The Geology of Early Humans in the Horn of Africa

edited by

Jay Quade
Department of Geosciences
University of Arizona
Tucson, Arizona 85721
USA

Jonathan G. Wynn
Department of Geology
University of South Florida
4202 E. Fowler Ave, SCA 528
Tampa, Florida 33620
USA

THE
GEOLOGICAL
SOCIETY
OF AMERICA®

Special Paper 446

3300 Penrose Place, P.O. Box 9140 • Boulder, Colorado 80301-9140, USA

2008

Copyright © 2008, The Geological Society of America (GSA). All rights reserved. GSA grants permission to individual scientists to make unlimited photocopies of one or more items from this volume for noncommercial purposes advancing science or education, including classroom use. For permission to make photocopies of any item in this volume for other noncommercial, nonprofit purposes, contact The Geological Society of America. Written permission is required from GSA for all other forms of capture or reproduction of any item in the volume including, but not limited to, all types of electronic or digital scanning or other digital or manual transformation of articles or any portion thereof, such as abstracts, into computer-readable and/or transmittable form for personal or corporate use, either noncommercial or commercial, for-profit or otherwise. Send permission requests to GSA Copyright Permissions, 3300 Penrose Place, P.O. Box 9140, Boulder, Colorado 80301-9140, USA. GSA provides this and other forums for the presentation of diverse opinions and positions by scientists worldwide, regardless of their race, citizenship, gender, religion, or political viewpoint. Opinions presented in this publication do not reflect official positions of the Society.

Copyright is not claimed on any material prepared wholly by government employees within the scope of their employment.

Published by The Geological Society of America, Inc.
3300 Penrose Place, P.O. Box 9140, Boulder, Colorado 80301-9140, USA
www.geosociety.org

Printed in U.S.A.

GSA Books Science Editors: Marion E. Bickford and Donald I. Siegel

Library of Congress Cataloging-in-Publication Data
The geology of early humans in the Horn of Africa / edited by Jay Quade, Jonathan G. Wynn.
    p. cm. — (Special paper ; 446)
  "The contributions to this volume stem from a topical session at the 2004 Geological Society of America Annual Meeting"—Pref.
  Includes bibliographical references.
  ISBN 978-0-8137-2446-1 (pbk.)
     1. Fossil hominids—Ethiopia—Awash River Valley—Congresses. 2. Paleoanthropology—Ethiopia—Awash River Valley—Congresses. 3. Geology—Ethiopia—Awash River Valley—Congresses. 4. Awash River Valley (Ethiopia)—Congresses. I. Quade, Jay, 1955–. II. Wynn, Jonathan Guy, 1969–. III. Geological Society of America. Meeting (116th : 2004 : Seattle, Wash.)

GN282.G47 2008
569.90963—dc22

2008031482

Cover, above (left to right): Gawis-cranium, Gona Paleoanthropologic Project area; fossil elephant molar; large cleaver from the Asbole Faunal Zone of the Busidima Formation in the Dikika Research Project area. **Below:** Ounda Gona landscape showing the Hadar Formation and Busidima Formation, Gona Paleoanthropologic Project area. **Back cover:** Badland exposures of the upper Sidi Hakoma Member of the Hadar Formation in the Dikika Research Project area, looking across the verdant Awash River Valley to the Hadar Research Area.

10 9 8 7 6 5 4 3 2 1

Dedication

Richard L. Hay
(1926–2006)

Pioneer in East African geology and a greatly valued mentor and colleague.

Contents

Preface vii

1. *The geology of Gona, Afar, Ethiopia* 1
   J. Quade, N.E. Levin, S.W. Simpson, R. Butler, W.C. McIntosh, S. Semaw, L. Kleinsasser,
   G. Dupont-Nivet, P. Renne, and N. Dunbar

2. *Stratigraphy and geochronology of the late Miocene Adu-Asa Formation at Gona, Ethiopia* 33
   L.L. Kleinsasser, J. Quade, W.C. McIntosh, N.E. Levin, S.W. Simpson, and S. Semaw

3. *Magnetostratigraphy of the eastern Hadar Basin (Ledi-Geraru research area, Ethiopia)
   and implications for hominin paleoenvironments* 67
   G. Dupont-Nivet, M. Sier, C.J. Campisano, J R. Arrowsmith, E. DiMaggio, K. Reed,
   C. Lockwood, C. Franke, and S. Hüsing

4. *Stratigraphy, depositional environments, and basin structure of the Hadar and Busidima
   Formations at Dikika, Ethiopia* 87
   J.G. Wynn, D.C. Roman, Z. Alemseged, D. Reed, D. Geraads, and S. Munro

5. *Composite tephrostratigraphy of the Dikika, Gona, Hadar, and Ledi-Geraru project areas,
   northern Awash, Ethiopia* 119
   D.C. Roman, C. Campisano, J. Quade, E. DiMaggio, J R. Arrowsmith, and C. Feibel

6. *Tephrostratigraphy of the Hadar and Busidima Formations at Hadar, Afar Depression,
   Ethiopia* 135
   C.J. Campisano and C.S. Feibel

7. *Correlation and stratigraphy of the BKT-2 volcanic complex in west-central Afar, Ethiopia* ... 163
   E.N. DiMaggio, C.J. Campisano, J R. Arrowsmith, K.E. Reed, C.C. Swisher III,
   and C.A. Lockwood

8. *Depositional environments and stratigraphic summary of the Pliocene Hadar Formation
   at Hadar, Afar Depression, Ethiopia* 179
   C.J. Campisano and C.S. Feibel

9. *Paleoenvironmental context of the Pliocene A.L. 333 "First Family" hominin locality,
   Hadar Formation, Ethiopia* 203
   A.K. Behrensmeyer

10. *Herbivore enamel carbon isotopic composition and the environmental context of
    Ardipithecus at Gona, Ethiopia* 215
    N.E. Levin, S.W. Simpson, J. Quade, T.E. Cerling, and S.R. Frost

Preface

The field of paleoanthropology has made significant advances in early human origins research during the past two decades. For example, the initial phases of human evolution in Africa between 8 and 2.5 Ma are now known to be far more complex than was thought prior to the 1990s. The newly revealed complexity at the base of human phylogeny derives from recent discovery of four new genera (of a total of seven genera) belonging to the upright-walking subtribe hominina (*Ardipithecus, Kenyanthropus, Orrorin,* and *Sahelanthropus*) and eight newly defined species of the hominina (*S. tchadensis, O. tugenensis, Ar. kadabba, Ar. ramidus, Australopithecus anamensis, Au. bahrelghazali, Au. garhi,* and *K. platyops*)—all new taxa published since 1994. While these and other finds have driven human origins research forward, the discoveries have vitally depended on collaborative research with geologists whose contribution has been not only to date fossil discoveries, but increasingly aimed at establishing a more complete understanding of the geological context of speciation and extinction during these early evolutionary phases.

The need for collaboration between paleoanthropological and geological research has been evident since Raymond Dart's discovery of the Taung Child (*Australopithecus africanus*) in southern Africa, through the work of Louis and Mary Leakey in Kenya and Tanzania, and the ensuing large, international and multidisciplinary projects undertaken during the 1960s to 1980s. Throughout these decades of early discoveries, the role for geology in documenting the context of human evolution has significantly expanded. With recent realization of the complexity of early human origins, it is now clear that documenting the *context* of human evolution may be as important, if not more so, as the fossil discoveries themselves. It is also widely recognized that various subfields within geology uniquely provide these contextual data and interpretations. For example, unraveling the phylogenetic relationships among the several competing early hominins depends on accurate estimates of the ages of fossils, whether the fossil ages are based on radiogenic, tephrostratigraphic, magnetostratigraphic, or biostratigraphic data. Furthermore, beyond these geochronological constraints, recent geological studies have elucidated the interplay between global climate, regional environments, and patterns of human evolution. These efforts require not only a sound geological framework of the region of interest, but also interdisciplinary collaboration between a variety of convergent research fields, such as paleoecology, paleontology, geochronology, sedimentary geology and stratigraphy, volcanology, and tectonics.

Geologists Maurice Taieb and Jon Kalb were among the first scientists to recognize and describe the rich paleontologic potential of the Awash Valley of the Horn of Africa, a potential that was soon realized with the discovery of "Lucy" (*Australopithecus afarensis*) in the Hadar region of the Awash River Valley (Fig. 1). In the decades since Lucy's discovery, the Awash Valley has become one of several centers of activity of human origins research in Ethiopia, with ongoing research in the Middle Awash region, renewed research at Hadar, and newly developed projects at Dikika, Gona, Ledi-Geraru, and Woranso-Mille. Sedimentary basin deposits now exposed in the Awash Valley are perhaps the richest source of fossil hominins in the world, thanks in part to a recent expansion of the number of expeditions operating in the area. Beyond this paleoanthropological interest, the Afar triple junction region, which encompasses the sedimentary basins of the Awash Valley, is a virtual natural laboratory for geological research of many fields—from active rift tectonics and geodynamics to volcanology—all of which contribute to a modern understanding of the geological context of human evolution in the Horn of Africa. All aspects of geological understanding of these basins have accordingly increased significantly in the past ten years.

This volume brings together recent developments in the geological context of human evolution in the Lower Awash Valley. The focus of attention for the present volume is on four research projects that surround the long-studied Hadar Research Area. These research projects—Dikika, Gona, Hadar, and Ledi-Geraru—cover the geology of most of the Lower Awash Valley. Jonathan Wynn and colleagues describe the geology of the southernmost Dikika Project, the location of the Dikika child, "Selam." Jay Quade, Naomi Levin, Lynnette Kleinsasser, and colleagues summarize the geology and paleoecology of the Gona research area, which contains the longest known archaeological record in one location in the world. Diana Roman leads a paper summarizing the tephrostratigraphy of all four project areas, providing vital chronologic links between

TABLE 1. ^{40}Ar/^{39}Ar DATES FROM GONA

Sample	L#	Irrad	Min	Plateau (step-heat) or mean (SCLF) age*					Isochron age			Comments
				n	%^{39}Ar	MSWD	Age (Ma, ±2σ)	n	MSWD	^{40}Ar/^{36}Ar (Ma, ±2σ)	Age (Ma, ±2σ)	
Sagantole Fm.												
GONNL 3	55438-01	NM-186J	Plagioclase	6	66.8	5.6	*5.14 ± 0.27*	6	6.8	297.8 ± 12.5	5.11 ± 0.34	Hada Tuff, As Duma Mbr.
GONASH-51#	31200		Plagioclase	23	NA	NA	*4.56 ± 0.23*	NA	NA	NA	NA	Tuff, middle Segala Noumou Mbr.
GONASH-52#	31202		Plagioclase	16	NA	NA	*4.60 ± 0.46*	NA	NA	NA	NA	Tuff, middle Segala Noumou Mbr.
GON05 275	55977-01	NM-192H	Plagioclase	7	89.5	3.9	*4.53 ± 0.18*	7	4.1	300.1 ± 11.1	4.42 ± 0.33	Hada Tuff, As Duma Mbr.
WMASH 15	56273-01	NM-196J	Plagioclase	5	66.0	3.4	*4.66 ± 0.11*	5	1.9	258.0 ± 32.2	4.91 ± 0.22	Purple tuff locality, Segala Noumou Mbr.
WMASH 16	56251-01	NM-196F	Plagioclase	9	92.6	1.0	*4.47 ± 0.04*	9	1.0	294.1 ± 3.5	4.48 ± 0.05	Purple tuff locality, Segala Noumou Mbr.
WMASH-25#	52560-01	NM-141	Groundmass	7	90.6	1.5	*4.17 ± 0.21*	9	3.3	297.7 ± 1.6	4.05 ± 0.17	Basaltic dike cutting flow below GWM-5
WMASH-27#	52558-01	NA	Groundmass	7	86.6	3.4	*4.06 ± 0.39*	7	7.5	298.0 ± 3.0	4.00 ± 0.30	Basaltic dike cutting flow below GWM-5
WMASH 43	56257-01	NM-196G	Plagioclase	11	99.9	1.3	*4.23 ± 0.05*	11	1.5	295.0 ± 4.3	4.23 ± 0.06	Tuff, top of Segala Noumou Mbr.
WMASH 47**	56252-01	NM-196G	Groundmass	0	0.0	0.0	*0.00 ± 0.00*	8	20.1	300.0 ± 11.4	1.77 ± 1.28	Basalt flow, As Duma Mbr.
WMASH 48	56253-01	NM-196G	Groundmass	6	80.8	1.4	*3.58 ± 0.31*	6	1.7	297.0 ± 5.7	3.44 ± 0.61	Basaltic dike cutting flow below GWM-5
WMASH 49	57138-01	NM-208D	Groundmass	8	93.2	0.9	*4.17 ± 0.12*	8	1.3	297.8 ± 6.5	4.13 ± 0.17	Basaltic dike cutting flow below GWM-5
WMASH 50	56255-01	NM-196G	Groundmass	7	93.9	2.5	*4.29 ± 0.26*	7	2.6	292.9 ± 5.5	4.43 ± 0.38	Basaltic dike cutting flow below GWM-5
WMASH 55	56254-01	NM-196G	Groundmass	8	99.7	2.4	*3.97 ± 0.33*	8	2.8	296.6 ± 6.8	3.88 ± 0.64	Basalt flow capping the Sagantole Fm.
WMASH 59	57140-02	NM-208D	Groundmass	9	100.0	1.4	*4.42 ± 0.22*	9	1.6	296.8 ± 4.7	4.23 ± 0.73	Corestone, Barsuli Hill
WMASH 62**	57135-01	NM-208C	Groundmass	6	54.9	1.4	*4.42 ± 0.27*	6	0.6	301.4 ± 5.2	3.45 ± 0.87	Corestone, Baruli Hill
WMASH 70	56256-01	NM-196G	Plagioclase	8	89.7	1.8	*4.35 ± 0.07*	8	1.9	293.8 ± 4.3	4.37 ± 0.09	Yellow tuff at GWM-31
WMASH 65	57175	NM-2085	Sanidine	12	NA	2.3	*4.10 ± 0.07*	12	2.3	500.0 ± 500.0	4.07 ± 0.12	Tuff, base of As Aela
Hadar Fm.												
GONNL 68	55992	NM-192K	Sanidine	9	NA	0.6	*3.27 ± 0.10*	9	0.8	296.0 ± 3.0	3.26 ± 0.20	Tuff, top of As Aela
BKT-2L†	7201		Anorthoclase	21	NA	NA	*2.94 ± 0.01*	NA	NA	NA	NA	Tuff, Kada Gona
Busidima Fm.												
GONASH-79	56259	NM-196H	Sanidine	7	NA	1.4	*1.90 ± 0.10*	7	1.4	320.0 ± 50.0	1.83 ± 0.18	Tuff, Ounda Gona south
GONASH-14§	NA		Sanidine	25	NA	NA	*2.53 ± 0.31*	NA	NA	NA	NA	Tuff, Fialu
GONASH-16§	NA		Plagioclase	14	NA	NA	*1.64 ± 0.03*	NA	NA	NA	NA	Tuff, Fialu
GONASH-21§	NA		Sanidine	26	NA	NA	*2.17 ± 0.18*	NA	NA	NA	NA	Tuff, Ounda Gona north
GONASH-39§	NA		Sanidine	17	NA	NA	*2.69 ± 0.06*	NA	NA	NA	NA	Tuff, Dana Aoule
GONASH-41§	NA		Plagioclase	22	NA	NA	*2.27 ± 0.28*	NA	NA	NA	NA	Tuff, Dana Aoule
AST-2.75†	8302		Plagioclase	23	NA	NA	*2.52 ± 0.15*	NA	NA	NA	NA	Tuff, Kada Gona

Notes: Ages were calculated relative to FC-2 Fish Canyon Tuff sanidine interlaboratory standard (28.02 Ma; Renne et al., 1998). Analyses were performed at New Mexico Geochronology Research Laboratory using an MAP 215-50 mass spectrometer on line with automated all-metal extraction system. All errors are reported at ±2σ, unless otherwise noted. Details of irradiation, analytical procedures, calculation methods, and analytical data are in Table DR2 and Figure DR11 (see text footnote 1).
*Single-crystal laser fusion (SCLF) dates are in italics.
†Published in Semaw et al. (1997).
§Published in Quade et al. (2004).
#Published in Semaw et al. (2005).
**Shaded text denotes unreliable dates.

Sagantole Formation

The Sagantole Formation crops out continuously in the western half of the Gona Paleoanthropological Research Project area, resting conformably(?) on the Adu-Asa Formation to the west, and bounded to the east by the As Duma fault (Fig. 1). The type area for the Sagantole Formation is the central complex of the Middle Awash Project area (see volume introduction), where deposits of overlapping age and similar lithology are present (Kalb et al., 1982; Renne et al., 1999). In all, the Sagantole Formation at Gona covers an area of ~3 km × 30 km. Reconnaissance of this large area reveals that sediments make up >70% of the Sagantole Formation, interbedded with the remnants of cinder cones such as those at Umele Delti and around fossil site GWM-5 (Fig. 4). To the north and south of Gona, sediments of the Sagantole Formation grade laterally into stacked basalt flows (Fig. 1).

Sedimentary strata in the Sagantole Formation are all deformed and tilted gently to the E/NE at ~5°–10°. The entire area is cut by normal faults with offsets varying from less than a meter to tens of meters. The faults are oriented mostly NW-SE, subparallel to the As Duma fault (Fig. 4), but dip either east or west. The numerous faults made mapping and correlation of strata between fault blocks very difficult. Moreover, Quaternary-age gravels eroded in part from the prominent basalt ridges of the Adu-Asa Formation to the east have covered large tracts of the Sagantole Formation, preventing tracing of outcrops between areas.

At least one major interformational normal fault, the Segala Noumou fault, cuts through the Sagantole Formation from NW to SE (Fig. 4). The fault juxtaposes largely pale-colored fluviolacustrine sediments, which we newly designate the Segala Noumou Member of the Sagantole Formation (Fig. 4, Tl and Tu), to the east against much redder lacustrine, volcaniclastic and locally fanglomeratic sediments of the As Duma Member to the west (Fig. 4, Tr). Previously, in Semaw et al. (2005), we referred to the As Duma Member as the "WM-5 block," and the Segala Noumou Member as the "WM-3 block." The sediments of the As Duma Member appear to be older than those in the Segala Noumou Member to the east, based on their redder color, greater induration, and abundance of interlayered volcanic rocks. However, as we discuss later, geochronologic evidence points to both younger and older ages for the As Duma Member. Resolution of this apparent contradiction is important, as the Sagantole Formation on both sides of the Segala Noumou fault contains abundant remains of *Ardipithecus ramidus*, but only those to the east of the fault (Segala Noumou Member) can be firmly dated. In this paper, we concentrate on fossil-bearing sediments of both the Segala Noumou and As Duma Members bounded by the Busidima River on the north and the Sifi River on the south. Outside this area, the Sagantole Formation remains largely unstudied geologically at Gona.

Segala Noumou Member

Sediments of the Segala Noumou Member are bounded by the Segala Noumou fault to the west and the As Duma fault to the east (Fig. 4). The section is a minimum of 40 m thick, to which we can add at least another 15–20 m of unmeasured section near the As Duma fault. The Segala Noumou Member consists of pale-brown to green laminated (Fl) to massive (Fp) siltstone and cross-bedded sandstone (St or Sr) interbedded with minor conglomerates (Gm), altered tephras, and densely cemented carbonate tufas (Fig. 5; see Miall [1978] for notation). Table 3 summarizes the features of these deposits and their environmental interpretation. In general, the laminated, fish fossil–rich siltstones can be viewed as lacustrine, the massive calcareous siltstones as paleosols, the sandstones as fluvial and possibly deltaic (containing aquatic mollusks *Bellamya* sp., *Cleopatra* sp., *Biomphalaria* sp., *Melanoides tuberculata*, and Unionids), and the conglomerate as fluvial. Three lacustrine intervals appear to be represented in the Segala Noumou Member, two lower intervals dominated by fissile shale and an upper one represented by gastropod-rich (*Cleopatra* sp.) tufas, which can be traced across the middle of the member, suggesting shallow lake or paludal conditions (Fig. 5). Mammalian fossils are common in these tufas (Fig. 5, GWM-3, -3sw) or occur in bedded siltstone (Fh) possibly representing overbank deposits (Fig. 5, GWM-67). The conglomerates and sandstones are generally less than a few meters thick, and the bed forms are small, implying small-scale rivers and creeks. The conglomerates contain a mix of volcanic clasts cemented by sparry calcite.

Ash-fall tephras within the Segala Noumou Member are yellow to white in the field and vary in thickness from a few centimeters to over a meter. All show complete alteration of glass, and some show alteration of plagioclase phenocrysts, when present. Unlike the As Duma Member, the sediments of the Segala Noumou Member are not volcaniclastic, except at the base, and are not interbedded with flows or coarse cinder.

Geochronologic evidence from the Segala Noumou Member suggests that it ranges in age from 4.6 to 4.2 Ma. Dates on plagioclase from four air-fall tephras in the lower lacustrine interval range from 4.47 ± 0.04 Ma to 4.66 ± 0.11 Ma (Table 1, WMASH-15, -16, GONASH-51, -52). A fifth tephra (Table 1, WMASH-70) near fossil site GWM-31 (Fig. 4) in fluvial sediments, perhaps in the upper part of the Segala Noumou Member, produced an age of 4.35 ± 0.07 Ma. A final date from a tephra in the stratigraphically highest position in the Segala Noumou Member next to the As Duma fault is 4.23 ± 0.05 Ma (Table 1, WMASH-43; Fig. 4). The exact stratigraphic thickness dividing this sample from dated intervals lower in the member (shown in Fig. 5) is not known but is probably on the order of 15–20 m. A basalt flow (Table 1, WMASH-55), found resting on sediments of the Segala Noumou Member and possibly a part of the As Duma volcanics (Fig. 4), produced an age of 3.97 ± 0.33 Ma.

We sampled for magnetostratigraphy from 14 overlapping sites covering ~20 m of sedimentary thickness (Fig. 5). Paleomagnetic polarity directions from all samples are reversed, consistent with $^{40}Ar/^{39}Ar$ dates, which span 4.66 ± 0.11–4.23 ± 0.05 Ma. This range falls largely within magnetozone C3n.1r, a period of reversed magnetic polarity spanning 4.51–4.32 Ma (Lourens et al., 2004). Since we did not

TABLE 2. AGES AND CONTEXTS OF MAJOR FOSSIL AND ARCHAEOLOGICAL LOCALITIES AT GONA

Site name	Site type (tool-making tradition)	Age (Ma)	Context
Adu-Asa Formation			
ESC-1	Fossil	5.5–6.0	Fluviolacustrine
ESC-2	Fossil	5.5–6.0	Fluviolacustrine
ESC-3	Fossil	5.5–6.0	Fluviolacustrine
ESC-8	Fossil	5.5–6.0	Fluviolacustrine
ESC-9	Fossil	5.5–6.0	Fluviolacustrine
BDL1	Fossil	6.2–6.4	Fluvial
BDL2	Fossil	6.2–6.4	Fluvial
ABD1	Fossil	6.2–6.4	Lacustrine
ABD2	Fossil	6.2–6.4	Lacustrine
HMD-1	Fossil	6.2–6.4	Lacustrine
HMD-2	Fossil	6.2–6.4	Fluvial
Sagantole Formation			
Segala Noumou Member			
GWM-3, -3w	Fossil	4.3–4.4	Paludal
GWM-67	Fossil	4.5	Fluvial
GWM-31	Fossil	4.3–4.4	Fluvial
As Duma Member			
GWM-1	Fossil	>4.6	Marginal lacustrine
GWM-2	Fossil	>4.6	Marginal lacustrine
GWM-5 series	Fossil	>4.6	Marginal lacustrine
GWM-9n	Fossil	>4.6	Marginal lacustrine
GWM-16	Fossil	>4.6	Marginal lacustrine
Member uncertain			
GWM-10	Fossil	As Duma Mbr.?	Marginal lacustrine
GWM-11	Fossil	As Duma Mbr.?	Marginal lacustrine/fluvial
GWM-45	Fossil	Segala Noumou Mbr. ?	Fluvial
GWMS-6	Fossil	Segala Noumou Mbr. ?	Paludal
GWMS-7	Fossil	Segala Noumou Mbr. ?	Paludal
GWMS-11	Fossil	Segala Noumou Mbr. ?	Paludal
Busidima Formation			
OGS 6/7	Archaeological (Oldowan)	2.5–2.6	Type I (Awash mainstem)
DAN 1,3	Archaeological (Oldowan)	2.5–2.6	Type I (Awash mainstem)
BSN-6	Archaeological (Oldowan)	2.5–2.6	Type I (Awash mainstem)
EG-10, -12, -13, -24	Archaeological (Oldowan)	2.5–2.6	Type I (Awash mainstem)
WG-5	Archaeological (Oldowan)	2.2	Type I (Awash mainstem)
OGN-3	Archaeological (Oldowan)	2.1	Type I (Awash mainstem)
OGS-3	Archaeological (Oldowan)	2	Type I (Awash mainstem)
BSN-12	Fossil and archaeological (Acheulian)	1.5–1.7	Type I–II
BSN-40	Fossil	1.5–1.7	Type I (Awash mainstem)
BSN-49	Fossil	0.9–1.4	Type II (Awash tributary)
BSN-65	Fossil	1.2–1.3	Type II (Awash tributary)
BSN-17	Archaeological (Acheulian)	1.5–1.7	Type I–II
OGS-12	Archaeological (Acheulian)	1.5–1.6	Type II (Awash tributary)
OGS-5	Archaeological (Acheulian)	1.4–1.5	Type II (Awash tributary)
DAN-5	Fossil and archaeological (Acheulian)	1.5–1.7	Type I (Awash mainstem)
DAN-16	Archaeological (Acheulian)	1.0–1.1	Type I (Awash mainstem)
ABE-1	Archaeological (Acheulian)	0.6	Type I (Awash mainstem)
GAN-10	Archaeological (Acheulian)	0.25	Type II (Awash tributary)
YAN-1	Fossil and archaeological (Acheulian?)	0.4	Type II (Awash tributary)
GWS-2	Archaeological (Acheulian?)	0.4	Type II (Awash tributary)
YAS-1	Archaeological (Late Stone age?)	<0.05?	Type I (Awash mainstem)

C3n.2r (4.80–4.63 Ma) or C3n.3r (4.90–5.00). The geologic, radiometric, and paleontologic evidence (Semaw et al., 1997) certainly precludes the As Duma Member from being younger (i.e., younger than 4.2 Ma) than the Segala Noumou Member. Nevertheless, we regard dating of the As Duma Member as work in progress. The key to resolving this question will be dating of more felsic tephras from the As Duma Member, and magnetostratigraphic sampling of longer sections.

As Duma Volcanics

The As Duma volcanics consist of a series of dikes, flows, and cinder cones that are aligned along the As Duma fault (Fig. 1). Virtually all of the volcanics are basaltic in composition. Palogonitic alteration of the basalt is very common, suggesting extensive interaction with surface water and groundwater during and after emplacement.

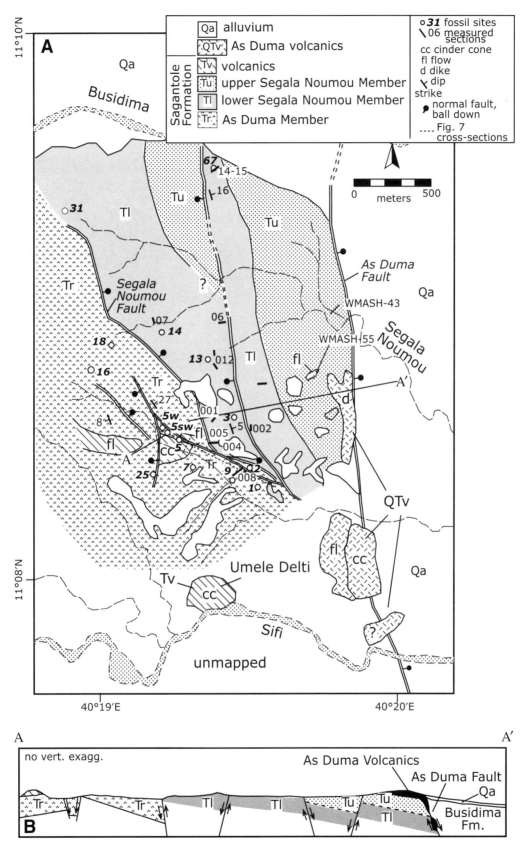

Figure 4. (A) Map and (B) cross section of the Sagantole Formation between the Busidima and Sifi. See Figure 1 for location of map area.

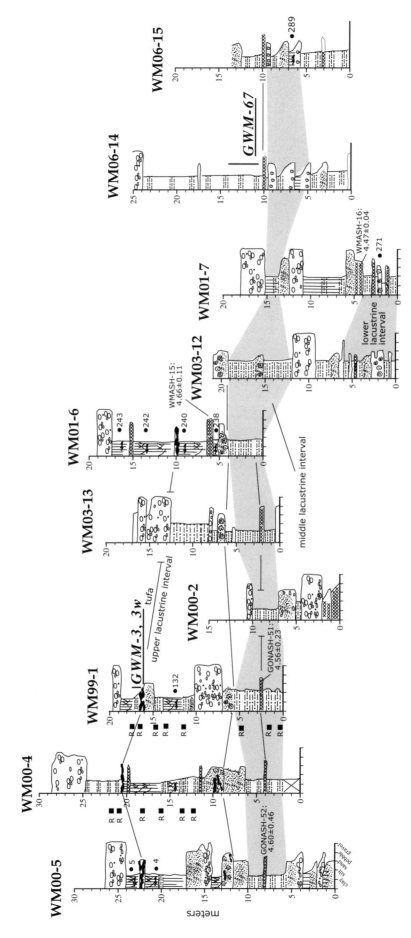

Figure 5. Measured stratigraphic sections within the Segala Noumou Member of the Sagantole Formation between the Busidima and Sifi. See Figure 4A for section locations and Figure 6 for legend.

TABLE 3. DEPOSITIONAL FEATURES OF THE SAGANTOLE FORMATION

Lithology	Symbol	Sedimentary features	Color	Fossils	Paleoenvironment
Silt or mudstone	Fl	Laminated, fissile	Pale green	Fish, plants, crocodiles, mollusks	Lacustrine
Silt or mudstone	Fh	Laminated to bedded, stained with hydrojarosite	Pale brown	Mammalian fossils	Floodplain overbank
Silt or mudstone	Fm	Slickensides, subhorizontal cracks, carbonate nodules and rhizoliths	Brownish red	Absent	Vertic paleosol
Sandstone	Sr or St	Ripple (Sr) or trough (St) cross-bedded	Light gray	Mollusks (*Bellamya* sp., *Biomphalaria* sp., *Melanoides tuberculata*, Unionids), reworked fossils	Fluvial or fluviodeltaic
Conglomerate	Gm	Massive	Dark gray	Absent	Fluvial
Tufa	Cm	Massive	Tan	Mollusks, mammalian fossils	Shallow lacustrine or paludal

Geologic relationships suggest that the As Duma volcanics all postdate 2.9 Ma, the age of the top of the Hadar Formation, but are not actively forming today. The main evidence for this is the alignment of the As Duma volcanics along the As Duma fault, which itself cuts the Sagantole Formation and Hadar Formations. The antiquity of the As Duma volcanics is supported by field evidence that suggests to us that volcanic activity has ceased or slowed along this reach of the As Duma fault. The cinder cones are deeply eroded and do not retain their original form. Dikes are cut by the As Duma fault, as apparent from their exposure only in the footwall of the fault. The presence of two (Table 4; samples BUSTASH-9/17 [BBT tephra] and ASASH-13) basaltic tephras in the younger Busidima Formation (2.7 to <0.15 Ma) hints at continued basaltic volcanic activity at some location in or close to the Gona Paleoanthropological Research Project area, perhaps from along the As Duma fault.

Hadar Formation

The Hadar Formation crops out in two areas of the Gona Paleoanthropological Research Project: One is along the Awash River in the easternmost portion of the project area, where the Awash River has cut down through the overlying Busidima Formation, exposing up to 125 m of the formation (Fig. 1). The other area of exposure is west of the As Duma fault in the As Aela area (Fig. 1), where the Hadar Formation rests directly on the Sagantole Formation.

The Hadar Formation has undergone extensive study over the past 30 yr (e.g., Aronson et al., 1977; Tiercelin, 1986; Walter, 1994; Yemane, 1997; Campisano et al., this volume, Chapter 8; Wynn et al., this volume), and the portions exposed along the lower Kada Gona have already been discussed in Quade et al. (2004). We present the full stratigraphic logs for that study in Figure DR2 (see footnote 1). In brief, the Hadar Formation at Gona and in the neighboring Hadar and Dikika areas consists of four members, in ascending order: the Basal, Sidi Hakoma, Denen Dora, and Kada Hadar Members. At Gona, deposition spans ca. 3.8–2.9 Ma, and the lower boundary exposed at As Aela is conformable with the Sagantole Formation, whereas the upper contact with the overlying Busidima Formation is marked by an angular unconformity.

Like the Sagantole Formation, the Hadar Formation along the Awash River is characterized by several laminated shale intervals rich in fossil mollusks and fish, representing at least three lacustrine transgressions into the Gona area between ca. 3.5 and 2.9 Ma. The lacustrine intervals are interbedded with fluvial-deltaic sands, overbank siltstones, and vertic paleosols (Quade et al., 2004).

At least 65 m of newly identified strata straddling the Sagantole-Hadar Formation contact are reported here for the first time. The As Duma fault turns to the northeast near the northern boundary of the project area, cutting upsection through the entire Sagantole Formation and exposing the lower and Hadar Formation (Fig. 1). Beds dip at ~7°–10° to the east.

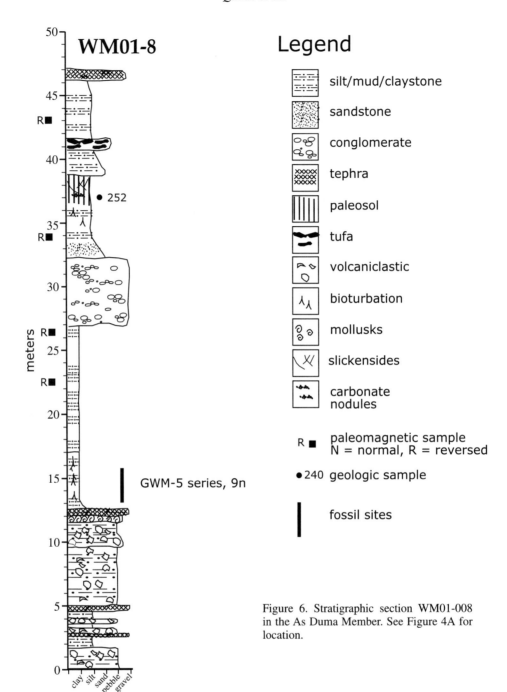

Figure 6. Stratigraphic section WM01-008 in the As Duma Member. See Figure 4A for location.

Most of the area is draped by younger gravels, but one badland area called As Aela by the local Afars exposes the continuous section that we present in Figure 8.

The section consists of interbedded siltstone, thin sands, gastropod-rich limestone, and a number of tephras. Sedimentologically, it is much like the underlying Segala Noumou Member of the Sagantole Formation. Siltstone (Fl) is laminated, greenish, and contains fish remains over a 7–8-m-thick interval in the middle of the section (Fig. 8, shaded area). A gastropod-rich limestone can be traced out along the base of this interval, which, combined with the fish fossil–rich siltstone, marks the transgression of a lake into the area. Above and below this interval, thin sandstone (Sr, St), conglomerate (Gm), and bedded (Fh) and massive (Fp) siltstone and mudstone characterize the section. We interpret all these latter lithologies as components of a small-scale alluvial system. The massive siltstones contain slickensides, subhorizontal cracks, and carbonate nodules, all features of modern Vertisols (Lynn and Williams, 1992).

The upper half of the As Aela section falls within the Basal and Sidi Hakoma Members of the Hadar Formation. A

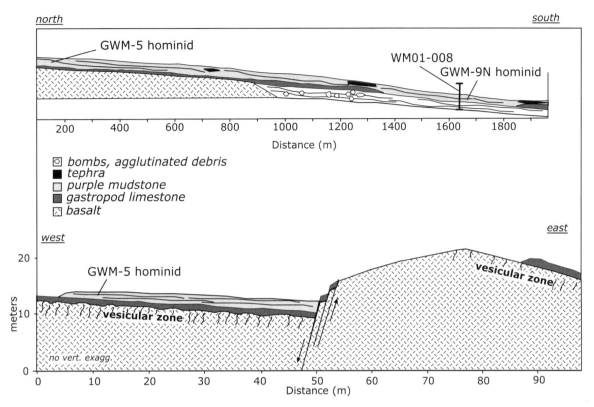

Figure 7. Detailed sections of fossil sites yielding *Ardipithecus ramidus* in the As Duma Member. See Figure 4A for locations and Figure 6 for legend.

conspicuous yellow pumice tephra (WMASH-65, Fig. 8) at the base of the exposed section at As Aela dates to 4.10 ± 0.07 Ma, whereas sanidine from a tephra (GONNL-68, Fig. 8) just below the top of the exposed section dates to 3.27 ± 0.10 Ma. In the type section of the Sagantole Formation in the Middle Awash Project area (see volume introduction), the top of the Sagantole Formation dates to 3.9 Ma. Wynn et al. (this volume) report on new findings from the Dikika area southeast of Gona where the base of the Basal Member of the Hadar Formation rests on a basalt and, near the base, contains the Ikini-Wargolo tephra (3.8 Ma). Glass is not preserved in tephras at As Aela with which we could test for the presence of the Ikini-Wargolo tephra.

Busidima Formation

The term Busidima Formation is a recent redesignation (Quade et al., 2004) of the upper portion of the Hadar Formation, which was formerly referred to as the upper portion of the Kada Hadar Member (e.g., Aronson et al., 1996). Outcrops of the Busidima Formation cover almost all of the central and eastern portions of Gona and rest unconformably on the underlying Hadar Formation, which crops out only along the Awash River and Kada Gona (Fig. 9). In areas largely outside Gona, the contact between the Busidima and Hadar Formations is visible as an angular unconformity where the Hadar Formation was locally faulted and tilted prior to deposition of the Busidima Formation (Wynn et al., this volume). The Busidima Formation at Gona is ~130 m thick and spans the period 2.7 Ma to <0.16 Ma.

In Quade et al. (2004), we described the stratigraphy and sedimentology of the Busidima Formation in some detail, as well as aspects of the geochronology. Since then, we have developed significant new information, especially for the upper half of the Busidima Formation. Our intent here is to expand upon the story presented in Quade et al. (2004), placing particular emphasis on the new stratigraphic and geochronologic data. This purpose is especially important for placing the abundant fossil hominid and archaeological remains in the Busidima Formation in a firm geochronologic framework. To achieve this, we measured a total of 56 stratigraphic sections in the Busidima Formation (Fig. 9), fully located and reproduced in Figures DR1 through DR9 (see footnote 1). From these sections, more generalized composite sections (Fig. 10) were constructed that summarize the key stratigraphic features and site locations in each area of Gona (Fig. 9).

Sedimentology and Stratigraphy

The Busidima Formation is composed of up to 130 m of sediment, but no more than ~50–70 m of thickness is measurable in any one area. Such thicknesses are visible in bluffs immediately along the Awash, where erosion has removed the upper half of the formation. The upper half of the formation is only preserved

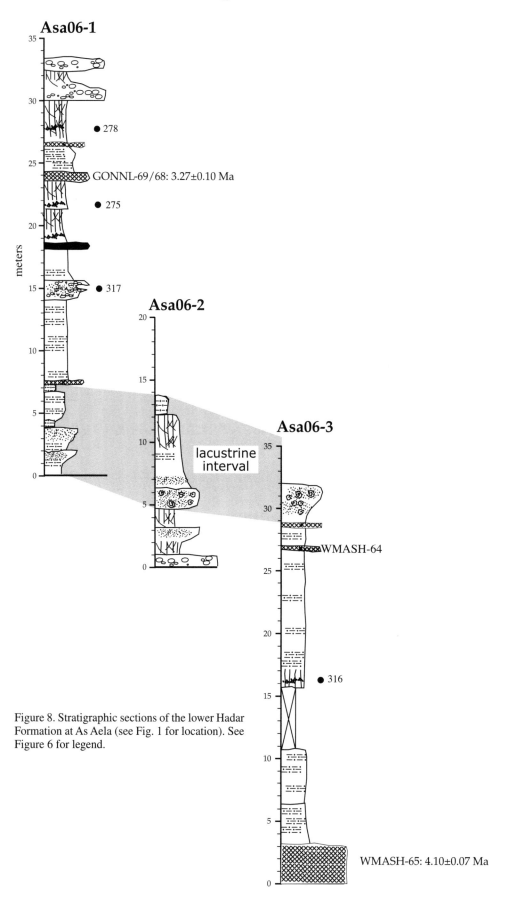

Figure 8. Stratigraphic sections of the lower Hadar Formation at As Aela (see Fig. 1 for location). See Figure 6 for legend.

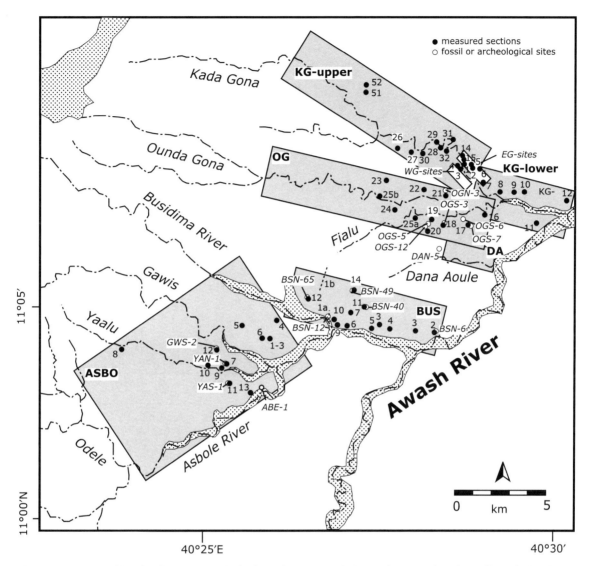

Figure 9. Location of key fossil and archaeological localities (open circles) and measured sections (filled circles) in the Busidima Formation. Individual sections are prefixed by area as follows in figure and in Figures DR1–DR9 (see text footnote 1): ASBO—Asbole area, KG—Kada Gona, OG—Ounda Gona, BUS—Busidima, and DA—Dana Aoule. Composite stratigraphic sections for each of these areas are presented in Figure 10.

in the headwaters of the Asbole, Busidima, Ounda, and Kada Gona drainages, far removed to the west of the Awash (Fig. 1). The Busidima Formation thins eastward into the Dikika Project Area as it approaches the hinge of the modern Awash half-graben (Wynn et al., this volume).

The base of the Busidima Formation in the Gona Paleoanthropological Research Project area is marked by a major unconformity cut into fluvial and lacustrine sediments of the underlying Hadar Formation (Figs. 3E and 3F). Above the unconformity, the Busidima Formation contains 1–3-m-thick conglomerates, in sharp contrast to the dominance of sand and mudstone in the underlying Hadar Formation. This makes the basal contact of the Busidima Formation easy to recognize and map at Gona (Fig. 1).

The Busidima Formation consists entirely of weakly consolidated sediments interlayered with very dispersed air-fall tephras (Quade et al., 2004). Lithologies include massive (Gm) to trough (Gt) cross-bedded conglomerate, rippled (Sr), trough cross-bedded (St), and massive (Sm) sandstone, and bedded/laminated (Fh/Fl) to massive (Fp) mudstone. In the lower half of the Busidima Formation, this succession of lithologies is organized in very regular 6–23-m-thick packages, fining upward from conglomerate at the base of each package to massive mudstones at the top. Quade et al. (2004) interpreted these fining-upward packages to represent deposition by a coarse-grained, meandering, ancestral Awash River. This contrasts with previous interpretation of these features of Yemane (1997), who viewed them as representing deposition on alluvial

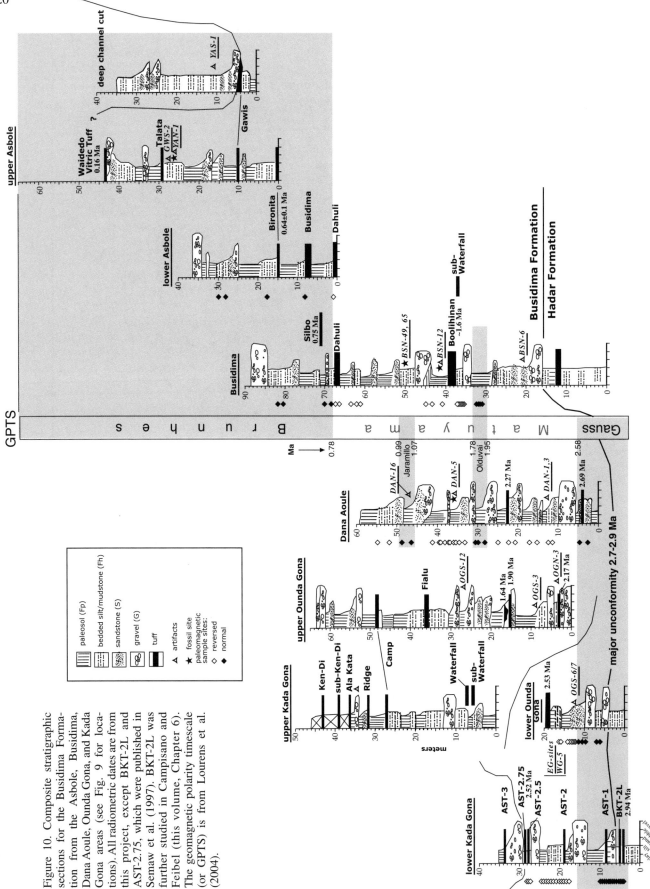

Figure 10. Composite stratigraphic sections for the Busidima Formation from the Asbole, Busidima, Dana Aoule, Ounda Gona, and Kada Gona areas (see Fig. 9 for locations). All radiometric dates are from this project, except BKT-2L and AST-2.75, which were published in Semaw et al. (1997). BKT-2L was further studied in Campisano and Feibel (this volume, Chapter 6). The geomagnetic polarity timescale (or GPTS) is from Lourens et al. (2004).

fans on the valley margins (rather than along its axis) by sheet-flooding. The evidence in favor of deposition by the ancestral Awash along the valley axis—rather than by its valley-flanking tributaries—is: (1) paleocurrent directions are N-NW, parallel to the flow of the modern Awash River at Gona today, (2) heterolithic gravels ("Type I" lithofacies gravels) in Busidima Formation are identical to those carried by the modern, coarse meandering Awash River, and (3) the fining-upward packages of the Busidima Formation contain clear fossil evidence (crocodiles, hippopotamus, marsh cane rats) of perennial water flow, much like the modern Awash, and unlike its modern tributaries, which are nearly all ephemeral (Quade et al., 2004).

We carefully mapped out the number and distribution of type I lithofacies gravels associated with the ancestral Awash River along a 1.5 km reach of the Busidima River (Fig. 1) in order understand its long-term behavior. Eight separate conglomerates in fining-upward packages were identified (Figs. DR5–DR7 [see footnote 1]), with an average thickness of 16 ± 5 m. The conglomerates are most numerous toward the east, in the area of the present position of the Awash River. Westward, the gravels feather out, disappearing near the confluence of the Busidima and Asbole Rivers. The gravels are distributed from the base of the formation (2.7 Ma) up to just below the Dahuli Tuff (0.81 Ma). This suggests a return time of the ancestral Awash River across Gona of eight conglomerates per 2 m.y., or once every ~250,000 yr.

Another important observation from the Busidima Formation is the presence of two distinct types of alluvial lithofacies: the type I lithofacies, containing the distinctive coarse gravels just described and associated with the ancestral Awash River, and type II lithofacies gravels, consisting of finer and less mature clasts in generally thinner conglomerate bodies than in type I (Quade et al., 2004). The type II lithofacies are interpreted to represent deposition by eastward-flowing paleotributaries to the ancestral Awash, tributaries that were in most, but not all, cases ephemeral.

The character of the Busidima Formation changes between 1.0 and 1.5 Ma, between the Dahuli Tuff (0.81 Ma) (Quade et al., 2004) and Boolihinan (1.6 Ma) Tuff. At about this level, the type I lithofacies gravels disappear (gravel 4.3, Fig. DR7 [see footnote 1]). After ca. 1.0 Ma, much finer sediments dominate, mostly thick paleosols (Fp) and their accompanying minor type II lithofacies gravels and sands (Fig. 3G). The density of mammalian fossils remains also drops off, an indication that the ancestral Awash River had moved eastward out of the Gona area. These gave way to a distal alluvial-fan setting characterized by broad surfaces that experienced deep pedogenesis, interspersed with small, largely ephemeral paleochannels filled by type II lithofacies.

Our surveys of the Asbole area (Fig. 1) since 2004 reveal that thick conglomerate bodies (type I) with coarse, well-rounded clasts reappear in the upper part of Busidima Formation (Fig. 3H). This transition occurs above the Bironita (0.64 ± 0.03 Ma) but below the Gawis (0.5 Ma) tephras. At least two very thick (17–30 m) fining-upward cycles are represented. The younger cycle contains abundant well-preserved aquatic mollusks, clearly attesting to perennial river flow.

Age Constraints

A rich variety of geochronologic constraints is available from the Busidima Formation, including over 35 tephras. Seven of these have been dated using $^{40}Ar/^{39}Ar$ (Table 1), all in the lower half of the Busidima Formation. Most tephras are aphyric but preserve fresh glass available for major-element analysis and hence potential tephrostratigraphic correlation. Some of these tephras can be confidently correlated with well-dated regional ash-fall events. The abundant well-bedded siltstone in the Busidima Formation yields robust magnetic polarities, providing key constraints on the age of the Busidima Formation and its rich archaeological and paleontological archive.

Tephra Occurrences

In the field, most of the tephras are either light gray or white. The white color nearly always denotes partial to complete alteration of glass to clay, whereas the gray color seems to be the primary color of unaltered glass, except where basaltic. Altered tephras are restricted in both time and space. The altered tephras are confined entirely to the lower half of the Busidima Formation (older than 1.6 Ma). Below this level, glass in most tephras shows some degree of alteration and, in some cases, complete conversion to secondary clays. The altered tephras are closely associated with the type I lithofacies of the paleo–Awash River, suggesting a causal connection. In this setting, the tephras would have been saturated shortly after burial by shallow groundwater on the ancestral Awash floodplain, probably leading to their alteration. By contrast, tephras in the levels from 0.5 to 1.5 Ma at Gona fell largely on distal alluvial fans and may have remained dry or only seasonally wetted even after burial due to the deep levels of the water table. As noted previously, the ancestral Awash River returned to Gona ca. 0.5 Ma, which would have placed almost all tephras at Gona in the saturated zone. However, such conditions may have been brief enough (<0.3 m.y.) for the glass in the tephras to be largely preserved.

Tephra occurrence is laterally very discontinuous in the Busidima Formation at Gona, in sharp contrast to the Hadar Formation, where tephras are readily traceable across long distances and between project areas (Roman et al., this volume; Wynn et al., this volume; Campisano et al., this volume, Chapter 6). For example, the Boolihinan and Dahuli Tuffs are major stratigraphic markers in the Busidima area, but they have not been found, despite intensive searching, along the Ounda Gona, Kada Gona, or Dana Aoule drainages. Conversely, the Fialu and Camp Tuffs are conspicuous and continuous over broad areas of Ounda and Kada Gona, and yet they have not been found along the Busidima. This pattern is an artifact of the cut-and-fill nature of the stratigraphy of the Busidima Formation by the paleo–Awash River. In many areas, this incision and backfilling by younger channels into older deposits are quite visible. For example, the channel containing site OGS-3 along the lower Ounda Gona (Fig. 10) is readily mappable and involves ~14 m of incision. The channel inset containing YAS-1 along the upper Asbole, involving at least 35 m of incision followed by backfilling, is another example (Fig. 10).

$^{40}Ar/^{39}Ar$ Dates

Seven $^{40}Ar/^{39}Ar$ dates are available from the Busidima Formation, all from the lower portion (Table 1). All the dates are on sanidine or plagioclase. They range from 2.69 ± 0.06 Ma to 1.641 ± 0.028 Ma.

Magnetostratigraphy

Paleomagnetic analyses are available from Semaw et al. (1997) and Semaw et al. (2003) for short intervals at the base of the Busidima Formation, and from data published here for the rest of the formation (Table DR1 [see footnote 1]). Over 100 sites are involved, and three to four samples were analyzed per site. Together, the data greatly refine our previous knowledge of the geochronology of the Busidima Formation.

Nearly all the major magnetic polarity intervals covering the last 2.7 m.y. appear to be represented at Gona, dated using the geomagnetic polarity timescale (GPTS) of Lourens et al. (2004). Our results show that sediments at the base of the Busidima Formation from Kada Gona (Semaw et al., 1997), at Ounda Gona (Semaw et al., 2003), and at Dana Aoule (Fig. 10) display normal polarity and belong to the Gauss chron. The Gauss-Matuyama transition (2.58 Ma) is very well fixed and occurs ~9 m below tephra AST-2.75 (2.517 ± 0.15) along the Kada Gona, 8 m below tephra GONASH-14 (2.534 ± 0.30) at Ounda Gona, and less than 5 m above tephra GONASH-39 (2.69 ± 0.06) at Dana Aoule (Figs. 3F and 10). Above these stratigraphic levels, only the Dana Aoule, Busidima, and lower Asbole areas have been sampled in detail. The Olduvai chron (1.78–1.95 Ma) has been detected at both Dana Aoule and Busidima, providing vital local geochronologic constraints on hominid remains and archaeological sites in those areas. The very short Jaramillo chron (0.99–1.07 Ma) only has been identified near the top of the Dana Aoule section (Fig. 10). The Brunhes-Matuyama (BM) boundary (0.78 Ma) occurs slightly (<1 m) above the Dahuli Tuff along the Busidima and the lower Asbole (Fig. 10). This is confirmed by the presence of the Silbo (0.75 Ma) and the Bironita (ca. 0.64 ± 0.03 Ma) Tuffs 4.5 and 15 m, respectively, above the Brunhes-Matuyama boundary (see next section).

Tephrostratigraphy

Mapping and major-element analyses reveal more than 35 chemically distinct tephras at Gona (Table 4). The distinctions are based on visual inspection of the glass morphology, on compositional distinctions (mainly CaO, FeO, TiO_2, followed by Al_2O_3, SiO_2, MgO, and MnO), and on stratigraphic level. All these criteria are important, because, in some cases, glass composition can be very similar (Fig. 11; Roman et al., this volume), but the tephras are clearly different based on stratigraphic position. Reliable analyses of alkalis were obtained for about two-thirds of the tephras

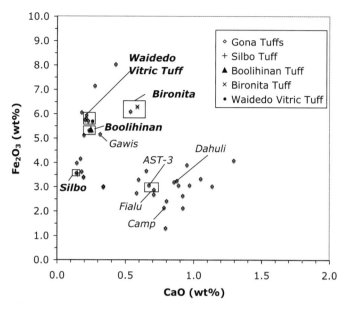

Figure 11. Plot of CaO versus Fe_2O_3 (wt% oxides) for all tephras (open diamonds) at Gona. The plots of some major tephra markers at Gona are noted, and the four Gona tephras with their regional tephra correlatives are indicated in boxes.

at Gona, allowing us to classify them chemically. Aside from two basaltic tephras (Table 4; BUSTASH 9/17 and ASASH-13), all other glass in tephras at Gona are rhyolitic and subalkaline (Roman et al., this volume).

Nearly all the tephras analyzed from the Busidima Formation are tied to measured stratigraphic sections, thus placing them in a firm local stratigraphic context. A combination of field evidence, radiometric and magnetostratigraphic dating, and extra-regional correlations allows us to place many of the tephras in an absolute chronologic framework (Fig. 10). Tephras with multiple occurrences and widespread distribution at Gona include: the sub-Waterfall (ca. 1.7 Ma), Boolihinan (1.6 Ma), AST-3/Fialu (~1.3 Ma), Camp (ca. 1.0 Ma), Dahuli (0.81 Ma), Silbo (0.75 Ma), Bironita (0.64 Ma), Gawis (ca. 0.55 Ma), Talata (ca. 0.38 Ma), and Waidedo Vitric (or WAVT) (0.16 Ma) Tuffs. More restricted in outcrop but stratigraphically important tephras include the Ridge (ca. 0.9 Ma), sub-Fialu A, B, C, and D (1.4–1.5 Ma), and Butte (<0.1 Ma?) Tuffs.

At least eight tephras can be correlated to tephras outside Gona, based on their major-element composition (Fig. 11 for CaO versus Fe_2O_3) and stratigraphic position. These include the Boolihinan, Silbo, Bironita, Dahuli, Ken-Di, Korina, Odele, and Waidedo Vitric Tuffs. The Boolihinan Tuff (Table 4: BUST-1, -10, BSN-12cr) is chemically very close to Deep Sea Drilling Project (DSDP) DEM-4–1 tephra at Ocean Drilling Program (ODP) Site 722 from the Gulf of Aden, dated at 1.6 Ma (Table 4; P. deMenocal et al., 2003, personal commun.). This correlation is supported by the position of the Boolihinan ~5 m above the top (1.78 Ma) of the Olduvai chron. The Dahuli Tuff, present but undated at Hadar (Yemane, 1997; Campisano et al., this volume,

Chapter 6), is found at Gona just below the Brunhes-Matuyama boundary; as such, we estimate its age to be 0.81 Ma. We correlate samples BUST-20 and -23 (Table 4) with the Silbo Tuff (0.75 Ma), another major tephra that fell over broad areas of East Africa and the Arabian Sea (McDougall, 1985; Haileab and Brown, 1994). This is consistent with its position ~5 m above the Brunhes-Matuyama (0.78 Ma) boundary. Sample ASASH-6 (Table 4) can be confidently correlated with the Bironita Tuff (ca. 0.64 ± 0.03 Ma; Clark et al., 1994 for the Middle Awash Project; Geraads et al., 2004, for the Dikika Project), at 14 m above the Brunhes-Matuyama boundary. The Ken-Di Tuff was found just south of the drainage divide separating the Gona and Hadar project areas and at the top of our measured section in that area. With the distinctively high %Fe_2O_3 of its main glass mode, the Ken-Di is also reported from the upper Busidima Formation at Hadar (Campisano et al., this volume, Chapter 6) and Dikika (Wynn et al., this volume). Two minor tuffs, the Korina and the Odele, are found in the uppermost Busidima Formation at both Gona and Dikika (Roman et al., this volume). Finally, a very prominent and continuous tephra sampled at six locations (Table 4) yields an excellent chemical match to the Waidedo Vitric Tephra (0.16 Ma; Clark et al., 2003) from the nearby Middle Awash Project area, consistent with its position very near the top of the section at Gona.

Archaeological Sites

The Busidima Formation contains almost the entire known (2.7 to <0.16 Ma) record of stone tool-making, arguably the longest documented archaeological record in one location in the world. Archaeological sites containing artifacts typical of the oldest recognized stone tool-making tradition, the Oldowan Industrial Complex, are numerous in the lowest stratigraphic levels at Gona (Semaw et al., 1997). They range in age from 2.5 to 2.6 Ma for the oldest sites (OGS-6/7, BSN-6, and the EG series sites) and 2.2–1.9 Ma (OGS-3, OGN-3, WG-5) for the youngest sites (Fig. 9; Table 2). As described in Quade et al. (2004), the tools are found in the sandy to silty, middle to upper portions of the fining-upward sequences so typical of the lower Busidima Formation. Moreover, gravels of the type I lithofacies lie at the base of these fining-upward sequences at all the known Oldowan sites. The sedimentologic evidence therefore consistently points to a floodplain setting for these sites adjacent to the channel of the ancestral Awash River. OGS-7 (Fig. 3F) is one such example, where dense concentrations of lithic debris occur on a sand lens marginal to a type I gravel (Semaw et al., 2003). These types of gravels are found largely in point bars of the modern Awash River. Aphanitic clasts from such point bars in the paleo–Awash River were carefully selected for by early hominids in Oldowan tool manufacture (Stout et al., 2005).

Archaeological sites containing implements belonging to the Acheulian Industrial Complex are also common at Gona, sometimes with associated hominid remains. The oldest sites (1.6–1.3 Ma) are represented by BSN-12, BSN-49, and BSN-65, OGS-12, and OGS-5. Younger examples of Acheulian sites include DAN-16 (1.0 Ma), ABE-1 (0.6 Ma), and GWS-2 (0.4 Ma) for the terminal Acheulian. As with the Oldowan sites, all these sites are found in the middle and upper portions of fining-upward sequences, indicative of a floodplain setting (Fig. 3G, BSN-49). Unlike the Oldowan sites, several important Acheulian sites, such as OGS-5, -12, BSN-49, and several sites at the YAN-1 stratigraphic level, are found in association with the type II lithofacies, a pattern seen elsewhere in East Africa (Rogers et al., 1994). Fossil evidence (aquatic snails, *Thryonomis swinderiansis* or marsh cane rat) suggests that these archaeological sites were located on large perennial tributaries to the ancestral Awash River.

Late Stone Age (?) sites such as YAS-1 in the upper part of the section are found in association with a deep (35 m) paleochannel filled with the type I lithofacies along the upper Asbole (Fig. 10). The detailed context of these youngest archaeological sites is work in progress.

REGIONAL GEOLOGICAL SYNTHESIS AND HOMINID PALEOENVIRONMENTS

The history of basin filling over the past ~6.4 m.y. at Gona is nearly continuously represented by volcanic but mainly sedimentary deposits (Fig. 12). In most respects, the geologic history of that basin filling at Gona follows the classic tectonosedimentary evolution of rift basins described by a number of studies (Lambiase, 1990; Gawthorpe and Leeder, 2000). In this progression, basins gradually expand, deepen, and interconnect, producing a very distinctive sedimentary succession. The basic structure of continental rifts is the half-graben (Bosworth, 1985); this is arguably the basin setting for Gona and surrounding project areas over the past 6.4 m.y., based on two general lines of evidence: The first is the increase in the eastward dip of beds in the Adu-Asa (5–20°E) through to the Sagantole (mostly 5–10°E) and Hadar Formations (1–5°E), which strongly suggests progressive rotation along a common normal fault or set of faults located on the east side of the area. This kind of "fanning" of dips is typical of half-graben growth structures in rifts (Schlische, 1991). The second is the eastward thickening of deposits in Hadar Formation (Dupont-Nivet et al., this volume; Wynn et al., this volume). This is presumed to occur in the direction of the hanging-wall depocenter of a half-graben downdipping to the east. By contrast, the Busidima Formation dips and thickens westward (Wynn et al., this volume), pointing to a change in the polarity of the half-graben, as the basin-bounding As Duma fault activated ca. 2.9 Ma.

Broadly speaking, there are four distinct phases represented at Gona that contrast in lithology, sediment accumulation rates, and depositional environment (Fig. 12). These include the "fault initiation" stage (after Gawthorpe and Leeder, 2000), represented by the Adu-Asa Formation (>6.4–5.2 Ma) in which basaltic volcanic rocks dominate over sediments in an incipient extensional basin. The second "interaction and linkage stage" is represented by the Sagantole (4.6–3.9) and Hadar (3.8–2.9) Formations. Here, local igneous activity decreased and finally disappeared except

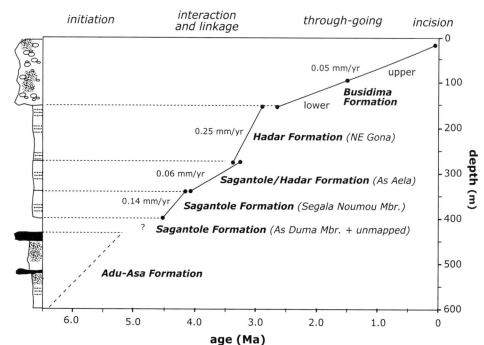

Figure 12. Sediment accumulation history and rates (in mm/yr) for the formations represented at Gona and the major tectonic stages (after Gawthorpe and Leeder, 2000).

along the As Duma fault, the basin enlarged and deepened, and sedimentation rates were very high (Fig. 12). Flow through the basin was likely partially confined at some downstream point north of Ledi-Geraru, impounding sediments and producing lakes and low-gradient rivers. The third "through-going fault stage" is represented by the Busidima Formation (2.7 to <0.2 Ma), when local depocenters were linked into a continuous drainage system by a single, large-scale meandering river. At the same time, sedimentation rates drastically decreased, gradients steepened, and sediments coarsened. In the final phase, basinwide sedimentation was brought to an end starting after ca. 0.16 Ma, when the Awash River incised rapidly and deeply, carving out the dramatic badlands seen today.

Adu-Asa Formation

The Adu-Asa Formation presents a sharp contrast to other periods of basin filling in that basaltic volcanism dominates over sedimentation. Accumulation rates appear to be fairly high (Fig. 12), but sedimentation occurred in small-scale rivers in narrow, probably north-south–oriented valleys. The dominantly basaltic composition of the Adu-Asa Formation (termed the Dahla Series Basalts by the volcanologic literature) is thought to mark the replacement of continental crust by incipient oceanic crust (Wolfenden et al., 2005) under the widening South Afar Rift.

The paleoenvironmental picture that emerges for the Adu-Asa Formation is one of multiple centers of eruptions, almost all basaltic but some felsic (including the Ogoti complex at Gona), covering most of the area with 3–10-m-thick flows and thin basaltic and felsic tephras (Fig. 13A). This must have been the general setting over a very large area to the south (>75 km), since the Adu-Asa Formation is lithologically similar in the Middle Awash area (WoldeGabriel et al., 2001). The presence of paleosols between many but not all flows (Fig. 3B) shows that >10^3–10^4 yr elapsed between basaltic eruptions in many cases, if soils developed at rates similar to those on basalt flows in the southwestern United States (Laughlin et al., 1994; van der Hoven and Quade, 2002). The active volcanic areas near Nazaret in the Main Ethiopian Rift south of Gona provide clues to how such basalt-dominated landscapes may have looked. Areas on older flows with continuous soil cover are densely vegetated with C_4 grasses and to a lesser extent C_3 trees and shrubs, whereas rugged younger flows carry mainly dispersed C_3 plants.

Paleorivers threaded their way along a narrow, probably north-south depression (Fig. 13A) during much of Adu-Asa Formation deposition. The rivers clearly had extralocal sources, given the mix of felsic and basaltic clasts present in river channels, but flow directions could not be established. Small lakes developed in the older (6.2–6.4 Ma) part of the sedimentary package, laying down greenish, fissile shales and diatomites. Well-developed paleosols are unsurprisingly lacking in this environment. The floodplain was narrow and probably very active, and at times locally covered by marshes and a shallow lake. Taken together, the Adu-Asa Formation may well represent the early stages of extension, the "rift initiation stage" of Gawthorpe and Leeder (2000), in which small half-grabens were filled by sediments from small rivers and lakes, and by ongoing basaltic volcanism. However, the limits of the basin containing the Adu-Asa Formation sediments are hard to establish. The pervasive eastward dip of the Adu-Asa Formation points to rotation along a west-dipping normal fault. This fault has not been identified but would have to be located to the east, probably now buried beneath younger deposits.

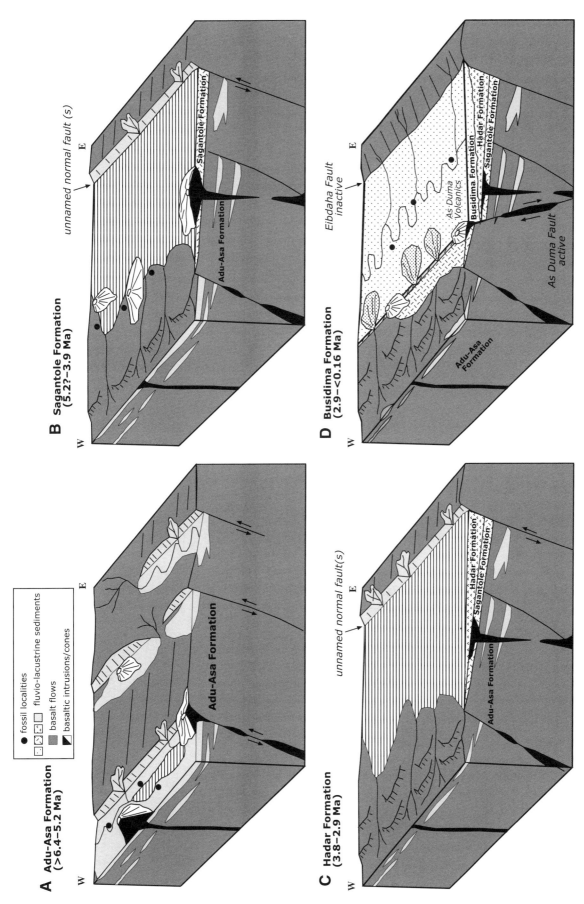

Figure 13. Tectono-stratigraphic evolution of the region for the (A) Adu-Asa, (B) Sagantole, (C) Hadar, and (D) Busidima Formations. The nature of the graben-bounding normal fault in B and C is speculative and is represented here as a single north-south-trending structure, identified as the Eibdaha fault by Dupont-Nivet et al. (this volume), or by Wynn et al. (this volume) as a series of unnamed normal faults. Solid dots show positions of hominid remains on paleolandscapes.

Additional key paleoenvironmental evidence for the Adu-Asa Formation comes from carbon isotopic evidence from fossil teeth (the intraflow paleosols lack soil carbonate for analysis). The carbon isotope evidence from teeth as presented in this volume (Levin et al., this volume) shows that C_4 plants (grasses) dominated the diets of most of the large herbivores (the hippopotamuses, horses, and elephants, for example), pointing to abundant grass cover somewhere in the vicinity of the water courses between basalt flows. The presence of C_3 (forest/shrubs) diet is also clearly indicated in some pigs, some bovids, and giraffes. These taxa fed in an open- rather than closed-canopy forest. Modern analogs near Nazaret would suggest that the narrow water courses were probably forested. Older basalt flows bordering the water courses were mantled by soil and covered mainly by C_4 grasses, whereas younger flows unsmoothed by pedogenesis were probably covered by very sparse C_3 trees and shrubs.

Sagantole Formation

We suggest that the Sagantole (<5.6–3.9 Ma) Formation reflects the next tectono-sedimentary stage in rifting: rapid and widespread sedimentation in a broadening and deepening of the basin. During this phase, graben-bounding faults extended and linked (Gawthorpe and Leeder, 2000), integrating the smaller Adu-Asa–age basins into a much larger one (Fig. 13B). Patterns of sedimentation and deformation give clear indications of the new basin geometry. The eastward dip of the Sagantole Formation suggests rotation along a west-dipping normal fault located somewhere east of Gona. Hence, deposition during Sagantole time was in an east-dipping half-graben. This would place Gona on the western side of the paleobasin, which is consistent with field evidence. Most of the Sagantole Formation is sedimentary at Gona, but to the north and south, these sediments interfinger with stacked basalt flows (Fig. 1). The fossil remains from the As Duma and Segala Noumou Members occur in a valley-margin setting, where lake deposits interfingered with small fluvial channels or lapped onto active basaltic cones and flows (Fig. 13B). This geometry suggests that Gona was situated at the western edge of a large lake and fluvial system in a basin that was centered to the east but is now down-faulted and buried under the Hadar Formation. Moreover, the Sagantole Formation at Gona is chopped by numerous small-scale normal faults, a distinguishing characteristic of hinge zones of half-grabens (Morley, 1995).

It is also highly likely that the lakes and rivers represented by the Sagantole Formation at Gona were linked to those farther south in the Middle Awash Project area, infilling one large "Sagantole" paleobasin or half-graben that was the immediate ancestor to the modern Awash half-graben. Sediments from the As Duma and Segala Noumou Members strongly resemble the Sagantole Formation exposed in the central complex of the Middle Awash. There, the Sagantole Formation dates between 5.6 and 3.9 Ma and is over 300 m thick (Renne et al., 1999), which is much thicker and overlaps the age of the As Duma (older than 4.6 Ma) and Segala Noumou (4.6–4.2 Ma) Members at Gona. Large areas of the Sagantole Formation west of the Segala Noumou fault are undated and may well fill much of the time gap back to ca. 5.0 Ma. The dated Segala Noumou Member (4.6–4.2 Ma) at Gona spans the upper Haradaso (5–4.4 Ma), Aramis (4.4–4.3 Ma), and Beidareem Members (4.3–4.19 Ma) of the Sagantole Formation. All three members are lacustrine and contain an ensemble of lithologies reminiscent of those in the Segala Noumou Member (Renne et al., 1999). Our view is that this portion of the Sagantole at Gona and in the Central Complex quite possibly was deposited in a single north-south–trending "Sagantole" paleovalley (half-graben), where Gona was on the western margins of the valley bordered by basalts, and the Central Complex was closer to the paleoaxis.

Carbon isotopic evidence from carbonates in the paleosols (Levin et al., 2004; Semaw et al., 2005) and from teeth (Levin et al., this volume) points to a mix of C_3 and C_4 vegetation in the western Gona Paleoanthropological Research Project area during deposition of the Segala Noumou Member. Ten paleosol profiles and 68 samples were taken within an ~2 km radius of hominid site GWM-3 (Fig. 4). The carbon isotopic composition of soil carbonates shows that the landscape was a mixed C_3/C_4 system, but C_3 plants (trees and shrubs) dominated in the area of sampling. This, therefore, was the paleoenvironmental context in which remains of *Ardipithecus ramidus* came to be preserved.

As with the Adu-Asa Formation, carbon isotope evidence from fossil teeth of the majority of large herbivores of the Sagantole Formation shows a clear preponderance of C_4 plants (grasses) in their diet (Levin et al., this volume). This contrasts with the dominance of C_3 plants indicated by the paleosol isotopic evidence in the vicinity of the hominid sites. This suggests that the low, well-watered areas represented by much of the Sagantole Formation were dominated by C_3 plants, whereas the surrounding, slightly higher elevation areas—whether volcanic or sedimentary—were grass covered, especially where soil cover was thick. Preservation in the geologic record is always biased toward the lowest topography, a setting that would therefore be strongly represented in our paleosol record.

Hadar Formation

Depositionally, the Hadar Formation appears to represent a continuation of the fluviolacustrine sedimentation that started in the Sagantole Formation time, and possibly within the same half-graben (Fig. 13C). The key difference is that the extent of the paleobasin is still visible despite faulting and younger sedimentation. Combined evidence from all the project areas in the region paints a clear picture of the Hadar-age basin. Both Dupont-Nivet et al. (this volume) and Wynn et al. (this volume) report significant thickening of the Hadar Formation eastward and northward, in the direction of what we assume to be hanging-wall depocenter. The Hadar Formation also dips gently eastward, from as much as 7°E to 10°E at As Aela, 1°E at Ledi-Geraru (Dupont-Nivet et al., this volume), and up to 2°E (but with much local variation depending on fault block) at Dikika (Wynn et al., this volume).

Dupont-Nivet et al. (this volume) identify the Eibdaha fault zone on the east side of Ledi-Geraru as a possible graben-bounding fault for the Hadar basin, juxtaposing older basalts in the hanging wall against Hadar Formation in the footwall. Wynn et al. (this volume), working at Dikika, recognize the Hadar graben as bounded to the east and northeast by a series of normal faults, rather than by a single fault. The location of this graben-bounding fault is a topic for further work.

Depositional patterns are the third key line of evidence in favor of a hanging-wall depocenter on the east and northeast side of the basin, and hence a west-dipping half-graben during Hadar deposition. Through much of Hadar Formation time, the Hadar and Gona project areas lay at the southern margin of a major lake system (Tiercelin, 1986; Wynn et al., this volume) that transgressed at least three times into the Gona area. Fluvio-lacustrine and deltaic sediments dominate the Hadar Formation in both areas (Campisano and Feibel, this volume, Chapter 8). Marginal lacustrine gastropod limestone crops out as far west as As Aela. The profundal facies, and hence the deepest part of the lake, occurs well to the north and east of Gona in the Ledi-Geraru project area (DiMaggio et al., this volume; DuPont-Nivet et al., this volume). In general, sediments in the Hadar Formation are finer grained than those in the underlying Sagantole Formation and in the overlying Busidima Formation, pointing to a low-gradient, low-energy system. However, very high sedimentation rates characterize the Hadar Formation at Gona (Fig. 12; 0.25 mm/yr) and central Hadar, and rates are as much a 1 mm/yr in central Ledi-Geraru (DuPont-Nivet et al., this volume). This indicates that the valley must have been partially structurally closed at some point north of Ledi-Geraru, and that the half-graben was very actively subsiding to accommodate the influx of sediment. Closure to the north and south may have been the result of hanging-wall highs along accommodation zones (Gawthorpe and Leeder, 2000). By deposition of the BKT-2U tephra (DiMaggio et al., this volume) at 2.94 Ma, the last Hadar-age lake had regressed, probably marking the filling of the half-graben to the height of the downstream accommodation zone.

Deposition older than ca. 4 Ma was strongly influenced by basaltic volcanism, whereas the Hadar and Busidima Formations are largely sedimentary at Gona, and the tephras are nearly all felsic, fine-grained, and probably extraregionally derived. To be sure, the As Duma fault created a conduit for the basaltic dikes, cinder cones, and flows of the As Duma volcanics, but these extrusions are very small in volume compared to the voluminous basaltic volcanism of the Sagantole and Adu-Asa Formations. Even this volcanism has slowed and apparently stopped, based on the deeply dissected nature of the cones and exhumation of the dikes. Limited basaltic volcanism also continued to the north in Hadar and Ledi-Geraru (Aronson and Taieb, 1981; Dupont-Nivet et al., this volume).

This gradual decline in volcanic activity along this sector of the southern Afar Rift is probably linked to the eastward shift and focusing of volcanism along the Wonji belt. The Wonji fault belt is a series of active volcanoes and east-stepping rift grabens located ~100 km east of Gona (Hayward and Ebinger, 1996; Tesfaye et al., 2003). The onset of faulting along the Wonji belt is unconstrained, but the age of the oldest dated volcanic rocks is ca. 1.6 Ma (Meyer et al., 1975). This process of eastward stepping and narrowing of volcanism is part of a longer pattern extending back to the early Miocene (Wolfenden et al., 2005). The volcanic "step" out of the Gona area occurred gradually, perhaps commencing during Segala Noumou Member time (4.6–4.2 Ma), and proceeding through sometime between the end of As Duma volcanism (younger than 2.9 Ma) and beginning of Wonji Belt volcanism (younger than 1.6 Ma). Although local volcanism had ended at Gona by Hadar time, tectonism had not. The high sedimentation rates (Fig. 12; 0.25 mm/yr) displayed by the Hadar Formation are probably in part linked to active slip along the as-yet-unidentified normal fault bounding the E-NE margin of the paleobasin.

Busidima Formation

The Busidima Formation in general marks an important shift in the style of basin filling. The Busidima Formation was laid down entirely by the ancestral Awash River and its tributaries (Fig. 13D). The filling pattern was one of local downcutting by 5–15 m into older sediment, creating shallowly dissected badlands, followed by backfilling and eventual overtopping of the eroded landscape by new sediment. The backfilling was accomplished by a combination of lateral channel accretion in a meandering fluvial system, and by significant overbank flooding (Quade et al., 2004). As locally dissected areas filled with sediment, the river axis migrated laterally and paleosols developed. Average paleosol development is in general weakest in the lower Busidima Formation, where the paleo–Awash River returned every 250,000 yr, and strongest when the Awash migrated entirely out of the Gona project area toward the east. Oldowan stone tools are confined to the three oldest (2.6–1.9 Ma) channel-filling events of the paleo–Awash River, the Acheulian to at least six paleo-Awash sequences above that (1.9–1.6; and ca. 0.6 Ma), and the Late Stone Age ? (younger than 0.05 Ma) to the last filling event at the top of the section.

We agree with Wynn et al. (this volume) that major westward thickening of the Busidima Formation clearly points to deposition in a west-dipping half-graben. This is the opposite polarity of the half-graben that accommodated the Sagantole, Hadar, and probably the Adu-Asa Formations, and, as such, it represents a major new stage in the basin evolution. Such shifts in basin polarity are common in the mature stages of rift-basin evolution, whereby normal faults along previously segmented half-grabens extend and merge, leading to sedimentation over former accommodation zones (Lambiase and Bosworth, 1995).

The western boundary of the newly formed half-graben in Busidima time was the As Duma normal fault. It runs along the western half of the Gona project area, juxtaposing the Busidima Formation against the Sagantole in the south and center (Fig. 1). Northward, the As Duma fault turns NE and cuts upsection through the Hadar Formation. This clearly shows that fault displacement commenced no earlier than 3.5 Ma, the age of the top of the As Aela section. On air photographs, the As Duma fault continues to cut upsection into the Hadar Formation. Combined with the rest of the evidence for westward thickening of the Busidima Formation (Wynn et al., this volume), the As Duma fault likely began to experience offset at some time late in Hadar Formation time. This led to the formation of the modern, west-dipping Awash graben and deposition of the Busidima Formation (Fig. 13D). Significantly, the Busidima Formation is thin to absent on the east side of Ledi-Geraru area, suggesting little or no movement along the E-NE bounding fault (the Eibdaha or other) after 2.9 Ma. We presume, therefore, that this area became the hinge zone of the Awash half-graben during Busidima time.

Other evidence strongly supports the concept of increasing interbasin linkage and sedimentary overtopping of Hadar-age accommodation zones (Gawthorpe and Leeder, 2000) in Busidima time. A fivefold decrease (0.25 versus 0.05 mm/yr) in sediment accumulation rates accompanied the conspicuous shift in depositional style from the Hadar to Busidima Formation. This indicates that much less sediment was arriving in the Awash half-graben than during Hadar time, or that more sediment was being exported from the graben than previously. In essence, the basin became a largely sediment-bypass zone, which is typical of rift basins that have overtopped accommodation zones (Lambiase, 1990). Slowing slip rates along the As Duma fault may have also contributed to lower sedimentation rates, as the most active extension shifted eastward by 1.6 Ma to the Wonji fault belt, as well as westward to the Borkena and the Dergaha-Sheket grabens. These areas today display the greatest extension rates (4.5 ± 0.1 mm/yr; Bilham et al., 1999) and seismic activity (Tesfaye et al., 2003). Having said this, a >10 m scarp marking the As Duma fault at most locations at Gona, except along active drainages, provides clear geomorphic evidence for recent movement (late Quaternary). Moreover, mid- to late Quaternary gravels resting on the As Duma volcanics are clearly offset by the As Duma fault by >4 m.

Mainly fluvial deposition, much lower sedimentation rates (<<0.05 mm/yr), and large-scale cutting-and-filling seem to characterize age-equivalent sediments of the Busidima Formation in other areas of the Awash basin. In the Hadar project area north of Gona, the Busidima Formation is also fluvial and only 45 m thick (Campisano et al., this volume, Chapters 6 and 8), whereas it is apparently not present at all in the middle Ledi-Geraru project area (DiMaggio et al., this volume). In the Middle Awash project area south of Gona, the Bouri Formation exposed in the Bouri Peninsula (a horst) spans about the same time period as the Busidima Formation, roughly 2.6–2.7 to 0.15 Ma (de Heinzelin et al., 1999; de Heinzelin, 2000; Clark et al., 2003). However, the total known thickness of the Bouri Formation is ~80 m, even less than the 130 m represented at Gona. Part of the reason is that up to 1.5 m.y. (2.5–1.0 Ma) of the sedimentary record appears to be missing at Bouri. The Bouri Formation is on average much finer grained than the Busidima Formation, and it is lacustrine for short periods, pointing to local structural impoundment (an accommodation zone?) such as that which occurs today by the Bouri horst. The dominantly fluvial nature of deposition at this time is seen closer to Gona in the eastern Middle Awash area (Williams et al., 1986).

In summary, the weight of the regional evidence during Busidima deposition shows many of the key features of the "through-going fault stage" of rift basins (Lambiase, 1990; Gawthorpe and Leeder, 2000). Consistent with this phase of evolution of most rifts: (1) lacustrine sedimentation in more localized basins during Hadar time and before was replaced with major axial drainage by a through-going river in Busidima time; (2) sedimentation rates sharply decreased due to decreased extension rates, possibly combined with partial hydrographic opening of the lower reaches of the basin north of Ledi-Geraru, draining the final Hadar-age lake dated at ca. or older than 2.94 Ma. This would have permitted greater net export of sediment from the basin. (3) Exposures show near-vertical stacking of axial channel belts adjacent to footwall fans, and (4) periodic (1.0–0.5 Ma) large-fan forcing of the axial river away from the footwall. The modern analog for this mature stage in rifting is the Rio Grande River, which flows along the Rio Grande Rift in the southwestern United States (Cavazza, 1986; Lambiase and Bosworth, 1995).

A very large sample of paleosols and over 285 carbon isotope analyses (Levin et al., 2004; Quade and Levin, 2008) from the lower Busidima Formation reveal a range of abundance of C_4 plants from nearly pure forest in a few settings to open grassland in others, for an average of ~50–50 mix of C_3 and C_4 plants for this portion of the formation. Detailed sampling along single paleosols reveals a vegetation pattern very similar to that seen across the floodplain of the modern Awash River, in which gallery forest grows in a narrow band along the active Awash channel, and swards of tall edaphic grasslands grading to *Acacia*-shortgrass savanna cover alluvial fans marginal to this. Isotopic evidence for nearly pure grassland first occurs at the base of the formation in a few samples, among the earliest evidence for a savanna ecosystem in East Africa. The proportion of C_4 grasses increases slightly to 60% in the upper Busidima Formation, in both the type I and type II lithofacies, similar to the mix of plant types seen across Gona today (Quade and Levin, 2008).

Late Quaternary Incision

Carving of the badlands at Gona by deep incision of the Awash River is a young event and arguably the most significant geomorphic event in the area since the beginning (2.7 Ma) of Busidima Formation time. The minor cutting-and-filling events (≤35 m) of the Busidima Formation bear no comparison to the deep (>250 m) incision that produced the modern valley. When and why did this incision occur?

The beginning of incision has yet to be worked out in detail, but we estimate that it started sometime in the last 160,000 yr. The main basis for this is the presence of the Waidedo Vitric Tephra (0.16 Ma) near the top of the Busidima Formation, along the upper Asbole. The presence of Late (?) Stone Age artifacts at YAS-1 (Fig. 10), in association with a paleo–Awash River high on the western side of the valley (Fig. 10), would suggest an even younger age (younger than 0.05 Ma) of incision, although detailed archaeological studies of the site have yet to be conducted. Another interesting clue to the downcutting history is the presence of the Butte tephra (Table 4) in inset paleochannels of the paleo–Awash River 40–75 m above the modern Awash River. Eventual dating of this tephra should fix the height of the recently incising paleo–Awash River at those times.

The causes for this unprecedented incision remain speculative. The explanation for incision that we favor is base-level drop of the Awash River in its lower reaches. Satellite photos suggest that the modern Awash River is deeply incised as far downstream as the Karrayu graben northeast of the Ledi-Geraru Project area (see Fig. 1 of volume introduction). Base-level lowering as this graben developed may be the key to the recent dramatic incision of the Awash River. The Karrayu graben is clearly an active part of the complex interaction of triple-junction tectonics, but its history is undocumented. Further downstream, the Awash River terminates in Lake Abhe, which occupies the Goba'ad graben, after passing through the Tendaho-Goba'ad discontinuity (see preface, this volume) bordering the southern side of the graben. The Goba'ad graben is a young feature dating to younger than 2.5 Ma (Gasse, 1990). Some studies (Courtillot et al., 1980; Tesfaye et al., 2003) connect the activation of Goba'ad and other east-west grabens with recent eastward propagation of the Gulf of Aden Rift into the southern Afar region. Major subsidence of these grabens sometime prior to ca. 200 ka may also have played a role in recent Awash River incision.

Tectonic versus Climatic Controls on Sedimentation

Gona provides a well-dated example of the tectono-sedimentary evolution of rifts. In accordance with models (Lambiase, 1990; Gawthorpe and Leeder, 2000) of this process, the record from Gona and surrounding areas displays a long-term coarsening-upward sequence, in which basal fluvial and minor lacustrine deposits (Adu-Asa Formation) grade upward into fine-grained lacustrine deposits (Sagantole and Hadar Formations), and into the coarse fluvial deposits of the Busidima Formation on top. Each stage in the rifting process lasted 1–3 m.y. The earliest "fault initiation" stage represented by the Adu-Asa Formation lasted a minimum of 1.2 m.y. (>6.4–5.2 Ma). The Sagantole (>4.6–3.9 Ma) and Hadar Formations (3.8–2.9), representing the "interaction and linkage stage," lasted 1.7–2.1 m.y. The final "through-going fault stage" exemplified by the Busidima Formation lasted ~2.5 m.y. (2.7 to <0.16 Ma). Models also describe a final "fault death" stage (Gawthorpe and Leeder, 2000) in which extension ceases, but this stage does not appear to have developed yet at Gona, given the clear evidence of significant recent activity along the As Duma fault. Rather, sedimentation appears to have ceased in the basin because of propagation of the rift triple junction westward, capturing the lower Awash and causing a base-level drop of the system that has only very recently (younger than 0.16 Ma) propagated into the Gona area.

From the foregoing, we clearly take the view that tectonics dominates over climate in determining the nature of basin deposition on long (10^6 yr) timescales. Recently, Trauth et al. (2005) suggested that East Africa experienced at least three wet phases during the last 2.7 Ma, based on the grouping of lake deposits at 2.7–2.5, 1.9–1.7, and 1.1–0.9 Ma in many sections across the region. This is the time spanned by the Busidima Formation, and, as we have pointed out, no lakes were present during this time in the northern Awash area. The reason is obvious: there must be a hydrographically closed basin to contain a lake. Development of a large, geologically conspicuous lake depends vitally on the tectonic stage of development of the basin, when extension rates are high and accommodation zones are not overtopped by sediment. The modern Awash valley is a cautionary example of the fundamental role of tectonics in lake formation. The Awash graben is terminated to the south in the Middle Awash Project area by the Bouri horst (see introduction), which represents an accommodation zone extending east from the Ayelu volcanic center. The only lake present in the valley, Yardi Lake, is dammed by uplift of the Bouri horst transverse to the main valley axis (de Heinzelin et al., 1999). It would be misplaced to use these deposits from a tectonically produced lake to deduce a wet climatic phase.

ACKNOWLEDGMENTS

The staff of the Gona Paleoanthropological Research Project would like to thank the Authority for Research and Conservation of Cultural Heritage of the Ministry of Culture and Tourism and the National Museum of Ethiopia for the research permit and general support. We appreciate K. Schick and N. Toth at the Stone Age Institute for their overall support. The L.S.B. Leakey Foundation, National Geographic Society, Wenner-Gren Foundation, and the National Science Foundation (NSF) funded this research, as did Tim White and the late Clark Howell through the Revealing Hominid Origins Initiative (RHOI-NSF) program (Division of Behavioral and Cognitive Sciences Award no. 0321893). We appreciate the hospitality of Culture and Tourism of the Afar Regional State at Semera and our Afar colleagues at Eloha. Fieldwork participants included Ali Ma'anda Datto, Ibrahim Habib (deceased), Asahamed Humet, and Yasin Ismail Mohamed. Melanie Everett, Steve Frost, Bill Hart, Lisa Peters, Mike Rogers, and Dietrich Stout are all warmly acknowledged for their help. We owe Matt Heizler a special thanks for providing so much help. We also acknowledge Berhane Asfaw, Yonas Beyene, Giday WoldeGabriel, Asahmed Humer, Alemu Admasu, and Menkir Bitew.

REFERENCES CITED

Aronson, J.L., and Taieb, M., 1981, Geology and paleogeography of the Hadar hominid site, Ethiopia, *in* Rapp, G., and Vondra, C.F., eds., Hominid Sites: Their Geologic Settings: Boulder, Colorado, Westview Press, p. 165–195.

Aronson, J.L., Schmitt, T.J., Walter, R.C., Taieb, M., Tiercelin, J.J., Johanson, D.C., Naeser, C.W., and Nairn, A.E.M., 1977, New geochronologic and paleomagnetic data for the hominid-bearing Hadar Formation of Ethiopia: Nature, v. 267, p. 323–327, doi: 10.1038/267323a0.

Aronson, J.L., Vondra, C.F., Yemane, T., and Walter, R., 1996, Character of the disconformity in the upper part of the Hadar Formation: Geological Society of America Abstracts with Programs, v. 28, no. 6, p. 28.

Bilham, R., Bendick, R., Larson, K., Mohr, P., Braun, J., Tesfaye, S., and Asfaw, L., 1999, Secular and tidal strain across the main Ethiopian Rift: Geophysical Research Letters, v. 26, no. 18, p. 2789–2792, doi: 10.1029/1998GL005315.

Bosworth, W.P., 1985, Geometry of propagating continental rifts: Nature, v. 316, p. 625–627, doi: 10.1038/316625a0.

Butler, R.F., 1992, Paleomagnetism: Magnetic domains to geologic terranes: Boston, Blackwell Scientific Publications, p. 83–104.

Campisano, C.J., and Feibel, C.S., 2008, this volume (Chapter 6), Tephrostratigraphy of the Hadar and Busidima Formations at Hadar, Afar Depression, Ethiopia, in Quade, J., and Wynn, J.G., eds., The Geology of Early Humans in the Horn of Africa: Geological Society of America Special Paper 446, doi: 10.1130/2008.2446(06).

Campisano, C.J., and Feibel, C.S., 2008, this volume (Chapter 8), Depositional environments and stratigraphic summary of the Pliocene Hadar Formation at Hadar, Afar Depression, Ethiopia, in Quade, J., and Wynn, J.G., eds., The Geology of Early Humans in the Horn of Africa: Geological Society of America Special Paper 446, doi: 10.1130/2008.2446(08).

Cavazza, W., 1986, Sedimentation pattern of rift-filling unit, Tesuque Formation (Miocene), Espanola Basin, Rio Grande Rift, New Mexico: Journal of Sedimentary Petrology, v. 59, p. 287–296.

Chernet, T., Hart, W.K., Aronson, J.L., and Walter, R.C., 1998, New age constraints on the timing of volcanism and tectonism in the northern Main Ethiopian Rift southern Afar transition zone (Ethiopia): Journal of Volcanology and Geothermal Research, v. 80, p. 267–280, doi: 10.1016/S0377-0273(97)00035-8.

Clark, J.D., de Heinzelin, J., Schick, K.D., Hart, W.K., White, T.D., WoldeGabriel, G., Walter, R.C., Suwa, G., Asfaw, B., Vrba, E., and Haile-Selassie, Y.T.D., 1994, African *Homo erectus*: Old radiometric ages and young Oldowan assemblages in the middle Awash Valley, Ethiopia: Science, v. 264, p. 1907–1909.

Clark, J.D., Beyene, Y., WoldeGabriel, G., Hart, W.K., Renne, P.R., Gilbert, H., Defleur, A., Suwa, G., Katoh, S., Ludwig, K.R., Boisserie, J.-R., Asfaw, B., and White, T.D., 2003, African *Homo erectus*: Old radiometric ages and young Oldowan assemblages in the middle Awash Valley, Ethiopia: Science, v. 423, p. 747–752.

Courtillot, V., Galdeano, A., and Le Mouel, J.L., 1980, Propagation of an accreting plate boundary: Discussion of new aeromagnetic data in the Gulf of Tadjdurah and the southern Afar: Earth and Planetary Science Letters, v. 47, p. 144–160, doi: 10.1016/0012-821X(80)90113-2.

de Heinzelin, J., 2000, Chapter 3: Stratigraphy, in de Heinzelin, J., Clark, J.D., Schick, K.D., and Gilbert, W.H., eds., The Acheulian and the Plio-Pleistocene Deposits of the Middle Awash Valley, Ethiopia: Royal Museum of Central Africa (Belgium) Annales-Sciences Géologiques, v. 104, p. 11–46.

de Heinzelin, J., Clark, J.D., White, T., Hart, W., Renne, P., WoldeGabriel, G., Beyene, Y., and Vrba, E., 1999, Environment and behavior of 2.5-million-year-old Bouri hominids: Science, v. 284, p. 625–629, doi: 10.1126/science.284.5414.625.

DiMaggio, E.N., Campisano, C.J., Arrowsmith, J.R., Reed, K.E., Swisher, C.C., III, and Lockwood, C.A., 2008, this volume, Correlation and stratigraphy of the BKT-2 Volcanic Complex in west-central Afar, Ethiopia, in Quade, J., and Wynn, J.G., eds., The Geology of Early Humans in the Horn of Africa: Geological Society of America Special Paper 446, doi: 10.1130/2008.2446(07).

Dupont-Nivet, G., Sier, M., Campisano, C.J., Arrowsmith, R., DiMaggio, E., Reed, K., Lockwood, C., Franke, C., and Hüsing, S., 2008, this volume, Magnetostratigraphy of the eastern Hadar Basin (Ledi-Geraru research area, Ethiopia) and implications for hominin paleoenvironments, in Quade, J., and Wynn, J.G., eds., The Geology of Early Humans in the Horn of Africa: Geological Society of America Special Paper 446, doi: 10.1130/2008.2446(03).

Fisher, R.A., 1953, Dispersion on a sphere: Proceedings of the Royal Society of London, ser. A, v. 217, p. 295–305, doi: 10.1098/rspa.1953.0064.

Gasse, F., 1990, Tectonic and climatic controls on lake distribution and environments in Afar from Miocene to present, in Kaz, B., ed., Lacustrine Exploration: Case Studies and Modern Analogues: American Association of Petroleum Geologists Memoir 50, p. 19–41.

Gawthorpe, R.L., and Leeder, M.R., 2000, Tectono-sedimentary evolution of active extensional basins: Basin Research, v. 12, p. 195–218, doi: 10.1046/j.1365-2117.2000.00121.x.

Gentry, A.W., 1981, Notes on Bovidae (Mammalia) from the Hadar Formation and from Amado and Geraru, Ethiopia: Kirtlandia, v. 33, p. 1–30.

Geraads, D., Alemsageged, Z., Reed, D., Wynn, J., and Roman, D.C., 2004, The Pleistocene fauna from Asbole, lower Awash Valley, Ethiopia, and its environmental and biochronological implications: Geobios, v. 37, p. 697–718, doi: 10.1016/j.geobios.2003.05.011.

Haileab, B., and Brown, F.H., 1994, Tephra correlations between the Gadeb prehistoric site and the Turkana Basin: Journal of Human Evolution, v. 26, p. 167–173, doi: 10.1006/jhev.1994.1009.

Hayward, N.J., and Ebinger, C.J., 1996, Variations in the along-axis segmentation of the Afar Rift system: Tectonics, v. 15, no. 2, p. 244–257, doi: 10.1029/95TC02292.

Hunt, J.B., and Hill, P.G., 1993, Tephra geochemistry: A discussion of some persistent analytical problems: The Holocene, v. 3, no. 3, p. 271–278, doi: 10.1177/095968369300300310.

Johanson, D.C., and Taieb, M., 1976, Plio-Pleistocene hominid discoveries in Hadar, Ethiopia: Nature, v. 260, p. 293–297, doi: 10.1038/260293a0.

Johanson, D.C., Taieb, M., and Coppens, Y., 1982, Pliocene hominids from the Hadar Formation, Ethiopia (1973–1977): Stratigraphic, chronologic, and paleoenvironmental contexts, with notes on hominid morphology and systematics: American Journal of Physical Anthropology, v. 57, p. 373–402, doi: 10.1002/ajpa.1330570402.

Kalb, J.E., Oswald, E.B., Tebedge, S., Mebrate, A., Tola, E., and Peak, D., 1982, Geology and stratigraphy of Neogene deposits, middle Awash Valley, Ethiopia: Nature, v. 298, p. 17–25, doi: 10.1038/298017a0.

Kimbel, W.H., Johanson, D.C., and Rak, Y., 1994, The first skull and other new discoveries of *Australopithecus afarensis*: Nature, v. 368, p. 449–451, doi: 10.1038/368449a0.

Kirschvink, J.L., 1980, The least-squares line and plane and the analysis of palaeomagnetic data: Geophysical Journal of the Royal Astronomical Society, v. 62, p. 699–718.

Kleinsasser, L.L., Quade, J., McIntosh, W.C., Levin, N.E., Simpson, S.W., and Semaw, S., 2008, this volume, Stratigraphy and geochronology of the late Miocene Adu-Asa Formation at Gona, Ethiopia, in Quade, J., and Wynn, J.G., eds., The Geology of Early Humans in the Horn of Africa: Geological Society of America Special Paper 446, doi: 10.1130/2008.2446(02).

Lambiase, J.J., 1990, A model for tectonic control of lacustrine stratigraphic sequences in continental rift basins, in Kaz, B., ed., Lacustrine Exploration: Case Studies and Modern Analogues: American Association of Petroleum Geologists Memoir 50, p. 265–276.

Lambiase, J.J., and Bosworth, W., 1995, Structural controls on sedimentation in continental rifts, in Lambiase, J.J., ed., Hydrocarbon Habitat in Rift Basins: Geological Society of London Special Publication 80, p. 117–144.

Laughlin, A.W., Poths, J., Healy, H.A., Reneau, S., and WoldeGabriel, G., 1994, Dating of Quaternary basalts using cosmogenic ^{3}He and ^{14}C methods with implications of excess ^{40}Ar: Geology, v. 22, p. 135–138, doi: 10.1130/0091-7613(1994)022<0135:DOQBUT>2.3.CO;2.

Levin, N.E., Quade, J., Simpson, S., Semaw, S., and Rogers, M., 2004, Isotopic evidence for Plio-Pleistocene environmental change at Gona, Ethiopia: Earth and Planetary Science Letters, v. 219, p. 93–100, doi: 10.1016/S0012-821X(03)00707-6.

Levin, N.E., Simpson, S.W., Quade, J., Cerling, T.E., and Frost, S.R., 2008, this volume, Herbivore enamel carbon isotopic composition and the environmental context of *Ardipithecus* at Gona, Ethiopia, in Quade, J., and Wynn, J.G., eds., The Geology of Early Humans in the Horn of Africa: Geological Society of America Special Paper 446, doi: 10.1130/2008.2446(10).

Lourens, L.J., Hilgen, F.J., Laskar, J., Shackleton, N.J., and Wilson, D.S., 2004, The Neogene Period, in Gradstein, F.M., Orgg, J.G., and Smith, A.G., eds., A Geologic Timescale 2004: Cambridge, Cambridge University Press, p. 409–440.

Lynn, W., and Williams, D., 1992, The making of a Vertisol: Soil Survey Horizons, v. 33, p. 45–50.

McDougall, I., 1985, K-Ar and ^{40}Ar/^{39}Ar dating of the hominid bearing Plio-Pleistocene sequence at Koobi Fora, Lake Turkana, northern Kenya:

Geological Society of America Bulletin, v. 96, no. 2, p. 159–175, doi: 10.1130/0016-7606(1985)96<159:KAADOT>2.0.CO;2.

McIntosh, W.C., and Chamberlin, R.M., 1994, ^{40}Ar/^{39}Ar geochronology of middle to late Cenozoic ignimbrites, mafic lavas, and volcaniclastic rocks in the Quemado region, New Mexico: New Mexico Geological Society Guidebook, v. 45, p. 165–185.

Meyer, W., Pilger, A., Rosler, A., and Stets, J., 1975, Tectonic evolution of the northern part of the Main Ethiopian Rift in southern Ethiopia, in Pilger, A., and Rosler, A., eds., Afar Depression of Ethiopia: Stuttgart, Schweizerbart, p. 352–362.

Miall, A.D., 1978, Lithofacies types and vertical profile models in braided river deposits: A summary, in Miall, A.D., ed., Fluvial Sedimentology: Canadian Society of Petroleum Geology Memoir 5, p. 597–604.

Morley, C.K., 1995, Developments in the structural geology of rifts over the last decade and their impact on hydrocarbon exploration, in Lambiase, J.J., ed., Hydrocarbon Habitat in Rift Basins: Geological Society of London Special Publication 80, p. 1–32.

Nielsen, C.H., and Sigurdsson, H., 1981, Quantitative methods for electron microprobe analysis of sodium in natural and synthetic glasses: The American Mineralogist, v. 66, p. 547–552.

Quade, J., and Levin, N., 2008, East African hominid paleoecology: Isotopic evidence from paleosols, in Sponnheimer, M., Reed, K., Lee-Thorp, J., and Ungar, P., eds., Early Hominin Paleoecology: Boulder, Colorado, University Press of Colorado (in press).

Quade, J., Levin, N., Semaw, S., Simpson, S., Rogers, M., and Stout, D., 2004, Paleoenvironments of the earliest toolmakers: Geological Society of America Bulletin, v. 116, p. 1529–1544, doi: 10.1130/B25358.1.

Renne, P.R., Swisher, C.C., III, Deino, A.L., Karner, D.B., Owens, T., and DePaolo, D.J., 1998, Intercalibration of standards, absolute ages and uncertainties in ^{40}Ar/^{39}Ar dating: Chemical Geology, v. 145, no. 1–2, p. 117–152, doi: 10.1016/S0009-2541(97)00159-9.

Renne, P.R., WoldeGabriel, G., Hart, W.K., Heiken, G., and White, T.D., 1999, Chronostratigraphy of the Mio-Pliocene Sagantole Formation, Middle Awash Valley, Afar Rift, Ethiopia: Geological Society of America Bulletin, v. 111, p. 869–885, doi: 10.1130/0016-7606(1999)111<0869:COTMPS>2.3.CO;2.

Rogers, M.J., Feibel, C.S., and Harris, J.W.K., 1994, Changing patterns of land use by Plio-Pleistocene hominids in the Lake Turkana Basin: Journal of Human Evolution, v. 27, p. 139–158, doi: 10.1006/jhev.1994.1039.

Roman, D.C., Campisano, C., Quade, J., DiMaggio, E., Arrowsmith, J R., and Feibel, C., 2008, this volume, Composite tephrostratigraphy of the Dikika, Gona, Hadar, and Ledi-Geraru project areas, northern Awash, Ethiopia, in Quade, J., and Wynn, J.G., eds., The Geology of Early Humans in the Horn of Africa: Geological Society of America Special Paper 446, doi: 10.1130/2008.2446(05).

Schlische, R.W., 1991, Half-graben basin filling models: New constraints on continental extensional basin development: Basin Research, v. 3, p. 123–141, doi: 10.1111/j.1365-2117.1991.tb00123.x.

Semaw, S., Renne, P., Harris, J.W.K., Feibel, C.S., Bernor, R.L., Fessaha, N., and Mowbray, K., 1997, 2.5-million-year-old stone tools from Gona, Ethiopia: Nature, v. 385, p. 333–335, doi: 10.1038/385333a0.

Semaw, S., Rogers, M.J., Quade, J., Renne, P., Butler, R.F., Dominguez-Roderigo, M., Stout, D., Hart, W.S., Pickering, T., and Simpson, S.W., 2003, 2.6-million-year-old stone tools and associated bones from OGS-6 and OGS-7, Gona, Ethiopia: Journal of Human Evolution, v. 45, p. 169–177, doi: 10.1016/S0047-2484(03)00093-9.

Semaw, S., Simpson, S.W., Quade, J., Renne, P.R., Butler, R.F., McIntosh, W.C., Levin, N., Dominguez-Rodrigo, M., and Rogers, M.J., 2005, Early Pliocene hominids from Gona, Afar, Ethiopia: Nature, v. 433, p. 301–305.

Simpson, S.W., Quade, J., Kleinsasser, L., Levin, N., MacIntosh, W., Dunbar, N., and Semaw, S., 2007, Late Miocene hominid teeth from Gona project area, Ethiopia: Program of the Seventy-Sixth Annual Meeting of the American Association of Physical Anthropologists, v. 132, no. S44, p. 219.

Stout, D., Quade, J., Semaw, S., Rogers, M., and Levin, N., 2005, Raw material selectivity of the earliest stone toolmakers at Gona, Afar, Ethiopia: Journal of Human Evolution, v. 48, p. 365–380, doi: 10.1016/j.jhevol.2004.10.006.

Taieb, M., Johanson, D.C., Coppens, Y., and Aronson, J.L., 1976, Geological and palaeontological background of Hadar hominid site, Afar, Ethiopia: Nature, v. 260, p. 289–293, doi: 10.1038/260289a0.

Tesfaye, S., Harding, D.J., and Kusky, T.M., 2003, Early continental breakup boundary and migration of the Afar triple junction, Ethiopia: Geological Society of America Bulletin, v. 115, no. 9, p. 1053–1067, doi: 10.1130/B25149.1.

Tiercelin, J.J., 1986, The Pliocene Hadar Formation, Afar Depression of Ethiopia, in Frostick, L.E., Renaut, R., Reid, I., and Tiercelin, J.J., eds., Sedimentation in the African Rifts: Geological Society of London Special Publication 23, p. 221–240.

Trauth, M.H., Maslin, M.A., Deino, A., and Strecker, M., 2005, Late Cenozoic moisture history of East Africa: Science, v. 309, p. 2051–2053, doi: 10.1126/science.1112964.

van der Hoven, S., and Quade, J., 2002, Tracing spatial and temporal variations in the sources of calcium in pedogenic carbonates in a semi-arid environment: Geoderma, v. 108, p. 259–276, doi: 10.1016/S0016-7061(02)00134-9.

Walter, R.C., 1994, Age of Lucy and the First Family: Laser ^{40}Ar/^{39}Ar dating of the Denen Dora Member of the Hadar Formation: Geology, v. 22, p. 6–10, doi: 10.1130/0091-7613(1994)022<0006:AOLATF>2.3.CO;2.

Walter, R.C., and Aronson, J.L., 1982, Revisions of K/Ar ages for the Hadar hominid site, Ethiopia: Nature, v. 296, p. 122–127, doi: 10.1038/296122a0.

Walter, R.C., and Aronson, J.L., 1993, Age and source of the Sidi Hakoma tephra, Hadar Formation, Ethiopia: Journal of Human Evolution, v. 25, p. 229–240, doi: 10.1006/jhev.1993.1046.

Watson, G.S., 1956, A test for randomness of directions: Monthly Notices of the Geophysical Journal of the Royal Astronomical Society, v. 7, p. 160–161.

Williams, M.A.J., Assefa, G., and Adamson, D.A., 1986, Depositional context of Plio-Pleistocene hominid-bearing formations in the Middle Awash valley, southern Afar Rift, Ethiopia, in Frostick, L.E., Renaut, R., Reid, I., and Tiercelin, J.J., eds., Sedimentation in the African Rifts: Geological Society of London Special Publication 25, p. 241–251.

WoldeGabriel, G., Aronson, J.L., and Walter, R.C., 1990, Geology, geochronology, and rift basin development in the central sector of the Main Ethiopian Rift: Geological Society of America Bulletin, v. 102, p. 439–458, doi: 10.1130/0016-7606(1990)102<0439:GGARBD>2.3.CO;2.

WoldeGabriel, G., White, T.D., Suwa, G., Renne, P., de Heinzelin, J., Hart, W.K., and Heiken, G., 1994, Ecological and temporal context of early Pliocene hominids at Aramis, Ethiopia: Nature, v. 371, p. 330–333, doi: 10.1038/371330a0.

WoldeGabriel, G., Haile-Selassie, Y., Renne, P.R., Hart, W.K., Ambrose, S.H., Asfaw, B., Heiken, G., and White, T., 2001, Geology and palaeontology of the late Miocene Middle Awash valley, Afar Rift, Ethiopia: Nature, v. 412, p. 175–178, doi: 10.1038/35084058.

Wolfenden, E., Ebinger, C., Yirgu, G., Renne, P.R., and Kelley, S.P., 2005, Evolution of a volcanic rifted margin: Southern Red Sea, Ethiopia: Geological Society of America Bulletin, v. 117, no. 7/8, p. 846–864, doi: 10.1130/B25516.1.

Wynn, J.G., Roman, D.C., Alemseged, Z., Reed, D., Geraads, D., and Munro, S., 2008, this volume, Stratigraphy, depositional environments, and basin structure of the Hadar and Busidima Formations at Dikika, Ethiopia, in Quade, J., and Wynn, J.G., eds., The Geology of Early Humans in the Horn of Africa: Geological Society of America Special Paper 446, doi: 10.1130/2008.2446(04).

Yemane, T., 1997, Stratigraphy and Sedimentology of the Hadar Formation, Afar, Ethiopia [Ph.D. thesis]: Ames, Iowa State University, 182 p.

MANUSCRIPT ACCEPTED BY THE SOCIETY 17 JUNE 2008

The Geological Society of America
Special Paper 446
2008

Stratigraphy and geochronology of the late Miocene Adu-Asa Formation at Gona, Ethiopia

Lynnette L. Kleinsasser*
Jay Quade
Department of Geosciences, University of Arizona, Tucson, Arizona 85721-0077, USA

William C. McIntosh
New Mexico Bureau of Geology and Mineral Resources, New Mexico Institute of Technology, 801 Leroy Place, Socorro, New Mexico 87801-4796, USA

Naomi E. Levin[†]
Department of Geology and Geophysics, University of Utah, 135 South 1460 East, Salt Lake City, Utah 84112-0111, USA

Scott W. Simpson
Department of Anatomy, School of Medicine, Case Western Reserve University, 10900 Euclid Avenue, Cleveland, Ohio 44106-4930, USA

Sileshi Semaw
Center for Research into the Anthropological Foundations of Technology, Stone Age Institute, P.O. Box 5097, Bloomington, Indiana 47407-5097, USA

ABSTRACT

The Gona area includes many rich fossil localities that are of great consequence to the study of human evolution. The Adu-Asa Formation, containing the oldest of these fossils, consists of nearly 200 m of fossil-bearing sedimentary rocks in thin (≤30 m), laterally variable sections interlayered with abundant basaltic lava flows. These volcanic and sedimentary rocks dip gently to the east and are repeated by north-northwest–trending, mostly west-dipping normal faults that accommodate extension in the Afar Rift.

The volcanic rocks in the Adu-Asa Formation are strongly bimodal. Basaltic lavas and tuffs are abundant, but we have also identified a rhyolite center and seven different silicic, or dominantly silicic, tuffs. Of these tuff units, we were able to identify four major tuffs across the Adu-Asa Formation at Gona by combining geochemical comparisons with detailed stratigraphic sections through fossil-bearing deposits: the Sifi, the Kobo'o, the Belewa, and the Ogoti Tuffs. New $^{40}Ar/^{39}Ar$ dates of these and other tuffs, as well as basalt flows, indicate that the formation spans the period from

---
*lkleinsasser@barrick.com
[†]Current address: Division of Geological and Planetary Sciences, California Institute of Technology, MC100-23, 1200 E. California Blvd., Pasadena, California 91125, USA.

Kleinsasser, L.L., Quade, J., McIntosh, W.C., Levin, N.E., Simpson, S.W., and Semaw, S., 2008, Stratigraphy and geochronology of the late Miocene Adu-Asa Formation at Gona, Ethiopia, in Quade, J., and Wynn, J.G., eds., The Geology of Early Humans in the Horn of Africa: Geological Society of America Special Paper 446, p. 33–65, doi: 10.1130/2008.2446(02). For permission to copy, contact editing@geosociety.org. ©2008 The Geological Society of America. All rights reserved.

34 Kleinsasser et al.

5.2 Ma to 6.4 Ma, although the oldest deposits within the Gona Paleoanthropological Research Project (GPRP) area have yet to be thoroughly surveyed. Known fossil localities within the Adu-Asa Formation at Gona are grouped into three temporal clusters, ranging in age from ca. 6.4 Ma to ca. 5.5 Ma.

Keywords: tephrostratigraphy, Gona, Adu-Asa Formation, $^{40}Ar/^{39}Ar$ dating, *Ardipithecus ramidus, Ardipithecus kadabba*.

INTRODUCTION

Genetic studies suggest that human and chimpanzee lineages diverged in Africa during the late Miocene–early Pliocene (Horai et al., 1992; Ruvolo, 1997; Chen and Li, 2001; Patterson et al., 2006). Patterson et al. (2006) estimated that the human-chimpanzee genome diverged permanently no earlier than 6.3 Ma, although they preferred a younger age of human-chimpanzee speciation, perhaps as young as 5.4 Ma. In contrast, recent hominid finds of *Ardipithecus kadabba* in Ethiopia, *Sahelanthropus tchadensis* in Chad, and *Orrorin tugenensis* in Kenya all date to this pre–5.4 Ma time period, suggesting that the hominid-chimpanzee divergence must have been earlier (WoldeGabriel et al., 2001; Vignaud et al., 2002; Sawada et al., 2002). Resolution of the apparent contradiction between genetic and fossil evidence, as well as separation of the phylogenetic relationships among the earliest hominids, rests upon the discovery and study of new, well-dated fossil material.

The Gona Paleoanthropological Research Project (GPRP), which has previously produced specimens of *Ardipithecus ramidus* in the early Pliocene Sagantole Formation (Semaw et al., 2005), has in recent field seasons contributed a number of discoveries in the older, and largely unstudied, deposits of the Adu-Asa Formation. These older hominids are assigned to the species *Ardipithecus kadabba* (Simpson et al., 2007). Secure dating of these and similar finds is crucial to illuminating the earliest chapter of our evolution.

GEOGRAPHIC AND GEOLOGIC SETTING

The Gona Paleoanthropological Research Project (GPRP), located ~300 km northeast of Addis Ababa, Ethiopia, contains a fossil-rich record of fluvial, lacustrine, and volcanic deposits spanning much of the last 6.5 m.y. (Fig. 1). The Gona Paleoanthropological Research Project lies within the Afar Rift and is 150–200 km west of the current triple junction between the Red Sea Rift, the Gulf of Aden Rift, and the Main Ethiopian Rift (Tesfaye et al., 2003). The Afar Rift is bounded on the north by the Danakil horst, on the east by the Southeast Ethiopian Highlands, and on the west by the Western Ethiopian escarpment (Quade and Wynn, this volume, Preface). Within this basin, the project area is bounded on the north by the Mille-Bati road, on the east by the Awash River, and on the south by the As Bole drainage. The western extent of the project area continues into the Western Ethiopian escarpment. These westernmost deposits have previously been referred to as the Dahla Series fissural basalts in the volcanological literature (Barberi et al., 1975; Wolfenden et al., 2005), but here we adopt the term Adu-Asa Formation. This term was coined by Kalb et al. (1982) and embraced, with some revisions, by later workers in the same area (WoldeGabriel et al., 2001) to encompass interbedded basalts and sedimentary rocks due south of the western part of Gona. Satellite photos strongly suggest north-south continuity of the Adu-Asa Formation in this region, a correlation confirmed by radiometric dates that we present in this paper.

The Adu-Asa Formation in the Gona Paleoanthropological Research Project area is composed of ~185 m of mostly basaltic lava flows intercalated with thin zones of volcaniclastic, fluvio-lacustrine sedimentary rocks. All fossil localities are confined to these sedimentary rocks. A rhyolite dome is exposed in the northern end of the project area and caps the Adu-Asa Formation there. Both rhyolitic and basaltic tuffs are common throughout the formation. However, the basaltic tuffs are generally too altered to use as geochemical markers, so we have mainly relied on the silicic tuffs to provide the necessary chronological control on the fossil localities (Fig. 2). In all but one case, the tuffs are ash-fall units that have been reworked to varying degrees and are interbedded with sedimentary rocks. The exception is a non-welded ash-flow tuff and its related surge deposit associated with the rhyolite dome in the northern end of the project area, although this tuff complex does have an ash-fall component as well.

Structurally, the Adu-Asa Formation at Gona is cut by numerous north-northwest–trending, west-dipping faults that have accommodated extension in the Afar Rift. Although these faults are too abundant to show in Figure 1, the topography strongly echoes the trend of this fault pattern. Dips on beds are gentle, generally to the east, and usually do not exceed 25°E. Thus, deposits in this formation tend to decrease in age to the

Figure 1. Locations of paleontological sites within the Adu-Asa Formation at Gona. Only sites mentioned in the text are shown. Site abbreviations are as follows: Hamadi Das (HMD), As Bole Dora (ABD), Bodele Dora (BDL), Henali (HEN), and Escarpment (ESC). Note the rhyolite dome in the northern end of the project area. Inset shows the location of the Gona Paleoanthropological Research Project area within Ethiopia. Satellite photo is an ASTER image (U.S Geological Survey [USGS] and Japan ASTER program, ASTER scene AST_L1A.003:2005991834, 1B, USGS, Sioux Falls, 24 December 2001).

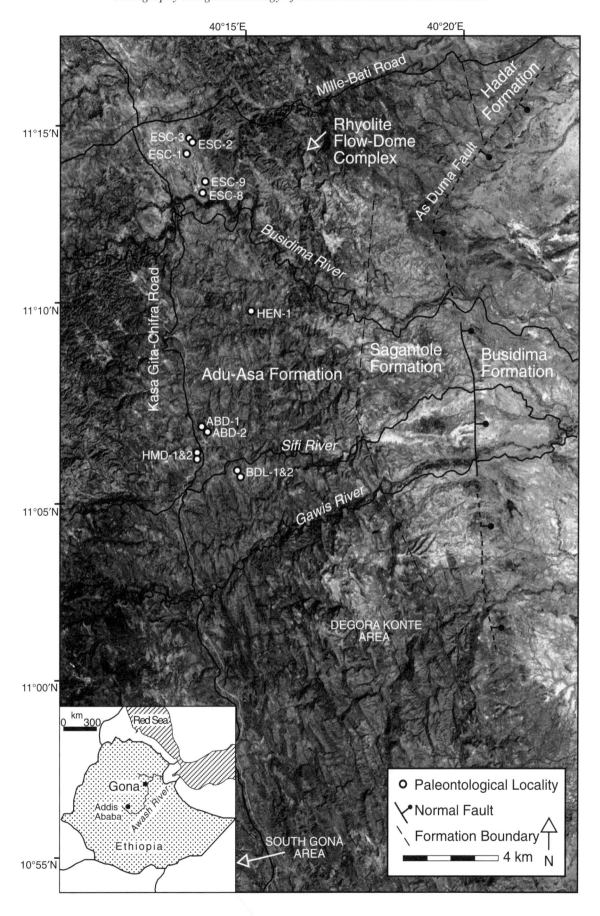

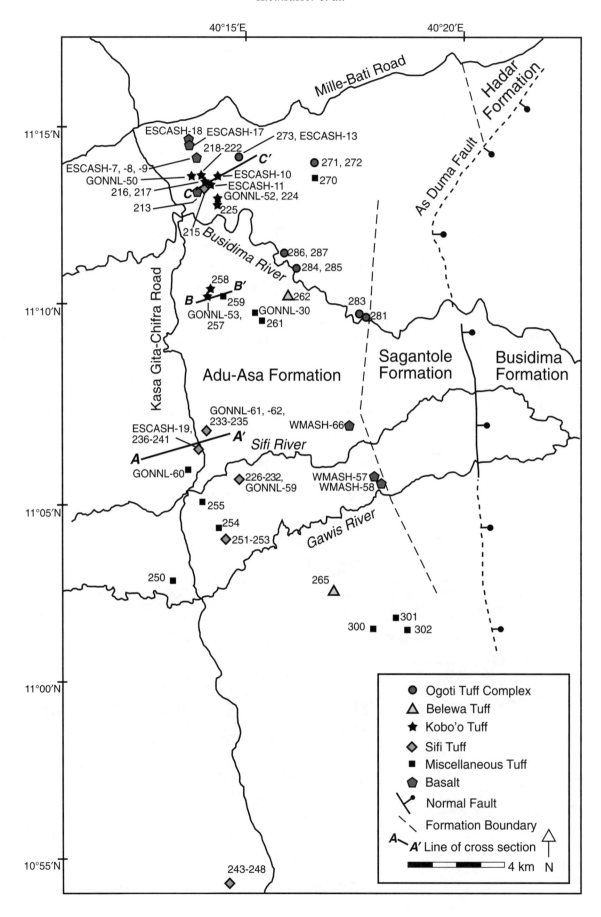

east. Repetitions of the stratigraphy by normal faults are common. The abundance of these faults, the similarity in appearance of basalt lava flows, and locally restricted exposures make correlations of outcrops between areas difficult. Here, the use of glass compositions from tuffs in correlations between areas—the central focus of this research—was vital to producing a coherent stratigraphic context for the fossil finds.

The Adu-Asa Formation is conformable with the younger Sagantole Formation, which is exposed to the east. The contact is characterized by a shift from mostly volcanic units with a sedimentary component in the Adu-Asa Formation to dominantly sedimentary units with a volcanic component in the Sagantole Formation. This lithologic contrast is strongly expressed topographically over much of the western Gona Paleoanthropological Research Project area. The Adu-Asa Formation is characterized by steep ridge (= basalt lavas)-and-swale (= intercalated sedimentary rocks) topography, whereas the sedimentary rocks that dominate the Sagantole Formation weather recessively. The top of the Adu-Asa Formation is marked by a final topographically high-standing basalt flow(s) (Fig. 1). Exceptions to this pattern can be found at the extreme northern and southern ends of the project area, where basalts instead of sedimentary rocks dominate the Sagantole Formation, and the transition between the Adu-Asa and Sagantole Formations is indistinct.

Several younger formations lie to the east of the Adu-Asa Formation within the Gona Paleoanthropological Research Project area. The Sagantole Formation at Gona is dominantly lacustrine volcaniclastic sedimentary rocks with intercalated basaltic lavas. While rich in fossils, the deposits of the Sagantole Formation at Gona are extensively faulted and contain only altered tuffs, making geochemical correlations difficult. It ranges in age from older than 4.6 Ma to 3.9 Ma (Semaw et al., 2005; Quade et al., this volume; Levin et al., this volume).

The Sagantole Formation is mostly bound on the east by the As Duma fault, a major north-south–trending, east-dipping normal fault that has been active subsequent to 4 Ma to present. At the surface, it juxtaposes the east-dipping Sagantole Formation to the west against the largely undeformed and much younger Busidima Formation to the east (Fig. 1). The exception is in the northern part of the Gona Paleoanthropological Research Project area, where the Sagantole Formation is in conformable contact with the Hadar Formation, and the As Duma fault separates the Hadar Formation from the Busidima Formation. At Gona, the Hadar and Busidima Formations span the period from 3.8 to <0.16 Ma (Quade et al., 2004; Quade et al., this volume).

Figure 2. Locations of samples from volcanic rocks (tuffs, basalts, obsidian) from the Adu-Asa Formation at Gona. For samples containing only a number, the prefix "GON05-" has been omitted. In cases where multiple samples were collected from the same locality, we have omitted markers for altered tuffs, obsidian samples, or basalts to aid clarity. Geochronological samples from the neighboring Sagantole, Hadar, and Busidima Formations are not shown.

METHODS

Fieldwork

The fieldwork and laboratory work for this study were conducted during 2003–2006. Fieldwork focused primarily on collecting volcanic units suitable for ^{40}Ar/^{39}Ar dating and/or major-element geochemical characterization using an electron microprobe. At most of the fossil localities, we also measured stratigraphic sections in order to document the relevant relationships between fossil-rich beds and the sampled volcanic units. Most of the geochronological samples taken in the Adu-Asa Formation at Gona are ash-fall tuffs, although several basalt samples were collected along with a few obsidian and ash-flow tuff samples. For ash-fall and ash-flow tuff samples, collection focused on obtaining both fresh glass shards and any juvenile phenocryst populations present in each unit, although almost every tephra unit encountered in the field was collected.

Many of the ash-fall deposits form multiple subunits in outcrop, which is at least in part a result of reworking. In these cases, we sampled each subunit in order to be sure that we had obtained a representative sample. Commonly, one subunit contained a greater density of phenocrysts and another contained a greater concentration of fresh glass shards. If any other populations were present, such as pumice lapilli or obsidian clasts, subunits containing these populations were also sampled in order to characterize the various components of the tuff. Thus, by sampling each subunit individually, we were able to account for the sedimentary sorting that may have separated different portions of a single tuff.

We also collected a few obsidian samples from the rhyolite dome in the northern end of the project area in order to characterize the composition of that silicic source. Samples collected included both glassy and spherulitic obsidian. We also sampled a basal pumice breccia.

For the basalts, collection efforts focused on obtaining samples that were as fresh and as little oxidized or hydrolyzed as possible. In addition, we looked for lava flows with holocrystalline groundmass and a small percentage of phenocrysts, but we collected hand samples of lavas at many stratigraphic levels throughout the Adu-Asa Formation at Gona. In practice, only outcrops that were pervasively argillized and friable were not sampled.

Laboratory Work

Tuffs

Samples were prepared by crushing and sieving each tuff into various size fractions, typically >500 mm, 500–250 mm, and 250–125 mm. If the sample contained unaltered glass shards, then a portion of the size fraction in which the shards were most abundant was used to make a microprobe mount. Most commonly, this was the 250–125 mm size fraction. Every tuff sample collected in the Adu-Asa Formation at Gona was processed for analysis on an electron microprobe. All suitable samples, whether glass shards, obsidian, or feldspar, were analyzed on a Cameca SX50 electron microprobe at the University of Arizona in the Department of Planetary Sciences Lunar and Planetary Laboratory.

For each component studied, we analyzed ~20 points, with each point on a different shard or crystal. In a typical suite of analyses, most shards/crystals proved chemically homogeneous. Often a few grain analyses were rejected prior to statistical analysis as contaminated if their compositions were different from the main compositional mode.

Many researchers have documented alkali mobility in glass as a result of electron bombardment during electron microprobe analysis (Hunt and Hill, 1993; Morgan and London, 1996; Nielsen and Sigurdsson, 1981). In determining the best analytical conditions to use, we followed some of the suggestions of Froggatt (1992) and Hunt and Hill (1993). Froggatt (1992) suggests that researchers use a beam defocused to at least 10 μm across, as well as a lower beam current when analyzing alkali elements. Hunt and Hill (1993) recommended that researchers analyze alkalis first, before a sample has time for significant mobilization to occur. After some experimentation, we settled on the analytical conditions in Table 1. We used these conditions for all analyses, whether glass or crystal.

Morgan and London (1996) described an optimal analytical setup for dealing with alkali mobility and the corresponding "grow-in" of Si and Al. As a comparison, we ran newly prepared mounts of selected samples using the setup conditions Morgan and London (1996) recommended and compared those results to the data obtained using the setup conditions in this study (Table 1).

Glass

If a size fraction contained fresh glass shards, then no further processing was necessary before creating a microprobe mount. Shards that had partially devitrified or were otherwise altered were ground away during polishing, as was any rind of clay alteration on otherwise well-preserved glass. We examined glass shards under a 10–40× binocular scope to establish their general morphology and followed the descriptive shard morphology system of Katoh et al. (2000), which is based on work by Ross (1928), Heiken (1972), and Yoshikawa (1976) (Fig. 3).

The few obsidian samples, either collected as corestones or picked out of a tuff sample as a clast, were prepared for the electron microprobe by lightly crushing them with a mortar and pestle and then mounting the pieces using the same process as that of the glass shards.

Phenocrysts

Many of the tuff samples collected also contained phenocrysts, usually feldspars. After separating a tuff sample into the various size fractions, if it was determined that feldspars were present, a small number (commonly 50–100) was extracted by hand and mounted.

Feldspars were analyzed not only to determine the suitability of the crystals for $^{40}Ar/^{39}Ar$ dating, but also as a check on any tuff correlations that were made based on volcanic glass chemistry. Crystals suitable for $^{40}Ar/^{39}Ar$ dating should be unaltered and contain ≥1% K_2O. Although plagioclase containing ≤1% K_2O was dated by the $^{40}Ar/^{39}Ar$ method, the large associated errors often compromised the utility of the sample. Extent of alteration was determined by examining backscattered electron (BSE) images of feldspars during electron microprobe work. If significant alteration was detected, it usually appeared as clay growth within cleavage planes of crystals.

If the glass composition of two tuff samples was identical but the phenocryst populations proved chemically distinct, then the two samples likely represent different eruptions. Because the possible range of feldspar compositions is less than in glass, however, comparisons based solely on the composition of feldspar populations are insufficient for firm correlation.

Tuff samples containing feldspars suitable for $^{40}Ar/^{39}Ar$ dating were sent to the New Mexico Geochronology Research Laboratory at the New Mexico Institute of Mining and Technology and were dated either by single-crystal laser fusion (SCLF) or incremental heating by resistance furnace or CO_2 laser. For details on analytical methods used in obtaining the $^{40}Ar/^{39}Ar$ dates on tuffs presented in this study, see GSA Data Repository Tables 1 and 2.[1]

Basalts

Basalt samples were also crushed and sieved into various size fractions. Generally, the 100–120 mm size fraction was further processed by placing the sample in an ultrasonic cleaner with a dilute HCl solution. Groundmass concentrates from this size fraction were obtained through further magnetic and hand-picking techniques.

Basalt samples were analyzed with an electron microprobe to determine their suitability for dating by the $^{40}Ar/^{39}Ar$ method. Backscatter electron microscopy (BSE) images of the basalts

---
[1]GSA Data Repository item 2008193, comprehensive electron microprobe and $^{40}Ar/^{39}Ar$ geochronology data, is available at www.geosociety.org/pubs/ft2008.htm, or on request from editing@geosociety.org, Documents Secretary, GSA, P.O. Box 9140, Boulder, CO 80301-9140, USA.

TABLE 1. ELECTRON MICROPROBE ANALYTICAL CONDITIONS

	Elements	Beam size (mm)	Accelerating voltage (kv)	Current (nA)	Time (s)
Condition A	Na, K	10	15	8	10
	Si, Mg, Al, Ca, Mn, Fe, Ti	1–3 (spot)	15	20	20
Condition B*	Na, Si, Al, K	20	15	2	20
	Mg, Ca, Mn, Fe, Ti	20	15	20	20

Notes: Unless otherwise noted, all probe data in this study were analyzed using condition A.
*Condition B is from Morgan and London (1996).

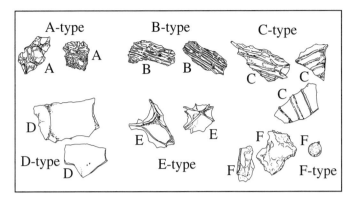

Figure 3. Morphological classification of glass shards used in this study: A-type is frothy glass shard with intrashard bubbles; B-type is glass shard composed of slender, fibrous threads; C-type is bubble-wall with stretched glass texture containing side walls of cylindrical vesicles; D-type is platy glass shard as part of a bubble wall much larger than the shard, may contain 1–2 ridges; E-type is platy glass shard with bubble-wall junctions; and F-type is miscellaneous glass shards including blocky shards and whole bubbles. In this study, shards identified as F-type are blocky and thick. Figure was modified from Katoh et al. (2000) and is based on previous studies by Ross (1928), Heiken (1972), and Yoshikawa (1976).

were useful in assessing the amount of clay alteration and glass content, and the compositional data obtained on the electron microprobe allowed characterization of the K content of the basalt. As with the tuffs, basalt samples were sent to the New Mexico Geochronology Research Laboratory. Groundmass concentrates were dated by incremental heating by either resistance furnace or CO_2 laser. Methodological details on $^{40}Ar/^{39}Ar$ dates obtained from basaltic groundmass are shown in GSA Data Repository Table 2 (see footnote 1).

Similarity Coefficients

We calculated the similarity coefficient for all possible pairings of glass analyses as well as feldspar pairs in order to statistically evaluate our geochemical correlations. The similarity coefficient, or SC, is a statistical measure first developed by Borchardt et al. (1972) and later refined by Rodbell et al. (2002). Created specifically for comparing the chemical compositions of glass in tuffs, it is a measure of the similarity of two tuffs based on a suite of geochemical analyses. If two samples have the same mean and standard deviation for every oxide included in the analysis, the SC would be equal to 1. In practice, an SC of 0.95 or greater is generally considered to be a valid correlation (Sarna-Wojcicki et al., 1980; Davis, 1985), whereas it is common for samples of the same tuff to produce SCs that are slightly lower. An SC of 0.92 is often taken as the lower limit for an acceptable correlation (Froggatt, 1992).

We used analyses of Na_2O, K_2O, SiO_2, MgO, Al_2O_3, CaO, MnO, FeO, and TiO_2; all measured Fe is expressed as FeO for these calculations. Although we analyzed for additional elements, we only used the nine oxides listed here in the SC calculations, because the other oxides were almost always present in amounts at or below the detection limit of the electron microprobe.

Following the equation as defined in Rodbell et al. (2002), the SC was calculated as:

$$d(A,B) = \left[\sum_i (R_i \times g_i) / \Sigma g_i \right],$$

where:

$d(A,B)$ = the similarity coefficient for samples A and B,
$R_i = Xi_A/Xi_B$ if $Xi_B > Xi_A$ and $R_i = Xi_B/Xi_A$ if $Xi_A > Xi_B$,
Xi_A = concentration of element i in sample A,
Xi_B = concentration of element i in sample B,
$g_i = 1 - \{([\sigma i_A/Xi_A]^2 + [\sigma i_B/Xi_B]^2)/E\}^{1/2}$,
σi_A = the standard deviation of element i in sample A,
σi_B = the standard deviation of element i in sample B, and
$E = 1 - $ (detection limit/[average of Xi_A, Xi_B]).

We calculated the average detection limit for every oxide in every sample, as the detection limit on each oxide can vary with every analysis. When determining the SC for samples A and B, we used whichever detection limit was larger.

RESULTS

Tuffs

The felsic glass composition from tuffs in the Adu-Asa Formation is primarily rhyolitic or dacitic in character (Fig. 4), although of course this may not be representative of the magma as a whole, since it does not take into account the contribution of phenocryst chemistry. For bimodal units, the mafic glass component plots as a basalt or basaltic andesite.

We calculated the similarity coefficient, or SC, for each sample pair of glass and phenocryst analyses (GSA Data Repository Tables 3 and 4 [see footnote 1]), except samples for which we did not obtain the detection limits of the microprobe analyses, which are necessary for the calculation.

In all, we identified four major tuffs in the Adu-Asa Formation at Gona, as well as three minor glassy tuffs and a crystal-rich series of related tuffs. Electron microprobe analyses, of both glass and feldspar, confirm many of the tentative correlations that were made in the field based on outcrop appearance and stratigraphic position (Tables 2 and 3; Figs. 5, 6, and 7). We named the four major glassy tuffs the Sifi Tuff, the Kobo'o Tuff, the Belewa Tuff, and the Ogoti Tuff Complex. The Hamadi Das crystal-rich sequence (HMDS) tuffs includes a number of altered, plagioclase-rich tuffs that lie stratigraphically below the Sifi Tuff. An important subunit of the Hamadi Das crystal-rich sequence tuffs is the Bodele Tuff, which is exposed at the Bodele Dora fossil localities and is an important constraint on the ages of those finds. Type localities/sections for each of these tuffs are presented in GSA Data Repository Figures 1 and 2 (see footnote 1), unless the type section is already included in Figures 6 or 7. Complete electron microprobe results on all glass and feldspar analyses are documented in GSA Data Repository

TABLE 2. SUMMARY OF ADU-ASA FORMATION GLASS ANALYSES

Marker tuff	Locality	Sample number	Number of shards	Major-element composition (wt%)												
				Na$_2$O	F	K$_2$O	SiO$_2$	MgO	Al$_2$O$_3$	ZrO$_2$	CaO	Cl	MnO	FeO	TiO$_2$	BaO	Total
Sifi	Below BDL sites	GONNL-59	23	2.61	0.12	2.99	73.18	0.01	11.87	0.04	0.47	0.03	0.07	1.84	0.16	0.04	93.44
Sifi	Above ABD sites	GONNL-61	16	2.59	0.07	2.99	72.99	0.01	11.88	0.05	0.46	0.04	0.06	1.85	0.17	0.04	93.21
Sifi	Near ESC-8	GON05-215a	20	2.48	0.04	2.44	73.64	0.03	12.05	0.05	0.41	0.04	0.06	1.74	0.14	0.06	93.18
Sifi	Near ESC-8	GON05-215b	17	2.37	0.04	2.38	72.98	0.02	12.31	0.05	0.36	0.04	0.06	1.73	0.14	0.02	92.50
Sifi	Near BDL sites	GON05-228	19	2.34	0.06	2.57	73.41	0.01	12.01	0.04	0.45	0.04	0.07	1.84	0.17	0.05	93.05
Sifi	Below BDL sites	GON05-231b	21	2.41	0.08	2.94	73.25	0.01	11.67	0.04	0.47	0.04	0.08	1.85	0.18	0.05	93.05
Sifi	Below BDL sites	GON05-231c	16	2.29	0.06	2.98	72.82	0.01	11.63	0.04	0.46	0.04	0.06	1.85	0.17	0.05	92.46
Sifi	Below BDL sites	GON05-231d	19	2.28	0.06	2.91	73.39	0.01	11.61	0.03	0.47	0.03	0.07	1.86	0.17	0.05	92.96
Sifi	Above ABD sites	GON05-234b	15	2.69	0.05	2.98	72.58	0.02	11.50	0.04	0.45	0.03	0.06	1.88	0.19	0.04	92.52
Sifi	Above ABD sites	GON05-234c	8	2.61	0.06	4.03	75.88	0.09	12.14	0.06	0.42	0.02	0.08	1.57	0.28	0.07	97.30
Sifi	Lateral to measured HMD section, above sites	GON05-238	20	1.96	0.03	2.82	72.75	0.01	11.88	0.05	0.49	0.03	0.08	1.94	0.19	0.05	92.30
Sifi	South Gona, lateral to measured section	GON05-243	19	2.68	0.05	1.82	72.23	0.01	11.91	0.05	0.48	0.03	0.07	1.93	0.19	0.05	91.50
Sifi	Near Gawis River	GON05-251a	20	1.23	0.02	2.72	73.42	0.01	11.94	0.05	0.49	0.03	0.09	1.95	0.19	0.06	92.19
Sifi	Near Gawis River	GON05-251b	19	1.32	0.03	2.26	73.74	0.01	12.01	0.04	0.49	0.03	0.08	1.98	0.18	0.06	92.21
Kobo'o (silicic A)	Near ESC sites	ESCASH-11a	21	2.24	0.05	2.26	70.32	0.01	11.48	0.09	0.69	0.04	0.13	2.51	0.22	0.08	90.11
Kobo'o (silicic A)	Near ESC sites	GONNL-52	19	2.05	0	2.05	70.26	0.01	12.02	0.06	0.66	0.04	0.11	2.50	0.23	0.04	90.08
Kobo'o (silicic A)	Between Kasa Gita-Chifra road and HEN sites	GONNL-53	17	2.54	0	1.77	70.12	0.00	12.23	0.07	0.61	0.05	0.10	2.42	0.21	0.06	90.23
Kobo'o (silicic A)	Above ESC-9	GON05-216a	9	2.07	0.07	1.94	72.36	0.02	12.10	0.08	0.70	0.04	0.12	2.60	0.25	0.04	92.39
Kobo'o (silicic A)	Above ESC-9	GON05-216b	10	2.08	0.01	1.86	71.98	0.02	11.95	0.08	0.69	0.05	0.11	2.66	0.22	0.05	91.76
Kobo'o (silicic A)	Above ESC-9	GON05-216d	16	2.31	0.04	2.26	73.29	0.01	12.18	0.07	0.62	0.06	0.09	2.45	0.22	0.07	93.68
Kobo'o (silicic A)	Near ESC sites	GON05-219a	17	0.82	0.04	1.90	71.77	0.02	12.13	0.08	0.68	0.04	0.11	2.52	0.21	0.08	90.41
Kobo'o (silicic A)	Near ESC sites	GON05-219b	20	0.86	0	1.93	71.62	0.02	12.15	0.09	0.71	0.05	0.12	2.65	0.23	0.06	90.54
Kobo'o (silicic A)	Near ESC sites	GON05-224a	18	1.57	0	1.91	71.19	0.02	12.12	0.07	0.67	0.04	0.11	2.49	0.23	0.09	90.53
Kobo'o (silicic A)	Near ESC-8, -9	GON05-224b	17	1.72	0.07	2.04	71.29	0.02	12.11	0.07	0.65	0.04	0.11	2.49	0.23	0.06	90.91
Kobo'o (silicic A)	Near ESC-8, -9	GON05-225	18	1.76	0.02	1.56	71.27	0.02	12.06	0.08	0.69	0.05	0.12	2.53	0.23	0.07	90.46
Kobo'o (silicic A)	Between Kasa Gita-Chifra road and HEN sites	GON05-257a	40	2.86	0.04	1.75	72.25	0.01	12.01	0.07	0.61	0.06	0.10	2.43	0.20	0.08	92.46
Kobo'o (silicic A)	Between Kasa Gita-Chifra road and HEN sites	GON05-257b	20	2.62	0.03	1.71	72.82	0.01	12.14	0.07	0.63	0.07	0.10	2.45	0.21	0.08	92.94
Kobo'o (silicic A)	Between Kasa Gita-Chifra road and HEN sites	GON05-258	17	2.22	0.05	1.99	71.10	0.01	12.21	0.07	0.62	0.06	0.10	2.40	0.23	0.06	91.11
Kobo'o (silicic B)	Near ESC sites	ESCASH-10	13	1.99	0.03	1.70	69.27	0.03	11.99	0.07	0.85	0.04	0.11	2.81	0.23	0.07	89.17
Kobo'o (silicic B)	Near ESC sites	ESCASH-11b	29	1.90	0.05	2.19	70.12	0.03	12.52	0.08	0.86	0.05	0.12	2.72	0.23	0.05	90.94
Kobo'o (silicic B)	Near ESC sites	GONNL-50	29	1.68	0.07	1.71	70.40	0.03	12.49	0.08	0.86	0.05	0.12	2.77	0.23	0.07	90.55
Kobo'o (silicic B)	Above ESC-9	GON05-216c	17	2.11	0.04	2.23	70.99	0.03	12.33	0.07	0.86	0.05	0.11	2.84	0.24	0.07	91.98
Kobo'o (mafic)	Near ESC sites	ESCASH-11b	12	2.11	0.13	1.26	52.70	3.34	12.98	0.15	6.99	0.02	0.36	12.58	2.94	0.00	95.55
Kobo'o (mafic)	Near ESC sites	GON05-219c	16	2.52	0.05	1.33	50.95	2.94	12.79	0.16	6.92	0.02	0.42	13.69	2.78	0.00	94.58

(continued)

TABLE 2. SUMMARY OF ADU-ASA FORMATION GLASS ANALYSES (continued)

Marker tuff	Locality	Sample number	Number of shards	Major-element composition (wt%)												
				Na$_2$O	F	K$_2$O	SiO$_2$	MgO	Al$_2$O$_3$	ZrO$_2$	CaO	Cl	MnO	FeO	TiO$_2$	BaO	Total
Belewa	In Belewa drainage	GON05-262a1	15	1.69	0.05	5.43	74.85	0.01	11.33	0.10	0.24	0.06	0.03	2.05	0.18	0.03	96.06
Belewa	In Belewa drainage	GON05-262a2	13	1.52	0.08	5.10	75.31	0.01	11.38	0.09	0.23	0.05	0.03	2.02	0.17	0.03	96.02
Belewa	In Belewa drainage	GON05-262b	13	2.06	0.07	5.65	74.83	0.01	11.44	0.08	0.21	0.06	0.03	1.81	0.15	0.02	96.43
Belewa*	In Belewa drainage	GON05-262b obs	15	2.48	0.09	5.58	75.53	0.01	11.63	0.05	0.23	0.06	0.03	1.77	0.15	0.02	97.64
Belewa	In Belewa drainage	GON05-262c	16	1.76	0.07	5.53	74.93	0.01	11.40	0.06	0.22	0.06	0.02	1.75	0.15	0.01	95.97
Belewa	In Belewa drainage	GON05-262d	32	1.82	0.10	5.67	74.30	0.01	11.42	0.08	0.23	0.06	0.03	1.89	0.16	0.02	95.78
Belewa*	In Belewa drainage	GON05-262d obs	16	2.20	0.04	5.77	75.10	0.00	11.46	0.07	0.22	0.06	0.02	1.81	0.15	0.02	96.94
Belewa	In Belewa drainage	GON05-262e	15	1.70	0.07	5.38	75.74	0.01	11.48	0.07	0.24	0.06	0.03	2.01	0.18	0.02	96.99
Belewa	In Belewa drainage	GON05-262f	15	1.66	0.08	5.41	75.58	0.01	11.46	0.07	0.23	0.06	0.03	1.82	0.15	0.00	96.55
Belewa*	In Belewa drainage	GON05-262f obs	12	2.20	0.08	5.68	75.34	0.01	11.46	0.07	0.23	0.06	0.02	1.91	0.16	0.03	97.26
Belewa	In Belewa drainage	GON05-262h	14	1.84	0.03	5.64	75.06	0.02	11.74	0.06	0.28	0.05	0.03	1.94	0.18	0.03	96.91
Belewa*	In Belewa drainage	GON05-262h obs	17	2.45	0.08	5.82	74.93	0.01	11.71	0.05	0.26	0.06	0.04	1.90	0.17	0.01	97.48
Belewa	Degora Konte	GON05-265a	20	1.94	0.04	4.91	72.20	0.00	10.92	0.08	0.21	0.06	0.04	1.78	0.14	0.03	92.36
Belewa	Degora Konte	GON05-265b	20	2.01	0.06	5.17	72.37	0.00	10.87	0.07	0.21	0.06	0.03	1.74	0.13	0.01	92.74
Belewa*	Degora Konte	GON05-265c obs	21	1.40	0.07	5.24	73.66	0.00	11.29	0.07	0.21	0.06	0.02	1.76	0.15	0.02	93.98
Ogoti†	W of rhyolite flow-dome complex	ESCASH-13 F1	21	1.56	0.08	5.37	73.51	0.02	11.43	0.02	0.44	0.04	0.04	1.57	0.14	0.05	94.29
Ogoti†	W of rhyolite flow-dome complex	ESCASH-13 M2	8	2.05	0.07	6.36	73.11	0.02	11.30	0.03	0.44	0.05	0.05	1.65	0.16	0.05	95.34
Ogoti†	W of rhyolite flow-dome complex	ESCASH-13 G1	18	2.07	0.13	6.19	73.34	0.02	12.04	0.04	0.45	0.05	0.04	1.65	0.15	0.03	96.20
Ogoti†	W of rhyolite flow-dome complex	ESCASH-13 F3	15	1.64	0.06	5.57	73.71	0.02	11.99	0.03	0.45	0.04	0.04	1.63	0.15	0.06	95.39
Ogoti	W of rhyolite flow-dome complex	GON05-273b	20	1.87	0.07	5.83	73.00	0.01	12.08	0.03	0.41	0.07	0.04	1.78	0.15	0.05	95.38
Ogoti	W of rhyolite flow-dome complex	GON05-273c	17	1.92	0.09	6.12	73.71	0.01	12.44	0.03	0.47	0.06	0.05	1.95	0.17	0.06	97.10
Ogoti#	In Busidima River	GON05-283	15	1.98	0.05	6.27	74.55	0.02	11.84	0.03	0.40	0.04	0.04	1.56	0.13	0.02	96.94
Ogoti#	In Busidima River	GON05-284b	19	2.05	0.09	5.90	74.35	0.02	12.12	0.03	0.48	0.04	0.04	1.76	0.15	0.03	97.06
Ogoti#	In Busidima River	GON05-284c	19	2.06	0.09	5.99	73.45	0.02	12.03	0.02	0.44	0.04	0.04	1.66	0.13	0.06	96.04
Ogoti#	In Busidima River	GON05-284d	20	1.84	0.07	5.97	72.98	0.02	11.98	0.03	0.43	0.04	0.04	1.71	0.13	0.07	95.32
Ogoti#	In Busidima River	GON05-286a	20	2.26	0.05	6.07	74.91	0.03	12.13	0.03	0.47	0.04	0.05	1.73	0.13	0.04	97.93
Ogoti*#	In Busidima River	GON05-286b	16	2.08	0.06	5.52	72.09	0.04	11.83	0.04	0.48	0.04	0.04	1.74	0.16	0.05	94.17
Ogoti†	W of rhyolite flow-dome complex	ESCASH-13 M2	13	2.28	0.07	0.82	50.45	5.05	12.94	0.02	9.77	0.02	0.23	12.49	2.52	0.01	96.71
Obsidian§	Rhyolite flow-dome complex	GON05-270	13	2.08	0.12	6.48	74.26	0.02	12.66	0.04	0.66	0.03	0.03	0.79	0.17	0.06	97.39
Obsidian§	Rhyolite flow-dome complex	GON05-272	18	2.90	0.10	5.39	72.61	0.02	11.39	0.04	0.40	0.04	0.04	1.67	0.12	0.07	94.79
Obsidian§	In Busidima River	GON05-281	20	2.49	0.08	5.70	70.90	0.03	11.94	0.04	0.56	0.04	0.04	1.88	0.17	0.07	93.95
Obsidian§	In Busidima River	GON05-285	13	2.43	0.07	5.99	71.93	0.15	11.57	0.04	0.57	0.04	0.04	1.85	0.25	0.03	94.96
Obsidian§	In Busidima River	GON05-287	20	2.35	0.08	5.90	72.13	0.03	11.32	0.03	0.42	0.04	0.04	1.50	0.13	0.04	94.02
Unnamed 1	Degora Konte	GON05-300	11	1.69	0.05	1.92	67.57	0.31	13.02	0.09	1.29	0.05	0.13	3.34	0.39	0.04	89.88
Unnamed 2	Degora Konte	GON05-301	18	3.20	0.08	3.88	71.90	0.02	13.09	0.03	1.02	0.04	0.06	2.00	0.13	0.06	95.51
Unnamed 3	Degora Konte	GON05-302	12	2.74	0.09	3.64	72.79	0.02	12.22	0.04	0.41	0.12	0.06	2.10	0.12	0.04	94.40

Notes: Summary of glass analyses from the Adu-Asa Formation at Gona; total Fe is expressed as FeO. Unless otherwise noted, all samples are of glass shards in ash-fall tuffs. Samples were analyzed on a Cameca SX50 electron microprobe at the Lunar and Planetary Laboratory, University of Arizona, using the setup conditions listed in Table 1. Sample locations are shown in Figure 2. Fossil localities are shown in Figure 1.
*Obsidian clast.
†Pumice clast.
§Glassy rhyolite.
#Ash-flow or surge deposit.

41

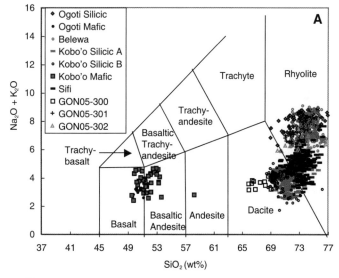

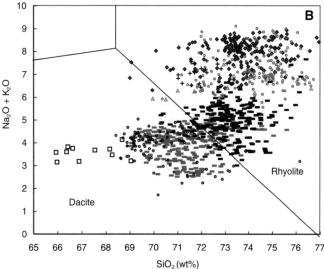

Figure 4. (A) Total alkali-silica diagram of tuff analyses from the Adu-Asa Formation. (B) Detail of silicic tuff analyses. Note that these analyses represent only the vitric ash component of the tuffs. Figure is after Le Bas et al. (1986).

Table 5 (see footnote 1). The $^{40}Ar/^{39}Ar$ dates obtained on these units are shown in Figures 8, 9, and 10 and summarized in Tables 4 and 5. Detailed results are included in GSA Data Repository Tables 1 and 2 (see footnote 1).

Sifi Tuff

The Sifi Tuff is a critical marker horizon because it is often associated with fossil localities (Figs. 1 and 2). Outcrops of this tuff are found along strike across much of the Gona project area (Fig. 2). The Sifi Tuff appears as lenses in fluvial sedimentary rocks in the southernmost part of the Gona Paleoanthropological Research Project area, and it is exposed at the As Bole Dora (ABD), Bodele Dora (BDL), and Hamadi Das (HMD) groups of fossil sites, as well as near the Escarpment (ESC) fossil localities (Figs. 1 and 6).

Fossil-rich beds lie both above and below the Sifi Tuff. At the ABD sites, fossil-bearing beds lie below both the Sifi Tuff and a diatomite bed, whereas at the BDL-2 site, the fossils derive from conglomerates above the Sifi Tuff. At the HMD sites, fossil-bearing deposits are exposed both above and below the level of the Sifi Tuff. There, the fossils can be traced to the siltstones below the level of the Sifi Tuff, as well as to a conglomerate unit above (Fig. 6).

In the central portion of the project area, sedimentary rocks containing the Sifi Tuff show evidence for a shift from a lacustrine to a more fluvial environment. Dark, laminated mudstone and diatomite beds are common in the lower part of the ABD, BDL, and HMD stratigraphic sections, whereas the BDL and HMD sections contain more sandstones and conglomerates above the level of the Sifi Tuff. The Sifi Tuff is heavily reworked into lenses, which vary in thickness from ~0.2 to 2 m. In places, the lenses are discontinuous. This large variation in thickness suggests that the transition from a lacustrine to a fluvial environment was completed by the time the Sifi Tuff was deposited.

Glass shards are rhyodacitic and typically ~0.5 mm in size with A-type morphology, although some B-type shards are present (Fig. 3). Glass in the Sifi Tuff is distinguished by a CaO content of 0.4%–0.5%, an MnO content of 0.06%–0.08%, and a K_2O content of ~2.5% (Table 2). We were not able to identify a homogeneous population of phenocrysts, so the Sifi Tuff is not suitable for radiometric dating.

Kobo'o Tuff

The Kobo'o Tuff is intercalated with fluvial sedimentary rocks in the northern half of the project area, although it may be present along strike in areas not well surveyed to the south (Fig. 2). The Kobo'o Tuff is reworked and varies in thickness from 0.5 m to ~3 m in paleochannels. Like the Sifi Tuff, exposures of the Kobo'o Tuff are repeated due to the abundant normal faults, and repetitions generally occur less than 1 km apart. Whereas the Kobo'o Tuff has only been identified at one fossil site, ESC-9 (Fig. 6), this tuff is repeatedly found near the ESC cluster of sites (Figs. 1 and 2). The fossils at ESC-9 were not in situ, but they likely derive from the sands and conglomerates below the level of the Kobo'o Tuff.

The sedimentary section associated with the Kobo'o Tuff is dominantly fluvial, but basalt flows are also common. In the measured sections containing the Kobo'o Tuff, sedimentary rocks were typically pale red claystone with interbedded volcaniclastic sandstone, conglomerate, and aphanitic basalts.

The Kobo'o Tuff is clearly felsic at the base and strongly bimodal toward the top in outcrop. In hand sample, the bimodal portion exhibits a "salt and pepper" appearance, consisting of ~60%–75% felsic shards and 25%–40% mafic shards. This pattern, as shown in Figure 7A, was noted at multiple sample collection sites. In some sample localities, a final felsic layer caps this felsic to bimodal sequence, but this uppermost layer is not always present. The striking bimodal nature of the Kobo'o Tuff

TABLE 3. SUMMARY OF ADU-ASA FORMATION FELDSPAR ANALYSES

Marker tuff	Locality	Stratigraphic position	Sample number	Number of grains	Major-element composition (wt%)													
					Na_2O	F	K_2O	SiO_2	MgO	Al_2O_3	ZrO_2	CaO	Cl	MnO	FeO	TiO_2	BaO	Total
HMDS	HMD	Below Sifi	GON05-236a	13	2.05	0.04	0.04	48.06	0.15	32.40	0.00	16.02	0.01	0.01	0.65	0.05	0.02	99.50
HMDS	HMD	Below Sifi	GON05-236b	18	2.57	0.04	0.08	49.21	0.15	30.21	0.00	14.63	0.02	0.01	0.67	0.05	0.03	97.66
HMDS	HMD	Below Sifi	GON05-239	17	5.99	0.02	0.28	56.89	0.04	26.96	0.00	8.88	0.01	0.01	0.32	0.05	0.05	99.51
HMDS	South Gona	Above Sifi	GON05-244	10	3.84	0.02	0.14	52.81	0.15	29.22	0.00	12.65	0.01	0.01	0.67	0.09	0.02	99.63
HMDS	South Gona	Below Sifi	GON05-246	15	5.98	0.05	0.28	57.51	0.03	27.49	0.00	8.90	0.01	0.01	0.35	0.04	0.03	100.70
HMDS	South Gona	Below Sifi	GON05-248	15	5.94	0.02	0.28	57.70	0.02	26.65	0.01	8.72	0.01	0.01	0.32	0.04	0.02	99.75
Bodele A	BDL	Below Sifi	GON05-229	13	3.76	0.03	0.13	51.70	0.12	29.46	0.00	12.90	0.02	0.01	0.79	0.06	0.02	99.01
Bodele B	BDL	Below Sifi	GON05-230	13	6.32	0.03	0.49	57.93	0.02	26.59	0.01	7.99	0.01	0.01	0.39	0.04	0.05	99.90
	ABD	Below Sifi	GONNL-62	20	7.67	0.03	1.80	63.02	0.02	22.95	0.00	4.07	0.01	0.01	0.55	0.05	0.19	100.38
	ABD	Below Sifi	GON05-233	18	8.22	0.03	0.58	62.79	0.01	23.66	0.00	4.70	0.01	0.01	0.19	0.02	0.12	100.37
Kobo'o	ESC-9	Kobo'o	GON05-216b	18	8.72	0.03	1.39	65.55	0.00	22.28	0.00	2.69	0.01	0.01	0.27	0.02	0.22	101.20
	ESC-9	Below Kobo'o	GON05-217	17	3.02	0.04	0.10	49.91	0.14	30.71	0.01	14.21	0.02	0.01	0.65	0.07	0.04	98.90
Belewa	In Belewa	Belewa	GON05-262a1	8	6.60	0.08	7.05	66.92	0.00	19.61	0.00	0.10	0.01	0.02	0.22	0.01	0.08	100.69
Belewa	In Belewa	Belewa	GON05-262a2	12	6.33	0.03	7.03	66.74	0.00	19.26	0.00	0.11	0.01	0.01	0.22	0.03	0.07	99.84
Belewa	In Belewa	Belewa	GON05 262i	6	6.31	0.03	6.62	65.04	0.00	19.53	0.01	0.43	0.01	0.01	0.19	0.01	0.81	99.00
Belewa	Degora Konte	Belewa	GON05-265c	18	6.60	0.04	6.31	65.26	0.01	19.94	0.01	0.37	0.00	0.01	0.20	0.02	0.40	99.17
Ogoti	Rhyolite dome	Ogoti	GON05-271	17	8.01	0.04	3.43	65.22	0.01	21.18	0.01	1.90	0.00	0.01	0.23	0.01	0.50	100.56
Ogoti[†]	W of dome	Ogoti	GON05-273b[§]	17	7.18	0.06	5.29	65.36	0.01	20.06	0.01	0.76	0.00	0.01	0.20	0.02	0.64	99.58
Ogoti[†]	In Busidima	Ogoti	GON05-283	22	7.54	0.04	4.04	65.37	0.00	20.95	0.01	1.44	0.01	0.01	0.23	0.02	0.51	100.17
Ogoti[†]	In Busidima	Ogoti	GON05-284a	17	7.90	0.05	3.84	65.71	0.01	21.69	0.00	1.71	0.01	0.01	0.23	0.01	0.47	101.64
Ogoti[†]	In Busidima	Ogoti	GON05-286a	18	7.96	0.03	3.15	65.73	0.00	21.56	0.00	2.06	0.01	0.00	0.23	0.02	0.46	101.23
Ogoti[†]	In Busidima	Ogoti	GON05-286c	19	8.23	0.06	3.05	63.47	0.01	21.04	0.01	2.18	0.01	0.01	0.25	0.02	0.39	98.71
	Degora Konte		GON05-302	4	8.46	0.00	3.17	66.99	0.01	20.27	0.01	1.06	0.02	0.01	0.32	0.00	0.41	100.71
	W of road*		GONNL-64	17	7.84	0.04	1.78	63.42	0.02	22.96	0.01	4.08	0.00	0.01	0.58	0.05	0.20	101.00
	Near HMD		GON05-241	13	4.26	0.02	0.20	51.96	0.12	29.56	0.00	12.45	0.02	0.01	0.82	0.09	0.00	99.53
HMDS	Near Gawis	Below Sifi	GON05-253	13	2.20	0.06	0.05	46.03	0.16	30.69	0.00	16.28	0.00	0.01	0.61	0.05	0.02	96.17
	HEN area		GON05-259a	4	6.71	0.04	0.38	57.23	0.03	24.78	0.02	7.75	0.00	0.01	0.42	0.02	0.04	97.43
	HEN area		GON05-259b	16	7.02	0.05	0.45	57.55	0.01	24.49	0.00	7.37	0.00	0.01	0.36	0.02	0.04	97.39
	HEN area		GON05-261	16	6.76	0.05	0.66	57.11	0.02	24.87	0.00	7.65	0.01	0.01	0.34	0.03	0.07	97.59

Notes: Summary of feldspar analyses from the Adu-Asa Formation at Gona; total Fe is expressed as FeO. Unless otherwise noted, all samples are of feldspar crystals in ash-fall tuffs. Samples were analyzed on a Cameca SX50 electron microprobe at the Lunar and Planetary Laboratory, University of Arizona, using the setup conditions listed in Table 1. Sample locations are shown in Figure 2. Fossil localities are shown in Figure 1.
*Sample GONNL-64 was collected outside of the area depicted in Figure 2 west of the Kasa Gita-Chifra road near the HMD sites and the Gawis River.
[†]Ash-flow or surge deposit.
[§]Resample of ESCASH-13.

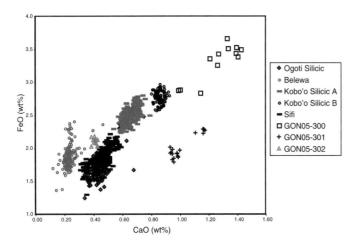

Figure 5. CaO versus FeO biplot of tuff analyses from the Adu-Asa Formation.

in outcrop is unique among the tuffs in the Adu-Asa Formation at Gona. The Ogoti Tuff Complex also has a mafic component to it, but it is not nearly as obvious in hand sample (see following).

Electron microprobe analysis reveals that the Kobo'o Tuff is actually polymodal, with two very similar rhyodacite phases in addition to the basaltic/basaltic andesite phase. The major differences between the two silicic components are the CaO and the FeO contents (Fig. 5; Table 2). In silicic mode A, the mean CaO content is 0.65% ($n = 265$), and in silicic mode B, the mean is 0.86% ($n = 93$). For the FeO content, silicic mode A contains ~2.5% FeO, whereas silicic mode B has ~2.7%–2.8% FeO. All other oxides are similar in both modes (Table 2). Silicic mode B is concentrated in the same beds as the mafic shards. Silicic mode A shards are 0.5–1 mm in diameter and are a mix of type A and B morphologies (Fig. 3). Silicic mode B shards are also 0.5–1 mm in diameter but display type D and E morphologies. The mafic shards are up to 1 mm in size and are dominantly type A morphology.

The Kobo'o Tuff is the only polymodal tuff identified in the Adu-Asa Formation at Gona (Fig. 5; Table 2). Both silicic modes contain ~0.22% TiO_2 and ~0.11% MnO, which are unique to the tuffs described here. The mafic component contains ~3% MgO and 7% CaO, which differs from the mafic component of the Ogoti Tuff Complex.

Unlike the Sifi Tuff, the Kobo'o Tuff contains chemically homogeneous populations of feldspars (Table 3). A combination of sanidine and plagioclase from sample GON05-216 (Fig. 7A) yielded a single-crystal $^{40}Ar/^{39}Ar$ age of 5.44 ± 0.06 Ma at the 2σ level (Table 4; Fig. 9).

We reanalyzed selected samples of the Kobo'o Tuff using the setup conditions recommended by Morgan and London (1996) and compared those analyses with the results obtained using the setup conditions in this study (Tables 1 and 6). Although the number of shards used in this comparison is small, the results highlight important differences in electron microprobe analytical conditions. For the mafic mode, there is no significant difference between the two sets of analytical conditions. For the silicic modes, however, the measured amounts of Na_2O and K_2O are lower and the measured amounts of SiO_2 and Al_2O_3 are higher for glass shards analyzed using the setup conditions in this study as compared to the analytical conditions suggested by Morgan and London (1996). This is a typical pattern during alkali mobilization, as electron bombardment causes alkalis, Si, and Al to migrate away from the electron beam (Hunt and Hill, 1993; Morgan and London, 1996; Nielsen and Sigurdsson, 1981).

The compositions reported in this study do reflect some alkali mobilization. However, the differences are largest in the Na_2O and Al_2O_3 contents of silicic analyses, and they are only significant at the 2σ level for Na_2O. This should be taken into account when considering tephrostratigraphic correlations between the tuffs presented here and those elsewhere in the region.

Belewa Tuff

The Belewa Tuff is known from two localities only (Fig. 2). The first is composed of 10 m of tephra interbedded with pinkish siltstone, sandstone, and conglomerate (Figs. 7B and 7C). The tuff at this locality, from which sample GON05-262 was collected, contains abundant perlitic obsidian clasts, lapilli-size pumice pieces, and ash containing glass shards and sanidine. Individual shards are ~1 mm in diameter, and associated pumice is 1–2 cm. Morphologically, shards were a mix of B-type with some A-type grains (Fig. 3).

The second locality, where sample GON05-265 was collected, is much finer grained than its chemical correlative GON05-262. This exposure is 1–2 m thick and contains coarse ash, perlitic obsidian fragments, and sanidine phenocrysts. Glass shards in this sample were typically 1 mm in diameter, and perlitic obsidian fragments up to 1 mm in diameter were also present. Shard morphologies in this outcrop are dominantly B-type, with a small amount of A- and F-type shards (Fig. 3).

In the proximal outcrop of the Belewa Tuff, there is some degree of soil development between a few of the tuffaceous layers (Figs. 7B and 7C). The degree of pedogenesis is slight and is indicated primarily by the angular, blocky jointing with slickensides that is prominent in the claystone units. Nevertheless, this demonstrates hiatuses between eruptions of chemically identical material. These younger layers are thinner and finer grained than those at the base of the outcrop and likely did not spread material far from the source. It is the thick, lapilli-sized deposits at the base of the first outcrop that likely correlate to the second, more distal outcrop where GON05-265 was sampled.

The Belewa Tuff is distinguished chemically by a CaO content of ~0.23%, which is lower than any of the other tuffs described here (Fig. 5; Table 2). Sanidine (with a K_2O content of 6%–7%) from sample GON05-265 yielded an $^{40}Ar/^{39}Ar$ age of 5.47 ± 0.04 Ma (2σ) for the Belewa Tuff (Table 4; Fig. 9).

Ogoti Tuff Complex

The Ogoti Tuff Complex is the only pyroclastic unit in the Adu-Asa Formation at Gona found to contain more than just an ash-fall component. Like the other tuffs, the ash-fall part of the

Ogoti Tuff Complex is interbedded with sedimentary rocks. The ash-flow overlies the glassy obsidian portion of a rhyolite flow, whereas the surge deposit directly underlies the basal pumice breccia of a rhyolite flow and overlies a basalt unit. This complex is located in the northern part of the study area (Figs. 1 and 2) and clearly originates from the only eruptive center we found in the Adu-Asa Formation (see following).

The Ogoti Tuff Complex is bimodal with a minor basaltic component (Table 2). The ash-fall tuff and surge deposits contain 1–2-mm-diameter glass shards as well as lapilli-sized pumice and centimeter-sized perlitic obsidian fragments. The ash-fall deposit has multiple tuffaceous beds separated by thin silty interbeds, for a total thickness of ~4 m. The surge deposit is ~2 m in thickness and is made up of cross-bedded layers and a channelized distribution of sublayers. The ash-flow tuff is nonwelded and contains 1–2-mm-diameter glass shards, and blocks of obsidian and pumice up to 25 cm in diameter are common.

Glass in the Ogoti Tuff Complex is characterized by a CaO content similar to the Sifi Tuff (~0.45%), coupled with a K_2O content of ~6% (Fig. 5; Table 2). The mafic component is distinguished by a CaO content of 9%–10%, a MgO content of ~5%, and a mean MnO content of 0.23% (Table 2).

Rhyolitic glass shards in the Ogoti Tuff Complex display both A- and B-type morphologies (Fig. 3). The basaltic component was identified only as pumice clasts within ash-fall tuff sample ESCASH-13 (Table 2). Additional mafic shards are present in the other samples but have devitrified and thus were not suitable for analysis.

Euhedral anorthoclase crystals, 1–2 mm in diameter, are abundant in the ash-flow and surge phases of the Ogoti Tuff Complex. A $^{40}Ar/^{39}Ar$ date on phenocrysts from ash-flow sample GON05-271 indicates that the Ogoti Tuff Complex is 5.80 ± 0.20 Ma (2σ) (Table 4). However, the age-probability plot is very broad and multimodal (Fig. 9), as reflected in the high mean square of weighted deviates (MWSD) of 10.96. We take this to indicate contamination by older feldspar populations, possibly incorporated from older material entrained in the ignimbrite during eruption. A recalculation using only the six youngest grains produced an age of 5.84 ± 0.07 Ma with a MWSD value of 0.86 (Fig. 9). While this age result is slightly older than the original calculation, the error range and MWSD are significantly lower. Thus, we view the recalculated result as the most accurate age determination of the ash flow, but it is a maximum age (Fig. 9).

A subsequent attempt to date the Ogoti Tuff Complex was more successful. Plagioclase from the ash-fall portion yielded a single-crystal $^{40}Ar/^{39}Ar$ date of 5.57 ± 0.15 Ma (2σ) (Table 4; Fig. 9). While broadly similar in composition, feldspars from the ash-fall component contained more K_2O and less CaO and Na_2O than the feldspars in the ash-flow tuff or the surge deposit (Table 3). Thus, we interpret the dominant feldspar population in the ash-fall portion of the Ogoti Tuff Complex as a juvenile population and the feldspars in the ash-flow and surge deposits as partly nonjuvenile.

Other Glassy Tuffs

Besides the four major tuffs in the Adu-Asa Formation, we analyzed three other geochemically distinct ash-fall deposits (Table 2), each of which is known from only a single outcrop. The paucity of other samples from these tuffs may be due to a lack of survey in the area and is not necessarily a reflection of the extent of deposits. All three units are exposed near the top of the Adu-Asa Formation, in the Degora Konte area just west of the boundary between the Adu-Asa Formation and the early Pliocene Sagantole Formation (Fig. 2).

Glass from sample GON05-300 is dacitic, while samples of glass from GON05-301 and GON05-302 are rhyolitic (Fig. 4). All three have an average shard size of ~0.5 mm. GON05-300 and GON05-301 have dominantly C-type morphology, while GON05-302 contains A-type shards with some B-type shards as well (Fig. 3). Sample GON05-300 is distinguished by a lower silica content (~68% SiO_2) than the other silicic tuffs in the Adu-Asa Formation at Gona, as well as higher FeO, CaO, MgO, and MnO contents (Fig. 5; Table 2). Sample GON05-301, with the exception of GON05-300, has the highest CaO content of any of the silicic tuffs in the formation at 1% CaO (Table 2). Finally, GON05-301 has a CaO content similar to that of the Sifi Tuff and the Ogoti Tuff Complex (~0.4% CaO); however, the FeO and Al_2O_3 contents of sample GON05-301 are higher than those of either the Sifi Tuff or the Ogoti Tuff Complex (Table 2). Only GON05-302 contains an obvious population of phenocrysts (Table 3), but the sample is as of yet undated.

Hamadi Das Crystal-Rich Tuff Sequence and the Bodele Tuff

A sequence of plagioclase-rich tuffs, which we refer to as the Hamadi Das Crystal-Rich Tuff Sequence (HMDS), is found at several locations below the Sifi Tuff, including the ABD, BDL, and HMD sites as well as at South Gona. In most cases, the Sifi Tuff lies 5–25 m above this tuff sequence, such as at the HMD sites (Fig. 6). The sole exception is at South Gona, where one of these tuffs (GON05-244) is found lateral to and at a level a few meters above the Sifi Tuff (Fig. 6). Many of the tuffs in this sequence are reworked and probably contain some nonjuvenile populations of plagioclase. Glass in all Hamadi Das tuffs is too altered for analysis, probably because many of the tuffs in this part of the Adu-Asa Formation were deposited in a lacustrine setting. Without glass analyses, it is difficult to sort out the exact stratigraphic relationships and correlations for each of these units.

However, we can in some cases develop a more precise stratigraphy given the stratigraphic closeness of the some of the Hamadi Das tuffs to the Sifi Tuff, combined with analyses of the feldspar compositions and detailed field observations. In particular, at the BDL sites, there are two altered plagioclase-rich tuffs that are exposed below both the level of the fossils (~9 m) and the Sifi Tuff (~4 m) (Fig. 6). These tuffs are a subset of the Hamadi Das tuffs and contain different populations of feldspars. We hereafter refer to as these units as the Bodele A Tuff (GON05-229) and the Bodele B Tuff (GON05-230) (Table 3).

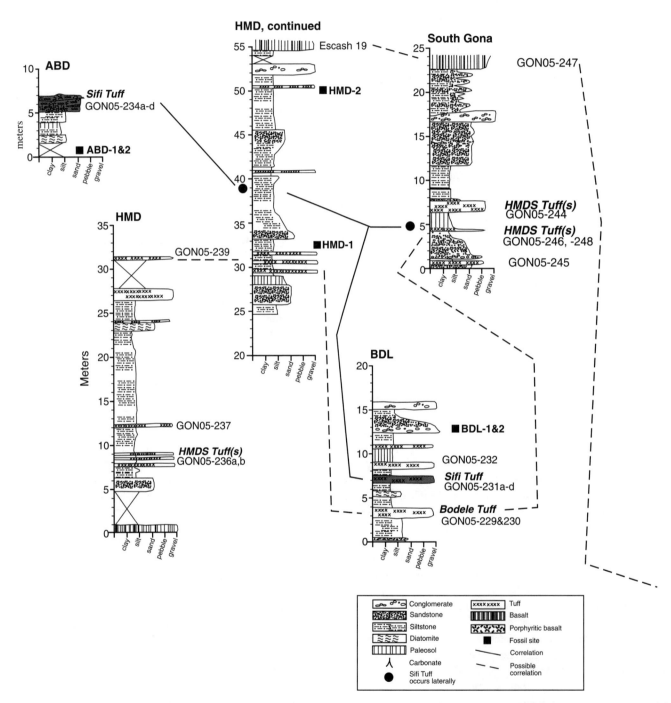

Figure 6 (*on this and following page*). Measured stratigraphic sections containing the Sifi, the Kobo'o, and the Hamadi Das crystal-rich sequence (HMDS) tuffs, including the Bodele Tuff, with fossil localities marked. Scale is in meters. A composite stratigraphic section with all of the major marker units is shown in Figure 11. All measured sections are assumed to be unfaulted.

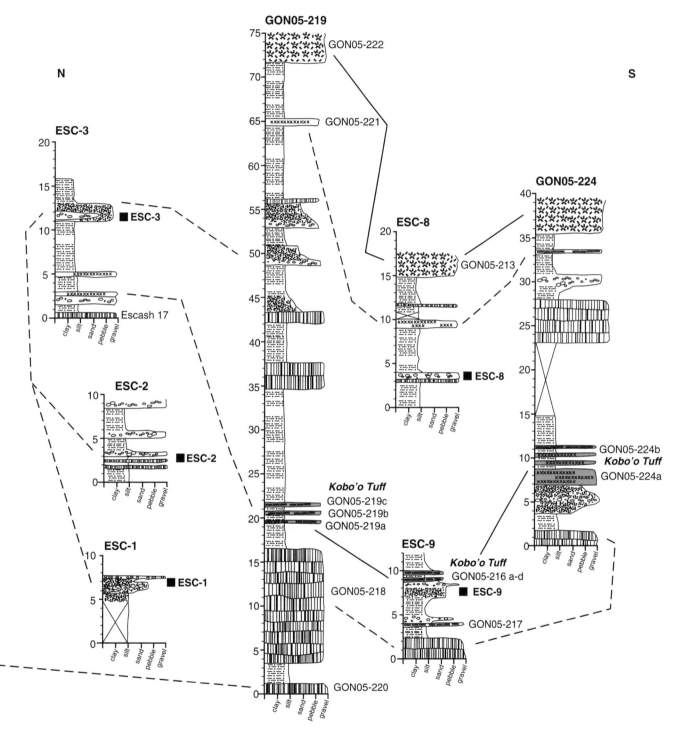

Figure 6 (*continued*).

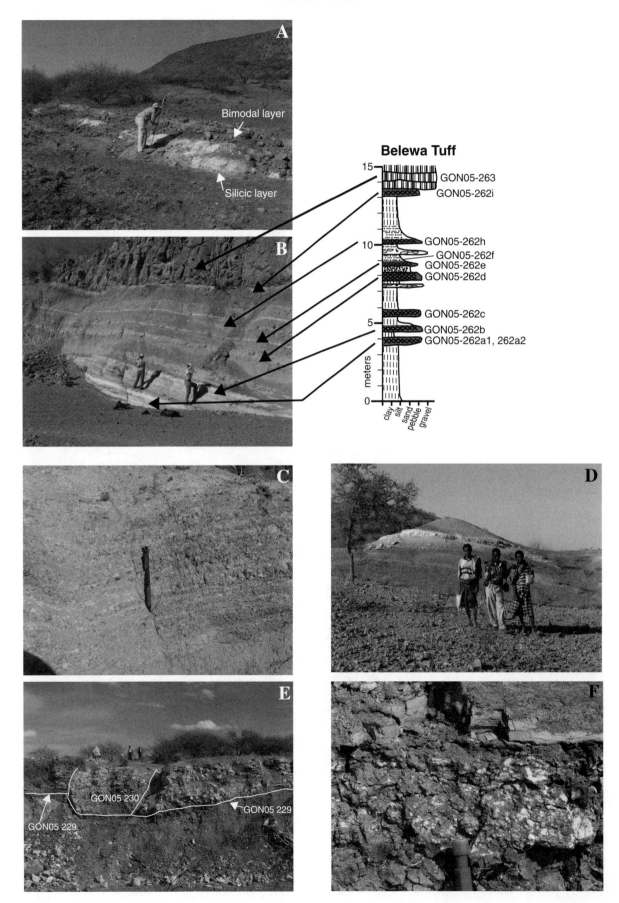

Figure 7. (A) Photograph of the Kobo'o Tuff type locality. Sample GON05-216b (Fig. 2) from this outcrop yielded a single-crystal ^{40}Ar/^{39}Ar date of 5.44 ± 0.06 Ma (2σ) on sanidine and plagioclase (Fig. 9; Table 4). The Kobo'o Tuff here and elsewhere is bimodal, it consists of mainly silicic ash layers at the base, and it is more mafic at the top. Person for scale. (B–C) Photographs and measured stratigraphic section of the type locality for the Belewa Tuff and where sample GON05-262 was collected (pencil in C is 15 cm). (D) The more distal correlate where GON05-265 (Fig. 2) was sampled and yielded a single-crystal ^{40}Ar/^{39}Ar date of 5.47 ± 0.04 Ma (2σ on sanidine; Fig. 9; Table 4). (E–F) Photographs of the Bodele A and B Tuffs at the BDL fossil localities (Figs. 1 and 2) in lacustrine mudstone. These phenocryst-rich, altered tuff units occur below both the Sifi Tuff and the level of the fossils at the BDL sites. The Bodele B Tuff (GON05-230) (shown in detail in F) is a slump deposit containing 0.5-cm-scale plagioclase crystals and lapilli-sized pumice pieces. The Bodele A Tuff (GON05-229) is composed of a double layer of millimeter-scale plagioclase that is disrupted by the slump deposit that contains the Bodele B Tuff (GON05-230). Hammer for scale in E is ~40 cm. End of pencil in F is ~1 cm.

The Bodele A Tuff is a doublet of thin tuffs in laminated lake beds. These beds are locally disrupted by small slumps composed of tuffaceous brown mud, from which we obtained the Bodele B Tuff (Fig. 7A). Material in the slump deposit is a mix of dark brown mud, altered pumice ≤2 cm in diameter, and very abundant feldspars ~0.5 cm in diameter (Fig. 7F). Based on the outcrop relationships, the slump must be younger than the doublet. The abundance and chemical homogeneity of Bodele B phenocrysts suggest that, although locally redeposited by slumping, incorporation of older phenocrysts, such as from the chemically distinct Bodele A Tuff, did not occur. The Bodele B Tuff yielded an ^{40}Ar/^{39}Ar plateau age of 6.48 ± 0.22 Ma (2σ) (GON05-230) on plagioclase (Table 5; Fig. 10).

The age of the Bodele B Tuff is supported by results from nearby South Gona. The composition of the plagioclase in sample GON05-246 is identical to that in the underlying tuff (GON05-248) and is strikingly similar to the plagioclase in the Bodele B Tuff at the BDL sites (Table 3). All three units are 3–8 m below the level of the Sifi Tuff in our measured sections (Fig. 6). Although the composition of phenocrysts alone is insufficient for a firm correlation, this observation combined with the stratigraphic constraint of the Sifi Tuff suggests that the tuffs from South Gona are likely additional examples of the Bodele B Tuff. Initial attempts to date plagioclase from sample GON05-246, collected from the South Gona section, yielded a single-crystal ^{40}Ar/^{39}Ar date of 5.64 ± 0.58 Ma (2σ) (Table 4). This determination, however, was made using only 3 of 15 analyses and is not reliable. A later attempt to date plagioclase from the same sample using incremental heating was more successful and yielded a

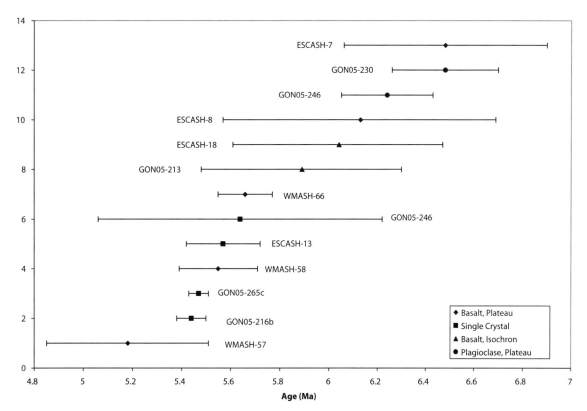

Figure 8. Summary graph of all ^{40}Ar/^{39}Ar dates from the Adu-Asa Formation. Locations of samples are given in Figure 2 and details are given in Tables 4 and 5. Full details are available in GSA Data Repository Tables 1 and 2 (see text footnote 1).

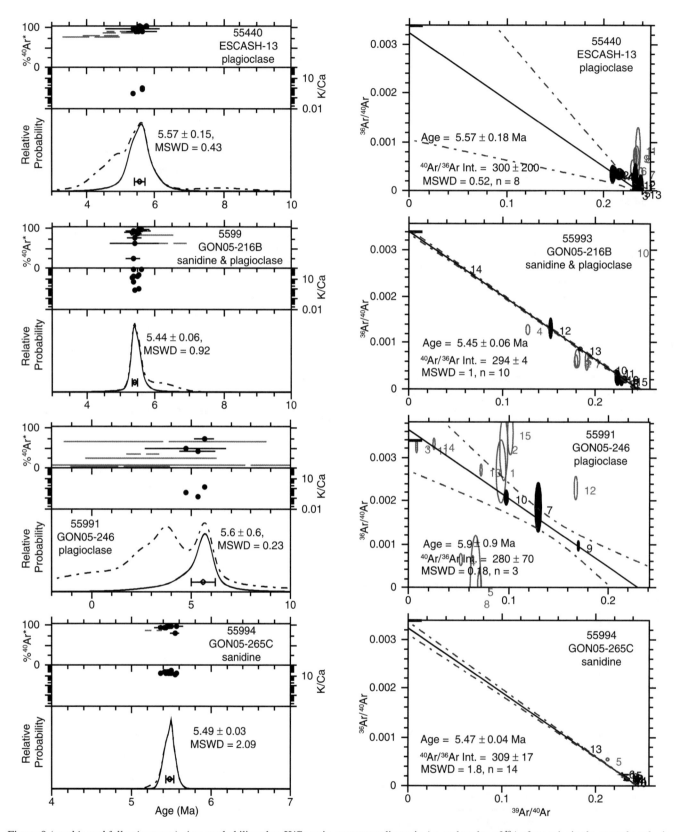

Figure 9 (*on this and following page*). Age-probability plot, K/Ca ratio, percent radiogenic Ar, and moles of ^{39}Ar for each single-crystal analysis. Details of irradiation, analytical procedures, calculation methods, and analytical data are given in GSA Data Repository Table 1 (see text footnote 1). The second set of graphs for GON05-271 was recalculated to include only the youngest population of feldspars. MSWD—mean square of weighted deviates.

reasonably flat age spectra with a plateau age of 6.24 ± 0.19 Ma (2σ) (Table 5; Fig. 10). This is very similar to the age of the Bodele B Tuff.

Obsidian and the Source of the Ogoti Tuff Complex and the Belewa Tuff

In the Gona Paleoanthropological Research Project area, the only volcanic source thus far identified in the Adu-Asa Formation was the silicic center in the northernmost part of the project area (Figs. 1 and 2). We analyzed the glassy obsidian portion of a rhyolite flow(s) from four different localities (GON05-270, -272, -281, and -287) as well as one sample from a basal pumice breccia (GON05-285) (Table 2). Samples GON05-272 and -287 were collected from directly beneath outcrops of the ash-flow component of the Ogoti Tuff Complex, while GON05-285 was collected from directly above the surge component of the Ogoti Tuff Complex. GON05-287 is a sample of glassy obsidian exposed above a spherulitic obsidian layer and below and conformable to sediments of the Sagantole Formation. The chemistry of these samples is very similar to, and in some cases indistinguishable from, that of the Ogoti Tuff Complex (Tables 2, 3, and 7). Similarity coefficients for the obsidian versus Ogoti Tuff Complex samples range from 0.77 to 0.95, with a median SC of 0.88. Thus, based on the similar chemistry and the spatial relationships between the silicic flows and outcrops of the Ogoti Tuff Complex, we can be certain that the source of the Ogoti Tuff Complex is this silicic center in the northern end of the Gona Paleoanthropological Research Project area.

The Ogoti Tuff Complex is not the only silicic tuff we can attribute to this source. Chemically, the Ogoti Tuff Complex is very similar to the Belewa Tuff. When comparing a sample of the Ogoti Tuff Complex to a sample of the Belewa Tuff, the average SC is 0.84, and one of the sample pairs gives an SC as high as 0.90 (Table 7). Moreover, grain-size contrasts between the two outcrops of the Belewa Tuff also point to a nearby silicic center as the source. The first outcrop of the Belewa Tuff (Figs. 2, 7B, and 7C) where sample GON05-262 was collected is much coarser-grained than the outcrop where the second sample was collected (GON05-265, Figs. 2 and 7D) and thus more proximal to the source. The first outcrop is located ~12 km north of the second, more distal outcrop. Thus, the source of the Belewa Tuff is almost certainly the rhyolite flow-dome identified in the northern end of the Gona Paleoanthropological Research Project area (Figs. 1 and 2).

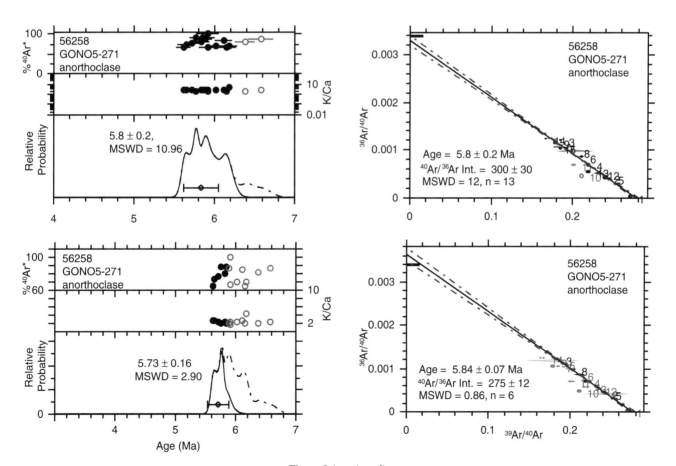

Figure 9 (continued).

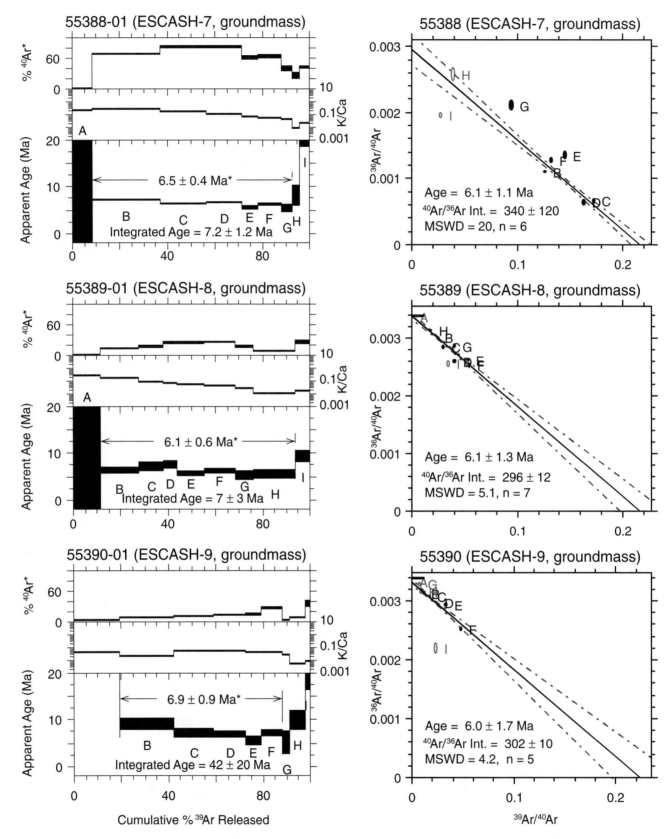

Figure 10 (*on this and following four pages*). Plateau and isochron ages for step-heated samples dated by the $^{40}Ar/^{39}Ar$ method. Details of irradiation, analytical procedures, calculation methods, and analytical data are given in GSA Data Repository Table 1 (see text footnote 1). MSWD—mean square of weighted deviates.

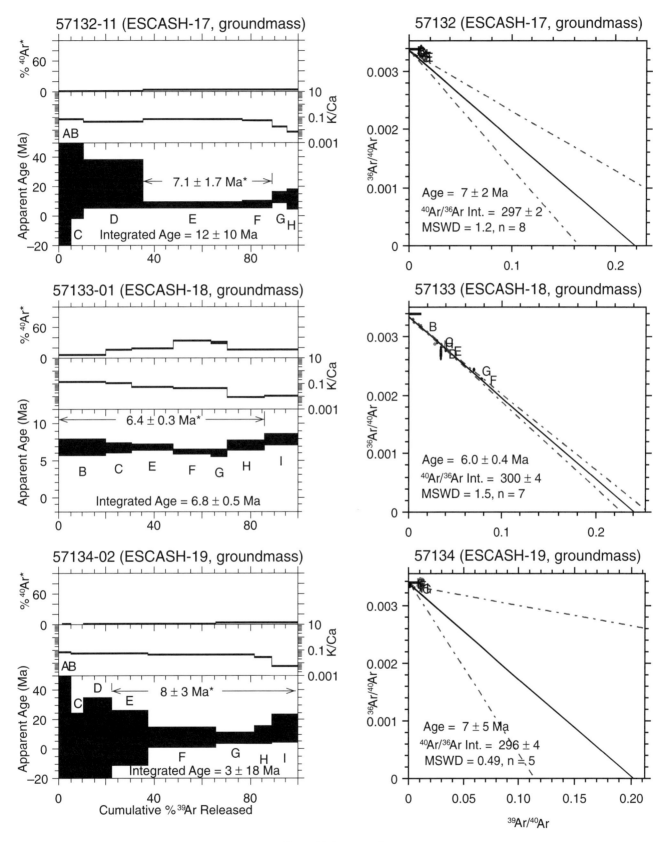

Figure 10 (*continued*).

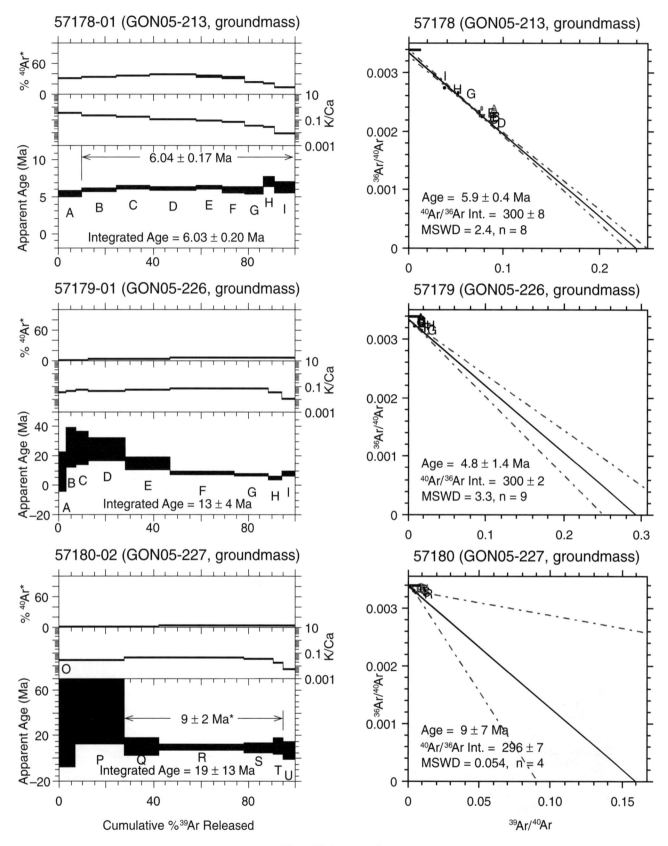

Figure 10 (*continued*).

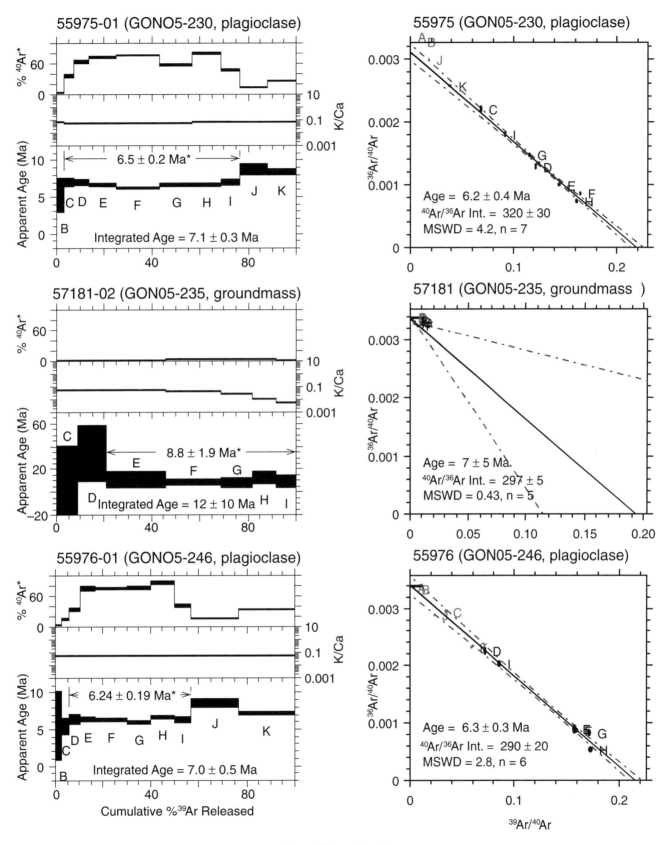

Figure 10 (*continued*).

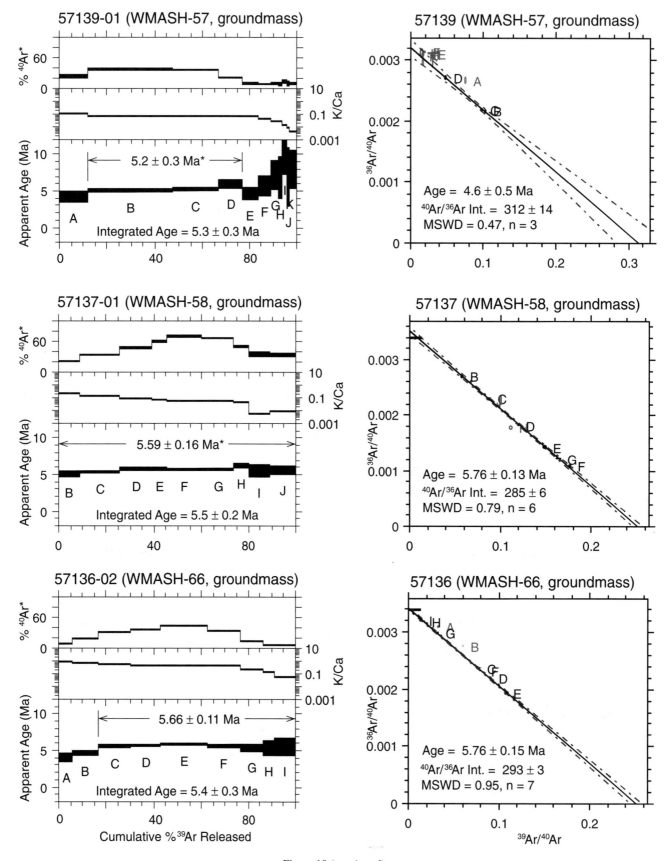

Figure 10 (*continued*).

TABLE 4. SUMMARY OF SINGLE-CRYSTAL ^{40}Ar/^{39}Ar RESULTS

Sample	Unit	Lab #	Irradiation #	Material	Mean age			Isochron age		Comments
					n	MSWD	Age (Ma, ±2σ)	n	MSWD	^{40}Ar/^{36}Ar (±2σ)	Age (Ma, ±2σ)	

Sample	Unit	Lab #	Irradiation #	Material	n	MSWD	Age (Ma, ±2σ)	n	MSWD	^{40}Ar/^{36}Ar (±2σ)	Age (Ma, ±2σ)	Comments
ESCASH-13	Ogoti ash-fall tuff	55440	NM-186K	Plagioclase	8	0.43	**5.57 ± 0.15**	8	0.52	300 ± 200	5.57 ± 0.18	Good
GON05-216b	Kobo'o Tuff	55993	NM-192K	Sanidine and plagioclase	10	0.92	**5.44 ± 0.06**	10	1.00	294 ± 4	5.45 ± 0.06	Very good
GON05-246	Bodele B Tuff	55991	NM-192K	Plagioclase	3	0.23	**5.64 ± 0.58**	3	0.18	280 ± 70	5.90 ± 0.90	Fair
GON05-265c	Belewa Tuff	55994	NM-192K	Sanidine	14	2.09	5.49 ± 0.03	14	0.38	309 ± 17	**5.47 ± 0.04**	Very good
GON05-271	*Ogoti ash-flow tuff*	56258	NM-196H	Anorthoclase	13	10.96	5.80 ± 0.20	13	12.00	300 ± 30	5.80 ± 0.20	
GON05-271	*Ogoti ash-flow tuff*	56258	NM-196H	Anorthoclase	6	2.80	5.73 ± 0.16	6	0.86	275 ± 12	5.84 ± 0.07	

Notes: Ages were calculated relative to FC-2 Fish Canyon Tuff sanidine interlaboratory standard (28.02 Ma; Renne et al., 1998). All errors are reported at ±2σ, unless otherwise noted. Details of irradiation, analytical procedures, calculation methods, and analytical data are in GSA Data Repository Tables 1 and 2 (see text footnote 1). Locations of samples are given in Figure 2. Analyses in italics indicate questionable accuracy. Bold denotes preferred ages. MSWD—mean square of weighted deviates.

TABLE 5. SUMMARY OF STEP-HEATED ^{40}Ar/^{39}Ar RESULTS

Sample	Unit	Location	Lab #	Irradiation #	Material	Plateau age				Isochron age			Comments
						n	%^{39}Ar	MSWD	Age (Ma) ± 2σ	n	MSWD	^{40}Ar/^{36}Ar ± 2σ	Age(Ma) ± 2σ	

Sample	Unit	Location	Lab #	Irradiation #	Material	n	%^{39}Ar	MSWD	Age (Ma) ± 2σ	n	MSWD	^{40}Ar/^{36}Ar ± 2σ	Age(Ma) ± 2σ	Comments
ESCASH-7	Porphyritic basalt	Caps N ESC sites	55388-01	NM-186B	Gm*	6	84.6	18.8	**6.48 ± 0.42**	6	20.3	338.6 ± 122.7	6.12 ± 1.06	Usable
ESCASH-8	Porphyritic basalt	Caps N ESC sites	55389-01	NM-186B	Gm	7	82.5	4.3	**6.13 ± 0.56**	7	5.1	295.6 ± 12.4	6.13 ± 1.30	Usable
ESCASH-9	Porphyritic basalt	Caps N ESC sites	*55390-01*	NM-186B	Gm	5	68.7	5.1	6.92 ± 0.88	5	4.2	301.8 ± 10.4	5.95 ± 1.78	
ESCASH-17	*Basalt flow*	Below ESC 3	57132-11	NM-208C	Gm	2	53.8	0.1	7.06 ± 1.69	8	1.2	297.4 ± 2.1	6.65 ± 2.19	
ESCASH-18	Porphyritic basalt	Caps N ESC sites	57133-01	NM-208C	Gm	7	86.2	2.1	6.37 ± 0.32	7	1.5	299.7 ± 4.1	**6.04 ± 0.43**	Good (isochron age)
ESCASH-19	*Basalt flow*	Caps HMD section	57134-02	NM-208C	Gm	5	77.5	0.4	8.31 ± 2.79	5	0.5	296.4 ± 3.7	7.15 ± 5.02	
GON05-213	Porphyritic basalt	Above ESC 8	57178-01	NM-208K	Gm	8	89.7	2.3	6.07 ± 0.17	8	2.4	299.6 ± 8.3	**5.89 ± 0.41**	Good (isochron age)
GON05-226	*Basalt flow*	Caps BDL sites	57179-01	NM-208K	Gm	0	0.0	0.0	0.00 ± 0.00	9	3.3	299.9 ± 2.3	4.81 ± 1.37	
GON05-227	*Basalt flow*	Caps BDL sites	57180-02	NM-208K	Gm	4	67.2	0.1	9.33 ± 2.21	4	0.1	296.1 ± 6.6	8.78 ± 6.69	
GON05-230	Bodele B tuff	Below BDL sites	55975-01	NM-192H	Plag.	7	73.3	5.7	**6.48 ± 0.22**	7	4.2	322.0 ± 28.0	6.18 ± 0.35	Fair
GON05-235	*Basalt flow*	Caps ABD sites	57181-02	NM-208K	Gm	5	72.0	0.4	8.76 ± 1.87	5	0.4	297.0 ± 4.6	7.24 ± 4.67	
GON05-246	Bodele B tuff	South Gona	55976-01	NM-192H	Plag.	6	51.0	2.1	**6.24 ± 0.19**	6	2.8	294.1 ± 23.6	6.26 ± 0.30	Fair
WMASH-57	Basalt flow	Caps Adu-Asa Form.	57139-01	NM-208D	Gm	3	65.0	3.3	**5.18 ± 0.33**	3	0.5	311.9 ± 14.0	4.60 ± 0.52	Fair
WMASH-58	Basalt flow	Caps Adu-Asa Form.	57137-01	NM-208C	Gm	6	73.5	3.3	**5.55 ± 0.16**	6	0.8	284.7 ± 5.7	5.76 ± 0.13	Good
WMASH-66	Basalt flow	Caps Adu-Asa Form.	57136-02	NM-208C	Gm	7	83.3	1.3	**5.66 ± 0.11**	7	1.0	293.1 ± 2.8	5.76 ± 0.15	Very good

Notes: Ages were calculated relative to FC-2 Fish Canyon Tuff sanidine interlaboratory standard (28.02 Ma; Renne et al., 1998). Analyses were performed at New Mexico Geochronology Research Laboratory using an MAP 215-50 mass spectrometer online with automated all-metal extraction system. All errors are reported at ±2σ, unless otherwise noted. Details of irradiation, analytical procedures, calculation methods, and analytical data are in GSA Data Repository Tables 1 and 2 (see text footnote 1). Locations of samples are given in Figure 2. Analyses in italics indicate low-radiogenic yield analyses with poor precision and questionable accuracy. Bold denotes preferred ages. High mean square of weighted deviate (MSWD) values are outlined.

*Groundmass concentrate.

TABLE 6. COMPARISON OF ANALYTICAL CONDITIONS

Kobo'o Tuff	Condition*	N	Na$_2$O	K$_2$O	SiO$_2$	MgO	Al$_2$O$_3$	CaO	MnO	FeO	TiO$_2$	Total[†]	Total[§]
Silicic A													
AVERAGE	A	263	2.04	1.91	71.54	0.01	12.06	0.65	0.11	2.49	0.22	91.27	91.03
σ			0.65	0.22	1.33	0.01	0.32	0.04	0.02	0.11	0.04	1.60	1.60
AVERAGE	B	21	3.44	2.10	71.44	0.01	11.56	0.68	0.12	2.51	0.23	92.28	92.08
σ			0.19	0.12	1.90	0.01	0.21	0.04	0.02	0.10	0.03	1.96	1.94
Silicic B													
AVERAGE	A	95	1.88	1.95	70.26	0.03	12.37	0.85	0.12	2.76	0.23	90.70	90.46
σ			0.52	0.32	1.32	0.01	0.34	0.04	0.02	0.10	0.05	1.46	1.45
AVERAGE	B	3	3.26	2.16	68.64	0.03	11.76	0.87	0.12	2.66	0.25	90.01	90.01
σ			0.02	0.12	2.08	0.01	0.13	0.01	0.02	0.10	0.09	2.19	2.19
Mafic													
AVERAGE	A	36	2.42	1.30	51.77	3.16	12.93	6.98	0.39	13.25	2.85	95.33	95.06
σ			0.61	0.08	1.57	0.25	0.29	0.25	0.04	0.76	0.15	1.79	1.76
AVERAGE	B	7	2.39	1.39	53.11	2.75	12.34	6.63	0.40	13.24	2.63	95.09	94.88
σ			0.63	0.05	2.08	0.34	0.27	0.41	0.05	0.63	0.18	1.79	1.79

Notes: Comparison of electron microprobe setup conditions and their effect on the apparent composition of samples.
*Conditions A and B are given in Table 1.
[†]Probe measured total.
[§]Summed total of oxides shown.

Basalts

The basalt flows in the Adu-Asa Formation at Gona are typically blue-gray in color, holocrystalline, and range in thickness from 1 to 10 m. In general, both the number and thickness of basalt lava flows increase between the level of the Sifi Tuff and the Kobo'o Tuff (Fig. 6; see also Quade et al., their Fig. 3B, this volume). This trend of increased volcanism and/or decreased sedimentation continues through to the top of the Adu-Asa Formation at Gona, although there is a shift to silicic volcanism as represented by the rhyolite dome in the northern end of the project area (Fig. 2). Further to the south, however, the top of the Adu-Asa Formation is still dominated by basalt flows. Although not well surveyed, it appears that below the level of the Hamadi Das tuffs, there is another large section of basalt lavas, which appears as an area of high topographic relief west of the Kasa Gita-Chifra Road and the HMD fossil sites in Figure 1. Alteration of basalt units can be substantial, especially near faults, where argillization of the matrix has resulted in a friable, sand-like texture. In many cases, relatively unweathered "corestones" can be found within an otherwise pervasively altered unit.

We focused on the tuffs as stratigraphic markers, since many different basaltic lava flows look similar in the field. We nonetheless still sampled many of the basaltic flows as a supplement to the geochronological information obtained from the tuffs. Complete details for the ^{40}Ar/^{39}Ar dates on basaltic groundmass are presented in GSA Data Repository Table 2 (see footnote 1).

Samples WMASH-57, -58, and -66 are from the easternmost basalt flows in the Adu-Asa Formation, and thus they cap the entire formation in the central Gona Paleoanthropological Research Project area (Figs. 1 and 2). Basaltic groundmass concentrates from these samples yielded ^{40}Ar/^{39}Ar plateau dates of 5.18 ± 0.33 Ma (2σ), 5.55 ± 0.16 Ma (2σ), and 5.66 ± 0.11 Ma (2σ), for samples WMASH-57, -58, and -66, respectively (Table 5; Fig. 10). Based on these dates, we consider 5.4 Ma to be a reasonable estimate for the age of the top of the Adu-Asa Formation in the central Gona Paleoanthropological Research Project area. While WMASH-57 yielded a younger age than 5.4 Ma, that sample has a fairly large associated error. WMASH-66, in contrast, yielded an age significantly older than 5.4 Ma, but this sample was collected west of samples WMASH-57 and -58, and it may represent a slightly older basalt flow. An upper boundary of 5.4 Ma is consistent with the plateau dates on both WMASH-57 and -58, as well as the dates on many of the tuffs within the Adu-Asa Formation.

Throughout the Adu-Asa Formation, there are many basalt flows in close stratigraphic association with fossil localities. Samples GON05-226 and -227 are from a blue-gray basalt with a fine-grained groundmass and occasional dispersed plagioclase phenocrysts up to 0.5 cm in diameter. This unit caps the sedimentary rocks at the BDL fossil localities and is similar both in description and stratigraphic placement to basalt sample GON05-235, which caps the ABD fossil localities (Figs. 1, 2, and 6). Similar blue-gray aphanitic flows, with no visible phenocrysts reported, are found capping the HMD fossil sites (ESCASH-19), underlying the ESC-3 site (ESCASH-17), and at the base of the stratigraphic section containing GON05-219 (GON05-218 and GON05-220) (Figs. 1, 2, and 6). The presence of a thin mudstone bed between samples GON05-218 and GON05-220 indicates the existence of at least two different blue-gray aphanitic flows (Fig. 6). Although the exact number of basalt units in this part of the Adu-Asa Formation is not clear, the relationship of these units to the fossil localities is unambiguous.

Attempts to date the aphanitic flows were unsuccessful. Samples GON05-226, GON005-227, GON05-235, ESCASH-17, and ESCASH-19 all had low radiogenic yields (generally <10%), poor precision of individual steps, and disturbed age spectra (Table 5; Fig. 10).

We were able to consistently identify one basalt flow in the field. This unit is porphyritic and has numerous plagioclase

TABLE 7. SELECTED SIMILARITY COEFFICIENTS (SC) (GLASS ANALYSES)

	262a1	262a2	262b	262c	262d	262e	262f	262h	262b obs†	262d obs†	262f obs†	262h obs†	265a	265b	265c obs†	ESC.-13 F1§	ESC.-13 M2*§	ESC.-13 G1§	ESC.-13 F3§	283**	286a**	286b†**	ESC.-13 M2*	270#	272#	281#	285#	287#
262a1	1.00																											
262a2	0.97	1.00																										
262b	**0.93**	0.91	1.00																									
262c	0.95	**0.93**	0.97	1.00																								
262d	0.95	**0.93**	0.97	0.96	1.00																							
262e	0.99	0.97	**0.92**	**0.94**	**0.94**	1.00																						
262f	0.96	0.95	0.96	0.98	0.96	0.96	1.00																					
262h	0.95	**0.92**	**0.93**	**0.94**	0.95	0.95	**0.93**	1.00																				
262b obs†	**0.92**	0.91	0.96	0.96	0.95	0.91	0.95	**0.92**	1.00																			
262d obs†	**0.92**	0.90	0.99	0.96	0.97	0.91	0.95	**0.92**	0.97	1.00																		
262f obs†	**0.94**	**0.92**	0.98	0.96	0.98	**0.93**	0.95	**0.94**	0.97	0.98	1.00																	
262h obs†	**0.92**	0.90	**0.94**	**0.92**	0.95	**0.92**	0.91	**0.94**	0.95	0.95	0.95	1.00																
265a	**0.94**	**0.94**	0.97	0.96	0.97	**0.92**	0.96	**0.93**	0.96	0.96	0.96	**0.93**	1.00															
265b	**0.93**	0.91	0.98	0.96	0.96	**0.94**	0.96	**0.92**	0.96	0.98	0.97	**0.93**	0.97	1.00														
265c obs†	**0.94**	**0.94**	0.95	0.96	**0.94**	**0.94**	0.98	**0.92**	**0.94**	**0.94**	**0.94**	0.91	0.95	0.95	1.00													
ESC.-13 F1§	0.86	0.86	0.85	0.89	0.86	0.87	0.89	0.89	0.86	0.87	0.85	0.85	0.85	0.86	0.88	1.00												
ESC.-13 M2*§	0.84	0.82	0.88	0.87	0.88	0.85	0.85	0.88	0.87	0.88	0.88	0.88	0.87	0.87	0.84	**0.93**	1.00											
ESC.-13 G1§	0.83	0.81	0.87	0.87	0.87	0.84	0.85	0.88	0.87	0.86	0.86	0.87	0.86	0.87	0.83	**0.93**	0.98	1.00										
ESC.-13 F3§	0.86	0.85	0.86	0.89	0.87	0.87	0.90	0.90	0.86	0.86	0.86	0.86	0.85	0.85	0.87	0.98	**0.93**	**0.93**	1.00									
283**	0.84	0.82	0.88	0.88	0.88	0.85	0.86	0.89	0.88	0.88	0.87	0.87	0.87	0.87	0.84	**0.92**	**0.94**	0.95	**0.92**	1.00								
286a**	0.83	0.81	0.87	0.86	0.88	0.85	0.84	0.88	0.88	0.88	0.88	0.88	0.85	0.86	0.83	**0.92**	0.95	0.95	**0.94**	**0.93**	1.00							
286b†**	0.85	0.83	0.89	0.87	0.88	0.86	0.86	0.90	0.89	0.89	0.88	0.90	0.87	0.88	0.85	**0.92**	**0.94**	0.97	**0.94**	0.91	0.96	1.00						
ESC.-13 M2*	0.39	0.37	0.41	0.38	0.41	0.39	0.40	0.38	0.39	0.41	0.42	0.39	0.40	0.42	0.38	0.36	0.38	0.38	0.38	0.40	0.38	0.42	1.00					
270#	0.73	0.73	0.76	0.75	0.76	0.75	0.74	0.78	0.75	0.76	0.76	0.78	0.75	0.75	0.72	0.80	0.86	0.86	0.81	0.83	0.83	0.85	0.39	1.00				
272#	0.84	0.81	0.86	0.86	0.85	0.83	0.84	0.87	0.89	0.86	0.86	0.88	0.85	0.86	0.84	0.90	0.90	0.91	0.90	0.90	**0.93**	0.91	0.38	0.77	1.00			
281#	0.83	0.81	0.86	0.82	0.85	0.83	0.81	0.88	0.86	0.85	0.86	0.89	0.83	0.83	0.80	0.85	0.89	0.89	0.86	0.86	0.90	**0.93**	0.38	0.85	0.88	1.00		
285#	0.82	0.80	0.84	0.82	0.85	0.82	0.82	0.83	0.86	0.85	0.85	0.88	0.84	0.84	0.81	0.81	0.85	0.85	0.86	0.87	0.87	**0.92**	0.41	0.79	0.85	0.90	1.00	
287#	0.82	0.80	0.86	0.85	0.85	0.82	0.83	0.86	0.88	0.87	0.86	0.89	0.84	0.85	0.82	**0.92**	**0.94**	0.95	0.91	0.95	0.95	**0.92**	0.39	0.82	**0.93**	0.89	0.88	1.00

Notes: Similarity coefficients (SC) for glass analyses from units attributed to the rhyolite dome in the north end of the GPRP area. SCs of 0.95 or greater are outlined, and SCs between 0.92 and 0.94 are shown in bold. Samples labeled with only a number (e.g., 265a) have had the sample prefix "GON05-" omitted. The prefix "ESC." is short for ESCASH. Unless otherwise noted, analyses are of glass shards in ash-fall tuffs. Average compositions of each sample are given in Table 2, along with name of the tuff, where applicable. SCs were calculated following the formula in Rodbell et al. (2002), which was modified from work by Borchardt et al. (1972).

*Sample contains both a felsic and a mafic population. The higher SCs denote the felsic split.
†Obsidian clast.
§Pumice lapilli.
#Glassy rhyolite.
**Ash-flow or surge deposit.

59

crystals over 2 cm in length and a black holocrystalline groundmass. It lies 25–50 m above the Kobo'o Tuff in our measured stratigraphic sections, assuming continuous stratigraphy unbroken by faults (Fig. 6, type locality given in GSA Data Repository Fig. 1 [see footnote 1]). Consistent identification of the porphyritic basalt is important because it is exposed prominently in section above many of the ESC fossil sites, including ESC-8 (GON05-213) and the ESC-1, -2, and -3 fossil localities (ESCASH-7, -8, -9, and -18) (Figs. 1 and 2). Except for ESC-9, the ESC sites are not in section with any unaltered tuffs, so this porphyritic basalt is the only available stratigraphic marker that can constrain the age of these sites. In all cases, the ESC sites are stratigraphically below the porphyritic basalt.

Our ^{40}Ar/^{39}Ar dating of the porphyritic basalt was more successful than for the aphanitic flows. Plateau ages for groundmass concentrates from ESCASH-7 and -8 were 6.48 ± 0.42 Ma (2σ) and 6.13 ± 0.56 (2σ), respectively (Table 5; Fig. 10). For samples GON05-213 and ESCASH-18, the isochron age is a better estimate of the eruption age than the plateau age, as these samples yielded isochrons with ^{40}Ar/^{39}Ar intercepts slightly higher than the atmospheric value (295.5). GON05-213 yielded an isochron age of 5.89 ± 0.41 Ma, while ESCASH-18 yielded an isochron age of 6.04 ± 0.43 Ma. Sample ESCASH-19, which is also from the porphyritic basalt, did not yield a usable date (Table 5; Fig. 10).

DISCUSSION

Geological History

We were able to construct a composite stratigraphic section for the upper (<6.4 Ma) part of the Adu-Asa Formation at Gona by combining the measured stratigraphic sections with outcrop patterns of the various volcanic units (Fig. 11). Our estimate of the composite thickness of the Adu-Asa Formation east of the Kasa Gita-Chifra Road (Figs. 1 and 2) is ~185 m. We also developed a geologic cross section in Figure 12 that reflects all of the tephrostratigraphic and structural constraints available.

The base of the stratigraphic sequence is dominantly lacustrine, as indicated by the presence of diatomite beds and laminated mudstone. This part of the formation contains many altered basaltic tuffs, which we collectively refer to as the Hamadi Das Crystal-Rich Tuff Sequence (HMDS). An important subunit of the Hamadi Das tuffs is the Bodele Tuff, which lies at the top of the Hamadi Das Crystal-Rich Tuff Sequence and forms an upper age limit on these ash-fall units. Plagioclase from the Bodele Tuff at the BDL fossil sites and a likely correlate in South Gona yielded ^{40}Ar/^{39}Ar ages of ca. 6.2–6.4 Ma (Tables 4 and 5; Figs. 9 and 10).

The Sifi Tuff is the oldest tuff with preserved glass in the Adu-Asa Formation at Gona and is an important stratigraphic marker due to its direct association with many fossil sites as well as its widespread occurrence. This tuff is exposed prominently 5–25 m above units of the Hamadi Das Crystal-Rich Tuff Sequence and the Bodele Tuff in multiple locations (Fig. 6).

While the stratigraphic sequence below the level of the Sifi Tuff is lacustrine, fluvial deposition had taken over by the time the Sifi Tuff was erupted, as indicated by outcrops of the Sifi Tuff that are channelized, reworked, and variable in thickness from 0 to 2 m. Fluvial sedimentation is dominant through the rest of the stratigraphic sequence, as the sedimentary units above the level of the Sifi Tuff are typically red or pinkish mudstone interbedded with cross-bedded sandstone and conglomerate. Because the Sifi Tuff does not contain a homogeneous population of phenocrysts, it is unsuitable for ^{40}Ar/^{39}Ar dating.

An aphanitic basalt flow caps the stratigraphic sequence in our measured sections of the BDL, ABD, and HMD fossil sites, as well as at South Gona, while a sequence of aphanitic basalt flows is exposed at the base of several measured sections containing the Kobo'o Tuff and the porphyritic basalt (Fig. 6). These aphanitic basalt exposures may not be from exactly the same flows, but it is probable that they represent the link between the lower part of the stratigraphic sequence, which contains the Hamadi Das Crystal-Rich Tuff Sequence and Sifi Tuff, and the sequence containing the Kobo'o Tuff. Multiple attempts to date these aphanitic units have proved unsuccessful because these analyses typically had low radiogenic yields, poor precision, and/or disturbed age spectra (Table 5; Fig. 10).

Like the Sifi Tuff, the Kobo'o Tuff is fluvially reworked and contains abundant well-preserved glass. It is exposed below a porphyritic basalt flow in two measured stratigraphic sections and above a series of aphanitic basalt flows (Fig. 6). This porphyritic unit is also an important stratigraphic marker because it caps several fossil localities. Sanidine and plagioclase crystals from the Kobo'o Tuff yielded an ^{40}Ar/^{39}Ar age of 5.44 ± 0.06 Ma (2σ), whereas the ^{40}Ar/^{39}Ar dates obtained for the porphyritic basalt flow were consistently closer to 6 Ma (Tables 4 and 5; Figs. 9 and 10). However, as the single-crystal analyses yielded the more precise ages on tuffs both above and below the level of the porphyritic basalt, we prefer the younger age of ca. 5.5 Ma for the upper part of the Adu-Asa Formation at Gona, even though the dates on this porphyritic basalt are significantly older.

Deposits above the level of the porphyritic basalt are not well surveyed, and, as a result, the upper portion of the composite stratigraphic section is loosely constrained and is likely more complex, but it does reflect all our observations and represents the area on Figure 1 east of the ESC sites, including the rhyolite dome. We observed rhyolite in the north end of the project area directly and conformably underlying sediments of the Sagantole Formation, so we can be sure that the composite stratigraphic and cross sections presented here contain the top of the formation, at least for the northern end of the project area (Figs. 11 and 12).

Figure 11. Composite stratigraphic section of the Adu-Asa Formation at Gona. The section is schematic and combines all measured stratigraphic sections (Figs. 6 and 7) as well as general geological observations. Only ^{40}Ar/^{39}Ar dates considered reliable are shown (Tables 4 and 5; Figs. 8, 9, and 10). HMDS—Hamadi Das crystal-rich sequence tuffs.

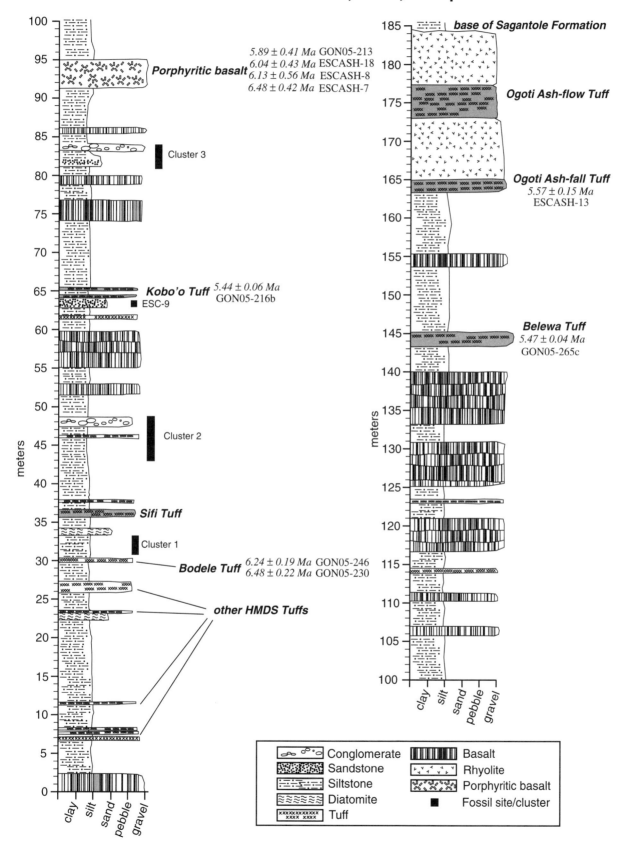

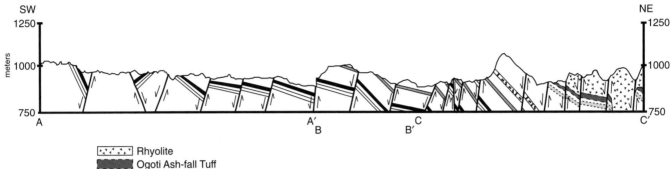

Figure 12. East-west composite cross section through the Adu-Asa Formation at Gona. Cross section is schematic and incorporates all known outcrops and orientations of the major marker units along the lines of cross section. See Figure 2 for locations of cross-section segments. Vertical exaggeration is ~3.1. HMDS—Hamadi Das crystal-rich sequence tuffs.

This rhyolite dome is also the likely source of both the Belewa and Ogoti Tuffs. Given that the deposits in the Adu-Asa Formation are generally east-dipping and thus become younger to the east, we have placed the Belewa and Ogoti Tuffs above the level of all known fossil sites in the Adu-Asa Formation as well as the Hamadi Das Crystal-Rich Tuff Sequence and Sifi and Kobo'o Tuffs. As exposures of the Belewa Tuff are consistently west of outcrops of the Ogoti Tuff, we interpret the Belewa Tuff to be below the level of the Ogoti Tuff in the composite stratigraphic section (Figs. 2 and 11). Single-crystal ^{40}Ar/^{39}Ar analyses of the Belewa and Ogoti Tuffs yielded ages of 5.47 ± 0.04 Ma (2σ) and 5.57 ± 0.15 Ma (2σ), respectively (Table 4; Fig. 9).

It is important to note that we use the term Ogoti Tuff Complex to include any tuff unit that displays the characteristic chemical composition, regardless of whether the units are directly equivalent temporally. This clarification is necessary because the ash-fall, ash-flow, and surge deposits sampled may not have all been deposited at precisely the same time. However, the amount of time lapsed was certainly smaller than the error range associated with even the most precise ^{40}Ar/^{39}Ar date.

It should also be noted that the ^{40}Ar/^{39}Ar dates on the Kobo'o, Belewa, and Ogoti ash-fall tuffs reflect the time of eruption, and not deposition, of the tuff, since sedimentary processes have clearly reworked each ash-fall tuff. However, the lag time between eruption and deposition is inconsequential, since this process would have occurred shortly after eruption, and there is very little dilution of tuffaceous material with other sedimentary components. In addition, given the similarity in the ^{40}Ar/^{39}Ar dates obtained on the Kobo'o, Belewa, and Ogoti Tuffs, we can infer that this part of the Adu-Asa Formation accumulated rapidly.

We have not included samples GON05-300, GON05-301, and GON05-302 in either the composite stratigraphic section or cross section because these units are tuffs that were encountered once each. It is likely that these units are younger than the Belewa Tuff, however, because they are exposed to the east of Belewa Tuff sample GON05-265 and 2–3 km west of sedimentary rocks of the Sagantole Formation. Thus, these units are near the top of the Adu-Asa Formation.

Age of Fossil Localities

The fossil localities in the Adu-Asa Formation at Gona fall into three temporal clusters. The oldest grouping (sites ABD-1, -2, HMD-1; Figs. 1, 2, 6, and 11) is stratigraphically below the Sifi Tuff, the second cluster is above the Sifi Tuff (sites HMD-2, BDL-1, -2; Figs. 1, 2, 6, and 11), and the third and youngest cluster (sites ESC-1, -2, -3, -8, and -9; Figs. 1, 2, 6, and 11) is around the level of the Kobo'o Tuff and the porphyritic basalt. We estimate the age of the oldest cluster to be ca. or younger than 6.4 Ma, the second cluster to be between 6.4 Ma and 5.5 Ma, and the third cluster to be ca. 5.5 Ma. While the second cluster can only be constrained to a rather large interval, it is closely associated with deposits toward the older end of this age range.

Sites included in the first temporal cluster are exposed directly below the Sifi Tuff, and we interpret them to be above the Bodele Tuff. This cluster includes the ABD-1, -2, and HMD-1 sites. At the ABD sites, the fossils are confined to the conglomerates and sandstones below a diatomite layer (Fig. 6), although given the low topographic relief at these sites, it is has not been possible to determine exactly which stratigraphic layer contains the fossils. At HMD-1, fossils have been traced to a siltstone ~7 m below the level of the Sifi Tuff and ~2 m above a series of tuff deposits that likely correlate to the Bodele Tuff (Fig. 6). As these fossil sites occur just above the level of the Bodele Tuff, we estimate their age to be slightly younger than 6.4 Ma.

A site in the Henali area (HEN-1) (Fig. 1) was not observed directly in section with any of the major stratigraphic markers discussed here. However, an altered, plagioclase-rich tuff (GON05-261) was collected near HEN-1, and the composition and outcrop appearance of this tuff are similar to the Hamadi Das Crystal-Rich Tuff Sequence tuffs, of which the Bodele Tuff is a subunit (Table 3; Fig. 2). Thus, the fossils from HEN-1 may be similar in age to the oldest temporal cluster of sites.

The second cluster contains fossils from the BDL-1, -2, and HMD-2 sites. At the BDL sites (BDL-1, -2), the conglomerates and sands ~6 m above the level of the Sifi Tuff and ~10 m above the Bodele Tuff contain the fossils (Fig. 6). At HMD-2, fossils are associated with the conglomerate unit ~10 m above the level of the Sifi Tuff (Fig. 6). These fossil-bearing conglomerate and sandstone units above the Sifi Tuff comprise the second temporal cluster.

The age of the second temporal cluster of sites is loosely constrained by the $^{40}Ar/^{39}Ar$ dates on the Bodele B Tuff, which is located below, and the Kobo'o Tuff, which we place above the level of the sites included in these clusters. The $^{40}Ar/^{39}Ar$ dates obtained on plagioclase from the Bodele B Tuff (GON05-230) and its likely correlate in South Gona (GON05-246) yielded ages of 6.48 ± 0.22 Ma (2σ) and 6.24 ± 0.19 Ma (2σ), respectively (Figs. 7E, 7F, 9, and 10; Tables 4 and 5). The sites in the first and second temporal clusters are closer to, and often occur near, exposures of the Sifi Tuff and the Hamadi Das Crystal-Rich Tuff Sequence–Bodele Tuff (Figs. 1, 2, and 6). These fossil sites are stratigraphically well below the Kobo'o Tuff (~17 m in our composite stratigraphic section), which yielded a date of 5.44 ± 0.06 Ma (2σ) on sanidine and plagioclase (Figs. 9 and 11; Table 4). Thus, we can constrain the ages of the first and second temporal clusters to between 6.4 Ma and ca. 5.5 Ma, but they are much closer to the 6.4 Ma than the 5.5 Ma age based on stratigraphic thicknesses.

The third and youngest temporal cluster of sites includes the ESC-1, -2, -3, -8, and -9 fossil localities. These sites are associated with the porphyritic basalt and/or the Kobo'o Tuff, and they are younger than the sites associated with the Sifi Tuff (Figs. 1, 2, 6, and 11). In all cases, sites included in this third cluster are below the level of the porphyritic basalt. We observed this relationship directly for sites ESC-1, -2, -3, and -8. ESC-9 is associated with sandstone directly below the Kobo'o Tuff, and as the Kobo'o Tuff is below the level of the porphyritic basalt, ESC-9 must also predate the porphyritic basalt (Fig. 6).

The fossils at ESC-1, -2, and -3 are from a conglomerate layer at least 7 m below the porphyritic basalt, and at ESC-3, there is an altered tuff at the base of the stratigraphic section. We speculate that this altered tuff is the Kobo'o Tuff, which would place the ESC-1, -2, and -3 sites between the level of the Kobo'o Tuff and the porphyritic basalt. Multiple outcrops of the Kobo'o Tuff occur near many of the ESC sites (Figs. 1 and 2), so it is plausible that this altered tuff is the Kobo'o Tuff. At ESC-8, the fossil-bearing units are derived from conglomerates and are below the porphyritic basalt and above an aphanitic basalt flow. For these reasons, we have placed the third temporal cluster of sites (specifically, sites ESC-1, -2, -3, and -8) in a conglomerate unit at ~83 m on the composite stratigraphic section (Fig. 11). We have included ESC-9 in the third temporal cluster of sites due to its association with the Kobo'o Tuff, but this site likely predates the ESC-1, -2, -3, and -8 sites, since the fossils from ESC-9 are below the Kobo'o Tuff, while we interpret the rest of these sites as being located above the level of the Kobo'o Tuff. We estimate the age of the fossil localities in the third temporal cluster to be ca. 5.5 Ma, based on the $^{40}Ar/^{39}Ar$ dates on the Kobo'o Tuff (Figs. 6 and 11; Tables 4 and 5).

Potential for Correlations with Other Paleoanthropological Projects

To date, late Miocene and early Pliocene deposits in the Afar region have only been studied in the Middle Awash project area, located ~90 km due south of Gona (Kalb et al., 1982; Renne et al., 1999; WoldeGabriel et al., 2001). The $^{40}Ar/^{39}Ar$ dates from the Adu-Asa Formation there are late Miocene in age and thus close to the age of the Adu-Asa Formation at Gona (WoldeGabriel et al., 2001). In the Middle Awash area, a tuff unit near the base of the Sagantole Formation yielded a $^{40}Ar/^{39}Ar$ date on plagioclase of 5.55 ± 0.1 Ma (Renne et al., 1999). Later work in the Middle Awash area constrained fossils in the Adu-Asa Formation there to between 5.54 ± 0.17 Ma and 5.77 ± 0.08 Ma, based on $^{40}Ar/^{39}Ar$ dates on groundmass from a basaltic lava flow and a basaltic tuff, respectively (WoldeGabriel et al., 2001). Basalt directly underlying the base of the Adu-Asa Formation in the Middle Awash yielded $^{40}Ar/^{39}Ar$ ages on groundmass of 6.33 ± 0.07 Ma and 6.16 ± 0.06 Ma (WoldeGabriel et al., 2001). These ages overlap with those from the Adu-Asa Formation as we have it mapped at Gona.

However, tuffs with published descriptions from the Middle Awash area are largely basaltic and thus are chemically dissimilar to the tuffs characterized here. It may be that the Hamadi Das Crystal-Rich Tuff Sequence tuffs, including the Bodele Tuff, are the same as those described in the Middle Awash area. However, the complete lack of unaltered glass in the Hamadi Das Crystal-Rich Tuff Sequence tuffs at Gona prevents the comparison.

There is one tuff from the Middle Awash, named the Witti Tuff, which shares some similarities to a tuff at Gona. Like the Kobo'o Tuff, the Witti Tuff is bimodal (Table 8; WoldeGabriel et al., 2001). However, low-K plagioclase from three different samples of the Witti Tuff yielded $^{40}Ar/^{39}Ar$ ages of 5.63 ± 0.12 Ma, 5.57 ± 0.08 Ma, and 5.68 ± 0.07 Ma, whereas the Kobo'o Tuff contains higher-K sanidine and yielded a slightly younger date of 5.44 ± 0.06 Ma (2σ). In addition, the mafic component of the Witti Tuff is significantly higher in CaO, MgO, FeO, and TiO_2 (Table 8). Although these are different tuffs, they may have come from the same source. If this is the case, then it is likely that the Adu-Asa Formation at Gona above the level of the Kobo'o Tuff postdates the published portions of the Adu-Asa Formation as described at the Middle Awash project (WoldeGabriel et al., 2001), and the Kobo'o Tuff is the product of a melt that had evolved since the eruption that produced the Witti Tuff.

TABLE 8. KOBO'O AND WITTI TUFF COMPARISON

	N	Na_2O	K_2O	SiO_2	MgO	Al_2O_3	CaO	MnO	FeO	TiO_2	P_2O_5	Total[†]
Kobo'o Tuff												
Silicic A	263	2.04	1.91	71.54	0.01	12.06	0.65	0.11	2.49	0.22	NA*	91.03
Silicic B	95	1.88	1.95	70.26	0.03	12.37	0.85	0.12	2.76	0.23	NA	90.46
Mafic	36	2.42	1.30	51.77	3.16	12.93	6.98	0.39	13.25	2.85	NA	95.06
Witti Tuff												
Silicic	38	2.59	4.79	70.00	0.00	12.00	0.90	0.10	2.38	0.20	0.02	92.97
Mafic	39	2.30	1.19	50.93	4.36	12.66	8.04	0.24	14.1	3.65	0.65	98.11

Notes: Comparison of glass analyses on the Kobo'o Tuff at Gona and the Witti Mixed Magmatic Tuff from the Middle Awash area. Analytical conditions for the Kobo'o Tuff are given in Table 1 (condition A). The Witti Tuff was analyzed at Los Alamos National Laboratory using a Cameca SX50 electron microprobe with a 15 nA current, accelerating potential of 15 kV, and a 10 μm beam size. All Fe is expressed as FeO. Data on the Witti Tuff are from WoldeGabriel et al. (2001).
*NA—not analyzed.
[†]Summed total of oxides shown.

CONCLUSIONS

The Adu-Asa Formation at Gona is ~185 m thick and is composed largely of stacked basalt flows interbedded with fluviolacustrine sediments and numerous ash-fall tuffs. Within the main sedimentary interval, environments shifted from lacustrine at the base to fluvial above. At the same time, the composition of the volcanic units also shifted, from basaltic lava flows and tuffs to a greater component of silicic material.

We have identified seven different silicic, or dominantly silicic, tuffs in the Adu-Asa Formation at Gona, as well as a series of altered, crystal-rich basaltic tuffs and a distinctive porphyritic basalt unit. Of the silicic tuffs, four form major stratigraphic markers, which in conjunction with the crystal-rich sequence of tuffs and a porphyritic basalt, have allowed us to correlate fossil-bearing deposits and clarify the overall stratigraphy of the deposits in the Adu-Asa Formation.

We have determined that the fossil localities in the Adu-Asa Formation at Gona are grouped into three major temporal clusters. The oldest and middle clusters of sites are associated with the Hamadi Das Crystal-Rich Tuff Sequence tuffs, including the Bodele Tuff, as well as the Sifi Tuff. The oldest cluster lies between the Hamadi Das Crystal-Rich Tuff Sequence tuffs and the Sifi Tuff, while the middle cluster is above the level of the Sifi Tuff. Localities included in these clusters are the ABD, BDL, and HMD groups of sites. The youngest cluster of fossil localities is associated with the porphyritic basalt unit that is stratigraphically above the Kobo'o Tuff. This group of sites includes sites ESC-1, -2, -3, and -8. Based on the rapid apparent deposition rates of the upper Adu-Asa Formation, we estimate the ages of the youngest fossils to be ca. 5.5 Ma. The older sites are constrained to between 6.4 Ma and ca. 5.5 Ma but likely date toward the older end of this range.

While we have yet to firmly correlate deposits of the Adu-Asa Formation at Gona with other paleoanthropological projects in East Africa, it is likely that they are contemporaneous with to slightly younger than the deposits of the Adu-Asa Formation as described in the Middle Awash study area. A test of this proposal awaits publication of the entire sections of the Sagantole and Adu-Asa Formations at the Middle Awash area and characterization of the tuffs they contain.

ACKNOWLEDGMENTS

We thank K. Schick and N. Toth at the Center for Research into the Archaeological Foundations of Technology (CRAFT) for their support of this project, and Ambacho Kebeda, Soloman Kebede, Haptewold Habtemichael, and Yonas Beyene for help with permits. We also thank Authority for Research and Conservation of Cultural Heritage of the Ministry of Culture and Tourism of Ethiopia for the field permit. Financial support was provided by the LSB Leakey Foundation, National Geographic, Wenner-Gren Foundation, and the Revealing Hominid Origins Initiative (RHOI)/National Science Foundation (SBR-9910974 and Behavorial and Cognitive Sciences [BCS] Award 0321893). Matt Heizler, Nelia Dunbar, Lisa Peters, Ariel Dickens, Melanie Everett, Steve Frost, Bill Hart, Mike Rogers, and Dietrich Stout are warmly acknowledged for all their help and interesting scientific exchanges. Our special thanks go to Asahmed Humet and many other Afars who in various ways facilitated this research. Kleinsasser also thanks the Department of Geosciences at the University of Arizona and the Bert Butler Foundation for funding, as well as Ken Domanik, Eric Seedorff, Joaquin Ruiz, and Christa Placzek for their generous assistance.

REFERENCES CITED

Barberi, F., Ferrara, G., and Santacroce, R., 1975, Structural evolution of the Afar triple junction, *in* Pilger, A., and Rösler, A., eds., Afar Depression of Ethiopia: Stüttgart, West Germany, Schweizerbart, p. 38–54.

Borchardt, G.A., Aruscavage, P.J., and Millard, H.T., Jr., 1972, Correlation of the Bishop Ash, a Pleistocene marker bed, using instrumental neutron activation analysis: Journal of Sedimentary Petrology, v. 42, p. 301–306.

Chen, F.-C., and Li, Wen-Hsiung, 2001, Genomic divergences between humans and other hominoids and the effective population size of the common ancestor of humans and chimpanzees: American Journal of Human Genetics, v. 68, p. 444–456, doi: 10.1086/318206.

Davis, J.O., 1985, Correlation of late Quaternary tephra layers in a long pluvial sequence near Summer Lake, Oregon: Quaternary Research, v. 23, p. 38–53, doi: 10.1016/0033-5894(85)90070-5.

Froggatt, P.C., 1992, Standardization of the chemical analysis of the ICCT Working Group: Quaternary International, v. 13–14, p. 93–96, doi: 10.1016/1040-6182(92)90014-S.

Heiken, G.H., 1972, Morphology and petrography of volcanic ashes: Geological Society of America Bulletin, v. 83, p. 1961–1988, doi: 10.1130/0016-7606(1972)83[1961:MAPOVA]2.0.CO;2.

Horai, S., Satta, Y., Hayasaka, K., Kondo, R., Inoue, T., Ishisa, T., Hayashi, S., and Takahata, N., 1992, Man's place in *Hominoidea* revealed by mitochondrial DNA genealogy: Journal of Molecular Evolution, v. 35, p. 32–43, doi: 10.1007/BF00160258.

Hunt, J.B., and Hill, P.G., 1993, Tephra geochemistry: A discussion of some persistent analytical problems: The Holocene, v. 3, p. 271–278, doi: 10.1177/095968369300300310.

Kalb, J.E., Oswald, E.B., Tebedge, S., Mebrate, A., Tola, E., and Peak, D., 1982, Geology and stratigraphy of Neogene deposits, Middle Awash valley, Ethiopia: Nature, v. 298, p. 17–25, doi: 10.1038/298017a0.

Katoh, S., Nagaoka, S., WoldeGabriel, G., Renne, P., Snow, M.G., Beyene, Y., and Suwa, G., 2000, Chronostratigraphy and correlation of the Plio-Pleistocene tephra layers of the Konso Formation, southern Main Ethiopian Rift, Ethiopia: Quaternary Science Reviews, v. 19, p. 1305–1317, doi: 10.1016/S0277-3791(99)00099-2.

Le Bas, M.J., Le Maitre, R.W., Streckeisen, A., and Zanettin, B., 1986, A chemical classification of rocks based on the total alkali-silica diagram: Journal of Petrology, v. 27, p. 745–750.

Levin, N.E., Simpson, S.W., Quade, J., Cerling, T.E., and Frost, S.R., 2008, this volume, Herbivore enamel carbon isotopic composition and the environmental context of *Ardipithecus* at Gona, Ethiopia, *in* Quade, J., and Wynn, J.G., eds., The Geology of Early Humans in the Horn of Africa: Geological Society of America Special Paper 446, doi: 10.1130/2008.2446(10).

Morgan, G., VI, and London, D., 1996, Optimizing the electron microprobe analysis of hydrous alkali aluminosilicate glasses: American Mineralogist, v. 81, p. 1176–1185.

Nielsen, C.H., and Sigurdsson, H., 1981, Quantitative methods for electron microprobe analysis of sodium in natural and synthetic glasses: The American Mineralogist, v. 66, p. 547–552.

Patterson, N., Richter, D.J., Gnerre, S., Lander, E.S., and Reich, D., 2006, Genetic evidence for complex speciation of humans and chimpanzees: Nature, v. 441, p. 1103–1108, doi: 10.1038/nature04789.

Quade, J., and Wynn, J.G., eds., 2008, this volume, Preface, *in* Quade, J., and Wynn, J.G., eds.,The Geology of Early Humans in the Horn of Africa: Geological Society of America Special Paper 446.

Quade, J., Levin, N., Semaw, S., Stout, D., Renne, P., Rogers, M., and Simpson, S., 2004, Paleoenvironments of the earliest stone toolmakers, Gona, Ethiopia: Geological Society of America Bulletin, v. 116, p. 1529–1544, doi: 10.1130/B25358.1.

Quade, J., Levin, N.E., Simpson, S.W., Butler, R., McIntosh, W.C., Semaw, S., Kleinsasser, L., Dupont-Nivet, G., Renne, P., and Dunbar, N., 2008, this volume, The geology of Gona, *in* Quade, J., and Wynn, J.G., eds., The Geology of Early Humans in the Horn of Africa: Geological Society of America Special Paper 446, doi: 10.1130/2008.2446(01).

Renne, P.R., WoldeGabriel, G., Hart, W.K., Heiken, G., and White, T.D., 1999, Chronostratigraphy of the Miocene-Pliocene Sagantole Formation, Middle Awash valley, Afar Rift, Ethiopia: Geological Society of America Bulletin, v. 111, p. 869–885, doi: 10.1130/0016-7606(1999)111<0869:COTMPS>2.3.CO;2.

Rodbell, D.T., Bagnato, S., Nebolini, J.C., Seltzer, G., and Abbott, M.B., 2002, A Late Glacial–Holocene tephrochronology for glacial lakes in southern Ecuador: Quaternary Research, v. 57, p. 343–354, doi: 10.1006/qres.2002.2324.

Ross, C.S., 1928, Altered Palaeozoic volcanic materials and their recognition: The American Association of Petroleum Geologists Bulletin, v. 12, p. 143–164.

Ruvolo, M., 1997, Molecular phylogeny of the hominoids; inferences from multiple independent DNA sequence data sets: Molecular Biology and Evolution, v. 14, p. 248–265.

Sarna-Wojcicki, A.M., Bowman, H.R., Meyer, C.E., Russell, P.C., Asaro, F., Michel, H., Rowe, J.J., Baedeker, P.A., and McCoy, G., 1980, Chemical Analyses, Correlations, and Ages of Late Cenozoic Tephra Units of East-Central and Southern California: U.S. Geological Survey Open-File Report 80-231, 52 p.

Sawada, Y., Pickford, M., Senut, B., Itaya, T., Hyodo, M., Miura, T., Kashine, C., Chujo, T., and Fujii, H., 2002, The age of *Orrorin tugenensis*, an early hominid from the Tugen Hills, Kenya: Compte Rendus: Comptes Rendus Paleovol, v. 1, p. 293–303, doi: 10.1016/S1631-0683(02)00036-2.

Semaw, S., Simpson, S.W., Quade, J., Renne, P.R., Butler, R.F., McIntosh, W.C., Levin, N., Dominguez-Rodrigo, M., and Rogers, M.J., 2005, Early Pliocene hominids from Gona, Ethiopia: Nature, v. 433, p. 301–305, doi: 10.1038/nature03177.

Simpson, S.W., Quade, J., Kleinsasser, L., Levin, N., McIntosh, W., Dunbar, N., and Semaw, S., 2007, Late Miocene hominid teeth from Gona project area, Ethiopia: American Association of Physical Anthropologists, 76th Annual Meeting, Philadelphia, Abstracts, v. 132, no. S44, p. 219.

Tesfaye, S., Harding, D.J., and Kusky, T.M., 2003, Early continental breakup and migration of the Afar triple junction, Ethiopia: Geological Society of America Bulletin, v. 115, p. 1053–1067, doi: 10.1130/B25149.1.

Vignaud, P., Duringer, P., Mackaye, H.T., Likius, A., Blondel, C., Boiserrie, J.-R., de Bonis, L., Eisenmann, V., Etienne, M.-E., Geraads, D., Guy, F., Lehmann, T., Lihoreau, F., Lopez-Martinez, N., Mourer-Chauviré, C., Otero, O., Rage, J.-C., Schuster, M., Viriot, L., Zazzo, A., and Brunet, M., 2002, Geology and palaeontology of the Upper Miocene Toros-Menalla hominid locality, Chad: Nature, v. 418, p. 152–155, doi: 10.1038/nature00880.

WoldeGabriel, G., Haile-Selassie, Y., Renne, P.R., Hart, W.K., Ambrose, S.H., Asfaw, B., Heiken, G., and White, T., 2001, Geology and palaeontology of the late Miocene Middle Awash valley, Afar Rift, Ethiopia: Nature, v. 412, p. 175–178, doi: 10.1038/35084058.

Wolfenden, E., Ebinger, C., Yirgu, G., Renne, P.R., and Kelley, S.P., 2005, Evolution of a volcanic rifted margin: Southern Red Sea, Ethiopia: Geological Society of America Bulletin, v. 117, p. 846–864, doi: 10.1130/B25516.1.

Yoshikawa, S., 1976, The volcanic ash layers of the Osaka Group: Journal of the Geological Society Japan, v. 82, p. 497–515.

MANUSCRIPT ACCEPTED BY THE SOCIETY 17 JUNE 2008

The Geological Society of America
Special Paper 446
2008

Magnetostratigraphy of the eastern Hadar Basin (Ledi-Geraru research area, Ethiopia) and implications for hominin paleoenvironments

Guillaume Dupont-Nivet*
Faculty of Geosciences, Paleomagnetic Laboratory "Fort Hoofddijk," Utrecht University, Budapestlaan 17, 3584 CD Utrecht, The Netherlands

Mark Sier
Faculty of Geosciences, Paleomagnetic Laboratory "Fort Hoofddijk," Utrecht University, Budapestlaan 17, 3584 CD Utrecht, The Netherlands, and *Faculty of Archaeology, Leiden University, P.O. Box 9515, 2300 RA Leiden, The Netherlands*

Christopher J. Campisano
Institute of Human Origins, School of Human Evolution and Social Change, Arizona State University, P.O. Box 874101, Tempe, Arizona 85287-4101, USA

J Ramón Arrowsmith
Erin DiMaggio
School of Earth and Space Exploration, Arizona State University, Tempe, Arizona 85287-1404, USA

Kaye Reed
Institute of Human Origins, School of Human Evolution and Social Change, Arizona State University, P.O. Box 874101, Tempe, Arizona 85287-4101, USA

Charles Lockwood[†]
Department of Anthropology, University College London, Gower Street, London WC1E 6BT, UK

Christine Franke[§]
Faculty of Geosciences (FB 5), University of Bremen, P.O. Box 330440, D-28334 Bremen, Germany

Silja Hüsing
Faculty of Geosciences, Paleomagnetic Laboratory "Fort Hoofddijk," Utrecht University, Budapestlaan 17, 3584 CD Utrecht, The Netherlands

THIS PAPER IS DEDICATED TO THE MEMORY OF CHARLES LOCKWOOD.

*Corresponding author: gdn@geo.uu.nl.
[†]Deceased.
[§]Current address: Laboratoire des Sciences du Climat et de l'Environnement, Avenue de la Terrasse, Bat. 12, F-91198 Gif-sur-Yvette Cedex, France.

Dupont-Nivet, G., Sier, M., Campisano, C.J., Arrowsmith, J R., DiMaggio, E., Reed, K., Lockwood, C., Franke, C., and Hüsing, S., 2008, Magnetostratigraphy of the eastern Hadar Basin (Ledi-Geraru research area, Ethiopia) and implications for hominin paleoenvironments, *in* Quade, J., and Wynn, J.G., eds., The Geology of Early Humans in the Horn of Africa: Geological Society of America Special Paper 446, p. 67–85, doi: 10.1130/2008.2446(03). For permission to copy, contact editing@geosociety.org. ©2008 The Geological Society of America. All rights reserved.

ABSTRACT

To date and characterize depositional environments of the hominin-bearing Hadar Formation, lacustrine sediments from the eastern part of the Hadar Basin (Ledi-Geraru research area) were studied using tephrostratigraphy and magnetostratigraphy. The Sidi Hakoma Tuff, Triple Tuff-4, and the Kada Hadar Tuff, previously dated by $^{40}Ar/^{39}Ar$ in other parts of the basin, were identified using characteristic geochemical composition and lithologic features. Paleomagnetic samples were collected every 0.5 m along an ~230-m-thick composite section between the Sidi Hakoma Tuff and the Kada Hadar Tuff. A primary detrital remanent magnetization mostly carried by (titano-) magnetites of basaltic origin was recognized. Consistent with existing data of the Hadar Basin, paleomagnetic directions show a postdepositional counterclockwise vertical-axis tectonic rotation (~5°–10°) and shallowing of paleomagnetic inclination (~5°–10°) related to sedimentation and compaction. Two normal-polarity intervals (chrons 2An.3n and 2An.2n) are recorded bracketing a reversed interval identified as the Mammoth event (chron 2An.2r). Resulting sediment accumulation rates (~90 cm/k.y.) are high compared to existing accumulation-rate estimates from the more western part of the Hadar Basin. The resulting eastward increasing trend suggests that deposition took place in an eastward-tilting basin. Sediment accumulations were constant throughout the basin from ca. 3.4 to 3.2 Ma. At 3.2 Ma, a regional and relatively short-lived event is indicated by significant change in depositional conditions and a large increase in accumulation rate. This disruption may have been related to increased climate variability due to astronomical climate forcing. It provides a possible explanation for changes in the Hadar faunal community and *Australopithecus afarensis* in particular.

Keywords: magnetostratigraphy, tephrostratigraphy, paleoanthropology, paleoenvironment, Pliocene, East African Rift.

INTRODUCTION

To understand how past environments may have influenced hominin evolution, temporal and causal relationships must be established between records of paleoenvironmental changes and the hominin fossil record. Recent advances arise from climate modeling and the availability of well-calibrated regional paleoclimatic records (deMenocal, 2004; Sepulchre et al., 2006), but accurately dated analysis of hominin-bearing sediments is needed to place fossils in their tectonic and climatic context. The Pliocene Hadar Basin in the Afar region of northern Ethiopia (Fig. 1) includes some of the world's most complete collections of hominins (Johanson and Taieb, 1976; Johanson et al., 1982; Kimbel et al., 2004; Alemseged et al., 2006). As a result, sediments of the Hadar Basin have been studied intensively since the 1970s. Successful dating has been achieved using the combination of radiochronologic methods (mainly $^{40}Ar/^{39}Ar$) on interlayered volcanic deposits and magnetostratigraphy (Aronson et al., 1977; Schmitt and Nairn, 1984; Renne et al., 1993; Tamrat et al., 1996; Walter and Aronson, 1993; Walter, 1994). Understandably, most geological work (e.g., Campisano and Feibel, this volume, and references therein) has focused on regions with high densities of fossil finds in the Hadar research area (Fig. 1). However, geologic work over the entire Hadar Basin is needed to fully constrain its paleoenvironment. The Dikika research area south of the Awash River and the Gona research area west of Hadar have been recently examined extensively (Quade et al., 2004, this volume; Wynn et al., 2006; Wynn et al., this volume). In the present study, we investigate and date the stratigraphy from a still poorly explored region of the eastern Hadar Basin that is referred to as the Ledi-Geraru research area, initially documented by the International Afar Research Expedition in the 1970s. The Ledi-Geraru area, east of the Hurda (= Ourda) wadi and north of the Awash River, provides the opportunity to better constrain the Hadar Basin history in space and time and to assess if environmental changes observed in other research areas are indeed regional and not related to local disturbance from faults, hiatuses, or variations in sediment accumulation. We present chronology of this previously undated stratigraphy using magnetostratigraphic analysis along with correlation of tuffaceous horizons found in the Ledi-Geraru area to their radiochronologically dated counterparts in the Hadar Formation. The results enable us to place paleontological finds in a well-calibrated chronostratigraphic framework and increase our understanding of the tectonic and climatic conditions that prevailed during their deposition.

GEOLOGICAL SETTING

Regional Geologic Setting

The Hadar Basin formed as an extensional subbasin of the subsiding Afar Depression at the triple junction of the Ethiopian Rift, the Aden Rift, and the Red Sea Rift (Fig. 1). Exposures of

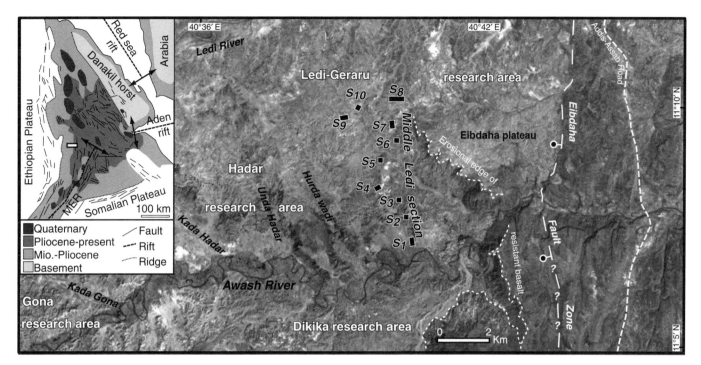

Figure 1. Locations of the subsections of the Middle Ledi section sampled for magnetostratigraphy (S_1–S_{10}) within the Ledi-Geraru research area. Map is overlain on grayscale version of Advanced Spaceborne Thermal Emission and Reflection Radiometer (ASTER) imagery and Shuttle Radar Topography Mission (SRTM) imagery. The Eibdaha fault zone delineation is based on field observations, analysis of 90 m digital elevation models (SRTM), and stereo aerial photography. Inset shows location of sampling area (white box) in the tectonic context of the Afar triple junction (simplified from Thurmond et al., 2006). MER—Main Ethiopian Rift.

the Hadar Formation defining the Hadar Basin are located along the Ethiopian Rift margin, just east of the Ethiopian escarpment, which separates the Afar Depression from the Ethiopian Plateau. The Ethiopian Plateau, probably present since the Oligocene (Pik et al., 2003), controls eastward river flow into the Afar Depression (in particular, the Awash River) but also northward flow into the Mediterranean through the Nile River. Atmospheric circulation modeling shows that the plateau obstructs zonal circulation and deflects Indian monsoon flow, such that enhanced Miocene-Pliocene uplift possibly induced aridification and environmental change in East Africa (Sepulchre et al., 2006; Gani et al., 2007; Spiegel et al., 2007). Rifting and subsidence of the Afar Depression commenced after the Oligocene (Ukstins et al., 2002). The most activity has occurred since 20 Ma, when the Gulf of Aden Rift joined the triple junction, driving widespread volcanism, faulting, and rotations of crustal blocks (Acton et al., 2000; Manighetti et al., 2001; Audin et al., 2004). Alternatively, it has been proposed that the Main Ethiopian Rift, propagating from the southwest, reached the Afar Depression and connected to the triple junction only 11 m.y. ago (Wolfenden et al., 2004). Plate kinematic and crustal thickness reconstructions of the Afar suggest that, although stretching rates remained constant, regional subsidence in the Hadar block was virtually complete by mid-Pliocene time (Redfield et al., 2003). Sustained deformation and volcanic activity during Pliocene time possibly shifted away from the margin—where the Hadar Formation was deposited—toward the center of the Main Ethiopian Rift. This was associated with a notable northward propagation of volcanism since 3 Ma from the Ethiopian Rift into the Afar Depression (Lahitte et al., 2003).

Hadar Basin Stratigraphy

Major sedimentary accumulation began in the late Miocene to early Pliocene with fluviolacustrine deposits of the Sagantole Formation, which are correlated to the Middle Awash region further south and have been dated from 5.6 Ma to 3.9 Ma (Renne et al., 1999; Quade et al., 2004). The focus of this study is the Hadar Formation, which is composed of fluvial, paludal, and lacustrine deposits that rapidly accumulated (30–90 cm/k.y.) between ca. 3.8 and 2.9 Ma after normal faulting of the Sagantole Formation (Wynn et al., 2006). The Hadar Formation is defined (Taieb et al., 1976) and described in detail in the Hadar research area (Campisano, 2007; Campisano and Feibel, this volume). It is divided into four members separated by radiometrically dated tephras. The members, from bottom to top are: (1) the Basal Member, below the Sidi Hakoma Tuff (ca. 3.42 Ma), (2) the Sidi Hakoma Member, between the Sidi Hakoma Tuff and Triple Tuff-4 (ca. 3.25 Ma), (3) the Denen Dora Member, between Triple Tuff-4 and the Kada Hadar Tuff (ca. 3.20 Ma), and (4) the Kada Hadar Member, between the Kada Hadar Tuff and the Busidima unconformity surface (BUS; Wynn and Roman, this volume), which is located just above the Bouroukie Tuff 2

(BKT-2 ca. 2.95 Ma). Additional marker horizons are described in Campisano and Feibel (this volume). The Sidi Hakoma Member includes a lignite layer, gastropods beds, and the Kada Damum Basalt. The Denen Dora Member includes a widespread fluvial sandstone (DD-3 sand) below the Kada Hadar Tuff. The middle and upper parts of the Kada Hadar Member comprise the laminated "Confetti Clay," with the Kada Hadar Tuff a few meters below, and the BKT-1 and BKT-2 tuffs. Dating and correlation to the Hadar Formation are also provided by magnetostratigraphy in the Gona research area (Quade et al., this volume) and the Hadar research area (Aronson et al., 1977; Schmitt and Nairn, 1984; Renne et al., 1993; Tamrat et al., 1996). The accurate stratigraphic positions of reversals from the Hadar studies have been recently synthesized (Campisano, 2007). The Hadar Formation at Hadar falls within the Gauss normal-polarity chron 2An (C2An) and includes the reversed-polarity Kaena chron 2An.1r (C2An.1r) and Mammoth chron 2An.2r (C2An.2r), providing four additional correlation points at the well-dated boundaries of these chrons. The Hadar Formation is separated from the overlying Busidima Formation (ca. 2.7 to <0.6 Ma) by a major area wide, ~200 k.y. angular unconformity (Quade et al., this volume; Wynn et al., this volume). Below this unconformity, there are fluviolacustrine deposits; above it, there are conglomeratic fluvial channel deposits interpreted either as the ancestral Awash River (Quade et al., 2004; Wynn et al., 2006) or as alluvial fans from the Ethiopian Plateau (Yemane, 1997).

Paleoenvironments of the Hadar Formation

The regional and local tectonic context of deposition during Hadar Formation time is still poorly understood. The present location of the Hadar basin exposure at the margin of the Main Ethiopian Rift suggests a simple east-west extensional configuration with subsidence controlled by syndepositional down-drop along normal faults parallel to the Ethiopian escarpment such as the east-dipping As Duma fault (Quade et al., 2004). This is substantiated by records of eastward to northward paleodrainages (Tiercelin, 1986; Quade et al., 2004; Behrensmeyer, this volume). However, the Hadar Formation thickness increases eastward (Tiercelin, 1986; Walter, 1994) to northeastward (Wynn et al., 2006), conspicuously suggesting an increasing rate of subsidence toward the eastern rather than the western margin of the basin. As a result, it was recently suggested that the Hadar Formation was deposited in response to northeast-southwest extension associated with the Red Sea Rift system, and then it shifted to east-west extension associated with the Ethiopian Rift during deposition of the Busidima Formation (Wynn et al., 2006).

Paleoenvironmental studies generally show that during deposition of the Hadar and Busidima Formations, global cooling was associated with more arid conditions in Africa, with some environmental variability at various localities (Quade et al., 2004). Analysis of the Hadar faunal assemblage indicates a range of available habitats including open and closed woodlands, gallery forests, edaphic grasslands, and shrublands. The paleontological record indicates slightly more xeric conditions beginning around 3.2 Ma and a distinct faunal turnover at ca. 3.0 Ma, with evidence of an influx of more arid-adapted taxa (Campisano and Reed, 2007; Reed, 2008). Stable carbon isotope values from pedogenic carbonates at Gona indicate a gradual shift from woodlands to grassy woodlands in the early Pliocene, to more open, but still mixed, environments in the late Pleistocene, possibly associated with increasing aridity and fluctuations in the timing and source of rainfall (Levin et al., 2004). Pollen assemblages suggest a large biome shift, up to 5 °C cooling, and a 200 to 300 mm/yr rainfall increase just before 3.3 Ma, consistent with a global marine $\delta^{18}O$ isotopic shift (M2) (Bonnefille et al., 2004). Relative to modern East African lakes, $\delta^{18}O$ values from lacustrine mollusk shell horizons of the Hadar Formation indicate large and high-frequency lake-level fluctuations that point to wetter and probably cooler summers, possibly attributable to the strengthening of the Atlantic-derived air mass component to the Ethiopian monsoon (Hailemichael et al., 2002). However, the timing and periodicity of the climate variations observed in these studies are still insufficiently precise to allow definite correlation to known regional and global climate conditions accurately constrained in the marine realm.

CHRONOSTRATIGRAPHIC ANALYSIS

Stratigraphic and Geochemical Correlations

Compared to Hadar and Gona, the Hadar Formation deposits in the Ledi-Geraru area are generally thicker and more distal from the source region, and there is a predominance of lacustrine over fluvial deposits. Therefore, they provide better preservation and resolution of the paleoenvironmental history. The stratigraphy is best exposed at the Middle Ledi section (Fig. 1) where sediments are relatively uninterrupted by faulting and essentially flat lying, with a low bedding dip of 0°–2° (<5°) to north-northwest. As a result, the stratigraphy is exposed along a south-to-north transect of several kilometers on strata well-exposed on small hills and along ridges. The Middle Ledi section is a composite of ten overlapping subsections correlated to each other using unambiguous marker beds. The composite section spans an ~230-m-thick continuous sedimentary interval including the complete Sidi Hakoma and Denen Dora Members (Fig. 2). The Sidi Hakoma Tuff, Triple Tuff-4, and the Kada Hadar Tuff are correlated between the Hadar and Ledi-Geraru research areas using the following stratigraphic relationships and lithologic descriptions.

Sidi Hakoma Tuff (SHT)

The Sidi Hakoma Tuff is typically preserved in the region as a widespread white bentonite, ~20 cm thick, overlying Basal Member fluvial-deltaic sands. Exposures immediately west and south of the Middle Ledi section are located a few meters above the level of the modern Awash floodplain along the course of the river. Vitric channel-fill and crevasse-splay deposits of the Sidi Hakoma Tuff are also preserved at both Hadar and Dikika (Wynn

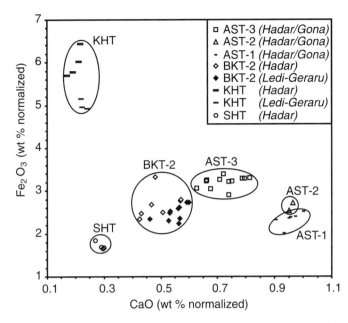

Figure 2. Normalized bivariate plot showing major-element composition from various tuffaceous horizons from the Hadar, Gona, and Ledi-Geraru research areas. Hadar/Gona data are from Hart et al. (1992) and Campisano (2007). Ledi-Geraru data are from DiMaggio et al. (this volume).

et al., 2006; Campisano and Feibel, this volume). A 25-cm-thick white bentonite located near the bottom of the Middle Ledi section (~4.15 m level) matches the description of the Sidi Hakoma Tuff bentonite exposed to the west and south. It is found above sands presumably of the Basal Member. The lack of a vitric facies in the Middle Ledi section precludes a definitive geochemical correlation to the Sidi Hakoma Tuff, but the tuff's lithology, position on the Awash floodplain, and associated context provide suitable evidence for a correlation. A unique lignite layer associated with a thin (~2 cm) white bentonite and a *Corbicula* shell horizon at approximately the 40 m level provides additional evidence that the basal bentonite represents the Sidi Hakoma Tuff. A similar lignite layer is known from the middle of the Sidi Hakoma Member in both central Hadar and along the border of the Hadar and Ledi-Geraru research areas (at Hurda wadi; Tiercelin, 1986). Furthermore, the lignite layer in central Hadar is also associated with thin bentonites and a *Corbicula* shell horizon, referred to as the Kada Me'e Tuff Complex (Campisano and Feibel, this volume).

Triple Tuff 4 (TT-4)

Although none of the Triple Tuffs preserves a vitric component, their unique lithologic context (specifically that of Triple Tuff-4), is diagnostic. At Hadar and Dikika, Triple Tuff-4 is a ubiquitous white bentonite encased in an olive-green laminated fissile-shale that preserves apatite nodules and an ostracodite unit in its lower portion (Tiercelin, 1986; Wynn et al., 2006). In the Middle Ledi section, a similar lithologic unit at the 151 m level (150.87 ± 0.675 m) is proposed as a correlate to the Triple Tuff-4/ostracod-shale unit. This laminated olive-green shale preserves a white bentonite, apatite nodules, and low concentrations of ostracods. Instead, the lower half of the shale contains a high concentration of *Melanoides* shells and carbonized plant pieces. However, *Melanoides* shells are occasionally preserved below the ostracodite at Hadar, and ostracod abundance was observed to be significantly lower in the Hurda Wadi Triple Tuff-4 exposures compared to central and western Hadar (Campisano, 2007, personal commun.). Based on its lithologic and interpreted chronostratigraphic position (see following), we propose this sequence as a correlate to the Triple Tuff-4/ostracod-shale from Hadar.

Kada Hadar Tuff (KHT)

At Hadar, the Kada Hadar Tuff is typically preserved as a beige, silty bentonite, 20–80-cm thick, but vitric channel-fill deposits (likely crevasse-splay channels) are occasionally preserved (Yemane, 1997; Campisano, 2007). A 1–2-m-thick vitric channel fill preserved at the top of the Middle Ledi section is similar in outcrop appearance to the Hadar Kada Hadar Tuff channel-fill deposit. This tephra deposit is not laterally continuous in the uppermost part of the section, and it is separated from better-exposed underlying strata by colluvial cover. Its stratigraphic position has been measured both by direct thickness measurements (yielding 228.34 m level) and by derivation from measured horizontal distances assuming constant dip from an exposed underlying outcrop (yielding 227.53 m level). A conservative 2 m uncertainty allowing for dip variations was thus attributed to the direct thickness measurements, yielding a 229.34 ± 3.00 m level within the 2-m-thick tuff. The BKT-2 complex was positively identified (DiMaggio et al., this volume) well above the proposed Kada Hadar Tuff correlate (perhaps by as much as 60–100 m, although the exact stratigraphic distance was not determined), confirming that the Ledi-Geraru research area includes almost the entire Hadar Formation, from at least Sidi Hakoma Tuff to BKT-2. The Kada Hadar Tuff correlation is confirmed by diagnostic major-element glass chemistry (Fig. 2; Table 1). Glass from the Kada Hadar Tuff is chemically distinct from all other Hadar Formation tephra in its exceptionally low concentrations of Al_2O_3 and CaO and high concentrations of Fe_2O_3. When compared to existing geochemical data from other vitric tephra horizons from the Hadar Formation (Hart et al., 1992; Walter, 1994; Campisano, 2007; DiMaggio et al., this volume), the Middle Ledi Kada Hadar Tuff is clearly distinct from the Sidi Hakoma Tuff and BKT-2 of the Hadar Formation as well as the AST series of tephra of the Busidima Formation (Table 1). The Middle Ledi sample displays the low Al_2O_3 and CaO and high Fe_2O_3 concentrations diagnostic of the Kada Hadar Tuff and is within 1σ uncertainty for virtually all major elements. The slightly larger spread in Fe_2O_3 values is within the expected error related to minor differences in analytical procedure. Based on these congruent analyses, we conclude that the vitric tuff horizon located at the top of the Middle Ledi stratigraphic section is confidently identified as the Kada Hadar Tuff.

TABLE 1. MAJOR-ELEMENT GEOCHEMISTRY OF THE KADA HADAR TUFF

Sample	n		SiO_2	TiO_2	Al_2O_3	MnO	Fe_2O_3	MgO	CaO	Na_2O	K_2O	Total
Middle Ledi												
AM06-1047	21	avg	73.96	0.27	9.69	0.18	4.51	0.05	0.21	0.92	1.16	90.95
		±1σ	1.03	0.04	0.21	0.05	0.29	0.07	0.02	0.17	0.16	1.20
AM06-1047	8	avg	72.54	0.30	9.89	0.15	4.61	0.01	0.20	0.89	0.90	89.50
		±1σ	0.63	0.05	0.32	0.08	0.16	0.02	0.03	0.48	0.30	1.14
AM06-1047	5	avg	74.08	0.28	9.69	0.15	4.48	0.03	0.22	0.90	1.27	91.09
		±1σ	0.24	0.03	0.18	0.03	0.25	0.04	0.02	0.36	0.31	0.39
Hadar												
E02-7413.AV	15	avg	70.55	0.26	8.98	0.20	5.40	0.02	0.19	2.40	1.57	89.76
		±1σ	2.41	0.04	0.33	0.04	0.37	0.01	0.03	0.49	0.20	3.04
E01-7341.AV	14	avg	69.21	0.27	8.78	0.19	5.63	0.01	0.19	1.61	1.38	87.49
		±1σ	1.66	0.03	0.16	0.06	0.31	0.01	0.03	0.39	0.15	1.49

Note: Samples were analyzed by electron microprobe; *n*—number of analysis; avg—average values listed as wt% with associated standard deviation (1σ).

Paleomagnetic Sampling

The Middle Ledi section was measured to within centimetric precision for 200.24 m total thickness starting 4.15 m below the base of the Sidi Hakoma Tuff. In the last 30-m-thick section below the Kada Hadar Tuff, sampling and logging were hindered by the poor exposure. Sampling of 2.5-cm-diameter cylindrical rock cores was performed with an electric drill mounted with a diamond-coated bit and cooled with an electric air compressor, both devices powered by a portable generator. Cores were oriented with a compass corrected for local declination (1.7°). One core was sampled at each of the 401 stratigraphic levels separated by an average interval of ~50 cm. All encountered lithologies were sampled. Meter-thick lenticular sandstone beds were avoided as much as was allowed by the outcrop configuration to avoid possible hiatuses associated with the erosional scouring typical of these high-energy deposits. Corrections of negligible bedding tilts were not applied due to the low magnitude of measured dips (<5°) and the potential error associated with the measurements of such low dips.

Paleomagnetic Analysis

Demagnetization and Characteristic Remanent Magnetization (ChRM)

Remanent magnetizations of samples were measured on a 2G Enterprises DC SQUID cryogenic magnetometer. Thermal demagnetization was performed at up to 18 thermal steps in a shielded furnace at the Utrecht University paleomagnetic laboratory. A first selection of pilot samples distributed every 4 m throughout the sampled stratigraphy was stepwise thermally demagnetized in great detail in order to determine characteristic demagnetization behavior, establish the most efficient demagnetization temperature steps, and localize stratigraphic intervals with potential paleomagnetic reversals. These results guided further processing of a second selection of samples at higher stratigraphic resolution from key parts of the section. This ultimately enabled us to distinguish between reliable and less reliable results and to confidently locate reversals in the stratigraphy. In total, 236 samples were thermally demagnetized.

Two main characteristic remanent magnetization (ChRM) components stand out clearly on vector end-point diagrams and stereographic projections. A low-temperature component (LTC) with only normal-polarity directions is typically demagnetized from ~100 °C to 250 °C. A high-temperature component (HTC) with normal- or reversed-polarity directions decays from ~350 °C to 525 °C (Fig. 3A). On average, 60% of the LTC remanence was quickly removed by 250 °C, leaving a slow decaying HTC up to 525 °C. Generally the remanence reached near-zero values above 550 °C, but in some samples, a higher-temperature component concordant with the HTC was recorded from 550 °C to >620 °C. In some of these samples, the contribution of the LTC was indicated by a normal direction overlapping on a reversed-polarity direction carried by the HTC (Fig. 3B). The LTC was found exclusively with a normal-polarity orientation, while HTC show normal and reverse polarity. This clearly identifies the LTC as a normal overprint. In a majority of samples, the HTC is well expressed, enabling straightforward isolation of normal- or reversed-polarity directions. However, in some samples, the HTC was poorly preserved and did not yield reliable directions (Fig. 3B). For these, overlapping LTC, although usually removed below 350 °C, could be found, extending in a few cases as high as 450 °C, thus indicating the maximum extent of the LTC contribution. This information was crucial for selecting reliable normal-polarity direction. All normal-polarity directions that did not extend beyond 450 °C were considered to be a possible overprint and thus unreliable for further analysis (Fig. 3C).

Rock Magnetism and Magnetic Mineralogy

Rock magnetic experiments were performed to assess the magnetic mineralogy and to better constrain the reliability of the ChRM. To test for potential artifacts due to mineral transformation during stepwise heating, alternating field (AF) demagnetization was performed on 29 pilot samples (Fig. 3D). These AF results show the same general behavior as the thermal demagnetization. This indicates that mineral transformation upon heating

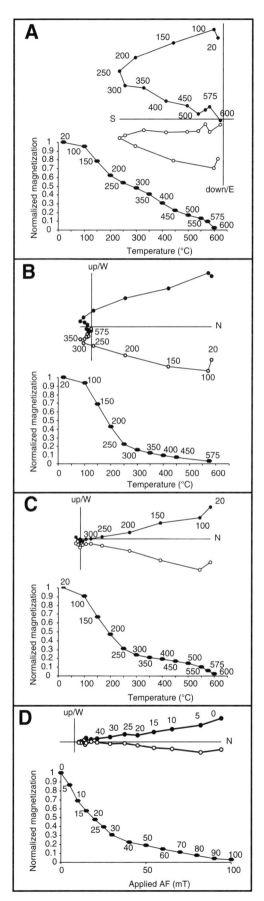

Figure 3. Typical demagnetization behaviors. Top diagrams: vector end point of typical thermal (A, B, C) or alternating field (AF) (D) demagnetizations. Full (open) symbols are projections on horizontal (vertical) plane. Bottom diagrams: normalized magnetization versus temperature or applied field. (A) Typical thermal demagnetization (quality 1). (B) Strong normal overprint (quality 3). (C) Full normal overprint (rejected). (D) Typical AF demagnetization.

did not significantly alter the results. However, AF treatment was less effective in separating the ChRM components.

High-temperature-range thermomagnetic experiments were performed on dry bulk sediments from five representative samples of various lithologies of the investigated stratigraphic section. Measurements using the KLY3-CS bridge show that (1) during heating, a slight susceptibility increase between room temperature and 300–400 °C is followed by a nearly linear decrease up to ~550 °C and a main drop between 550 and 580 °C; and (2) during cooling, susceptibility curve is irreversible below 500 °C (Fig. 4A, lower panel). This irreversibility at lower temperatures suggests the initial presence of maghemite or titanomaghemite altered during heating and therefore not apparent during cooling. The main drop between 550 and 580 °C is characteristic of fine-grained magnetite or Ti-poor titanomagnetite. These results are consistent with high-field thermomagnetic runs performed on a Curie balance, which also show a quasi-linear decrease during ~350–500 °C heating with a slightly sharper moment decrease above 500 °C and an irreversible cooling curve (Fig. 4A, upper panel). These thermomagnetic results, along with the thermal and AF demagnetization behavior and typically high values of NRM and susceptibility (Table DR1[1]), rule out iron sulfides, goethite, or hematite as important contributors to the remanence. However, the results strongly suggest the predominance of fine-grained magnetite and/or Ti-poor titanomagnetite for the HTC, as well as the influence of maghemite and/or titanomaghemite on the LTC.

Low-temperature-range thermomagnetic experiments were performed on dry bulk sediments from five other representative samples using a Quantum Design Magnetic Properties Measurement System (MPMS) at the Faculty of Geosciences at the University of Bremen (Fig. 4B). All zero-field-cooled and field-cooled curves (temperature range 5–300 K, applied field 5 T) lack the indication of stoichiometric magnetite usually shown by the Verwey structural phase transition at ~120 K (Franke et al., 2007). The shift to a lower-temperature transition would be indicative of (1) nonstoichiometric, slightly oxidized magnetite (the Verwey transition can be totally suppressed at a sufficiently high degree of maghematization) or (2) sufficiently high Ti-rich magnetite (the Verwey transition can be totally suppressed at Ti contents > 0.04; Kakol et al., 1994). The presence of Ti-bearing magnetic mineral phases in the samples is further supported by the characteristics of the curves of the low-temperature thermomagnetic

---

[1]GSA Data Repository item 2008194, which includes tabulated characteristic remanent magnetization directions, is available at www.geosociety.org/pubs/ft2008.htm, or on request from editing@geosociety.org, Documents Secretary, GSA, P.O. Box 9140, Boulder, CO 80301-9140, USA.

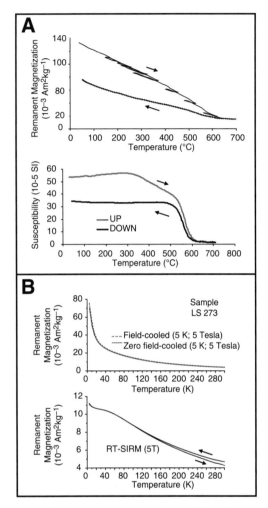

Figure 4. (A) Thermomagnetic experiments. Top diagram: typical high-field magnetic moment upon heating and subsequent cooling in successive temperature ranges measured on Curie balance. Bottom diagram: typical low-field susceptibility upon heating and subsequent cooling measured on kappabridge KLY3-CS. (B) Low-temperature Magnetic Properties Measurement System (MPMS). Top diagram: zero-field-cooled (solid line) and field-cooled (dotted line) remanence (5 T applied field) monitored between 5 and 300 K. Bottom diagram: Low-temperature cycle (between 300 and 5 K) of the room-temperature saturation isothermal remanent magnetization (RT-SIRM, 5 T). Black arrows mark the cooling and subsequent warming curve.

thermomagnetic curves and thermal demagnetization behavior. In addition, they provide hysteresis measurements and three-component IRM experiments showing a dominant low-coercivity component demagnetized below 600 °C. These further suggest a remanence dominated by pseudo–single-domain titanomagnetite clearly associated with the HTC, and a minor contribution from a higher-coercivity mineral such as titanomaghemite.

To further test these suggestions, scanning electron microscopic (SEM) analyses were performed on polished sections of the five representative samples also used for the MPMS measurements described above, using a FEI XL30 SFEG SEM at the Utrecht Electron Microscopy facility at 15 kV acceleration voltage. A thin carbon coating of a few nanometers was applied on the polished sections to avoid surface charging. Backscattered electron imaging was used for visualizations, and energy-dispersive X-ray spectroscopy was used to examine the elemental composition. Recorded elemental spectra were normalized to their respective oxygen maxima. EDAX PhiRhoZ processing software was used to (semi-) quantify the obtained elemental spectra. The micrographs and element spectra show detrital grains of iron (-titanium) mineral phases within the spatial resolution of the SEM within a 0.1–100 μm grain-size range (Fig. 5). Element spectra show magnetite composition (Fig. 5A, 1) and Fe:Ti ratios indicative of (Ti-rich) titanohematites ($Fe_{3-x}Ti_xO_4$, $0 \leq x \leq 1$) (Fig. 5B, 1, 3, and 5) and Ti-poor titanomagnetites ($Fe_{2-y}Ti_yO_3$, $0 \leq y \leq 1$)) (Fig. 5B, 2, 4, and 6), which were calculated according to their stoichiometric formulas (Table 2). The magnetic grains contain minor amounts of metal ions other than Fe and Ti, because substitution of Fe by Mg or Al is a typical phenomenon in detrital (titano-) magnetite. To account for this effect, the total of these three elements (Fe_Σ) was calculated according to Dillon and Franke (2008). The calculated titanomagnetite compositions therefore range between TM09 and TM54, while the titanohematite grains have compositions close to pure ilmenite (TH89 to TH98). According to Buddington and Lindsley (1964), the ilmenite content in a titanohematite phase coexists with a titanomagnetite phase of complementary composition. This intergrowth would be typical for low-temperature exsolution processes of slow-cooling basaltic rocks. The resulting domain size of the titanomagnetite is therefore expected to be much smaller than the observed overall particle sizes because it is restricted by the lamellae formed by the titanohematite. The detrital character of these (titano-) magnetites and titanohematites is evident by their fragmental appearance, relatively smooth surfaces, and sharp, curved edges related to shrinkage cracks (Fig. 5B). These indicate variable degrees of maghematization of the grains and thus further support the detrital basaltic origin of the magnetic mineralogy.

Basaltic grains were most probably derived from the numerous basaltic flows available in the drainage basin of the Afar Rift. Recent weathering of these titanomagnetites during the Brunhes normal interval may have produced a chemical remanent magnetization in the titanomaghemite that overprinted, sometimes completely, the original detrital remanent magnetization.

runs (cycled between 300 and 5 K, 5 T applied field at room temperature; RT-SIRM, Fig. 4B, lower panel), which indicate a mixed signal of Ti-poor titanomagnetites and Ti-rich titanohematites (Dillon and Franke, 2008) since the presence of notable amounts of pure hematite or goethite has already been excluded. This characteristic mineral combination supports a basaltic origin as the main source of the magnetic material (e.g., Dunlop and Özdemir, 1997; Krása et al., 2005). Parallel conclusions have been reached by Tamrat et al. (1996) based on extensive rock magnetic experiments on samples from comparable lithologies collected from the Hadar research area. They show similar

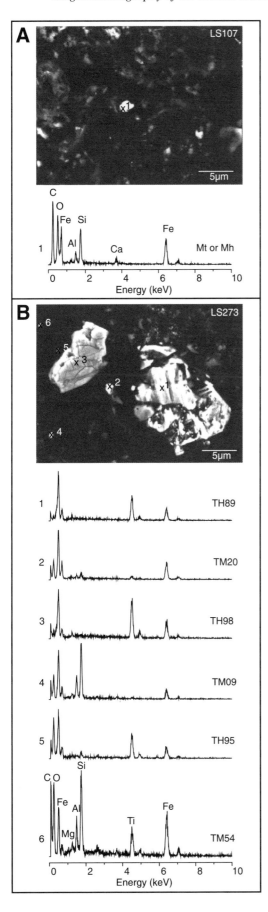

Figure 5. Backscatter electron micrographs and associated energy-dispersive X-ray spectra from spot analyses marked by cross symbols in samples (A) LS107 and (B) LS273. Whiteish particles correspond to opaque minerals, rich in Fe and Ti (Mh—maghemite, Mt—magnetite, TH—titanohematite, TM—titanomagnetite).

ChRM Direction Analysis

ChRM directions were calculated using least square analysis (Kirschvink, 1980) on a minimum of four consecutive steps of the HTC. Line fits were not anchored to the origin. ChRM directions with maximum angular deviation (MAD) above 30° were systematically rejected. Careful selection of ChRM directions (especially of normal polarity directions) is required by the recognized occurrences of a normal overprint potentially extending to relatively high demagnetization temperatures (450 °C). For 93 samples, normal- and reversed-polarity directions were clearly identified on the HTC, where linear decay extended several temperature steps above 450 °C (quality-1 directions, Table DR1 [see footnote 1]). For 69 samples, normal or reversed polarities are indicated but are less reliable (quality-2 directions) because of directional scatter and/or LTC overlap on HTC, resulting in directions slightly divergent from ideal univectorial decay toward the origin. For 57 unreliable samples (quality-3 directions), normal- or reverse-polarity determination is ambiguous due to a strong normal-polarity LTC extending up to 450 °C that partially or fully overprints the HTC. For these samples, unambiguous polarity determination was virtually impossible, especially for HTC of normal polarity. However, for some HTC of reversed-polarity direction, a ChRM direction could be calculated through line fit forced to the origin or through great circle analysis (using the mean of quality-1 reversed directions as set reference point according to the methods of McFadden and McElhinny, 1988), and these were graded as quality-3. To avoid misinterpreting normal overprints as primary record of the magnetic field, which would ultimately result in constructing an erroneous magnetostratigraphic record, we opted for the conservative approach and rejected quality-2 and quality-3 directions altogether from the following magnetostratigraphic record. The remaining set of 93 quality-1 ChRM directions cluster around

TABLE 2. CATION ELEMENTAL CONTENTS					
Sample	Particle	Ti (%)	Fe_Σ (%)	Fe_Σ/Ti	Mineral
LS273	1	14.36	17.93	1.25	TH89
LS273	2	1.12	15.82	14.13	TM20
LS273	3	18.5	19.37	1.05	TH98
LS273	4	0.3	9.69	32.30	TM9
LS273	5	7.07	7.84	1.11	TH95
LS273	6	3.28	14.8	4.51	TM54
LS107	1	0	11.14	—	Mt or Mh

Note: Particle—Label of particle on Figure 5; Ti—element atomic % from semiquantitative energy-dispersive spectroscopy (Fig. 5); Fe_Σ—sum of Fe, Al, and Mg element atomic %; Mineral—mineral phase calculated according to Dillon and Franke (2008); TM—titanomagnetite; TH—titanohematite; Mt—magnetite; Mh—maghemite.

antipodal normal- and reverse-polarity mean directions (Fig. 6A) and pass the reversals test, suggesting that our selection procedure was successful in excluding directions biased by the normal overprint (Tauxe, 1998).

Virtual geomagnetic poles (VGP) were calculated from the remaining quality-1 directions, and the Vandamme (1994) criterion was applied to the VGPs. Eleven widely outlying VGPs were excluded by the Vandamme procedure and downgraded to quality-2 directions. The resulting set of 82 quality-1 ChRM directions provides reliable paleomagnetic polarity at an average interval of 2.4 m throughout the sampled section.

Rotation and Flattening of Paleomagnetic Directions

To compare our results to expected paleomagnetic poles, we computed the mean direction using quality-1 ChRM directions with MAD below 15° only (74 directions). The mean of combined normal and reverse directions was compared to the expected direction from the geocentric axial dipole (GAD) or to the expected direction calculated from the apparent polar wander path (APWP) of stable Africa (computed from 25 African poles with mean ages of 3.1 Ma; Besse and Courtillot, 2002). In both cases, results show, at 95% confidence level, that the mean inclination is significantly too shallow by ~5°–10°, while the mean declination is significantly rotated by ~5°–10° in a counterclockwise sense (Table 3). These values can be considered as minimum estimates because bedding tilt correction of the observed low-magnitude dips (0°–5° in northerly direction) was not performed on these data and would increase rotation and flattening by 0°–5°. Our results compare well with previously reported paleomagnetic directions from the same age at the Hadar section (Tamrat et al., 1996), which also show significant flattening and ~5° counterclockwise rotation when compared to the APWP of Africa (although rotation is statistically insignificant if compared to the GAD).

Paleomagnetic results suggest that a small counterclockwise rotation (~5°–10°) has affected the region after deposition of these rocks since ca. 3 Ma. This may reflect subsequent deformation of the Hadar Basin, possibly associated with the extension of the Afar Rift during the 3–1 Ma northward propagation of Stratoid Series volcanism (Manighetti et al., 2001; Lahitte et al., 2003; Wolfenden et al., 2004). Similar to Tamrat et al. (1996), we observe ~5°–10° flattening of inclination interpreted to result from sedimentary processes during deposition and compaction. This is supported by successful correction of flattening using the method of Tauxe (2005). The observed shallow inclination of 11.5° ± 4.9° is corrected to 21.3° (95% confidence interval is 17–26°; Fig. 6B). This compares well with the expected inclination (16.4° ± 4.9° for the APWP or 21.5° for the GAD). In addition to the reliability tests and quality selection of the ChRM directions, the rotation and successful correction of flattening further support a primary and detrital origin of the remanent magnetization. Thus, the selected directions are suitable for constructing a reliable magnetostratigraphic section.

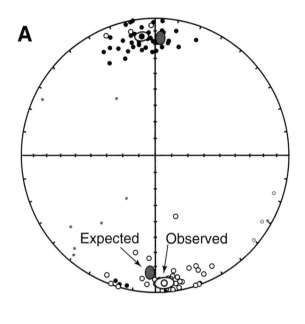

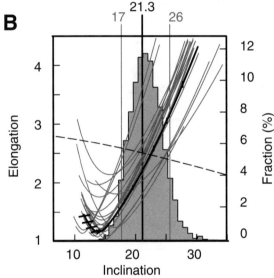

Figure 6. (A) Characteristic remanent magnetization (ChRM) directions projected in lower (full symbol) and upper (open symbol) hemispheres of stereogram. Smaller symbols are 11 directions rejected by Vandamme (1994) criterion. The observed means of normal and of reversed ChRM directions are compared to the expected direction calculated from the apparent polar wander path (APWP) of Africa at 3.1 Ma (Besse and Courtillot, 2002) at the site location (Table 3). The flattened distribution of the ChRM directions contrasts with the north-south–elongated distribution expected for geomagnetic directions at these low latitudes. (B) Correction of observed (Obs.) inclination error using the method developed in Tauxe (2005) applied to the quality-1 ChRM directions with maximum angular deviation (MAD) < 15° ($n = 72$). Black curve shows variation of the elongation of the data set distribution with respect to mean inclination when affected by a flattening factor ranging from 0.35 to 1.00; light-gray curves are the same for generated data sets from bootstrap analysis. The corrected inclination is given by the intersection with the expected elongation (dotted curve) from the geomagnetic model. Background histograms indicate distribution of corrected inclinations with mean of 21.3° and 95% confidence interval (17°–26°).

TABLE 3. MEAN PALEOMAGNETIC DIRECTIONS

Data set	Observed mean direction					Rotation		Flattening		Age	Reference pole			Expected I		Expected D	
	D_s (°)	I_s (°)	α_{95} (°)	k	n	R (°)	± ΔR (°)	F (°)	± ΔF (°)		Lat. (°N)	Long. (°E)	A_{95} (°)	I_x (°)	± δI (°)	D_x (°)	± δD (°)
Middle Ledi (this study)																	
Q-1	354.4	11.4	2.9	30.5	82	−8.3	± 3.2	5.0	± 4.5	3.1	86.2	176.9	2.6	16.4	± 4.9	2.7	± 2.6
Q-1	354.4	11.4	2.9	30.5	82	−5.6	± 2.4	10.1	± 2.3	GAD	90	180.0	0	21.5	± 0.0	0.0	± 0.0
Q-1 MAD<15	354.8	11.5	3.1	29.4	74	−7.9	± 3.3	4.9	± 4.6	3.1	86.2	176.9	2.6	16.4	± 4.9	2.7	± 2.6
Q-1 MAD<15	354.4	11.5	3.1	29.4	74	−5.6	± 2.5	10.0	± 2.5	GAD	90	180.0	0	21.5	± 0.0	0.0	± 0.0
Hadar (Tamrat et al., 1996)																	
Tamrat, 1996	358.6	7.0	4.0	17.9	72	−4.1	± 3.8	9.4	± 5.1	3.1	86.2	176.9	2.6	16.4	± 4.9	2.7	± 2.6
Tamrat, 1996	358.6	7.0	4.0	17.9	72	−1.4	± 3.2	14.5	± 3.2	GAD	90	180.0	0	21.5	± 0.0	0.0	± 0.0

Note: Data set—averaged paleomagnetic data set of ChRM directions (Q-1 indicates mean of quality-1 ChRM directions; Q-1 MAD < 15 indicates mean of quality-1 ChRM directions with maximum angular deviation [MAD] less than 15°; Tamrat et al., 1996). ChRM directions are from Tamrat et al. (1996). Observed mean directions: D_s—mean declination; I_s—mean inclination; α_{95}—angular radius of 95% confidence; k—concentration parameter; n—number of averaged ChRM directions. Rotation R ± ΔR (Flattening F ± ΔF)—difference and 95% confidence between the observed mean declination (inclination) and the declination (inclination) expected from the apparent polar wander path (APWP) of Africa at 3.1 Ma (Besse and Courtillot, 2002) at the site location (11.13°N, 40.67°E).

Magnetostratigraphic Analysis

Correlation to the Paleomagnetic Polarity Time Scale

Three distinct paleomagnetic polarity intervals separated by two reversals are clearly identified by the distribution of normal and reverse quality-1 directions (Fig. 7). The consistency of the polarity of quality-1 directions within those polarity intervals further supports the reliability of our data-selection procedure. At both reversals, careful demagnetization of all collected samples was performed to obtain the best possible resolution for the stratigraphic position of these reversals, resulting in positions of 99.05 ± 4.00 m for the bottom reversal and 191.11 ± 3.40 m for the top reversal. The thickness of these inversion intervals and the fewer quality-1 directions within them are consistent with the typical duration of weak transitional geomagnetic field during reversals (Clement et al., 2004; Valet et al., 2005).

Correlation to the paleomagnetic polarity time scale is indicated by the age of the tuff horizons recognized in the Middle Ledi sections and correlated to dated composite sections from the Hadar research area (Fig. 7). Similar to the pattern found in these other sections, the Sidi Hakoma Tuff and the Kada Hadar Tuff bracket a reversed interval that includes Triple Tuff-4 and is therefore unmistakably recognized as the Mammoth C2An.2r. The two identified reversals thus provide two age constraints to the Middle Ledi section in addition to the three dated tie points supplied by the correlation of the Sidi Hakoma Tuff, Triple Tuff-4, and the Kada Hadar Tuff.

Age References

To derive sediment accumulation rates from these five age reference tie points, we present below a review of the present level of precision and accuracy on these age estimates. The reference ages for the Mammoth C2An.2r boundaries have been revised since the last publications of magnetostratigraphic results from the Hadar Formation (Renne et al., 1993; Tamrat et al., 1996). These previous ages (3.330 Ma and 3.220 Ma for the bottom and top of the Mammoth C2An.2r, respectively) relied on the astronomically calibrated Pliocene time scale established by Hilgen (1991) based on the correlation of sedimentary cycles from Mediterranean marine successions to the precession time series of the astronomical solution of Berger and Loutre (1991). However, application of recent astronomical solutions (Laskar et al., 2004), which provide a better fit to the Mediterranean record, yields an updated time scale with 3.330 Ma and 3.207 Ma ages for the bottom and top of the Mammoth C2An.2r, respectively (Lourens et al., 1996, 2004). More recently, Lisiecki and Raymo (2005) revised these ages to 3.319 Ma and 3.210 Ma, respectively, based on a stack of 57 globally distributed benthic $\delta^{18}O$ records. Errors in these astronomical solutions may arise from uncertainties of tidal dissipation and changes in global ice volume altering Earth's dynamical ellipticity (Lourens et al., 2001), but the best fit to the Mediterranean record is obtained with present-day values of tidal dissipation and dynamical ellipticity, suggesting that these values have not significantly altered the Pliocene-Pleistocene record (Lourens et al., 1996). However, remaining uncertainties in the time lag among astronomical forcing, climate response, and registration in the stratigraphic record, and the assumption of a constant sedimentation rate between astronomically calibrated points, must be accounted for when constructing the time scale. Although these parameters are not readily resolvable, we apply a reasonable bracket for the related error previously proposed by Kuiper et al. (2004), by assigning a ±5 k.y. uncertainty to all the astronomically calibrated ages.

The ages of the three identified tuffaceous layers within the Middle Ledi section are provided by stratigraphic correlations to the Sidi Hakoma Tuff (SHT), Triple Tuff-4 (Triple Tuff-4), and the Kada Hadar Tuff (KHT), previously dated by $^{40}Ar/^{39}Ar$ at Hadar to 3.397 ± 0.029 Ma, 3.220 ± 0.012 Ma, and 3.175 ± 0.012 Ma, respectively; ages are given with their analytical error (Walter and Aronson, 1993; Walter, 1994). Since the publication of these ages, new intercalibration of the neutron fluence monitor standard (Fish Canyon Tuff sanidine age increased from 27.84 to 28.02 Ma by Renne et al., 1998) yields revised ages of 3.419 ± 0.029 Ma, 3.241 ± 0.012 Ma, and 3.196 ± 0.012 Ma for the Sidi Hakoma Tuff, Triple Tuff-4, and Kada Hadar Tuff, respectively. For the Sidi Hakoma Tuff, an independent age estimate of 3.41 ± 0.01 Ma is provided via geochemical correlation to

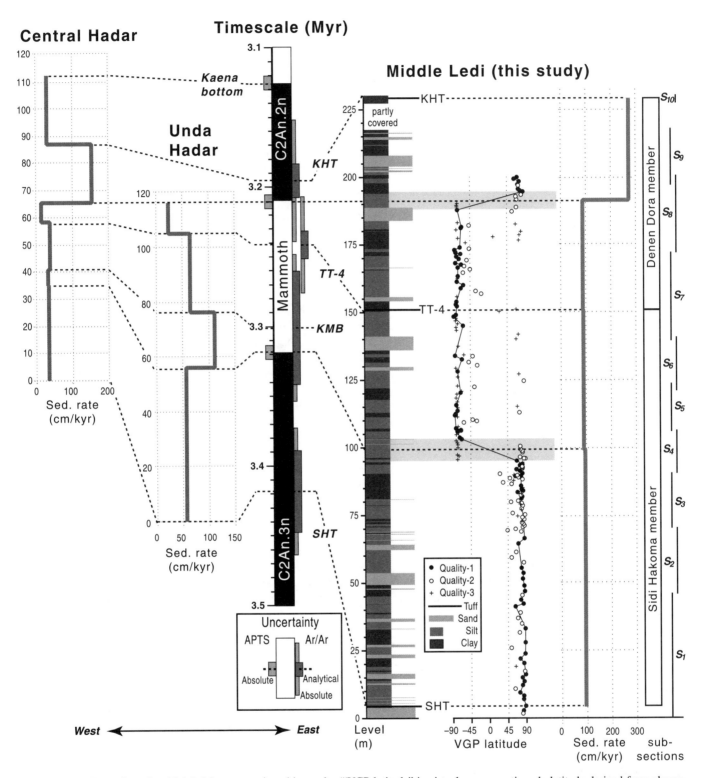

Figure 7. Middle Ledi section (right): Magnetostratigraphic results ("VGP latitude" is virtual geomagnetic pole latitude derived from characteristic remanent magnetization [ChRM] directions) with associated sediment accumulation rates (Sed. rate). Time scale (center): Correlations of tuffaceous and basaltic horizons (KHT—Kada Hadar Tuff, TT-4—Triple Tuff 4, KMB—Kada Damum Basalt, SHT—Sidi Hakoma Tuff) and chron boundaries to reference ages from calibrated $^{40}Ar/^{39}Ar$ dates on associated horizons and from the astronomically tuned polarity time scale (APTS), with their associated uncertainties (see Table 4). Unda Hadar and central Hadar (left): Stratigraphic levels of tuff horizons and chron boundaries and associated sediment accumulation rates from the Unda Hadar section (Renne et al., 1993) and the central Hadar section (Campisano, 2007) of the Hadar research area.

a tephra within the astronomically calibrated Gulf of Aden deposits (deMenocal and Brown, 1999). This slightly younger age is not significantly different from the $^{40}Ar/^{39}Ar$ result, but it provides a more precise age with lower absolute uncertainty. However, it should be noted that the tuning is based on the solution of Berger and Loutre (1991) not yet updated to the Laskar et al. (2004) solutions. A new age estimate for Triple Tuff-4 by Campisano (2007) of 3.256 ± 0.016 Ma is within one standard deviation of the previously reported Triple Tuff-4 age, but it suggests that Triple Tuff-4 could be slightly older. However, this estimate relies on only 10 single-crystal analyses in contrast to 20 single grains for Walter (1994). An independent check on age estimates of the Mammoth C2An.2r boundaries, the Sidi Hakoma Tuff, and the Triple Tuff-4 is provided later using our magnetostratigraphic results.

Sediment Accumulation Rates

For the five tie points provided by the three identified tuffaceous horizons and the two reversal boundaries of the Mammoth C2An.2r, the various proposed ages and associated uncertainties are plotted against stratigraphic level (Fig. 8).

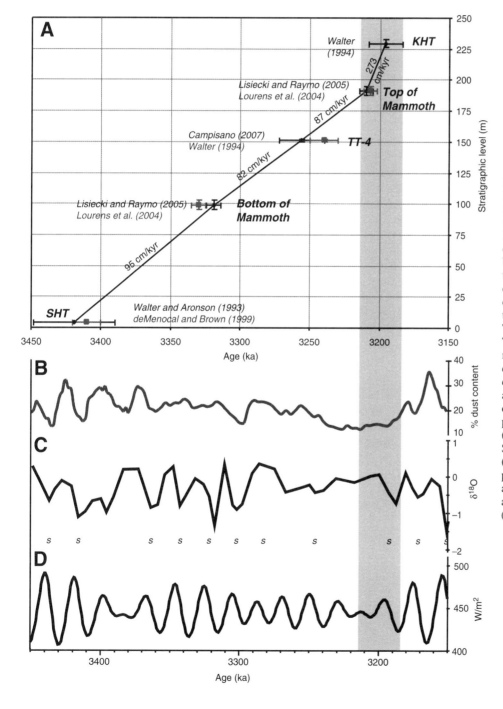

Figure 8. (A) Age versus stratigraphic level for the Middle Ledi section. Age estimates of tuffaceous horizons ($^{40}Ar/^{39}Ar$ date calibrated according to Renne et al. [1998] and given with 1σ error on age) are compared to chron boundary ages (from astronomically tuned polarity time scales with 5 k.y. estimated uncertainty on age) with associated uncertainties on stratigraphic position. KHT—Kada Hadar Tuff, TT-4—Triple Tuff 4, SHT—Sidi Hakoma Tuff. The best fit to constant sediment accumulation rates (indicated on the curve in cm/k.y.) is provided by age estimates depicted with black diamonds. Gray area indicates time span of the interval of high accumulation rate, which is compared to climate proxies and parameters: (B) Terrigenous dust off East Africa Deep Sea Drilling Project (DSDP) Site 721/722 (deMenocal, 1995), (C) Mediterranean planktonic $\delta^{18}O$ (Lourens et al., 1996) and sapropels (s) (Eimes et al., 2000), and (D) 21 June insolation at 11°N (Laskar et al., 2004).

Sediment accumulation rates are estimated by linear interpolation between the five successive tie points. An excellent linear fit is obtained below the top of the Mammoth C2An.2r down to the Sidi Hakoma Tuff. In this interval, interpolations of chron boundary ages fall within the analytical error of the ^{40}Ar/^{39}Ar ages. This supports the consistency between the calibrated ^{40}Ar/^{39}Ar ages (Renne et al., 1998) and the astronomically tuned polarity time scales (Lisiecki and Raymo, 2005; Lourens et al., 2005). The age estimates that minimize sediment accumulation rate variations are the Sidi Hakoma Tuff age of Walter and Aronson (1993), the Mammoth C2An.2r boundary ages of Lisiecki and Raymo (2005), and the Triple Tuff-4 age of Campisano (2007). Accordingly, rates are nearly constant (average 88 cm/k.y.) from the Sidi Hakoma Tuff at the base of the section to the top of the Mammoth C2An.2r, but a threefold increase (273 cm/k.y.) is indicated between the top of the Mammoth C2An.2r and the Kada Hadar Tuff (Table 4; Fig. 8).

This large increase in accumulation rates remains significant regardless of the proposed tie-point age estimate, and it is robust beyond uncertainties on ages and stratigraphic positions of reversals and tuffaceous horizons. Keeping the rate constant above the top of the Mammoth C2An.2r is not viable. It would require either an impossibly low stratigraphic position for the Kada Hadar Tuff at the 203 m level, or the unlikely option that the age of the Kada Hadar Tuff is too old by 3σ, even though this ^{40}Ar/^{39}Ar age with 26 sanidine analyses is one of the better constrained in the Hadar region (Walter, 1994). An error caused by invalid calibration of the fluence monitor standard is also unlikely because it would imply offsetting not only all the other otherwise concordant ^{40}Ar/^{39}Ar ages issued from the same laboratory conditions (i.e., the Berkeley Geochronology Center), but also the Sidi Hakoma Tuff astronomically calibrated in ocean records (deMenocal and Brown, 1999). Alternatively, constant rates would require shifting the age of the top of the Mammoth C2An.2r to a precession cycle older than reported. An error in the tuning of all the Mediterranean type sections (Lourens et al., 1996, 2005) and the 57 benthic δ^{18}O records (Lisiecki and Raymo, 2005) is unlikely. It would require not only changing the position of one precession cycle in these records, but also shifting positions of well-established 100 and 400 k.y. eccentricity cycles. Additionally, an error in the stratigraphic position of the paleomagnetic reversal in the type sections (e.g., through delayed acquisition of remanent magnetization) is unlikely given the detailed resolution of sampling and demagnetization of these sections (van Hoof and Langereis, 1991). Thus, we conclude from sediment accumulation rates in the Middle Ledi that deposition was near constant at ~90 cm/k.y. since ca. 3.4 Ma, and it increased significantly at ca. 3.2 Ma.

Comparison to Hadar Basin Stratigraphy

When compared to the Hadar and Gona areas, with typical 20–30 cm/yr accumulation rates (Campisano and Feibel, 2007; Quade et al., this volume), the ~90 cm/k.y. rates in the Ledi-Geraru area clearly indicate that it was positioned close to or within the depocenter of the Hadar Basin. When compared to the Unda Hadar (eastern) and central Hadar sections, an eastward thickening of the stratigraphy is immediately apparent (Fig. 7). The thickness between the Sidi Hakoma Tuff up to the top of the Mammoth C2An.2r in the central Hadar section is almost doubled in the Unda Hadar and more than tripled in the Middle Ledi. This eastward thickening is nearly identical among the central Hadar, Unda Hadar, and Middle Ledi sections at ~12 m/km (ratio of thickness increase in meters to E-W distance in kilometers between sections), and it translates into regional 1° easterly tilt of the Sidi Hakoma Tuff with respect to the horizon representing the top of the Mammoth C2An.2r.

Sediment accumulation rates derived between tie points compare well between the three sections (Fig. 7; Table 4). In the Unda Hadar, although the precise position of the Kada Hadar Tuff in relation to the top of the Mammoth interval is not readily available from published records, extrapolation of constant accumulation yields an unlikely stratigraphic position for the Kada Hadar Tuff (119.35 m) only a few meters above the Mammoth C2An.2r and much lower than the Confetti Clay horizon shown 15–25 m above the Mammoth C2An.2r in Schmitt and Nairn (1984). This strongly suggests that sediment accumulation must also have increased in this interval at Unda Hadar, because the Kada Hadar Tuff is only a few meters below the Confetti Clay (Campisano and Feibel, this volume). Below the Kada Hadar

TABLE 4. AGES, STRATIGRAPHIC LOCATIONS, AND ACCUMULATION RATES

Rock unit or reversal	Reference				Central Hadar		Unda Hadar		Middle Ledi	
	Age (ka)	±	δAge (ka)	ΔAge (ka)	Level (cm)	Rate (cm/k.y.)	Level (cm)	Rate (cm/k.y.)	Level (cm)	Rate (cm/k.y.)
SHT*	3419	±	43	45	0	–	0	–	432	–
Mam_b†	3319	±	5	5	3500	35	5600	56	9905	95
KMB§	3301	±	40	52	4081	32	7600	111	–	–
TT-4#	3256	±	16	36	5809	38	10,500	64	15,087	82
Mam_t†	3210	±	5	5	6509	15	11,600	24	19,111	87
KHT**	3196	±	12	43	8671	154	*11,935*	*extrapol.*	22,934	273
Kaena_b†	3127	±	5	5	11,201	37	–	–	–	–

Note: Reference ages of rock units are from: *—Walter and Aronson (1993), §—Renne et al. (1993), #—Campisano (2007), and **—Walter (1994). Reversal ages (_b and _t for chron bottom and top, respectively) are according to time scales of: †—Lisiecki and Raymo (2004). δAge—analytical error; ΔAge—absolute uncertainty; Level—stratigraphic levels, in italic if linearly extrapolated from rate at underlying tie point; Rate—sediment accumulation rate derived from linear interpolation.

Tuff, the chronostratigraphy at Hurda shows less constant rates (average of 64 cm/k.y.), possibly related to the larger uncertainty on the age of the Kada Damum Basalt (KMB) and on the reversals positions (using AF demagnetization only). In central Hadar, a pattern strikingly similar to that of the Middle Ledi is indicated by near constant rates (average of 30 cm/k.y.) from the Sidi Hakoma Tuff to the top of the Mammoth C2An.2r, followed by a fivefold increase (154 cm/k.y.) between the top of the Mammoth C2An.2r to the Kada Hadar Tuff. Interestingly, above the Kada Hadar Tuff, rates return to near average values (34 cm/k.y.) between the Kada Hadar Tuff and the base of the Kaena C2An.1r interval, indicating that the increase in accumulation rate was a short-lived event. Finally, the regional consistency of the results signifies that processes responsible for the accumulation increase operated at the scale of the entire Hadar Basin and cannot be attributed to inaccurate stratigraphic positions of reversals, to a local fault, or to a local change in the drainage pattern that would be restricted to only one section.

DISCUSSION

Tectonic Configuration of Hadar Basin

By extending the stratigraphic record further east into the Ledi-Geraru area, our results clearly show eastward thickening of the Hadar Formation. This result supports previous work that suggested thickening in eastward (Tiercelin, 1986) or northeastward (Wynn et al., 2006) directions (our results constrain the east-west component of thickening but not the north-south component). An eastward thickening pattern indicates that sediment accumulation space was more important on the eastern flank than on the western flank of the basin, suggesting an asymmetric eastward-tilting graben configuration. This basin geometry may have resulted from normal faulting on the eastern margin of the basin or eastward-increasing thermal subsidence (Einsele, 2000). We rule out the possibility that the dip formed before deposition to be later filled in by sediments because that would result in progradational stratigraphic successions (coarsening upward) in opposition to the observed aggradational patterns observed in the Hadar Formation.

To account for the observed 1° eastward dip accumulated in ~200 k.y. between the Sidi Hakoma Tuff and the top of Mammoth C2An.2r, significant subsidence is required toward the eastern margin of the basin (at least the ~300 m total thickness of the Hadar Formation). Normal faulting is present at the eastern end of the Ledi-Geraru research area. Basaltic lavas become progressively thicker and a more significant portion of the stratigraphy from west to east. Toward the west, the termination of these flows is expressed by an erosional escarpment defining the present irregular western edge of the Eibdaha Plateau because of their greater resistance to erosion (Fig. 1). Toward the east, these flows are cut by numerous north-south–trending normal faults. A major break (east of the Eibdaha Plateau) drops basalts with a thin Hadar sedimentary cover down to the west against the basalts to the east. This boundary continues to the south and is manifest as an ~50-m-high west-facing escarpment that we refer to as the Eibdaha fault zone. The Eibdaha fault zone forms the eastern boundary of the Hadar Formation stratigraphy, and the Eibdaha Plateau is tilted eastward both north and south of the Awash River. These relationships, observed in the field, on digital elevation models (90 m Shuttle Radar Topography Mission [SRTM] imagery), and on aerial photography, suggest a down-to-the-west normal displacement. However, the Eibdaha fault zone may postdate Hadar Formation deposition because it offsets the Hadar Formation, and rocks coeval with the Hadar Formation (Afar Stratoid Basalts) are exposed in the footwall (Kidane et al., 2003). Further investigation is required to constrain the age and displacement history on these faults.

Alternatively, the eastward thickness increase may result from increasing thermal subsidence, as suggested by the steadiness of the accumulation rates observed throughout the Hadar Basin during a relatively long time interval (ca. 3.4 Ma to ca. 3.2 Ma). We observe, however, that these rates (30–90 cm/k.y.) are greater than the low regional subsidence expected in the Pliocene by thermal and flexural modeling based on kinematic and crustal thickness reconstructions (Redfield et al., 2003). Redfield et al. (2003) estimated crustal thinning and thermal weakening at the scale of an entire 180-km-long "Hadar block" from the Ethiopian escarpment to the rift center since 20 Ma. In detail, Redfield et al. (2003) suggested that the logarithmic subsidence decrease since 6.2 Ma was negligible regionally, but that the elastic lithosphere became so thin that stretching accommodation was localized rather than regional. As a result, a high thermal gradient would be expected between the Ethiopian escarpment (elastic thickness up to 60 km) and the adjacent stretched lithosphere (elastic thickness down to 5 km). Consistently, the observed Hadar Formation thickening away from the Ethiopian escarpment may be related to the high gradient of flexural rigidity producing riftward downwarping through differential thermal subsidence, possibly enhanced by sediment loading as described on the western margin of the Afar Depression (Beyene and Abdelsalam, 2005), the Adama basin just south of the Afar Depression (Wolfenden et al., 2004), or the Gulf of Aden passive margin in Oman (Gunnell et al., 2007; Petit et al., 2007).

Interestingly, the observed sediment transport directions measured in the Hadar Formation are directed eastward, with proximal deposits (alluvial and fluvial sands and gravels) along the Ethiopian escarpment and distal lacustrine deposits toward the Afar Depression (Aronson and Taieb, 1981; Tiercelin, 1986; Quade et al., 2004). This suggests that the higher, eroding topography that provided the sediment source—such as the Ethiopian escarpment—was already present west of the basin as indicated by thermochronologic results from the Ethiopian Plateau (Pik et al., 2003; Gani et al., 2007). It should be noted that rift escarpments do not require active faulting to be sustained, but they are usually described as long-lasting (over 10s of m.y.) geomorphic features inherited from the main base-level drop of the initial rifting (e.g., Braun and van der Beek, 2004). We therefore argue that, although the Hadar Formation sediment originated from the existing rift escarpment to the west, the accumulation was not primarily controlled by faulting along the rift escarpment but rather by increased subsidence toward the Afar Depression.

Environmental Change Ca. 3.2 Ma

The remarkably constant sediment accumulation rates from ca. 3.4 Ma to ca. 3.2 Ma suggest that the Hadar Basin was at close to dynamic aggradational equilibrium during that period. Perturbation of this steady state is required by the conspicuous sediment accumulation increase occurring between the top of the Mammoth C2An.2r interval and the Kada Hadar Tuff. Interestingly, the most distinctive sedimentologic feature of this particular interval in the Middle Ledi section is a 65% increase in the amount of sandstone. This increase is also noted at Hadar, where this interval includes a set of sandy layers referred to as the Denen Dora-2 and -3 sands (DD-2s, DD-3s). This interval has been intensely studied because it preserves a large proportion of the hominin fossil record from the Hadar Basin (Behrensmeyer, this volume). In addition to the large number of hominins from the DD-2 and DD-3 interval, this sequence also preserves by far the greatest number of paleontological specimens from Hadar. Of the 11 submember divisions at Hadar, specimens from the DD-2 and DD-3 submembers account for more than 40% of the Hadar paleontological assemblage but represent less than 8% of the time recorded (Campisano, 2007). Although the abundance of preserved specimens may be a result of ecological or taphonomic factors related to DD-2 and DD-3 depositional environments, it may also be a function of increased preservation potential associated with the exceptionally high rate of sediment deposition documented in this interval.

Whereas the DD-2 sand is laterally discontinuous and only 2.5 m in maximum thickness, the DD-3 sand is extensive and laterally persistent. This upward-fining sand can reach up to 10 m in thickness and displays an erosional base, well-developed cross-bedding, and lateral accretion surfaces (Tiercelin, 1986; Campisano and Feibel, this volume). The DD-3 forms a time-transgressive progradational sequence typical of a dissecting meandering fluvial system coming from the western Hadar region, where it is often conglomeratic (Quade et al., 2004). The sudden occurrence of this high-energy fluvial system prograding onto more distal sediments suggests three possible mechanisms forcing the drainage basin configuration: (1) water-level lowering, (2) decrease in basin subsidence, and/or (3) higher detrital influx (Keighley et al., 2003). At Hadar, the Kada Hadar Tuff is typically interbedded within the overbank silts of the DD-3 system and subsequently transitions to low-gradient fluvial, paludal, and lacustrine deposition similar to that found below the DD-3. This suggests that the basin quickly returned to dynamic equilibrium after the event, as also shown by a return to average values of accumulation rates in the central Hadar region (Campisano, 2007; Table 4).

For the Hadar Basin, variations in sediment accumulation rates have usually been attributed to tectonically controlled variations in accumulation space through fault-slip and subsidence (Aronson and Taieb, 1981; Tiercelin, 1986). Although this may be the case for the widespread ca. 2.9 Ma unconformity separating the Busidima and Hadar Formations (Quade et al., 2004), a major regional tectonic event at ca. 3.2 Ma, with associated long-term subsidence and uplift, can be ruled out by the sudden and short-lived nature of the observed accumulation increase. Also, increased subsidence would more likely result in transgression of distal sediments rather than the observed prograding DD-type sands. Taking the ages and associated uncertainties at face value, the accumulation increase found between the top of the Mammoth C2An.2r interval and the Kada Hadar Tuff would have occurred in less than 30 k.y. Within such a short time span, the basin can be assumed to have maintained steady-state conditions in relation to tectonic subsidence and relief. The observed accumulation increase and progradation of the DD sands are thus more likely related to landscape equilibration after (1) a sudden change in drainage configuration and/or (2) climatic fluctuations (Einsele and Hinderer, 1998). Deciding between these two possible causes is difficult. The drainage configuration may have been suddenly and significantly altered by mechanisms such as a regional tuff deposition, suddenly increasing erosional efficiency, or minor local faulting opening the basin outlet, draining the lake, and resulting in a base-level drop that could have induced increased sediment transport to the basin from up-river catchments. At present, no evidence for the occurrence of such mechanisms has been reported from the Hadar region. Alternatively, climatic fluctuations at or below the scale of precession cycles (20 k.y.) may have produced lowering of the lake level and/or increased erosional efficiency and sediment transport to the basin, without necessitating a change in drainage configuration.

This climate-related hypothesis, previously proposed for the DD-3 sand (Tiercelin, 1986), can be substantiated by comparison to existing climate proxies (Fig. 8). Locally, pollen and stable isotope analysis have been gathered from the Hadar Basin (Bonnefille et al., 2004; Levin et al., 2004; Quade et al., 2004; Wynn et al., 2006). However, the age resolution of these studies is still insufficient to decipher a short event at ca.3.2 Ma. Regionally relevant climate indicators are available with higher time resolution from the Gulf of Aden and the Mediterranean. They show that the ca. 3.2 Ma event occurs at the end of a relatively "quiet" period expressed by low variability in the amount of terrigenous material transported to the Gulf of Aden (deMenocal, 1995) as well as in Mediterranean planktonic $\delta^{18}O$ values (Lourens et al., 1996), and by fewer sapropel layers in the eastern Mediterranean basin (sapropel layers are tied to low-latitude summer insolation driving East African monsoon rainfall on the Ethiopian Highlands where both the Blue Nile and the Awash River originate; Emeis et al., 2000). This quiet period corresponds to a peculiar configuration in Earth's orbital parameters arising from a combination of low eccentricity and low-amplitude obliquity variation, reducing the precession influence on Earth insolation and favoring damped seasonality (Laskar et al., 2004). We propose that the ca. 3.2 Ma event may have been related to the onset of increased climate variability, perturbing the landscape by increasing sediment discharge and increasing sediment accumulation after a relatively long period of low variability. Testing this hypothesis will require further work focusing on increasing temporal resolution of climatic proxies directly from the sediments of the Hadar Basin.

Relationship to Hominin Evolution and the Paleontological Record

The Hadar Basin contains abundant fossils of *Australopithecus afarensis* between 3.45 Ma and 2.95 Ma (Kimbel et al., 2004). These fossils sample an evolving lineage that can be traced back at least to the first appearance of an earlier species, *A. anamensis*, at ca. 4.2 Ma (Kimbel et al., 2006). Evolutionary changes occurred at various points in the *A. anamensis-afarensis* lineage. The last change was an increase in lower jaw size that probably indicates a general increase in body size (Lockwood et al., 2000; Kimbel et al., 2004). This increase in size took place at some point above the Kada Hadar Tuff (ca. 3.2 Ma) and before ca. 3.0 Ma, but the paucity of *A. afarensis* fossils between these ages makes it impossible to be more precise.

Among other mammalian fauna, there are an increase and influx of more arid-adapted taxa, particularly bovids, at Hadar during the interval of 3.2 Ma to 3.0 Ma (Campisano and Reed, 2007; Reed, 2008). The coincidence of a dramatic change in local rates of sediment accumulation with a shift in some records of global environmental change (discussed already) raises the possibility that the Afar faunal community was directly affected by these events. In particular, the onset of high-amplitude climate oscillations and increased aridity in eastern Africa between 3.15 and 2.95 Ma may have been linked to the size-related morphological changes within the *A. afarensis* lineage and the increasing proportion of more arid-adapted mammalian taxa during this period (Campisano and Feibel, 2007). These suggestions remain speculative, but they highlight the importance of well-calibrated records for fossils, chronology, and paleoenvironmental indicators to the study of variation within species as well as patterns of species turnover. In addition, more work is necessary in South Africa and other parts of the continent to determine whether the events in Hadar record are paralleled elsewhere.

ACKNOWLEDGMENTS

This work was funded by the Leakey Foundation, the Wenner Gren Foundation for Anthropological Research, the Institute of Human Origins, the School of Human Evolution and Social Change at Arizona State University, and the Netherlands Organisation for Scientific Research (NWO). Permission to conduct fieldwork was granted by the Authority for Research and Conservation of Cultural Heritage (of the Ethiopian Ministry of Youth, Sports, and Culture) with local permissions and assistance from the Culture and Tourism Bureau of the Afar Regional State government. We thank our field crew directed by Mesfin Mekonen, Bruno Paulet for help in the field and plane tickets, Klaudia Kuiper for $^{40}Ar/^{39}Ar$ issues, and Mark Dekkers and Andy Biggins for rock magnetic issues. Three anonymous reviews considerably improved the original manuscript, and we thank Jay Quade and Jonathan Wynn for their constructive comments.

REFERENCES CITED

Acton, G.D., Tessema, A., Jackson, M., and Bilham, R., 2000, The tectonic and geomagnetic significance of paleomagnetic observations from volcanic rocks from central Afar, Africa: Earth and Planetary Science Letters, v. 180, p. 225–241, doi: 10.1016/S0012-821X(00)00173-4.

Alemseged, Z., Spoor, F., Kimbel, W.H., Bobe, R., Geraads, D., Reed, D., and Wynn, J.G., 2006, A juvenile early hominin skeleton from Dikika, Ethiopia: Nature, v. 443, p. 296–301, doi: 10.1038/nature05047.

Aronson, J.A., and Taieb, M., 1981, Geology and paleogeography of the Hadar hominid site, Ethiopia, in Rapp, G., and Vondra, C.F., eds., Hominid Sites: Their Geologic Setting: Boulder, Colorado, Westview Press, p. 165–195.

Aronson, J.L., Schmitt, T.J., Walter, R.C., Taieb, M., Tiercelin, J.J., Johanson, D.C., Naeser, C.W., and Nairn, A.E.M., 1977, New geochronologic and palaeomagnetic data for the hominid-bearing Hadar Formation of Ethiopia: Nature, v. 267, p. 323–327, doi: 10.1038/267323a0.

Audin, L., Quidelleur, X., Coulie, E., Courtillot, V., Gilder, S., Manighetti, I., Gillot, P.Y., Tapponnier, P., and Kidane, T., 2004, Palaeomagnetism and K-Ar and Ar-40/Ar-39 ages in the Ali Sabieh area (Republic of Djibouti and Ethiopia): Constraints on the mechanism of Aden Ridge propagation into southeastern Afar during the last 10 Myr: Geophysical Journal International, v. 158, p. 327–345, doi: 10.1111/j.1365-246X.2004.02286.x.

Behrensmeyer, A.K., 2008, this volume, Paleoenvironmental context of the Pliocene A.L. 333 "First Family" Hominin Locality, Hadar Formation, Ethiopia, in Quade, J., and Wynn, J.G., eds., The Geology of Early Humans in the Horn of Africa: Geological Society of America Special Paper 446, doi: 10.1130/2008.2446(09).

Berger, A., and Loutre, M.F., 1991, Insolation values for the climate of the last 10,000,000 years: Quaternary Science Reviews, v. 10, p. 297–317, doi: 10.1016/0277-3791(91)90033-Q.

Besse, J., and Courtillot, V., 2002, Apparent and true polar wander and the geometry of the geomagnetic field in the last 200 million years: Journal of Geophysical Research, v. 107, doi: 10.1029/2000JB000050, 2300.

Beyene, A., and Abdelsalam, M.G., 2005, Tectonics of the Afar Depression: A review and synthesis: Journal of African Earth Sciences, v. 41, p. 41–59, doi: 10.1016/j.jafrearsci.2005.03.003.

Bonnefille, R., Potts, R., Chalie, F., Jolly, D., and Peyron, O., 2004, High-resolution vegetation and climate change associated with Pliocene *Australopithecus afarensis*: Proceedings of the National Academy of Sciences of the United States of America, v. 101, p. 12,125–12,129, doi: 10.1073/pnas.0401709101.

Braun, J., and van der Beek, P., 2004, Evolution of passive margin escarpments: What can we learn from low-temperature thermochronology?: Journal of Geophysical Research, v. 109, F04009, doi: 10.1029/2004JF000147.

Buddington, A.F., and Lindsley, D.H., 1964, Iron-titanium oxide minerals and synthetic equivalents: Journal of Petrology, v. 5, p. 310–357.

Campisano, C.J., 2007, Tephrostratigraphy and Hominin Paleoenvironments of the Hadar Formation, Afar Depression, Ethiopia [Ph.D. thesis]: New Brunswick, New Jersey, Rutgers, 601 p.

Campisano, C.J., and Feibel, C.S., 2007, Connecting local environmental sequences to global climate patterns: Evidence from the hominin-bearing Hadar Formation, Ethiopia: Journal of Human Evolution, v. 53, p. 515–527, doi: 10.1016/j.jhevol.2007.05.015.

Campisano, C.J., and Feibel, C.S., 2008, this volume, Depositional environments and stratigraphic summary of the Pliocene Hadar Formation at Hadar, Afar Depression, Ethiopia, in Quade, J., and Wynn, J.G., eds., The Geology of Early Humans in the Horn of Africa: Geological Society of America Special Paper 446, doi: 10.1130/2008.2446(08).

Campisano, C.J., and Reed, K.E., 2007, Spatial and temporal patterns of *Australopithecus afarensis* habitats at Hadar, Ethiopia: Philadelphia, Paleoanthropology Society Annual Meeting Abstracts, p. A6.

Clement, A.C., Hall, A., and Broccoli, A.J., 2004, The importance of precessional signals in the tropical climate: Climate Dynamics, v. 22, p. 327–341, doi: 10.1007/s00382-003-0375-8.

deMenocal, P.B., 1995, Plio-Pleistocene African Climate: Science, v. 270, p. 53–59, doi: 10.1126/science.270.5233.53.

deMenocal, P.B., 2004, African climate change and faunal evolution during the Pliocene-Pleistocene: Earth and Planetary Science Letters, v. 220, p. 3–24, doi: 10.1016/S0012-821X(04)00003-2.

deMenocal, P.B., and Brown, F.H., 1999, Pliocene tephra correlations between East African hominid localities, the Gulf of Aden and the Arabian Sea, *in* Agusti, J., Rook, L., and Andrews, P., eds., The Evolution of Terrestrial Ecosystems in Europe: Cambridge, Cambridge University Press, p. 23–54.

Dillon, M., and Franke, C., 2008, Diagenetic alteration of natural magnetic Fe-Ti oxides identified by energy dispersive spectroscopy (EDS) and low-temperature remanence and hysteresis measurements: Physics of the Earth and Planetary Interior, doi:10.1016/j.pepi.2008.08.003.

DiMaggio, E.N., Campisano, C.J., Arrowsmith, J R., Reed, K.E., Swisher, C.C., III, and Lockwood, C.A., 2008, this volume, Correlation and stratigraphy of the BKT-2 volcanic complex in west-central Afar, Ethiopia, *in* Quade, J., and Wynn, J.G., eds., The Geology of Early Humans in the Horn of Africa: Geological Society of America Special Paper 446, doi: 10.1130/2008.2446(07).

Dunlop, D., and Özdemir, Ö., 1997, Rock Magnetism: Fundamentals and Frontiers: Cambridge, Cambridge University Press, 573 p.

Einsele, G., 2000, Sedimentary Basins—Evolution, Facies, and Sediment Budget (2nd edition): Berlin, Springer-Verlag, 792 p.

Einsele, G., and Hinderer, M., 1998, Quantifying denudation and sediment-accumulation systems (open and closed lakes): Basic concepts and first results: Palaeogeography, Palaeoclimatology, Palaeoecology, v. 140, p. 7–21, doi: 10.1016/S0031-0182(98)00041-8.

Emeis, K.-C., Sakamoto, T., Wehausen, R., and Brumsack, H.-J., 2000, The sapropel record of the eastern Mediterranean Sea—Results of Ocean Drilling Program Leg 160: Palaeogeography, Palaeoclimatology, Palaeoecology, v. 158, p. 371–395, doi: 10.1016/S0031-0182(00)00059-6.

Franke, C., Frederichs, T., and Dekkers, M.J., 2007, Efficiency of heavy liquid separation to concentrate magnetic particles: Geophysical Journal International, doi: 10.1111/j.1365-246X.2007.03489.x, v. 170, p. 1053–1066.

Gani, N.D., Gani, M.R., and Abdelsalam, M.G., 2007, Blue Nile incision on the Ethiopian Plateau: Pulsed plateau growth, Pliocene uplift, and hominin evolution: GSA Today, v. 17, no. 9, p. 4–11, doi: 10.1130/GSAT01709A.1.

Gunnell, Y., Carter, A., Petit, C., and Fournier, M., 2007, Post-rift seaward downwarping at passive margins: New insights from southern Oman using stratigraphy to constrain apatite fission-track and (U-Th)/He dating: Geology, v. 35, p. 647–650, doi: 10.1130/G23639A.1.

Hailemichael, M., Aronson, J.L., Savin, S., Tevesz, M.J.S., and Carter, J.G., 2002, $\delta^{18}O$ in mollusk shells from Pliocene Lake Hadar and modern Ethiopian lakes: Implications for history of the Ethiopian monsoon: Palaeogeography, Palaeoclimatology, Palaeoecology, v. 186, p. 81–99, doi: 10.1016/S0031-0182(02)00445-5.

Hart, W.K., Walter, R.C., and WoldeGabriel, G., 1992, Tephra sources and correlations in Ethiopia: Application of elemental and neodymium isotope data: Quaternary International, v. 13–14, p. 77–86, doi: 10.1016/1040-6182(92)90012-Q.

Hilgen, F.J., 1991, Astronomical calibration of Gauss to Matuyama sapropels in the Mediterranean and implication for the geomagnetic polarity time scale: Earth and Planetary Science Letters, v. 104, p. 226–244, doi: 10.1016/0012-821X(91)90206-W.

Johanson, D.C., and Taieb, M., 1976, Plio-Pleistocene hominid discoveries in Hadar, Ethiopia: Nature, v. 260, p. 293–297, doi: 10.1038/260293a0.

Johanson, D.C., Taieb, M., and Coppens, Y., 1982, Pliocene hominids from the Hadar Formation, Ethiopia (1973–1977): Stratigraphic, chronological, and paleoenvironmental contexts, with notes on hominid morphology and systematics: American Journal of Physical Anthropology, v. 57, p. 373–402, doi: 10.1002/ajpa.1330570402.

Kakol, Z., Sabol, J., Kozlowski, A., and Honig, J.M., 1994, Influence of titanium doping on the magnetocrystalline anisotropy of magnetite: Physical Review B: Condensed Matter and Materials Physics, v. 49, p. 12,767–12,772.

Keighley, D., Flint, S., Howell, J., and Moscariello, A., 2003, Sequence stratigraphy in lacustrine basins: A model for part of the Green River Formation (Eocene), southwest Uinta Basin, Utah: Journal of Sedimentary Research, v. 73, p. 987–1006, doi: 10.1306/050103730987.

Kidane, T., Courtillot, V., Manighetti, I., Audin, L., Lahitte, P., Quidelleur, X., Gillot, P.Y., Gallet, Y., Carlut, J., and Haile, T., 2003, New paleomagnetic and geochronologic results from Ethiopian Afar: Block rotations linked to rift overlap and propagation and determination of a similar to 2 Ma reference pole for stable Africa: Journal of Geophysical Research–Solid Earth, v. 108, doi: 10.1029/2001JB000645.

Kimbel, W.H., Rak, Y., and Johanson, D.C., 2004, The Skull of *Australopithecus afarensis*: New York, Oxford University Press, 254 p.

Kimbel, W.H., Lockwood, C.A., Ward, C.V., Leakey, M.G., Rak, Y., and Johanson, D.C., 2006, Was *Australopithecus anamensis* ancestral to *A. afarensis*? A case of anagenesis in the hominin fossil record: Journal of Human Evolution, v. 51, p. 134–152, doi: 10.1016/j.jhevol.2006.02.003.

Kirschvink, J.L., 1980, The least-square line and plane and the analysis of paleomagnetic data: Geophysical Journal of the Royal Astronomical Society, v. 62, p. 699–718.

Krása, D., Shcherbakov, V.P., Kunzmann, T., and Petersen, N., 2005, Self-reversal of remanent magnetization in basalts due to partially oxidized titanomagnetites: Geophysical Journal International, v. 162, p. 115–136, doi: 10.1111/j.1365-246X.2005.02656.x.

Kuiper, K.F., Hilgen, F.J., Steenbrink, J., and Wijbrans, J.R., 2004, $^{40}Ar/^{39}Ar$ ages of tephras intercalated in astronomical tuned Neogene sedimentary sequences in the eastern Mediterranean: Earth and Planetary Science Letters, v. 222, p. 583–597, doi: 10.1016/j.epsl.2004.03.005.

Lahitte, P., Gillot, P.Y., Kidane, T., Courtillot, V., and Bekele, A., 2003, New age constraints on the timing of volcanism in central Afar, in the presence of propagating rifts: Journal of Geophysical Research, v. 108, doi: 10.1029/2001JB001689, 2123.

Laskar, J., Robutel, P., Joutel, F., Gastineau, M., Correia, A.C.M., and Levrard, B., 2004, A long-term numerical solution for the insolation quantities of the Earth: Astronomy and Astrophysics, v. 428, doi: 10.1051/0004-6361:20041335, p. 261–285.

Levin, N.E., Quade, J., Simpson, S.W., Semaw, S., and Rogers, M., 2004, Isotopic evidence for Plio-Pleistocene environmental change at Gona, Ethiopia: Earth and Planetary Science Letters, v. 219, p. 93–110, doi: 10.1016/S0012-821X(03)00707-6.

Lisiecki, L.E., and Raymo, M.E., 2005, A Pliocene-Pleistocene stack of 57 globally distributed benthic $\delta^{18}O$ records: Paleoceanography, v. 20, p. PA1003, doi: 10.1029/2004PA001071.

Lockwood, C.A., Kimbel, W.H., and Johanson, D.C., 2000, Temporal trends and metric variation in the mandibles and teeth of *Australopithecus afarensis*: Journal of Human Evolution, v. 39, p. 23–55, doi: 10.1006/jhev.2000.0401.

Lourens, L.J., Antonarakou, A., Hilgen, F.J., Van Hoof, A.A.M., Vergnaud Grazzini, C., and Zachariasse, W.J., 1996, Evaluation of the Plio-Pleistocene astronomical timescale: Paleoceanography, v. 11, p. 391–413.

Lourens, L.J., Wehausen, R., and Brumsack, H.J., 2001, Geological constraints on tidal dissipation and dynamical ellipticity of the Earth over the past three million years: Nature, v. 409, p. 1029–1033, doi: 10.1038/35059062.

Lourens, L.J., Hilgen, F.J., Laskar, J., Shackelton, N.J., and Wilson, D.S., 2004, The Neogene Period, *in* Gradstein, F.M., Ogg, J.G., and Smith, A.G., eds., A Geologic Time Scale 2004: Cambridge, Cambridge University Press, p. 409–440.

Manighetti, I., Tapponnier, P., Courtillot, V., Gallet, Y., Jacques, E., and Gillot, P.Y., 2001, Strain transfer between disconnected, propagating rifts in Afar: Journal of Geophysical Research–Solid Earth, v. 106, p. 13,613–13,665, doi: 10.1029/2000JB900454.

McFadden, P.L., and McElhinny, M.W., 1988, The combined analysis of remagnetization circles and direct observations in palaeomagnetism: Earth and Planetary Science Letters, v. 87, p. 161–172, doi: 10.1016/0012-821X(88)90072-6.

Petit, C., Fournier, M., and Gunnell, Y., 2007, Tectonic and climatic controls on rift escarpments: Erosion and flexural rebound of the Dhofar passive margin (Gulf of Aden, Oman): Journal of Geophysical Research, v. 112, B03406, doi: 10.1029/2006JB004554.

Pik, R., Marty, B., Carignan, J., and Lave, J., 2003, Stability of the Upper Nile drainage network (Ethiopia) deduced from (U-Th)/He thermochronometry: Implications for uplift and erosion of the Afar plume dome: Earth and Planetary Science Letters, v. 215, p. 73–88, doi: 10.1016/S0012-821X(03)00457-6.

Quade, J., Levin, N., Semaw, S., Stout, D., Renne, P., Rogers, M., and Simpson, S., 2004, Paleoenvironments of the earliest stone toolmakers, Gona, Ethiopia: Geological Society of America Bulletin, v. 116, p. 1529–1544, doi: 10.1130/B25358.1.

Quade, J., Levin, N.E., Simpson, S.W., Butler, R., McIntosh, W.C., Semaw, S., Kleinsasser, L., Dupont-Nivet, G., Renne, P., and Dunbar, N., 2008, this volume, The geology of Gona, Afar, Ethiopia, in Quade, J., and Wynn, J.G., eds., The Geology of Early Humans in the Horn of Africa: Geological Society of America Special Paper 446, doi: 10.1130/2008.2446(01).

Redfield, T.F., Wheeler, W.H., and Often, M., 2003, A kinematic model for the development of the Afar Depression and its paleogeographic implications: Earth and Planetary Science Letters, v. 216, p. 383–398, doi: 10.1016/S0012-821X(03)00488-6.

Reed, K.E., 2008, Paleoecological patterns at the Hadar hominin site, Afar Regional State, Ethiopia: Journal of Human Evolution, v. 54, no. 6, p. 743–768.

Renne, P.R., Walter, R.C., Verosub, K.L., Sweitzer, M., and Aronson, J.L., 1993, New data from Hadar (Ethiopia) support orbitally tuned time scale to 3.3 Ma: Geophysical Research Letters, v. 20, p. 1067–1070, doi: 10.1029/93GL00733.

Renne, P.R., Swisher, C.C., III, Deino, A.L., Karner, D.B., Owens, T.L., and DePaolo, D.J., 1998, Intercalibration of standards, absolute ages and uncertainties in $^{40}Ar/^{39}Ar$ dating: Chemical Geology, v. 145, p. 117–152, doi: 10.1016/S0009-2541(97)00159-9.

Renne, P.R., WoldeGabriel, G., Hart, W.K., Heiken, G., and White, T.D., 1999, Chronostratigraphy of the Mio-Pliocene Sagantole Formation, middle Awash Valley, Afar Rift, Ethiopia: Geological Society of America Bulletin, v. 111, p. 869–885, doi: 10.1130/0016-7606(1999)111<0869:COTMPS>2.3.CO;2.

Schmitt, T.J., and Nairn, A.E.M., 1984, Interpretations of the magnetostratigraphy of the Hadar hominid site: Ethiopia, v. 309, p. 704–706.

Sepulchre, P., Ramstein, G., Fluteau, F., Schuster, M., Tiercelin, J.J., and Brunet, M., 2006, Tectonic uplift and eastern Africa aridification: Science, v. 313, p. 1419–1423, doi: 10.1126/science.1129158.

Spiegel, C., Kohn, B.P., Belton, D.X., and Gleadow, A.J.W., 2007, Morphotectonic evolution of the central Kenya rift flanks: Implications for late Cenozoic environmental change in East Africa: Geology, v. 35, p. 427–430, doi: 10.1130/G23108A.1.

Taieb, M., Johanson, D.C., Coppens, Y., and Aronson, J.L., 1976, Geological and palaeontological background of Hadar hominid site, Afar, Ethiopia: Nature, v. 260, p. 289–293, doi: 10.1038/260289a0.

Tamrat, E., Thouveny, N., and Taieb, M., 1996, Magnetostratigraphy of the lower member of the Hadar Formation (Ethiopia): Evidence for a short normal event in the mammoth subchron: Studia Geophysica et Geodaetica, v. 40, p. 313–335, doi: 10.1007/BF02300746.

Tauxe, L., 1998, Paleomagnetic Principles and Practice: Dordrecht, Kluwer Academic Publisher, 299 p.

Tauxe, L., 2005, Inclination flattening and the geocentric axial dipole hypothesis: Earth and Planetary Science Letters, v. 233, p. 247–261, doi: 10.1016/j.epsl.2005.01.027.

Thurmond, A.K., Abdelsalam, M.G., and Thurmond, J.B., 2006, Optical-radar-DEM remote sensing data integration for geological mapping in the Afar Depression, Ethiopia: Journal of African Earth Sciences, v. 44, p. 119–134, doi: 10.1016/j.jafrearsci.2005.10.006.

Tiercelin, J.J., 1986, The Pliocene Hadar Formation, Afar Depression of Ethiopia, in Frostick, L.E., Renaut, R.W., Reid, I., and Tiercelin, J.J., eds., Sedimentation in the African Rifts: Oxford, Blackwell Scientific, p. 221–240.

Ukstins, I.A., Renne, P.R., Wolfenden, E., Baker, J., Ayalew, D., and Menzies, M., 2002, Matching conjugate volcanic rifted margins: $^{40}Ar/^{39}Ar$ chronostratigraphy of pre- and syn-rift bimodal flood volcanism in Ethiopia and Yemen: Earth and Planetary Science Letters, v. 198, p. 289–306, doi: 10.1016/S0012-821X(02)00525-3.

Valet, J.-P., Meynadier, L., and Guyodo, Y., 2005, Geomagnetic dipole strength and reversal rate over the past two million years: Nature, v. 435, p. 802–805, doi: 10.1038/nature03674.

Vandamme, D., 1994, A new method to determine paleosecular variation: Physics of the Earth and Planetary Interiors, v. 85, p. 131–142, doi: 10.1016/0031-9201(94)90012-4.

van Hoof, A.A.M., and Langereis, C.G., 1991, Reversal records in marine marls and delayed acquisition of remanent magnetization: Nature, v. 351, p. 223–224, doi: 10.1038/351223a0.

Walter, R.C., 1994, Age of Lucy and the First Family: Single-crystal $^{40}Ar/^{39}Ar$ dating of the Denen Dora and lower Kada Hadar Members of the Hadar Formation: Geology, v. 22, p. 6–10, doi: 10.1130/0091-7613(1994)022<0006:AOLATF>2.3.CO;2.

Walter, R.C., and Aronson, J.L., 1993, Age and source of the Sidi Hakoma Tuff, Hadar Formation, Ethiopia: Journal of Human Evolution, v. 25, p. 229–240, doi: 10.1006/jhev.1993.1046.

Wolfenden, E., Ebinger, C., Yirgu, G., Deino, A., and Ayalew, D., 2004, Evolution of the northern Main Ethiopian Rift: Birth of a triple junction: Earth and Planetary Science Letters, v. 224, p. 213–228, doi: 10.1016/j.epsl.2004.04.022.

Wynn, J.G., Alemseged, Z., Bobe, R., Geraads, D., Reed, D., and Roman, D.C., 2006, Geological and palaeontological context of a Pliocene juvenile hominin at Dikika, Ethiopia: Nature, v. 443, p. 332–336, doi: 10.1038/nature05048.

Wynn, J.G., Roman, D.C., Alemseged, Z., Reed, D. Geraads, D., and Munro, S., 2008, Stratigraphy, depositional environments, and basin structure of the Hadar and Busidima Formations at Dikika, Ethiopia, in Quade, J., and Wynn, J.G., eds., The Geology of Early Humans in the Horn of Africa: Geological Society of America Special Paper 446, doi: 10.1130/2008.2446(04).

Yemane, T., 1997, Stratigraphy and Sedimentology of the Hadar Formation [Ph.D. thesis]: Ames, Iowa State University, 182 p.

MANUSCRIPT ACCEPTED BY THE SOCIETY 17 JUNE 2008

Stratigraphy, depositional environments, and basin structure of the Hadar and Busidima Formations at Dikika, Ethiopia

Jonathan G. Wynn*
Diana C. Roman
Department of Geology, University of South Florida, Tampa, Florida 33620, USA

Zeresenay Alemseged
Department of Anthropology, California Academy of Sciences, San Francisco, California 94118, USA

Denné Reed
Department of Anthropology, University of Texas at Austin, Austin, Texas 78712, USA

Denis Geraads
Centre National de la Recherche Scientifique UPR 2147, 44 Rue de l'Amiral Mouchez, 75014, Paris, France

Stephen Munro
School of Archaeology and Anthropology, Australian National University, Canberra, ACT 0200, Australia

ABSTRACT

Sediments exposed in the Dikika Research Project area form a nearly continuous sequence spanning the period from older than 3.8 Ma to younger than 0.15 Ma. By developing a stratigraphic framework of sedimentary basins, we are able to reconstruct a regional geological history that illuminates environmental changes resulting from tectonic events in the Afar triple junction region. The sequence begins with the Basal Member of the Hadar Formation, which was deposited on a dissected and deeply weathered surface of the uppermost flow of Dahla Series Basalt (8–4 Ma). This contact signals an increase in sediment accumulation rate due to active extension along faults parallel to the Red Sea Rift system. Sediments of the Hadar Formation indicate the progressive infilling of the Hadar Basin and migration of the shoreline northward or northeastward toward the axial depocenter, with several brief transgressions southward. After 2.9 Ma, the Dikika Research Project area was uplifted, and the Hadar Formation was faulted and eroded on an angular unconformity. Subsequent to 2.7 Ma, sedimentation returned, although the character and position of the newly developed Busidima half-graben had changed. This basin was formed by

---

*jwynn@cas.usf.edu

Wynn, J.G., Roman, D.C., Alemseged, Z., Reed, D., Geraads, D., and Munro, S., 2008, Stratigraphy, depositional environments, and basin structure of the Hadar and Busidima Formations at Dikika, Ethiopia, *in* Quade, J., and Wynn, J.G., eds., The Geology of Early Humans in the Horn of Africa: Geological Society of America Special Paper 446, p. 87–118, doi: 10.1130/2008.2446(04). For permission to copy, contact editing@geosociety.org. ©2008 The Geological Society of America. All rights reserved.

the rotation of an asymmetric marginal half-graben around a border fault that paralleled the western escarpment of the Ethiopian Rift. The Busidima Formation deposited in this basin records the migration of the paleo–Awash River across its floodplain in response to a changing tectonic setting. These local paleoenvironmental changes are primarily the response to regional tectonics and are superimposed on the global and regional records of climate change.

Keywords: Hadar Formation, Busidima Formation, hominin evolution, paleoenvironment, Ethiopia.

INTRODUCTION

In 1999, the Dikika Research Project (DRP) began reconnaissance and survey of a little-explored but fossiliferous area in the northern Awash Valley of Ethiopia (Fig. 1), and this expanded to a multidisciplinary research project in 2002. Prior to the work of the Dikika Research Project, brief geological research in the area had been done during paleontological prospecting from the Hadar area during the 1970s (Taieb et al., 1972, 1976; Taieb and Tiercelin, 1979; Aronson and Taieb, 1981; Tiercelin, 1986) and during the Rift Valley Research Mission in Ethiopia (RVRME) during the 1970s and 1980s (Kalb et al., 1982a, 1982b; Kalb, 1993). To date, the geological research of the Dikika Research Project has been published in accounts of hominin and other paleontological reports (Alemseged and Geraads, 2000; Geraads et al., 2004; Geraads, 2005; Alemseged et al., 2005, 2006; Wynn et al., 2004, 2006).

The purpose of this paper is to present a full and updated account of the current state of geological research in the Dikika Research Project area, including but not limited to the area of interest for paleoanthropological and paleontological research. We start with current knowledge of the stratigraphy exposed in the Dikika Research Project area, mainly identified using tephrostratigraphic marker horizons mapped throughout the region. From this data, we interpret the regional record of sedimentary basin geometry, basin formation processes, and spatial patterns of paleoenvironments, elucidating new aspects of the tectonic history of this part of the Ethiopian Rift, made famous for its fossil hominins.

TECTONIC HISTORY OF THE AFAR TRIPLE JUNCTION REGION

Currently, the Dikika Research Project area lies near the intersection of three extensional tectonic regimes (Fig. 1A), where continental rifting of the East African Rift system meets the oceanic spreading centers of the Red Sea Rift system and Gulf of Aden Rift system. Curiously, Pliocene-Pleistocene sedimentary deposits of the Dikika Research Project area and the majority of strata containing Neogene hominin sites in the Awash River Valley are presently exposed on the shoulders of the Ethiopian Rift (Fig. 1). The fossiliferous exposures are not found in the actively extending and depositing sedimentary basins that track the volcanic and rifting axis of the Wonji fault belt (i.e., Yangudi and Adda-Do Basins in the Southern Afar Rift; Fig. 1). Rather, the current axial depositional basins are ~100 km to the east of fossiliferous exposures in the Awash River Valley, and the exposures are uplifted with respect to the central rift axis and sedimentary basins. This geographical situation implies some tectonic changes since deposition of the Pliocene-Pleistocene fossiliferous strata. These changes are part of the longer-term history of the development of the rift-rift-rift (R-R-R) triple junction in the Afar region (Wolfenden et al., 2004). One of our aims in this study is to elucidate the tectonic and depositional history during the period represented by the deposits exposed in the Dikika Research Project area.

The history of R-R-R tectonism in the Afar began with the eruption of ~1.5 million km^3 of flood basalts and other felsic volcanic rocks between 33 and 25 Ma (but principally during a 1 m.y. interval centered around 30 Ma), likely in response to the surfacing of a mantle plume (Ebinger et al., 1993; Hofmann et al., 1997). This early period of volcanism also marked the onset of continental extension, rift-flank uplift, and basin sedimentation in the Red Sea Rift system and Gulf of Aden Rift system. These events can be correlated across these rift systems (Hofmann et al., 1997; Ukstins et al., 2002; Watchorn et al., 1998). The major phase of extension in the Red Sea Rift system began sometime later, between 25 and 20 Ma (Menzies et al., 1997). Crustal extension, rift-flank uplift, and basin sedimentation did not reach the Ethiopian segment of the East African Rift system until much later, ca. 18–15 Ma in the southern segment of the Main Ethiopian Rift, well after East African Rift system extensional tectonics had already developed further south, in the Gregory Rift of Kenya (Ebinger et al., 1993). Furthermore, rift-flank uplift, extension, and basin sedimentation did not spread to the northern segment of the Main Ethiopian Rift until later still—sometime after 11 Ma (Wolfenden et al., 2004). Thus, the R-R-R junction, frequently discussed as typical of the tectonic setting of the Afar region, could not have formed until late in its tectonic history—some 18 m.y. after initiation of the mantle plume.

Much less is known regarding the extension and sedimentation history of the northern Main Ethiopian Rift than the southern and central rift (WoldeGabriel et al., 2000), and even less is known regarding the development of the Southern Afar Rift, which contains the Dikika Research Project area. Early stages of volcanism are evident in the central Main Ethiopian Rift from the

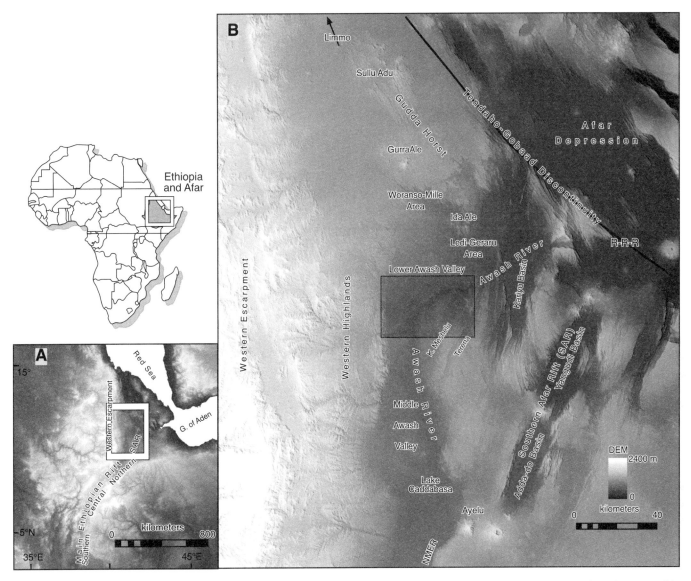

Figure 1. Location map and digital elevation model (DEM) of the Dikika Research Project area, showing major tectonic and topographic features of the Afar rift-rift-rift (R-R-R) junction region. (A) Afar triangle region, where the Ethiopian Rift system meets the Red Sea and Gulf of Aden Rift systems. Rectangle shows area detailed in B. (B) Detail of the Awash River Valley between the western escarpment and the active depositional basins of the Southern Afar Rift (SAR). Shaded rectangle shows area mapped in Figure 3. NMER—Northern Main Ethiopian Rift.

late Oligocene to early Miocene, but the central Main Ethiopian Rift flanks did not begin to uplift until ca. 10 Ma (WoldeGabriel et al., 1990), coincident with sedimentation in the central Main Ethiopian Rift (Wolfenden et al., 2004). Early rift-flank uplift and sedimentation on the southeastern flank of the northern Main Ethiopian Rift is evident at 11–10 Ma, marked by the fossil-bearing strata of the Chorora Formation (Geraads et al., 2002). A synthesis of recent geochronological data from hominid sites of the Awash Valley suggests larger-scale initiation of sedimentation on the western flank of the northern Main Ethiopian Rift somewhat later at ca. 8 Ma (White et al., 1993; WoldeGabriel et al., 1994, 2001; White et al., 2006).

During the period from 8 to 4 Ma, minimal sedimentary basin development occurred throughout much of the Southern Afar Rift, although thin patches of sediment attributed to the Adu Asa and Sagantole Formations are interspersed with basalt in the western margin at Gona (Quade et al., this volume). Volcanic deposits of this episode include the pre-Stratoid volcanics such as the Dahla Series Basalts (8–4 Ma; Audin et al., 2004). The Dahla Series is widespread in the eastern Afar Depression, but it is also exposed in the northwest-southeast–trending Gudda horst (Lahitte et al., 2003a, 2003b; Audin et al., 2004; Haile-Selassie et al., 2007), below the Afar Stratoid Series (<3.5 Ma; Kidane et al., 2003), and in the uplifted plateaus east of the

Dikika Research Project area (Tefera et al., 1996). These pre-stratoid basalts either correlate to, or interfinger with roughly coeval basalts, interspersed with the thin sediment wedges of the Adu-Asa and Sagantole Formations in the western margin of the Awash River Valley at Gona (Quade et al., this volume; Kleinsasser et al., this volume). The accommodation space providing for accumulation of the Adu-Asa and Sagantole Formations must indicate some minimal development of rift-flank uplift and sedimentation between 8 and 4 Ma, but it appears that this was mostly limited to the western escarpment region.

The period between 4 and 3 Ma was marked by major reorganization of the R-R-R junction and changes in tectonic patterns throughout the Afar region. Patterns of well-dated faults in the Adama Basin indicate a shift in extension direction during this interval in the northern Main Ethiopian Rift, from ~130/310° to the modern extension direction of ~105/285° (Wolfenden et al., 2004). This shift is consistent with global plate kinematic models for this period (Calais et al., 2003) and is roughly coincident with the initiation of seafloor spreading in the Red Sea Rift system, the eruption of the Afar Stratoid Series flood basalts (mostly <3.5 Ma; Kidane et al., 2003), and a predicted increase in extension rates in the Afar between 4 and 3 Ma from plate kinematics (Eagles et al., 2002). These tectonic events were likely the regional response to larger-scale tectonic patterns such as the propagation (sensu Hey, 1977) of the Red Sea and Gulf of Aden Rift systems into the Afar and the readjustment of plate boundaries (Audin et al., 2004; Lahitte et al., 2003a, 2003b; Manighetti et al., 1997).

Currently, the Ethiopian Rift system is separated from the distinctly different Red Sea Rift system and Gulf of Aden Rift system by an incipient plate boundary called the Tendaho-Goba'ad Discontinuity (Fig. 1). This left-lateral, oblique-slip fault zone marks the boundary between the continental extension of the Ethiopian Rift system and the marine extension of the Red Sea and Gulf of Aden Rift systems. Thus, the R-R-R currently lies along this incipient plate boundary (Wolfenden et al., 2004). The present extension direction of the southern Main Ethiopian Rift is defined by the motions of the Nubian and Somali plates and has a mean extensional azimuth of 128/308° ± 20° (Acocella and Korme, 2002; Bilham et al., 1999). It is clear that the current kinematic setting of the Afar has not been static over the past 30 m.y. Rather, the tectonic setting has evolved during the formation and evolution of the R-R-R triple junction such that plate boundaries and junctions, as well as extension directions, have migrated, primarily as a generalized response to rift propagation of the Red Sea and Gulf of Aden Rift systems into the Afar Depression (Courtillot et al., 1984; Manighetti et al., 1997). This history of rift propagation must leave a record in the patterns of sedimentary basins, which are formed due to isostatic forces and stretching of the continental lithosphere (McKenzie, 1978).

These and other major events in the tectonic history of the Afar region were unfolding during the time of deposition of sediments now exposed in the Dikika Research Project area, and it is our hypothesis that these events are recorded in sedimentation patterns of the Hadar and Busidima Formations. Our analysis of sedimentation patterns in the Dikika Research Project area is thus intended to promote interpretation of these events in light of the overall sedimentation and basin formation pattern in this part of the Afar R-R-R junction region.

STRATIGRAPHIC FRAMEWORK OF THE DIKIKA RESEARCH PROJECT AREA

Figure 2 summarizes the current stratigraphic definition of the Awash Group strata, which includes the Hadar and Busidima Formations and key marker beds exposed in the Dikika Research Project area. The formally defined stratigraphy of the Hadar Formation has evolved somewhat since the original stratotype definition (Taieb et al., 1972). The formation did not have definitive upper and lower bounds as late as the time of the discovery and publication of "Lucy," which gave the formation its notoriety (Johanson and Taieb, 1976; specimen A.L. 288-1; *Australopithecus afarensis*, 3.1 Ma). Ensuing geological work since the 1970s in the Hadar, Gona, and surrounding hominid paleontological areas has expanded the definition and modified the boundaries of the Hadar Formation. The initial definition of Taieb et al. (1972) defined only 80 m of deposits belonging to a more encompassing "Central Afar Group," which included strata now simply attributed to the Hadar Formation. A more complete definition of the Hadar Formation was provided by Taieb et al. (1976), which included >140 m of deposits but still lacked well-defined upper and lower boundaries. This definition continued through the work of the RVRME (Kalb et al., 1982a, 1982b), and later work of the Hadar Project (e.g., Kimbel et al., 1996). The Busidima Formation (stratotype defined by Quade et al., 2004; modified by Quade et al., this volume) absorbed some of the "upper" Hadar Formation and provided a definitive boundary with unconformable deposits overlying the central Hadar exposures, giving the Hadar Formation a definitive upper boundary (Fig. 2).

Radiometric age data for the Hadar Formation cited here refer to the work of Campisano (2007), which synthesizes new geochronological results with prior chronostratigraphic definitions (Aronson et al., 1977; Aronson and Taieb, 1981; Brown and Cerling, 1982; Kimbel et al., 1996; Walter, 1981, 1994; Walter and Aronson, 1982, 1993; Walter et al., 1996), recalibrated to modern standards. Other sources of dates are cited individually throughout. For the Hadar Formation, we mainly follow the stratigraphic nomenclature of Taieb et al. (1976), which divided the formation into four members: (1) Basal (below the Sidi Hakoma Tuff [SHT], 3.42 ± 0.03 Ma), (2) Sidi Hakoma (between the Sidi Hakoma Tuff and Triple Tuffs [TT]; we use TT-4 as the most distinctive, radiometrically dated and widespread marker, TT-4 = 3.256 ± 0.018 Ma), (3) Denen Dora (between TT and Kada Hadar Tuff [KHT] = 3.20 ± 0.01 Ma), and (4) Kada Hadar (above Kada Hadar Tuff). The Busidima Formation is defined by the stratotype of Quade et al. (2004; refined by Quade et al. this volume), but it is currently not subdivided at the member level.

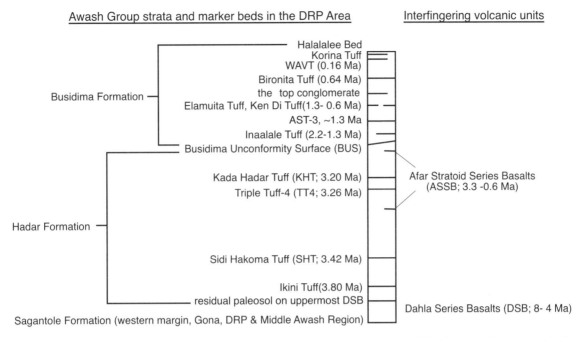

Figure 2. Stratigraphic summary of Awash Group strata exposed in the Dikika Research Project area. Left: marker beds of the Hadar and Busidima Formations. Right: volcanic units that interfinger with the Awash Group. WAVT—Waidedo Vitric Tuff.

Upper Boundary of the Hadar Formation

Despite the longstanding Hadar Formation stratotype and refinements to its chronostratigraphy, critical stratigraphic details of the upper and lower boundaries of the formation were missing from early publications that focused on documenting hominids. Thus, the upper and lower boundaries of the Hadar Formation remained unclear through the work of the RVRME (Kalb, 1993; Kalb et al., 1982a, 1982b) and later work of the Hadar Research Project (Aronson et al., 1977; Aronson and Taieb, 1981; Kimbel et al., 1996). Later work at Hadar began to identify the unconformable nature of what was then called the "upper" Kada Hadar Member (Vondra et al., 1996) and the associated stratigraphic problems of the upper part of the Hadar Formation. More recent work (Quade et al., 2004; Wynn et al., 2006) acknowledges a widespread and mappable disconformity or unconformity between the Hadar and Busidima Formations. This, and other recent observations, has more clearly defined the upper and lower boundaries of the Hadar Formation stratotype.

The unconformity between the formations separates the strata into two distinct depositional units that were deposited in distinctly different sedimentological, tectonic, and paleoenvironmental circumstances. This boundary has been described as a disconformity and has been called the Major Disconformity Surface (MDS), previously recognized based on a chronostratigraphic gap above a prominent tuff complex, referred to as the BKT-2 complex (2.94–2.96 Ma; Kimbel et al., 1996). However, within the Dikika Research Project area, the BKT-2 complex is not extensively exposed and is not everywhere below the unconformity, as is apparently the case at Hadar and Ledi-Geraru (DiMaggio et al., this volume). Furthermore, it is evident from mapping of the strata in the Dikika Research Project area that the boundary between the Hadar and Busidima Formations is the first of a series of unconformable surfaces, and it is observed to be an angular unconformity that cuts through different levels of the Hadar Formation (Figs. 3 and 4I). This surface is gently inclined and dips westward from outcrops at ~600 m elevation at Ikini to 560 m at Kada Gona and Degewin to 525 m at the lowest exposed surface at Busidima (Fig. 3). Strata of the Busidima Formation between Inaalale and the Awash River are uniformly tilted to the west by a similar dip of ~1° (measured with a laser theodolite on the surface of the AST-3 and confirmed by mapping of marker tuff beds; Fig. 3). Although the Busidima Formation is relatively undeformed and has a generally uniform dip, the Hadar Formation beneath the unconformity is variably oriented depending on the individual fault block. Although dip angles are very shallow, the most tilted block (exposures of the Sidi Hakoma Tuff in upper Arbosh) dips to the east by about ~2° (Fig. 3). Given the evidence of differences in outcrop orientation and differences in the degrees of faulting prior to erosion, we refer to this first unconformable surface above the BKT-2 Tuff complex as the Busidima unconformity surface (BUS), rather than describing it as a disconformity (Figs. 5–10). The Busidima unconformity surface truncates the Hadar Formation at the level of the BKT-2 in the Kada Gona area (Kimbel et al., 1996) but cuts much deeper in the Hadar Formation near Gor Gore, nearly to the level of the Sidi Hakoma Tuff (Figs. 3, 7–8). At Busidima, much younger sediments overlie the Busidima

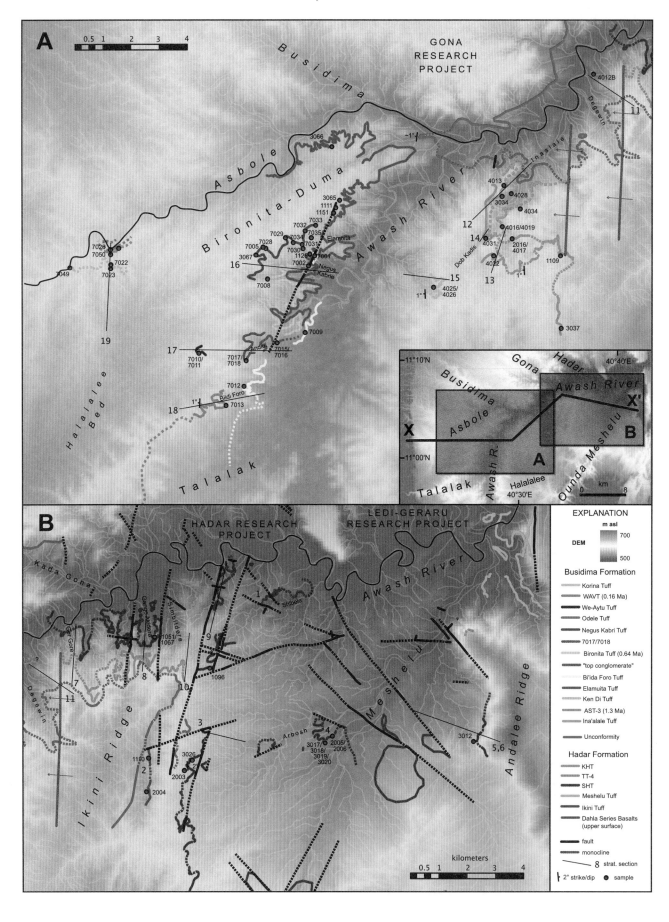

unconformity surface without the presence of Hadar Formation marker beds such as the BKT and AST (Bouroukie Tuffs and Artifact Site Tuffs).

Much of the observed lateral difference in the level at which the Busidima unconformity surface cuts into the Hadar Formation is due to structural deformation of the Hadar Formation along the series of N-NE–trending faults that formed prior to the onset of deposition of the Busidima Formation (at least 30 m vertical offset on some faults; Fig. 10). The Hadar Formation is mostly exposed in the eastern Dikika Research Project area, where the orientation and age of exposures are controlled by the intricacies of the intersection of several fault trends (Fig. 3B). Meanwhile, the younger Busidima Formation is mostly exposed in the western Dikika Research Project area, is relatively little faulted, and the age of exposures is consistently younger toward the As Duma border fault at the western margin of Awash River Valley (Fig. 3A).

Lower Boundary of the Hadar Formation

The lower boundary of the Hadar Formation was also undefined in previous work at Hadar, where the Basal Member was described as all strata below the Sidi Hakoma Tuff in a section at Ounda Leita (Taieb et al., 1976). This previous type section of the Basal Member began simply at the base of the observed exposure, leaving an open question about the definition of the lower contact. Jon Kalb and coworkers noted that the Hadar Formation was underlain by an undefined basalt in early work (Kalb et al., 1982b), but he rescinded this observation in later refinements (Kalb, 1993). We confirmed their original observation in Unduble, where a sequence of at least two deeply weathered paleosol surfaces was developed on two columnar basalt flows that directly underlie the sediments at the base of the Basal Member (Wynn et al., 2006; Figs. 3, 4B, 4D, and 6). This contact provides a well-defined basal boundary for the Basal Member of the Hadar Formation and completes the stratotype definition.

For several reasons, we attribute these basalts at the base of the section at Unduble (Fig. 6) to the Dahla Series (8–4 Ma), rather than the (Lower) Afar Stratoid Series (3.3–0.6 Ma; ages from Audin et al., 2004; Kidane et al., 2003). The Afar Stratoid Series includes the Kada Damomou Basalt (KMB, 3.3 Ma) and other flows that interfinger with the Sidi Hakoma Member at Hadar (Aronson et al., 1977) and elsewhere in the Dikika Research Project area (Fig. 4C). The basalts at Unduble clearly underlie the Hadar Formation, and the Hadar Formation contains tephras such as the Ikini (correlated to the Wargolo Tuff in the Turkana Basin and Gulf of Aden, 3.8 Ma) and Sidi Hakoma Tuff (correlated to the Tulu Bor Tuff in the Turkana Basin and Gulf of Aden, 3.4 Ma). These tephras predate the Afar Stratoid Series but postdate the Dahla Series. Furthermore, the basalts at Unduble were weathered during a period of in situ soil formation directly on the basalt that predates the formation of the Hadar Basin, and they required roughly 100 k.y. to form (Fig. 4D). Finally, the basalts at Unduble are deformed by faults that are syndepositional with the Hadar Formation, and that define

Figure 3. Geological map of the Dikika Research Project area. (A) The western Dikika Research Project area, where predominantly the Pliocene-Pleistocene Busidima Formation is exposed above the Busidima unconformity surface. Inset in A shows location of the two panels of the geological map, and cross-section line X–X′ shown in Figure 10. (B) The eastern Dikika Research Project area, in which the Pliocene Hadar Formation below the Busidima unconformity surface (BUS) is predominantly exposed. WAVT—Waideedo Vitric Tuff; KHT—Kada Hadar Tuff; SHT—Sidi Hadoma Tuff; TT-4—Triple Tuff-4.

Figure 4 (on following two pages). Depositional features of the Hadar Formation. (A) Growth fault in the Basal Member of the Hadar Formation. Dark-colored mudstone between two light-colored altered tephra beds thickens (left to right ~SW to NE) across a series of small-scale normal faults. (B) Exposure of a columnar flow unit of the Dahla Series Basalt, below the Hadar Formation. Paleosol in D is below this flow. Similar C horizon of a second paleosol is developed on the upper surface of this flow. (C) Interfingering basalt flows of the Afar Stratoid Series Basalt (ASSB) between laminated clays of the lower Hadar Formation (HF). These basalts show hyaloclastic features such as pillow structures and amygdaloidal and granular, slag-like textures. (D) Paleosol formed on flows of Dahla Series Basalts. C horizon shows spheroidally weathering basalt. (E) Gango Akidora amphitheater section of upper Denen Dora and Kada Hadar Member sediments of the Hadar Formation. At base of photo, planar cross-bedded fine-grained sands overlie the ostracod-bearing clays of the Denen Dora lacustrine sequence. Several lenticular, trough cross-bedded channels are evident in upper part of photo (one outlined). A lenticular channel of the Kada Hadar Tuff (KHT) is near the top of exposure, followed by the confetti clay (cc). (F) Typical Vertisol in the Hadar Formation, with angular, blocky ped structure, disseminated discrete carbonate nodules, numerous intersecting slickensided ped surfaces, and sand-filled crack network extending from overlying unit. (G) Photomicrograph of transverse cross section through calcified plant stem from the Basal Member of the Hadar Formation showing ribbed outer surface, calcified cellular structure, and aeration cavities (large sparry calcite filled, or light-colored circles) typical of submerged aquatic plants such as reeds, sedges, cattails, etc. Organic matter stains the outer surfaces black. (H) Photomicrograph of transverse cross section through calcified plant stem from the Busidima Formation showing (in order of exterior to interior): outer layer of calcite-cemented sand grains, micritic calcified outer surface, network of packed adventitious roots (light circles), and central cavity. Adventitious root pores and central cavity are lined with brightly colored sparry calcite; remainder of calcified root is micritic. (I) Busidima unconformity surface in upper Gor Gore. Paleochannel of Kada Hadar Tuff (Hadar Formation) dips to east. Overlying Busidima Formation is nearly horizontal, although the formation in general dips ~1° toward the As Duma border fault to the west. (J) Exposure of the Halalalee Bed paleosol on horizontal plateau between Asbole and Talalak. Dark brown calcareous paleosol is exposed over several tens of square kilometers. (K) Thenardite (Th) weathering out from sediments above the "top conglomerate" in exposures at Asbole. The loosely consolidated sequence of cumulative paleosols with thin granular surface horizons overlies the well-cemented "top conglomerate." (L) Detail of fine calcareous rhizoliths in the Bk horizon of the Halalalee Bed paleosol (see J).

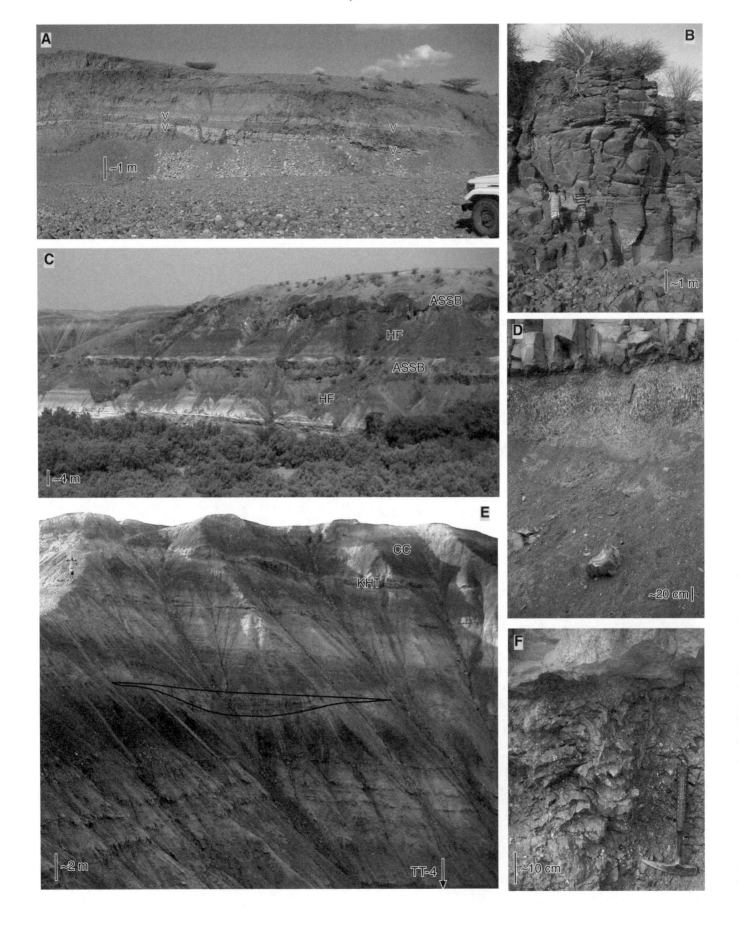

the structure and boundaries of the Hadar sedimentary basin (Fig. 10B). Meanwhile, the Dahla Series basalts are not down-dropped below the surface in areas to the northeast of a series of northwest-southeast–trending faults (Fig. 3), and the Hadar Formation thickens dramatically across this series of faults. To the north-northeast of the Dikika Research Project area, in the Woranso-Mille region, Haile-Selassie et al. (2007) similarly found a basement of basalt that underlies hominin-bearing sediments attributable to the Basal Member of the Hadar Formation (with a preliminary age of 3.5–3.8 Ma; Haile-Selassie et al., 2007). The stratigraphic relationship between the Basal Member exposed in the Dikika Research Project area and sediments attributed to the Sagantole Formation (>5.2–3.9 Ma; Quade et al., this volume) at the western margin of the Gona area remains unclear. However, ongoing geochronological work on the Dahla Series Basalt at Unduble may eventually result in the correlation of strata between widely separated exposures from the Dikika Research Project area to those of the Sagantole Formation in the western margin at Gona, and/or the lowermost Hadar Formation at Woranso-Mille (Haile-Selassie et al., 2007).

These observations establish the complete stratotype definition of the Hadar Formation, which includes all sediments deposited in the Hadar Basin. Thus, the type section of the Hadar Formation extends from the upper surface of the uppermost Dahla Series Basalts exposed in Unduble to the first unconformable surface above the BKT-2 complex (Busidima unconformity surface) that marks the boundary to the overlying Busidima Formation (Fig. 2).

Upper Boundary of the Busidima Formation

Similar to the Hadar Formation, the Busidima Formation, as it was defined in the initial publication, did not have a well-defined upper boundary (Quade et al., 2004). With continued work in the Dikika Research Project and Gona areas, this volume now contains a complete stratotype definition of the Busidima Formation (see also Quade et al., this volume). A continuous sequence of sediments is observed from the Busidima unconformity surface to a widespread paleosol that forms plateaus throughout the Gona, Dikika Research Project, and Middle Awash regions, originally mapped by Kalb et al. (1982b) as the "Halalalee Bed." This plateau surface is obvious on topographic maps, aerial and satellite imagery, and is horizontal (as opposed to the slight ~1° west dip of the underlying strata of the Busidima Formation). The ~1-m-thick bed is a paleosol that consists of dark brown illuvial clays with fine interlaced rhizocretions (sensu Klappa, 1980) overlain by a dark brown granular surface horizon (Figs. 4J and 4L). These observations similarly complete the stratotype definition of the Busidima Formation (Fig. 2), which includes all sediments between the regional unconformity (Busidima unconformity surface) exposed at Gona, Dikika, and Hadar up to the Halalalee Bed (Halalalee Bed as defined by Kalb et al., 1982b). These lithostratigraphic definitions will allow the Hadar and Busidima Formations to be mapped and extended into the Middle Awash area, where the lithostratigraphy remains informal and preliminary, despite very precise chronostratigraphy (White et al., 1993; Clark et al., 1994, 2003; White et al., 2006).

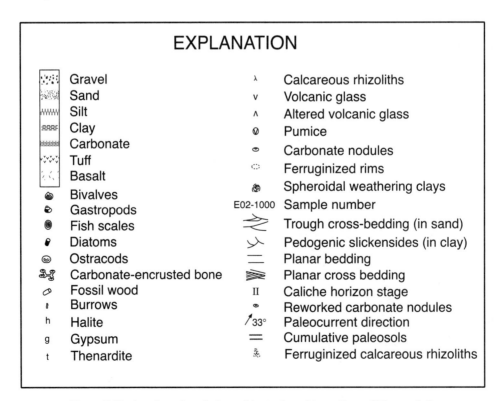

Figure 5. Explanation of symbols used in stratigraphic sections of Figures 6–9.

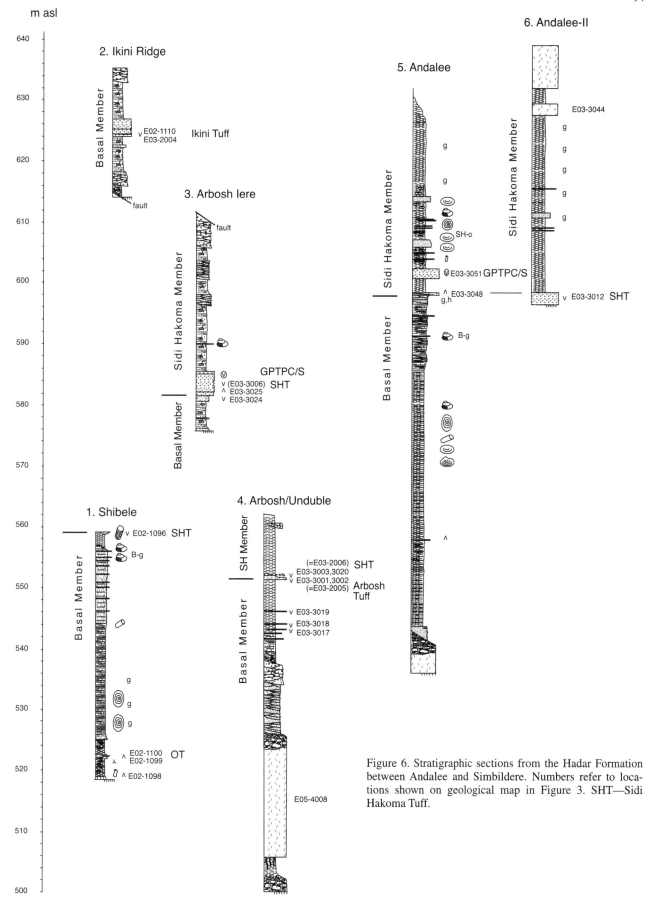

Figure 6. Stratigraphic sections from the Hadar Formation between Andalee and Simbildere. Numbers refer to locations shown on geological map in Figure 3. SHT—Sidi Hakoma Tuff.

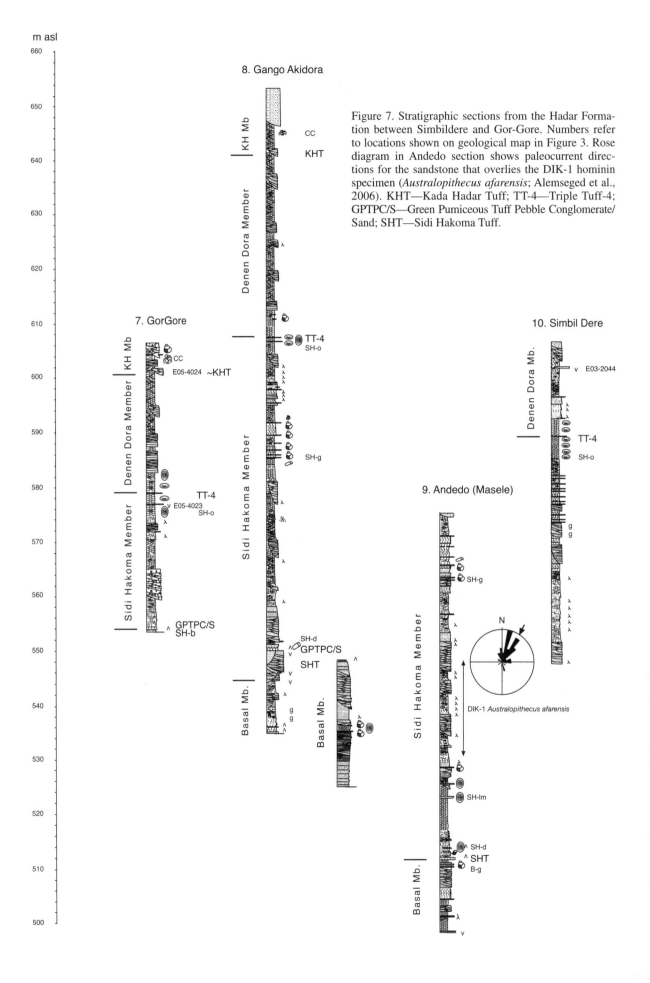

Figure 7. Stratigraphic sections from the Hadar Formation between Simbildere and Gor-Gore. Numbers refer to locations shown on geological map in Figure 3. Rose diagram in Andedo section shows paleocurrent directions for the sandstone that overlies the DIK-1 hominin specimen (*Australopithecus afarensis*; Alemseged et al., 2006). KHT—Kada Hadar Tuff; TT-4—Triple Tuff-4; GPTPC/S—Green Pumiceous Tuff Pebble Conglomerate/Sand; SHT—Sidi Hakoma Tuff.

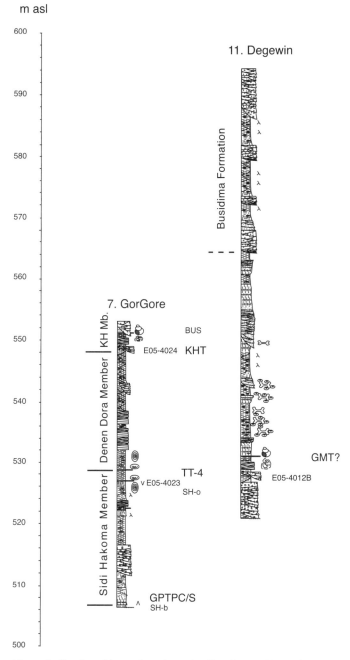

Figure 8. Stratigraphic sections from the Gor-Gore to Degewin areas. Numbers refer to locations shown on geological map in Figure 3. KHT—Kada Hadar Tuff; TT-4—Triple Tuff-4; GPTPC/S—Green Pumiceous Tuff Pebble Conglomerate/Sand; SHT—Sidi Hakoma Tuff.

TEPHROSTRATIGRAPHY

Tephra beds within the Hadar and Busidima Formations have proven to be useful isochronous markers that can establish correlation within regions of poor and discontinuous outcrop (Fig. 3), between research project areas and between sedimentary basins (i.e., Brown and Cerling, 1982), and indeed as distant as between the East African Rift and marine basins such as the Gulf of Aden (Brown et al., 1992). Chemical analyses of tephra units within the Dikika Research Project area analyzed by Diana C. Roman have been reported in publications of the geology (Geraads et al., 2004; Alemseged et al., 2005, Wynn et al., 2006) and are fully summarized here with new analyses up to 2007 (Tables 1 and 2). Further details of analytical methods and correlations are provided by Roman et al. (this volume), who also provide a more comprehensive description of all tephrostratigraphy from the Hadar, Gona, Ledi-Geraru, and Dikika projects in the Lower Awash Valley.

Hadar Formation

In the eastern Dikika Research Project area, we have identified and mapped three widespread tephras already known and documented from the type area of the central Hadar Formation (Figs. 6–8; Sidi Hakoma Tuff, one of the Triple Tuffs, specifically TT-4, and the Kada Hadar Tuff). The Sidi Hakoma Tuff is widespread, both locally and regionally, and is ubiquitous throughout any exposure of the Hadar Formation in the entire Dikika Research Project area. Its widespread exposure makes this an excellent marker of the presence or absence of the Hadar Formation. If the boundary between the Basal to Sidi Hakoma Member is exposed, the Sidi Hakoma Tuff is always present in one of a variety of sedimentary facies that range from fluvial to deltaic to lacustrine (Figs. 3, 6–8). In the badlands of the Awash River surrounding the Simbildere graben, a relatively complete and continuous section of the Hadar Formation is exposed that extends from a level just below the Sidi Hakoma Tuff to above the Kada Hadar Tuff. Local correlation of the Sidi Hakoma Tuff, based on its chemical composition (Roman et al., this volume), has extended our mapping from the previously documented exposures near the Awash River to other areas within the Dikika Research Project area. This effort has been crucial to correlating between several areas of exposures separated by faults of unknown offset. For example, by geochemically identifying an exposure of the Sidi Hakoma Tuff along the western slopes of Andalee Ridge, we could document another, more complete section of the Hadar Formation, extending from the Dahla Series Basalts up through the Sidi Hakoma Tuff, and continuing up to the Kada Hadar Tuff (Fig. 3). Also, between these two sets of exposures (Andalee and Simbildere) on isolated hilltops in the Arbosh drainages, we identified the Sidi Hakoma Tuff exposed in a couplet with the Arbosh Tuff (described later). Finally, in the upper Meshelu and Arbosh drainages, the Sidi Hakoma Tuff is exposed in a line of gently tilted cuestas (dipping ~2° to the east) of nearly 5 km in length, in an area previously mapped as belonging the Pleistocene Weihatu Formation (Kalb et al., 1982b). In contrast, exposures of the Triple Tuff-4 and Kada Hadar Tuff generally lack unaltered glass and were recognized based on their distinctive lithostratigraphy and stratigraphic position (compared to detailed descriptions of Walter, 1994).

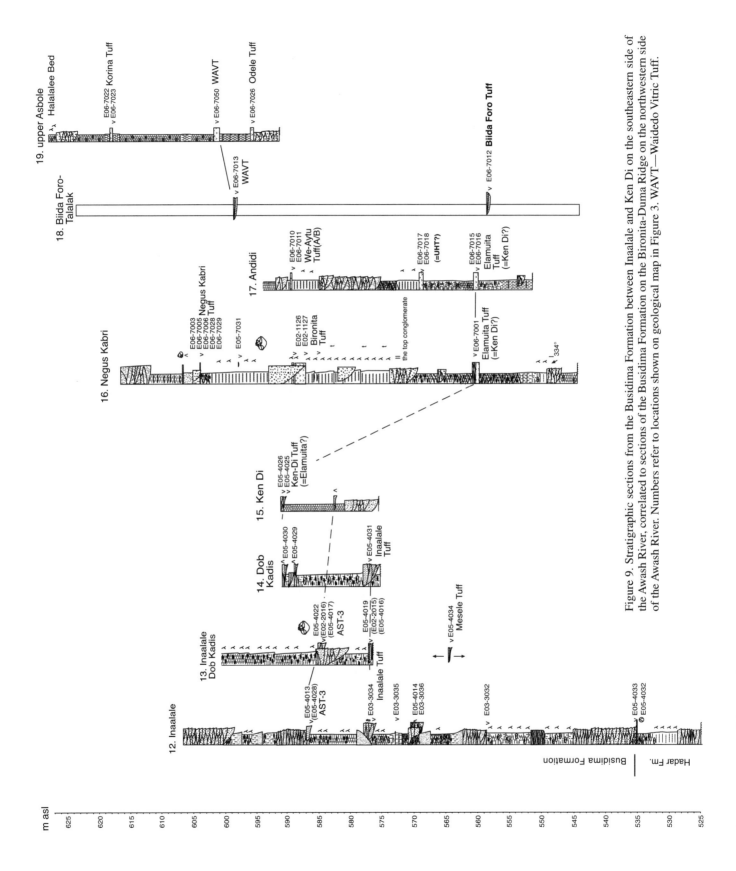

Figure 9. Stratigraphic sections from the Busidima Formation between Inaalale and Ken Di on the southeastern side of the Awash River, correlated to sections of the Busidima Formation on the Bironita-Duma Ridge on the northwestern side of the Awash River. Numbers refer to locations shown on geological map in Figure 3. WAVT—Waidedo Vitric Tuff.

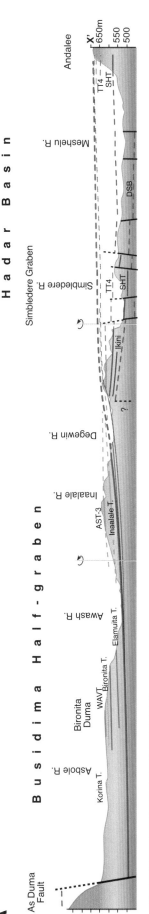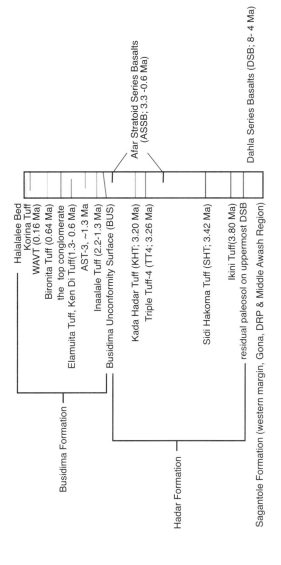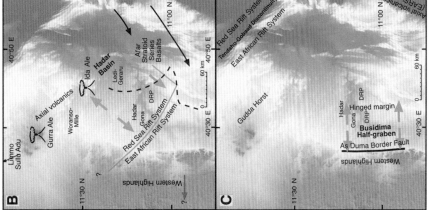

Figure 10. Cross section of the Hadar and Busidima Formation strata between the western highlands and Andalee Ridge (A), and tectonic and sedimentary basin map for the Hadar (B) and Busidima (C) Formations. Opposing arrows show extension directions. DRP—Dikika Research Project; WAVT—Waidedo Vitric Tuff.

TABLE 1. MAJOR-ELEMENT GEOCHEMICAL DATA (WT% OXIDE, EPMA) FOR TEPHRA GLASS SHARDS OF THE BUSIDIMA AND HADAR FORMATIONS[§]

Tuff Name	Derivation	Reference	Sample	Beam conditions	SiO_2	TiO_2	Al_2O_3	Fe_2O_3	MnO	MgO	CaO	Na_2O	P_2O_5	K_2O	Total	OT	Age (Ma)
Busidima Formation																	
AST-3	Artefact Site Tuff	Locals	E02-1109	10, 5, 5–10	73.58	0.22	14.37	2.97	n/a	0.08	0.83	3.88	n/a	4.08	100.00	94.38	1.3
AST-3	Artefact Site Tuff	Locals	E03-3037	12, 5, 10	73.69	0.27	13.30	3.48	0.11	0.04	1.12	3.70	n/a	4.29	100.00	95.19	1.3
AST-3	Artefact Site Tuff	Locals	E04-4041	12, 5, 5	76.43	0.15	13.12	2.44	0.33	0.00	0.75	2.50	0.02	4.27	100.00	94.52	1.3
AST-3	Artefact Site Tuff	Locals	E04-4017	12, 5, 5	74.70	0.20	13.19	2.85	0.31	0.05	0.57	4.37	0.03	3.72	100.00	95.5	1.3
AST-3	Artefact Site Tuff	Locals	E04-4028	12, 5, 5	73.24	0.16	13.57	3.15	0.35	0.07	0.75	4.72	0.03	3.96	100.00	95.92	1.3
AST-3	Artefact Site Tuff	Locals	E04-4022	12, 5, 5	73.54	0.21	13.58	2.93	0.32	0.06	0.69	4.58	0.06	4.02	100.00	95.45	1.3
AST-3	Artefact Site Tuff	Locals	E04-4013	12, 5, 5	74.30	0.15	13.18	3.07	0.31	0.05	0.62	4.51	0.03	3.77	100.00	95.27	1.3
Bi'idi Foro	Stream	Locals	E06-7009	10, 5, 5	75.46	0.38	11.15	3.29	n/a	0.01	0.24	4.74	0.04	4.69	100.00	91.43	
Bi'idi Foro	Stream	Locals	E06-7012	10, 5, 5	76.13	0.23	11.41	2.48	n/a	0.04	0.25	4.75	0.03	4.68	100.00	89.70	
–Bironita	Ridge	Locals	E03-3066	12, 5, 10	73.20	0.30	12.51	6.39	0.26	0.03	0.99	3.33	n/a	2.97	100.00	91.64	0.64
Bironita	Ridge	Locals	E02-1111*	10, 5, 5–10	72.64	0.38	12.72	6.45	n/a	0.02	0.62	3.88	n/a	3.29	100.00	93.02	0.64
Bironita	Ridge	Locals	E02-1127*	10, 5, 5–10	71.93	0.36	12.54	6.37	n/a	0.02	0.59	4.39	n/a	3.80	100.00	93.94	0.64
Bironita	Ridge	Locals	E02-1151*	10, 5, 5–10	74.08	0.38	12.91	5.98	n/a	0.02	0.58	3.07	n/a	2.98	100.00	91.64	0.64
Bironita	Ridge	Locals	E02-1126*	10, 5, 5–10	71.75	0.45	12.05	6.43	n/a	0.02	0.57	5.10	n/a	3.63	100.00	92.64	0.64
Bironita	Ridge	Locals	E03-3065	12, 5, 10	72.97	0.39	12.58	6.50	0.21	0.02	0.64	3.56	n/a	3.13	100.00	91.3	0.64
Bironita	Ridge	Locals	E06-7002	10, 5, 5	73.61	0.40	12.28	7.21	n/a	0.01	0.67	2.75	0.00	3.07	100.00	89.88	0.64
Bironita	Ridge	Locals	E06-7008	10, 5, 5	72.05	0.36	12.19	6.62	n/a	0.03	0.58	4.59	0.03	3.54	100.00	90.76	0.64
Unnamed	Sample	Locals	E06-7017	10, 5, 5	72.25	0.31	14.60	3.42	n/a	0.10	0.74	4.46	0.01	4.12	100.00	89.45	
Elamuita	Stream	Locals	E06-7001	10, 5, 5	75.00	0.32	10.41	7.87	n/a	0.08	0.40	3.03	0.03	2.86	100.00	86.30	
Elamuita	Stream	Locals	E06-7015	10, 5, 5	74.25	0.42	10.26	7.87	n/a	0.03	0.41	4.18	0.03	2.54	100.00	86.24	
Elamuita	Stream	Locals	E06-7016	10, 5, 5	76.47	0.39	10.16	7.73	n/a	0.06	0.39	2.58	0.06	2.16	100.00	87.25	
Inaalale	Stream	Locals	E02-2015	12, 5, 10	75.08	0.32	13.32	4.10	0.17	0.08	0.31	3.16	n/a	3.45	100.00	90.68	
Inaalale	Stream	Locals	E03-3034	12, 5, 10	75.85	0.26	12.91	3.81	0.13	0.08	0.27	3.25	n/a	3.44	100.00	92.95	
Inaalale	Stream	Locals	E04-4031	12, 5, 5	75.20	0.23	12.61	4.10	0.41	0.08	0.28	3.69	0.03	3.38	100.00	94.09	
Ken Di (mode 1)	Region	Locals	E04-4025, -4026	12, 5, 5	76.44	0.35	9.81	7.81	0.54	0.02	0.28	2.50	n/a	2.22	100.00	92.71	
(mode 2)					71.66	0.48	13.29	7.39	0.55	0.07	0.79	2.96	n/a	2.73	100.00	92.60	
Korina	Region	Topo	E06-7022	10, 5, 5	73.40	0.21	13.26	2.76	n/a	0.02	0.92	5.02	0.05	4.37	100.00	92.33	<0.15
Korina	Region	Topo	E06-7023	10, 5, 5	73.93	0.22	13.47	2.27	n/a	0.03	0.82	4.98	0.04	4.24	100.00	89.11	<0.15
Korina	Region	Topo	E06-7049	10, 5, 5	73.80	0.19	13.39	2.29	n/a	0.02	0.86	5.05	0.04	4.39	100.00	92.36	<0.15
Mesele Stream (= alt. Inaalale)		Topo	E04-4034	12, 5, 5	78.16	0.29	9.83	5.77	0.44	0.10	0.36	2.47	0.04	2.55	100.00	93.05	
Negus Kabri	Stream	Locals	E03-3067	12, 5, 10	73.20	0.25	13.92	2.98	0.12	0.04	1.16	4.47	n/a	3.86	100.00	95.45	
Negus Kabri	Stream	Locals	E06-7003	10, 5, 5	72.99	0.10	13.98	2.72	n/a	0.01	1.20	4.73	0.01	4.26	100.00	94.00	
Odele	Region	Topo	E06-7026	10, 5, 5	71.48	0.25	13.86	3.24	n/a	0.02	0.60	5.60	0.02	4.93	100.00	93.16	
WAVT	Waidedo Vitric Tuff	Locals	E06-7050	10, 5, 5	74.33	0.32	9.75	5.35	n/a	0.04	0.28	5.20	0.00	4.72	100.00	88.91	0.16
WAVT	Waidedo Vitric Tuff	Topo	E06-7013	10, 5, 5	75.96	0.36	9.62	6.15	n/a	0.03	0.28	3.90	0.04	3.67	100.00	88.48	0.16
WeAytu-A (mode 1)	Region	Topo	E06-7010	10, 5, 5	72.95	0.10	13.91	3.50	n/a	0.01	0.72	5.78	0.00	4.87	100.00	94.04	
(mode 2)					71.11	0.10	13.91	3.50		0.01	0.72	5.78	0.00	4.87	100.00		
WeAytu-B	Region	Topo	E06-7011	10, 5, 5	77.44	0.25	11.17	2.50	n/a	0.06	0.26	4.30	0.01	4.00	100.00	88.09	
E06-7017	Sample		E06-7017	10, 5, 5	72.25	0.31	14.60	3.42	n/a	0.10	0.74	4.46	0.01	4.12	100.00	89.45	

(*continued*)

TABLE 1. MAJOR-ELEMENT GEOCHEMICAL DATA (WT% OXIDE, EPMA) FOR TEPHRA GLASS SHARDS OF THE BUSIDIMA AND HADAR FORMATIONS[§] (continued)

Tuff Name	Derivation	Reference	Sample	Beam conditions	SiO_2	TiO_2	Al_2O_3	Fe_2O_3	MnO	MgO	CaO	Na_2O	P_2O_5	K_2O	Total	OT	Age (Ma)
Hadar Formation																	
Sidi Hakoma	Stream	Hadar	E02-1067[†]	10, 5, 5–10	78.30	0.16	13.69	1.64	n/a	0.06	0.31	3.60	n/a	2.23	100.00	93.39	3.42
Sidi Hakoma	Stream	Hadar	E02-1096[†]	10, 5, 5–10	78.20	0.17	12.88	1.66	n/a	0.07	0.32	4.30	n/a	2.41	100.00	91.8	3.42
Sidi Hakoma	Stream	Hadar	E02-1051[†]	10, 5, 5–10	78.00	0.15	12.89	1.58	n/a	0.07	0.31	4.30	n/a	2.68	100.00	91.42	3.42
Sidi Hakoma	Stream	Hadar	E02-2003[†]	12, 5, 10	77.71	0.13	13.23	1.67	0.07	0.07	0.31	3.33	n/a	3.49	100.00	91.58	3.42
Sidi Hakoma	Stream	Hadar	E02-2006[†]	12, 5, 10	76.60	0.17	13.12	1.60	0.07	0.07	0.31	3.79	n/a	4.28	100.00	92.48	3.42
Sidi Hakoma	Stream	Hadar	E03-3017[†]	12, 5, 10	77.20	0.14	12.92	1.52	0.06	0.06	0.26	3.84	n/a	3.99	100.00	93.99	3.42
Sidi Hakoma	Stream	Hadar	E03-3018[†]	12, 5, 10	77.68	0.18	13.24	1.76	0.06	0.06	0.31	2.95	n/a	3.76	100.00	93.39	3.42
Sidi Hakoma	Stream	Hadar	E03-3019[†]	12, 5, 10	78.60	0.13	13.37	1.55	0.08	0.06	0.27	2.98	n/a	2.96	100.00	90.67	3.42
Sidi Hakoma	Stream	Hadar	E03-3020[†]	12, 5, 10	77.95	0.23	13.19	1.51	0.06	0.06	0.29	3.18	n/a	3.55	100.00	90.34	3.42
Sidi Hakoma	Stream	Hadar	E03-3012[†]	12, 5, 10	78.42	0.22	14.19	1.52	0.07	0.10	0.41	2.70	n/a	2.37	100.00	88.76	3.42
Arbosh	Stream	Hadar	E02-2005[†]	12, 5, 10	75.00	0.14	13.66	2.09	0.11	0.00	0.67	4.37	n/a	3.97	100.00	91.49	3.42
Ikini	Ridge	Locals	E02-1110[†]	10, 5, 5–10	77.18	0.08	12.21	2.11	n/a	0.03	0.18	3.60	n/a	4.61	100.00	93.95	3.8
Ikini	Ridge	Locals	E02-2004[†]	12, 5, 10	77.45	0.11	12.31	2.45	0.06	0.02	0.16	3.57	n/a	3.85	100.00	91.99	3.8
Other																	
Mt. Ayelu		Mapped	E03-3026	12, 5, 10	72.24	0.20	13.68	3.38	0.13	0.01	0.51	4.71	n/a	5.14	100.00	96.4	
Sagantole-a	Region	Topo	E06-7036	10, 5, 5	76.73	0.37	11.31	2.20	n/a	0.10	0.16	2.88	0.02	6.24	100.00	92.67	
Sagantole-b	Region	Topo	E06-7037a	10, 5, 5	71.82	0.39	14.35	2.01	n/a	0.04	1.24	4.92	0.02	5.21	100.00	92.57	
Sagantole-a	Region	Topo	E06-7037b	10, 5, 5	76.19	0.18	11.72	2.62	n/a	0.08	0.19	2.92	0.06	6.05	100.00	91.55	

Note: n/a—not analyzed. EPMA—electron probe microanalysis.
*Geraads et al. (2004).
[†]Wynn et al. (2006).
[§]Further details are provided in Roman et al. (this volume).

TABLE 2. TRACE-ELEMENT AND RARE EARTH ELEMENT GEOCHEMICAL DATA (PPM; XRF & ICP-MS) FOR GLASS SEPARATES OF TEPHRA FROM THE BUSIDIMA AND HADAR FORMATIONS

Tuff Name	Sample	Ba	Nb	Rb	Sr	Y	Zn	Zr	F	Sc	Hf	Ta	Th	U	La	Ce	Nd	Sm	Eu	Tb	Yb	Lu
Busidima Formation																						
Negus Kabri	E03-3067	692	90.1	99	131	71	115	488	n/a	5	n/a	n/a	17	n/a	72	136	n/a	n/a	n/a	n/a	n/a	n/a
Bironita	E02-1111	950	117	80	50.8	92.2	198	713	687	n/d	16.7	8.1	12.3	3.27	81	162	77.9	15.6	3834	2.91	9.56	1.66
Bironita	E02-1127	965	121	82	55.1	94.2	197	714	706	2	16.4	8.1	12.5	3.22	80.8	163	77.8	15.4	3719	n/a	9.25	1.62
Bironita	E02-1151	875	112	77	71.2	85.4	188	695	n/d	4	16.6	8.2	12.2	2.96	77.9	156	73.8	15.4	3610	2.84	8.79	1.54
AST-3	E02-1109	798	114	89	73	83.7	120	714	790	3	18.8	8	17.6	4.62	107	218	96.2	18.8	2606	2.94	8.07	1.4
AST-3	E03-3037	783	103.2	93	159	93	138	794	n/a	6	n/a	n/a	19	n/a	89	178	n/a	n/a	n/a	n/a	n/a	n/a
Inaalale	E02-2015	205	183.7	149	12	151	191	1299	n/a	3	n/a	n/a	24	n/a	134	291	n/a	n/a	n/a	n/a	n/a	n/a
Hadar Formation																						
Sidi Hakoma	E02-1051	298	82.1	100	55.3	72.8	80	450	267	4	12.3	5.8	17.2	4.4	84.4	169	73.5	14	1282	2.2	6.84	1.13
Sidi Hakoma	E03-3017	196	81	102	44	73	102	466	n/a	n/a	11.9	n/a	18	0.5	96	182	76.7	n/a	n/a	n/a	n/a	n/a
Sidi Hakoma	E03-3012	444	81	65	105	64	76	364	n/a	n/a	8.8	n/a	17	0.8	86	166	65.0	n/a	n/a	n/a	n/a	n/a
Arbosh	E02-2005	731	85	89	71	93	123	378	n/a	n/a	10.8	n/a	13	0.7	116	221	94.4	n/a	n/a	n/a	n/a	n/a
Ikini	E02-1110	45	115	210	34.1	98.2	126	752	1251	3	20.6	8.7	29.6	6.06	69.4	147	62.6	14.8	575	2.8	10.5	1.77
Ikini	E02-2004	51	139.3	242	33	117	339	801	n/a	2	n/a	n/a	36	n/a	70	128	n/a	n/a	n/a	n/a	n/a	n/a
Other																						
Mt. Ayelu	E03-3026	787	118	113	14	98	167	890	n/a	n/a	20.4	n/a	17	0.60	106	200	86.40	n/a	n/a	n/a	n/a	n/a

Note: n/a—not analyzed, n/d—not detected, XRF—X-ray fluorescence spectrometry, ICP-MS—inductively coupled plasma–mass spectrometry. Further details are provided in Roman et al. (this volume).

We also mapped three new tephras in the Hadar Formation of the Dikika Research Project area, two of which have dated extrabasinal correlates (Fig. 3). The Ikini Tuff is exposed in discontinuous and lenticular channel bodies on the eastern slopes of Ikini Ridge (Figs. 3 and 6). The section containing the Ikini Tuff is uplifted with respect to much of the Hadar Formation exposures on at least two sets of intersecting faults (Fig. 3). Electron microprobe analyses (EMPA) major-element data on individual glass shards, combined with minor- and trace element-data by x-ray fluorescence (XRF) show that this unit correlates to the Wargolo Tuff of the Turkana Basin, Kenya, and to its correlate the VT-3 (Vitric Tuff 3) from the Middle Awash Region of Ethiopia (Haileab and Brown, 1992; the Wargolo in the Gulf of Aden has an orbitally tuned date of 3.80 ± 0.01 Ma; deMenocal and Brown, 1999). The correlation of the Ikini and VT-3 Tuffs may help to solve an uncertain relationship between the Sagantole and Hadar Formations (Haileab and Brown, 1992). Sections containing the VT-3 (= Ikini; extrabasinal correlates are signified in this paper with equalities in parentheses) in the Middle Awash had previously been attributed to the Aramis Member of the Sagantole Formation (Haileab and Brown, 1992; Kalb, 1993; Kalb et al., 1982a, 1982b), although this was based on distant lithological correlations, and the Aramis Member is now dated at ca. 4.4–4.3 Ma (Renne et al., 1999). Later revisions of the chronostratigraphy in the Middle Awash region placed sections containing VT-1 to VT-3 in the Belohdelie Member of the Sagantole Formation (previously formation "X"), and the still informally designated Formation "W" (White et al., 1993; Renne et al., 1999; although it is unclear what features differentiate formation "W" from the underlying Sagantole Formation, besides its age). Our finding of a correlate to the VT-3 in the Dikika Research Project area means that the formally defined lithostratigraphic unit of the Hadar Formation may absorb what has previously been attributed to formation "W" in the Middle Awash region. These interpretations are however problematic because the basin geometry is unclear from descriptions of strata in the Middle Awash region, and it is difficult to assess whether they are deposited in contiguous stratigraphy. Further mapping and stratigraphy in the Kada Meshelu and Terena areas may resolve some of the uncertainty of the relationship between the Sagantole and Hadar Formations and help to define patterns of depositional environments necessary to understand temporal and spatial patterns of paleoenvironments throughout the Awash River Valley.

Another new tuff, which we call the Arbosh Tuff, is exposed immediately below the Sidi Hakoma Tuff, capping small hills in the lacustrine badlands of Arbosh (Figs. 3 and 6). Our correlation of this unit to the Maka Sand Tuff from Maka in the Middle Awash Valley is based on similar tephrostratigraphic chemical analyses (Wynn et al., 2006; Tables 1 and 2), and this confirms the lithological, biostratigraphic, and uncertain tephrochemical correlation upon which the interpreted age of *Australopithecus afarensis* from Maka depends (White et al., 1993).

Finally, in the northeastern Meshelu area, there is a very continuous and widespread altered tephra that is exposed in entirely lacustrine deposits of uncertain age, which we now call the Meshelu Tuff (Figs. 3 and 6). The ubiquitous nature of this unit, in addition to lithological features, might suggest that this may be another correlate of the Sidi Hakoma Tuff, although its deposition in entirely lacustrine environments typical of this northeast corner of the Dikika Research Project area has altered any primary glass beyond the capability for chemical analysis, except perhaps of melt inclusions with phenocrysts (method proposed by Haileab, 1994). Other similarly altered tephras occur throughout the Dikika Research Project area, such as the Ounda Leita Tuff (Tiercelin, 1986), and other bentonite units within the "triple tuffs," although these are not of regional extent, and are thus nonideal stratigraphic markers.

Busidima Formation

Tuff layers within the Busidima Formation have also aided in stratigraphic reconstruction and provided chronological constraints on the strata (Fig. 9). In fact, recent work in the Gona, Hadar, and Dikika Research Project areas has revealed a much larger number of tephras within the Busidima Formation than in the underlying Hadar Formation, making tephrostratigraphic correlation between regions very promising (Quade et al., this volume; Campisano and Feibel, this volume, Chapter 6; Roman et al., this volume). Although the Busidima Formation contains more abundant tephra units than the Hadar Formation, interpretations of the tephrostratigraphy in this part of the section have been problematic due to the slow, episodic, and laterally discontinuous nature of tephra deposition in discontinuous and predominantly fluvial settings showing episodes of erosional scour, and hence discontinuous exposure. Furthermore, secure correlation has been problematic because of similarities in chemical composition between what are known to be stratigraphically distinct units (Roman et al., this volume). Whereas some of the Hadar Formation tephra are deposited continuously over hundreds of square kilometers, many of the tephra layers in the Busidima Formation are exposed in lenticular channel bodies a few tens of meters in width, or at most in cut-and-fill channel deposits sometimes only a few meters in extent. Although many new tuffs have been recognized in the Busidima Formation, we formally name only those with well-defined and novel chemical compositions. In addition to the Bironita Tuff (0.64 Ma, described by Geraads et al., 2004), we have recently mapped a large number of tuffs in the Busidima Formation having as-yet unknown absolute chronostratigraphic positions. Many of those can be placed in a relative stratigraphic sequence on the basis of lateral mapping or chemical correlation to other exposures within the Dikika Research Project area or to the adjoining Gona and Hadar project areas (Roman et al., this volume). Here, we provide a "tuff-by tuff" description of each tuff layer in the Busidima Formation exposed in the Dikika Research Project area, in approximate stratigraphic order.

Inaalale Tuff

The Inaalale Tuff is a newly described tuff layer that is widely exposed below the AST-3 in the eastern Dikika Research Project area (Fig. 3) and is now recognized in both the Hadar and Gona research areas (Campisano and Feibel, this volume, Chapter 6; sample HE01-177 and sample E93-5035 were recognized in early tephrostratigraphic work in the Gona area; Feibel and Wynn, 1997; Campisano, 2007). In its type area, the Inaalale Tuff is below the AST-3 (Fig. 9), providing some stratigraphic constraint on the samples from Hadar and Gona. Also, the Inaalale Tuff from Hadar lies above the Salal Me'e Tuff (ca. 2.2 Ma) and is estimated to lie below the AST-3 (Campisano and Feibel, this volume, Chapter 6). The Inaalale Tuff in the Gona area lies in a short section in upper Gara'a Dora, a tributary of the Kada Gona. The correlation of the Inaalale Tuff across regions provides a definitive sequence among the Salal Me'e, Inaalale, and AST-3 Tuffs as three regionally extensive markers. Two of these tuffs can now be found across all three project areas in the Lower Awash Valley (Inaalale Tuff and AST-3). The Inaalale Tuff is not dated, but based on its stratigraphic position and correlations here, it must be between ca. 2.2 and ca. 1.3 Ma. Furthermore, several potential extrabasinal correlates can be identified based on major-element composition, including the Kangaki Tuff (Nachukui Formation, between 2.32 and 1.90 Ma), the Kay Behrensmeyer Site (KBS) (= H-2) Tuff (Turkana Basin, 1.88 ± 0.02 Ma), Tuff J (Shungura Formation, between 1.86 and 1.65 Ma), and the Okote Tuff (Koobi Fora Formation, between 1.62 and 1.48 Ma). On the basis of trace elements, the Inaalale Tuff is most similar to Tuff J. However, only limited trace-element data are available for the KBS tuff and Kangaki and do not exist to our knowledge for the Okote Tuff. Of these potential extrabasinal correlates, only the KBS has provided an absolute radiometric age determination of 1.88 ± 0.02 Ma (McDougall, 1981), but all fall within the possible stratigraphic interval of the Inaalale Tuff (between the Salal Me'e Tuff and AST-3). We leave the potential correlations to the Inaalale Tuff unresolved with the currently available data and estimate its age to be between ca. 2.2 and ca. 1.3 Ma.

Mesele Tuff

The Mesele Tuff lies at approximately the same level as the Inaalale Tuff, and it is exposed in a narrow lenticular channel collected during reconnaissance mapping (Fig. 9). The Mesele Tuff has a distinct chemical composition from the Inaalale, and it is similar to the Boolihinan, Gawis, and Waterfall Tuffs at Gona (Quade et al., this volume). Other potential chemical correlates include an unnamed tuff from the Turkana Basin (86-2491 of Haileab, 1995), the upper Nalukuwoi Tuff (between 2.52 and 2.34 Ma), the K-1 Tuff (between 1.65 and 1.39 Ma), and the Kimre Tuff (between 1.86 and 1.65 Ma). Further stratigraphic constraints are required to provide a unique identity for the Mesele Tuff.

AST-3

The AST-3 is widely exposed in the western Dikika Research Project area (Figs. 3 and 9; this tuff was previously noted as the Dob Kadis Tuff/AST-3 by Wynn et al. [2006]; we refer to it here simply as the AST-3). The AST-3 was first identified in what is now the Gona area by Roche and Tiercelin (1977), the third of what were then identified as the "Artefact Site Tuffs." The AST-3 is similar to a likely correlate, the Fialu Tuff, also at Gona (Quade et al., this volume). The AST-3 has also been noted in the upper Busidima Formation exposures in the Hadar area (Campisano and Feibel, this volume, Chapter 6), making this another excellent widespread stratigraphic marker across the Dikika, Gona, and Hadar research areas. Although it a useful marker bed, the age of the AST-3 remains poorly constrained. Quade et al. (this volume) estimate the age of the AST-3/Fialu at ca. 1.3 Ma on the basis of magnetostratigraphy and stratigraphic position above a 1.64 Ma tuff (GONASH-16). Meanwhile, Walter et al. (1996) obtained a fission-track age of 2.7 ± 0.6 Ma for the AST-3. The cut-and-fill structure in which this extensive tuff was deposited also suggests a period of widespread erosion prior to its deposition, which appears to have occurred predominantly across the upper of the two hinged surfaces of erosion that bound the Busidima half-graben. For this reason, the AST-3/Fialu may provide an appropriate boundary between subdivisions of the Busidima Formation, and an excellent stratigraphic reference point across the Lower Awash River Valley.

Ken Di Tuff

Like the AST-3, the Ken Di Tuff can be correlated across the Dikika, Hadar, and Gona research areas (cf. Campisano and Feibel, this volume, Chapter 6; Quade et al., this volume). Although its chemical composition is distinctly bimodal (Roman et al., this volume), there are not yet any potential extrabasinal correlates identified for this tuff. In the Dikika Research Project area, the Ken Di Tuff lies above the AST-3 (ca. 1.3 Ma) and is the uppermost stratigraphic marker horizon observed on the southeast side of the Awash River (Fig. 9). As such, it can be mapped below the extensive Bironita Tuff (0.64 Ma). At Hadar, the Ken Di Tuff lies below the Dahuli Tuff (0.8 Ma) and above the Burahin Dora and AST-3 (Campisano and Feibel, this volume, Chapter 6), while at Gona, the Ken Di is above a sequence of tuffs including the Camp and Ridge Tuffs (Quade et al., this volume). We note a similar chemical composition and stratigraphic position between the Ken Di Tuff from the southeast side of the Awash River to the Elamuita Tuff from the northwest side of the Awash in the Asbole area of the Dikika Research Project (see following description), but this correlation remains tentative. Like the Inaalale and AST-3 Tuffs, the widespread nature of this tuff makes it an excellent stratigraphic reference point across the Lower Awash River Valley.

Elamuita Tuff

The Elamuita Tuff, which lies below the Bironita Tuff and below Tuff E06-7016 and E06-7017, is a likely correlate to at least one mode of the bimodal Ken Di Tuff. The Ken Di Tuff from both Hadar and the Dikika Research Project areas contains two to three chemical modes, while our analysis of the Elamuita Tuff contains only one, matching the dominant Ken

Di Tuff chemical mode. Since only a small number of analyses were made of the Elamuita Tuff, it is likely that the minor modes were simply missed, or that the secondary mode is not present in some exposures. Additional analyses are expected to confirm this correlation and secure the correlation between the Ken Di and Elamuita Tuffs. This internal correlation, and further stratigraphic constraints between the Ken Di and Elamuita areas, may not only provide an extremely useful tie between two distinct areas of the Dikika Research Project strata (across the Awash River), but will link a well-constrained section in the Dikika Research Project stratigraphy to the poorly exposed upper portions of the stratigraphy at Hadar and the associated Acheulean stone tools (Campisano and Feibel, this volume, Chapter 6).

Bi'ida Foro Tuff

The Bi'ida Foro Tuff lies below the Bironita Tuff and is exposed in low hills near the Awash River from Talalak to Andedi (Figs. 3 and 9). As such, it is likely above the AST-3. This new tuff from the 2006 season was estimated in the field to be similar to the Elamuita Tuff, although not directly traceable through any outcrop. For this reason, it was not placed in a well-defined stratigraphic section, and only later chemical analysis confirmed the separate identity of the Bi'ida Foro and Elamuita Tuffs. The Bi'ida Foro Tuff has a large number of potential extrabasinal correlates but no identified correlates in the Awash Valley. At this point, we cannot identify a favored correlate, and we await further stratigraphic information.

E06-7017 and E06-7018

We also mapped and identified a tuff during the 2006 field season (samples E06-7017 and E06-7018) that is chemically similar to the Upper Hurda Tuff from Hadar (Campisano and Feibel, this volume, Chapter 6). However, the few analyses of this sample are insufficient to uniquely confirm this correlation. Samples E06-7017 and E06-7018 lie above the Elamuita Tuff and below the We-Aytu Tuff, but neither of these has a uniquely confirmed stratigraphic position. The stratigraphic position of the Upper Hurda Tuff at Hadar is also uncertain, although it occurs near Acheulean stone tools. Clearly, further work regarding this potential correlation is necessary and may help to resolve a number of stratigraphic relationships.

Bironita Tuff

Because it is very widespread, the Bironita Tuff is a very useful stratigraphic reference point for the Busidima Formation (in addition to the AST-3; Fig. 9). The Bironita Tuff is exposed mainly across the Bironita-Duma Ridge (Fig. 3), between the Awash and Asbole Rivers, but it is also exposed in the Gona area to the north (Quade et al., 2004, this volume). Major- and trace-element compositional data obtained by EPMA and XRF for the Bironita Tuff (Geraads et al., 2004) support a correlation to a tuff from the Bodo, Dawaitoli, and Hargufia areas of the Middle Awash that which has previously been dated at 0.64 ± 0.03 Ma (Clark et al., 1994). The Bironita Tuff has also been identified in the adjacent Gona Research area, where it is found low in the Brunhes chron (<0.78 Ma), and above a correlate to the 0.74 Ma Silbo Tuff (Quade et al., this volume), confirming its age and correlations. The Bironita Tuff occurs in conspicuous outcrop above a continuously mappable tabular conglomerate bed, which also makes a good lithostratigraphic marker (the "top conglomerate" of Geraads et al., 2004). This conglomerate likely continues as a single mappable conglomerate in the Gona area, just above site BSN-12 and the Boolihinan Tuff (Quade et al., this volume; Quade et al., 2007, personal commun.).

Negus Kabri Tuff

A regionally significant tephra layer, which we call the Negus Kabri Tuff, occurs near the top of the Busidima Formation in Bironita-Duma Ridge (Figs. 3 and 9). The Negus Kabri Tuff overlies the Bironita Tuff and is highly distinctive in both its major-element chemistry (specifically, its elevated CaO content) and its diagnostic appearance in the field. It is the lowermost in a package of approximately four vitric tuffs, one of which is sanidine-rich. This quadruplet is mappable over a wide area at the crest of the Bironita-Duma Ridge. The Negus Kabrie Tuff shows a strong chemical similarity to the Dahuli Tuff, near the top of the stratigraphy at Hadar, which also occurs below the Bironita Tuff at Gona (Campisano and Feibel, this volume, Chapter 6; Quade et al., this volume; the Gona outcrop of this tuff was formerly referred to as the Kuseralee Tuff by Quade et al., 2004). Radiometric dating of sanidines from within the Negus Kabrie Tuff, along with trace-element analysis of this sample and of the Dahuli Tuff, will help to not only place this tuff into an absolute chronology, but potentially further resolve Hadar-Gona-Dikika correlations.

Odele Tuff

The Odele Tuff lies stratigraphically above the Bironita Tuff and below the Waidedo Vitric Tuff (discussed later; Figs. 3 and 9). Based on major-elemental analysis by EMPA, the Odele Tuff is similar to several stratigraphically distinct tuffs, which include the AST-3/Fialu from Hadar and Gona, the Salal Me'e Tuff from Hadar, and the Upper Hurda Tuff (Hadar), as well as the KBS (= H2) Tuff from the Turkana Basin. The Odele Tuff in upper Asbole lies stratigraphically above the Bironita Tuff (0.64 Ma). This would leave only the Upper Hurda Tuff from Hadar as the remaining possible correlate. The stratigraphic position of the Upper Hurda Tuff at Hadar is uncertain due to poor exposure in the upper reaches of Hurda (Campisano and Feibel, this volume, Chapter 6), although it lies near or below the Ken Di and below the Dahuli Tuff. We leave any tentative correlation to the Odele Tuff uncertain. Meanwhile, further stratigraphic constraints on the Odele Tuff, in addition to a more robust chemostratigraphic correlation, would help to resolve the age of Acheulean stone tools from the upper Hurda region, as well as the composite stratigraphy of the Busidima Formation in these three contiguous project areas.

We-Aytu A/B Tuffs

A couplet of two chemically and physically distinct tephra layers, the We-Aytu A/B Tuffs, lies above the E06-7017 and E06-7018 and Elamuita Tuffs in the Andidi area (Fig. 9). This height in the section is likely similar to or above the level of the Bironita Tuff, but an exact stratigraphic relationship cannot be confirmed. We-Aytu A has a distinctive bimodal major-element chemistry, and there are currently no potential correlates for this tuff, although the Gawis Tuff from Gona is similar in its chemistry and stratigraphic position (Quade et al., this volume). There are no potential regional correlates and a large number of tephra with similar chemistry to the We-Aytu B, although most are likely stratigraphically below the We-Aytu Tuffs.

Waidedo Vitric Tuff

Another widespread stratigraphic marker throughout the Lower Awash Valley has been identified as a correlate to the Waidedo Vitric Tuff (0.16 Ma; Clark et al., 2003; Quade et al., this volume). Within the Dikika Research Project, this tuff is widely exposed in the upper Asbole sections (Fig. 9), as it is on the adjacent side at Gona (Quade et al., this volume). A second section is exposed in the Talalak region (Fig. 3), in which the Waidedo Vitric Tuff lies above the Bi'ida Foro Tuff and extends southward across the Talalak River into the Middle Awash region. The Waidedo Vitric Tuff in the Dikika Research Project area lies stratigraphically below the Korina Tuff (described next) and above the Odele Tuff and is thus above the Bironita Tuff (0.64 Ma). Our samples of the Waidedo Vitric Tuff in the Dikika Research Project area are also chemically similar to the Boolihinan Tuff (ca. 1.6 Ma) at Gona, as well as to two other tuffs from the Konso Formation (Katoh et al., 2000; an unnamed tuff, sample 9711-66, ca. 1.45 Ma; and the Droplet Tuff [DRT], between ca. 1.5 and 1.43 Ma), differing slightly from these in iron, sodium, and potassium content. We exclude correlation to the Konso Tuffs and the Boolihinan Tuff based on the position above the Bironita (0.64 Ma). The identification of the Waidedo Vitric Tuff in the Lower Awash Valley makes an excellent stratigraphic tie between widely separated sedimentary basins. This tuff is now found at Gona, the Dikika Research Project area, Herto (Middle Awash region), and in the Konso Formation (Quade et al., this volume; Clark et al., 2003), and thus it makes an extremely useful tephrostratigraphic marker for the Middle Pleistocene of the Horn of Africa.

Korina Tuff

The Korina Tuff is the uppermost tuff of the Busidima Formation in the Dikika Research Project area, and it is located above the Waidedo Vitric Tuff (0.15 Ma). The Korina Tuff has a large number of potential regional correlates within the Lower Awash Valley but no potential extrabasinal correlates. Given its stratigraphic position high above the Waidedo Vitric Tuff, we exclude most of the possible chemical correlates. Although it is a useful stratigraphic marker, based on its stratigraphic position, we are not able to identify a plausible tephrostratigraphic correlate at this time.

SEDIMENTARY BASIN STRUCTURE

Clearly, much remains to be done to formulate a complete tephrostratigraphic framework for the Dikika Research Project area and the Awash Valley in general. However, the stratigraphy and mapping described here are detailed enough to begin to formulate a model of the geometry and structure of the sedimentary basins in which the Hadar and Busidima Formation sediments were deposited. This model is based on the mapping of widespread marker horizons (mostly, but not exclusively, tephra) and the spatial patterns of sedimentary facies observed within the region. Due to active tectonics, most of the deposits of the Hadar Basin have likely either been uplifted and eroded or buried under modern alluvium to the northeast of the Dikika Research Project area. Despite this difficulty, we have a complete enough stratigraphy to interpret the geometry and patterns of depositional environments of the Hadar Basin and Busidima half-graben.

In general, we follow Taieb and Tiercelin (1979), who separated Ethiopian Rift basins into two types: axial and marginal. Axial basins are generally formed in full graben structures between volcanoes of the rift-axial volcanic chains. Examples include the Yangudi and Adda-do Basins of the Southern Afar Rift and axial basins parallel to the Red Sea in the Afar Depression (Fig. 1). Axial basins typically show much higher and more continuous sediment accumulation rates than marginal basins. Marginal basins on the other hand are formed in half-graben structures at the base of rift escarpments (e.g., at the western margin of the Awash Valley in Fig. 1). These asymmetric basins are typically bounded on the escarpment side by the sharp contact to a steep normal border fault, and on the axial side by a broad and gently sloping hinged margin. Examples of marginal half-grabens include the basin into which the Chorora Formation was deposited (11–10 Ma; Geraads et al., 2002) and marginal basins formed at the base of the western escarpment of the northern Ethiopian Rift (Beyene and Abdelsalam, 2005). Marginal basins typically show much slower and more episodic deposition than axial basins. As discussed next, we interpret the Hadar Basin to be an axial basin, consistent with tectonics of the Red Sea Rift system, while we interpret the Busidima half-graben to be a marginal basin formed by rift-flank uplift of the As Duma border fault, consistent with the modern tectonic setting of the Ethiopian Rift system.

Hadar Basin

Our regional mapping of sedimentary facies and thicknesses of the Hadar Formation suggests a well-developed and rapidly subsiding axial rift basin, with extension directions parallel to the Red Sea Rift system (Wynn et al., 2006; Fig. 10B). The thickening of the Hadar Formation toward the basin center in the northeast is observable across a series of northwest-southeast–trending normal faults that have overall syndepositional offset down to the northeast. The thickness of the Basal Member (older than 3.42 Ma) increases across this series of faults, as does sediment between distinct marker units (altered tephra beds in the growth structures of Fig. 4A). Likewise, the relative abundance of sedimentary facies indicating shallow lacustrine conditions increases across this series of faults (Figs. 6–8). In the extreme case, in far northeast of the Dikika Research Project area (lower Meshelu), the entire section from below the Meshelu Tuff to the Kada Hadar Tuff consists almost entirely of laminated mudstones, with only brief indications of soil formation in poorly drained conditions.

This evidence strongly suggests that the lake system and basin center were both to the north or northeast of the Dikika Research Project area during deposition of the Hadar Formation. Support for this basin paleogeography during later Hadar Formation time (ca. 3.2 Ma) is provided by Behrensmeyer (this volume), who traced a single paleochannel within the Denen Dora Member at Hadar along a 1.5 km stretch of its north-south length to a series of braided distributary systems. These results have the significant implication that the depocenter of the Hadar Basin paleolake existed to the north or northeast of Hadar during this time. Likewise, Haile-Selassie et al. (2007) reported lake-marginal deposits at 3.5–3.8 Ma in an area to the north-northeast of the Dikika Research Project area, in what was likely a contiguous basin with the deposits of the Basal Member of the Hadar Formation in the Dikika Research Project area. Finally, DiMaggio et al. (this volume) report on what may be the final sequence of extensive lacustrine deposits of the Hadar Basin at ca. 2.9 Ma. They interpret these deposits to be near the basin depocenter during their deposition, which is again to the northeast of the Dikika Research Project area (Fig. 1). Our interpretations contrast with previous reconstructions of the basin paleogeography during Denen Dora time, which have inferred a north-south elongate lake to the east of Hadar, formed by an north-south–elongated extensional axial graben (east-west extension direction), with border faults parallel to the western escarpment (Tiercelin, 1986; Yemane, 1997; Campisano, 2007).

During the initial period of deposition of the Hadar Formation (ca. 3.8 Ma), the chain of axial volcanoes north of the Dikika Research Project area, including Ida Ale, Gurra Ale, Sullu Adu, and Limmo, initiated silicic volcanism (Fig. 10). This trend and activity suggested to Lahitte et al. (2003a) that this interval is marked by the propagation of the Red Sea Rift into the Afar (see also Audin et al., 2004; Hall et al., 1984; Lahitte et al., 2003b). Later, beginning at ca. 3.3 Ma and synchronous with deposition of the Sidi Hakoma Member, flows of the Afar Stratoid Series Basalts (3.3–0.6 Ma; Lahitte et al., 2003b) interfingered with the predominantly lacustrine sediments of the lower Hadar Formation. In many cases, the Afar Stratoid Series Basalts show hyaloclastic features indicating deposition in the Hadar lake, a consistent feature of the basin paleogeography during deposition of the Hadar Formation (Fig. 4C).

During the deposition of the Hadar Formation, the As Duma border fault, which bounds the later structure of the Busidima half-graben, may not have yet been activated, as evidence suggests its activity coincided with the Busidima Formation. It is thus likely that the highlands of the western margin of the Western Escarpment were further west than their current position. This suggests that the currently exhumed exposures of pre–Hadar Formation basalts in the rift-flank (attributed to the Dahla Series in Semaw et al. [2005] but to the Adu-Asa Formation in Quade et al. [this volume]) may not yet have been exposed at the western margin of the Gona and Asbole areas until the later activation of the As Duma border fault at ca. 2.7 Ma. Rather, the rapid deposition of the Hadar Formation was likely to have increased dramatically north of the northwest-southeast–trending fault that bounded the Hadar Basin (Fig. 10B).

The configuration of the Hadar Basin depicted in Figure 10B is consistent with our observations of the stratigraphy and sedimentary facies of the Hadar Formation, as it is currently defined from older than 3.8 to ca. 2.9 Ma. Between 2.9 and 2.7 Ma, however, we find that the basin configuration of the Lower Awash Valley changed markedly to that of the Busidima half-graben, described next. In contrast to these interpretations, Campisano and Feibel (2007) suggested that minor sediment accumulation persisted in the Hadar Basin into the early Pleistocene, presumably based on the presence of ~40 m of Busidima Formation deposits that are exposed in the Hadar area, and as such, in what they attribute to the Hadar Basin. However, our tectonosedimentary model and their description of these sediments (Campisano and Feibel, this volume, Chapter 6) are more consistent with these exposures more simply being deposited in the Busidima half-graben, which existed in the northwest corner of the Hadar area (Fig. 10C), but not further east at Hadar.

Busidima Half-Graben

The strata of the Busidima half-graben are comparatively undeformed by active tectonics, and the basin geometry is evident from a geological cross section through the Dikika Research Project badland exposures (Fig. 10A). The sediments of the Busidima Formation thicken toward the border fault at the western margin across at least two gently dipping monoclines, characteristic of the hinged margins of rift half-grabens. Such asymmetric structures are typical of half-graben rift basins throughout the East African and other continental rift systems (Ebinger, 2002; Oliver Withjack et al., 2002). While the sediments of the Busidima Formation characteristically thicken toward the basin-bounding fault identified at As Duma (Quade et al., 2004), surficial outcrops become generally younger in this direction due to the ~1° westward dip of the strata (Fig. 3). The Busidima half-graben is likely contiguous with unconformable Pleistocene strata that overlie Pliocene deposits further south in the Middle Awash Valley, such as those at Andalee and the Weihatu Formation mapped throughout the region (Kalb et al., 1982a, 1982b). To the north, it appears that the Busidima half-graben does not extend much further than the northern extent of the As Duma border fault area and northwestern corner of the Hadar area. The fault and adjacent basin terminate at the intersection with a younger normal fault and graben that have a northeast-southwest orientation (Quade et al., this volume). Our interpreted basin structure of the Busidima half-graben is characteristic of a linear group of marginal half-grabens currently formed at the western margin of the Ethiopian Plateau in the Northern Afar region. These marginal structures are formed at the rift flanks, on the distal shoulders of the fully developed axial rift basins of the Southern Afar Rift and Afar Depression (Beyene and Abdelsalam, 2005). The latter basins are generally interpreted to be full grabens with an axial chain of central volcanoes (Lahitte et al., 2003a).

DEPOSITIONAL ENVIRONMENTS

Hadar Formation

Depositional environments of the central exposures of the Hadar Formation (Sidi Hakoma–lower Kada Hadar Member) have been extensively described and studied (Taieb et al., 1976; Taieb and Tiercelin, 1979; Aronson and Taieb, 1981; Tiercelin, 1986; Yemane, 1997; WoldeGabriel et al., 2000; Campisano and Feibel, this volume, Chapter 8). Here, we summarize and expand these studies focusing on new evidence from the Dikika Research Project area. The Basal Member is well exposed in the Dikika Research Project area, but it is much less so in the Hadar and Gona areas. Hence, we focus on these exposures, which provide crucial details to understanding the development of the Hadar Basin. It is also in these sections where we clearly see the basal contact of the Hadar Basin with the underlying deeply eroded basement surface of Dahla Series Basalts, a contact which is not evident elsewhere in the Awash River Valley, except perhaps in the newly described sections in the Woranso-Mille area (Haile-Selassie et al., 2007).

The type section of the Basal Member as proposed in previous work (Taieb et al., 1976; Tiercelin, 1986) was described in the Ounda Leita or Shibele River, near the Awash River (Fig. 3). Here, a prominent fault-bounded ridge is capped by the Sidi Hakoma Tuff, and it exposes some 40 m of laminated clays with abundant partially decomposed plant fragments, as well as fossil gastropods, fish, and ostracods. Two green and altered tuffs occur near the base of the exposure, called the Ounda Leita Tuffs (Tiercelin, 1986), although these could not be dated by radiometric or tephrostratigraphic means. As discussed already, a more complete section of the Basal Member is exposed in the Unduble, Arbosh, and Andalee areas in the far eastern Dikika Research Project area, where the Basal Member rests directly on the weathered surface of the Dahla Series Basalt basement. At the base of this section, two deeply weathered residual paleosols formed on the upper surface of two columnar-jointed basalt flows. The uppermost paleosol is erosionally truncated, and only the C horizon remains. The lowermost paleosol exhibits a complete profile of soil horizons (A-Bt-C-R), overlain directly by basalt flow (Fig. 4D). The Bt horizon (which is defined by clay transported from the A horizon) is composed mainly of smectite (with postburial sparry calcite in veins) and has a blocky ped structure with abundant clay skins (X-ray diffraction [XRD] data, from A.C. Calder, 2005, personal commun., confirm the identification of smectite). These features, combined with the lack of a Bk horizon (the accumulation of pedogenic calcium carbonate in the subsurface), is evidence of an extensive period of soil formation (on the order of 100 k.y.) in a relatively humid environment (>1000 mm mean annual rainfall; Retallack, 2001). The transition from both paleosol C horizons to bedrock is marked by increasingly fresh spheroidally weathering cobbles of highly fractured basalt, with increasing primary rock textures, grading into fresh basalt. These paleosols indicate at least two periods of nondeposition and extensive in situ soil formation on the stable geomorphic surfaces provided by basalt flows of the Dahla Series exposed at the base of the Hadar Formation in the Dikika Research Project area, which also likely occur in Woranso-Mille to the north (Haile-Selassie et al., 2007). These features strongly suggest an extended period of slow to negligible sediment accumulation prior to 3.8 Ma in this part of the Lower Awash Valley (Fig. 11A). Although sediment accumulation was slow to negligible in the Dikika Research Project area, slow and episodic accumulation is evident from deposits that are attributed to the Sagantole and Adu Asa Formations (>5.2–3.9 Ma) at the western margin of the Gona area (Quade et al., this volume), indicating some form of sedimentary basin, likely a basin associated with rift-flank uplift of the ancient western margin, which was further west from the current position (Fig. 11A).

The uppermost paleosol C horizon on the Dahla Series Basalt is overlain by white to light gray, poorly sorted, and massive medium- to coarse-grained sands to grits, occasionally with pebble bases. The clasts of these sands are composed predominantly of basaltic rock fragments and detrital sparry calcite grains in a matrix of smectitic clay and abundant clinoptilite (both fine-grained minerals identified with XRD data from A.C. Calder, 2005, personal commun.). The grits contain ovoid sparry calcite clasts, which appear to be the freshly eroded fragments of calcite-filled vesicles that are abundant in the underlying Dahla Series Basalts. These features, combined with the sedimentary structures, fresh basalt rock fragments, and redeposited zeolites, suggest that these beds represent the juvenile weathering products produced locally by the initial uplift and erosion of the stable surfaces of the Dahla Basalt and paleosol surfaces and, hence, the onset of sedimentation in the Hadar Basin. Most of the remainder of the Basal Member exposures in the eastern Dikika Research Project area fine upward to an extensive sequence of laminated clays with abundant plant fragments, sometimes preserving calcified casts of aquatic plant stems with diagnostic aerating structures (Fig. 4G). Many beds also contain fish scales and other fish bones, ostracods, and discontinuous bands of septarian nodules, and occasional thin bentonites likely derived from alteration of primary tephra deposits in the lake.

Throughout the eastern Dikika Research Project area, several white to very light gray, poorly sorted, medium-grained zeolite-bearing sand beds (similar to those described already) project from the southwest into the laminated clays that dominate to the northeast. The basaltic basement is below the horizon in the northeast Dikika Research Project area (around lower Meshelu), where thick sections of the Basal Member are downdropped by a

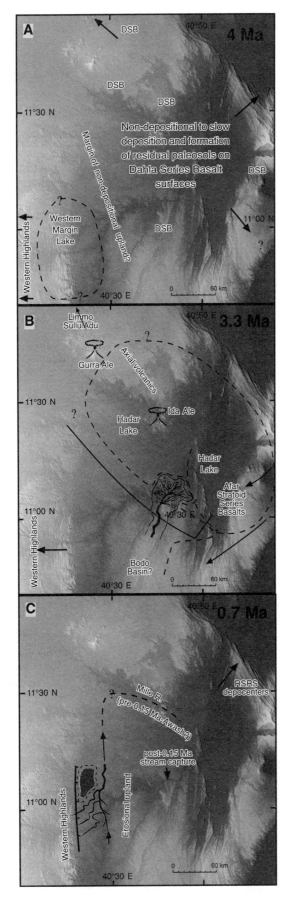

Figure 11. Paleogeographic maps of depositional environments at key intervals of deposition of the Hadar and Busidima Formations: (A) Depositional patterns just prior to onset of deposition in the Hadar Basin (specifically, ca. 4 Ma), a period marked by nondeposition to slow deposition in the north and east, with a marginal basin depositing predominantly lacustrine sediments at the western margin of the present-day Awash Valley. The uplifted western highlands were likely much further west during this interval. DSB—outcrops of Dahla Series Basalts. (B) Depositional patterns typical of the Hadar Formation (specifically at 3.3 Ma), and sediment accumulation in the Hadar Basin. Sediments interfinger with increasing thicknesses of the Afar Stratoid Series Basalts toward the northeast. (C) Depositional patterns of the Busidima Formation (specifically at 0.7 Ma). The As Duma border fault defines the Busidima half-graben and newly formed western highlands composed of uplifted Miocene and Pliocene sediments. RSRS—Red Sea Rift system.

series of northwest-southeast–trending faults. Several exposures of this series of faults show features indicating syndepositional offset such as growth fault structures (thickening of sediment between marker units; Fig. 4A) and general thickening of the Basal Member laminated clays on the hanging-wall blocks in the northeast. Similar features of syndepositional offset were suggested by Aronson and Taieb (1981) at Hadar, although their interpretations were of growth-fault structures on the north-northeast–trending fault group. These features of the lower Basal Member suggest the onset of a rapidly developing sedimentary basin that formed during active syndepositional faulting, with a widespread, but shallow lake to the northeast of the Dikika Research Project and a progradational shoreline to the southwest.

The upper Basal Member is marked by a transition from the laminated clays of the lower Basal Member through a generally coarsening-upward sequence of silty clays interbedded with light yellowish brown, cross-bedded, and moderately well-sorted medium- to fine-grained sands. The latter are more typical of the Sidi Hakoma Member, and they contrast with those of the lower Basal Member. At least two of these sands are capped by massive to horizontally bedded flaggy bioclastic sandstones with abundant gastropods (mostly *Melanoides tuberculata*; Basal gastropodites of Taieb et al., 1976), which can be identified across the Hadar Formation exposures of the Dikika Research Project area (Figs. 6–7). These beds represent shallow-water deposition in swampy shorelines and beaches, locally interbedded with fluvial sediments, all of which indicate a general trend of regression of the Basal Member lake from its southwestern shoreline.

The regression of the Basal Member lake continues through much of the lower Sidi Hakoma Member. From the uppermost Basal Member to the uppermost Sidi Hakoma Member, there is an alternating sequence of thick tabular, trough cross-bedded, medium- to coarse-grained sandstones with intervening brown to dark brown massive to pedogenically turbated clays with abundant small calcium carbonate nodules and rhizoform carbonate nodules. The latter are typical of poorly drained cracking-clay soils (fossil Pellic Vertisols; Fig. 4F; Retallack, 2001). This sedimentary facies is typified by the exposures at the DIK-1 hominin locality (Alemseged et al., 2006; Wynn et al., 2006; Table 3, Fig. 12).

TABLE 3. SEDIMENTOLOGICAL FEATURES AND INTERPRETATION OF FACIES EXPOSED IN RIDGE WHERE DIK-1-1 HOMININ WAS RECOVERED

Unit	Description	Interpretation
1	Lower contact is planar but irregular. Irregular and diffuse ledge-forming horizons of very large, commonly diffuse $CaCO_3$ nodules (10YR 8/1 white) within clay similar to Unit 2. Upper contact is gradational, with decreasingly abundant, smaller and more diffuse nodules.	Paleosol carbonate formed at phreatic-vadose zone contact in seasonally drained overlying paleosol of Unit 2.
2	10YR 3/1 & 10YR 4/1 very dark gray to dark gray fine-grained massive clay with abundant small to fine $CaCO_3$ nodules (<1 mm to 2 cm), generally with discrete boundaries, and rounded to subrounded shapes, very infrequently rhizoform. Very abundant slickensided surfaces forming network through entire unit. Very common MnO_2 dendrites and small nodules (<2 mm). Common Fe_2O_3 coatings on some ped surfaces. Also common Fe_2O_3 glaebules up to 1 mm with weakly concentric structure.	Seasonally drained, low topographic position paleosol (Pellic Vertisol), with evidence of seasonal saturation and reducing conditions (flooded), and seasonally oxidizing conditions. Facies association indicates a poorly drained subaerial delta-plain environment in topographic depressions. Likely herbaceous vegetation tolerant to these drainage conditions. Stable isotopes indicate C3 vegetation.
3	Planar and somewhat gradational contact (over 10s of cm) to 10YR 3/2 very dark grayish brown clay to clay with minor silt. Common slickensided surfaces not networked as in Unit 2. Common $CaCO_3$ nodules (up to 2 cm) with diffuse to discrete boundaries. No Fe/Mn nodules or concretions.	Moderately well-drained paleosol in relatively elevated topographic position with respect to Unit 2. Evidence of seasonal wetting but not extensive saturation. Facies association indicates moderately well-drained subaerial delta-plain environment on shallow slopes, but not in depressions (as in Unit 2). Likely dominated by herbaceous vegetation, but there is some evidence for woody vegetation—all likely C3.
4	Erosional and irregular contact to 5Y 6/2 light olive gray medium-grained sand. Contact with underlying clay is marked by irregular horizon of 5–15 cm nodular $CaCO_3$ (2.5 YR 5/6 red) with septarian structure and Fe_2O_3 Liesegang banding. Occasional vertical, white $CaCO_3$ rhizoliths extend from lower surface into underlying clay. Many anastomosing cross-sets with weak development of soil surface. Sand units generally fine upward from coarse grained (cL) to medium grained (mU), with occasional grit lags at bases of lowermost cross-sets. Individual cross-sets fine upward from mU to mL and range in size from ~5 cm to 25 cm and several meters in breadth with dips of up to 22°. Unit attains maximum thickness of 145 cm in southernmost exposure of DIK-1 ridge but thins to ~60 cm at intersection with fault. Upper 22–25 cm shows planar cross-beds of 1–3 cm thickness consisting of silty fine-grained sand.	Braided stream with broad shallow, sediment-choked anastomosing channels. Facies association indicates a subaerial delta channel, with low topographic relief and evidence of abundant flooding and poor drainage during wet season, and periodic drying, exposure, and erosion during dry seasons. Evidence in channels of woody vegetation adapted to moderately poor drainage. Stable isotopes indicate mostly C3.
5	Planar and regular gradational boundary to 10YR 5/3 brown silt with fine-grained sand. Planar cross-bedded to planar bedded. Frequent massively cemented horizons of 10 cm thick root mats of horizontal rhizoliths (~1–4 cm diameter) cemented by massive septarian $CaCO_3$ spar and abundant Fe_2O_3 staining.	Transitional to Unit 4, without erosional scouring and channelization, and lower energy setting within the braided plain of delta channel, possibly crevasse splay deposits from more active braided channel of Unit 4. Facies association indicates proximal to subaerial delta channels, with low topographic relief. Poor drainage during wet season restricts root development. Likely correlates to SH-2 sandstone at Hadar.
6	Planar and irregular contact marked by frequent Fe_2O_3-stained $CaCO_3$ nodules and clay clasts to 2.5Y 6/2 light yellowish brown fine- (fU) to medium-grained (mL) sand. Massive to planar bedded and weakly lithified and abundant Fe_2O_3 and MnO_2 staining and nodules.	Continuation of Unit 5, proximal to subaerial delta channel. Poor drainage and saturation indicated by reducing conditions.
7	Gradational fining-upward contact to 10YR 3/1 very dark gray clay as in Unit 2, although with coarser $CaCO_3$ nodules.	Transition to delta plain, poorly drained topographic depressions as in Unit 2.
8	Sharp erosional contact with up to 1 m local relief to 5Y 6/2 light olive gray medium-grained (mL) to coarse-grained (cU) sand similar to Unit 4, although generally coarser grained, especially at base of unit (which has a locally pebbly erosional surface). Cross-sets are 15–40 cm thick and average ~20 cm. Local dip angles of cross-sets are up to 15°. Occasional thin lenses of ~10 cm silty clay with weak granular pedogenic structure and ~1 cm rhizoliths extend into underlying cross-sets, notably at 30 cm from base. Unit 1: Paleosol carbonate formed at phreatic-vadose zone contact in seasonally drained overlying paleosol (Unit 2).	Braided stream as in Unit 4. Delta channel choked with sediment on a plain with low topographic relief and poor drainage. Seasonally to periodically flooded with extended periods of exposure. Erosional scour indicated at some levels, particularly at base of exposure in DIK-1. Likely correlates to SH-3 sandstone at Hadar, an extensively mappable geologic unit.

Note: Unit numbers correspond to Figure 12.

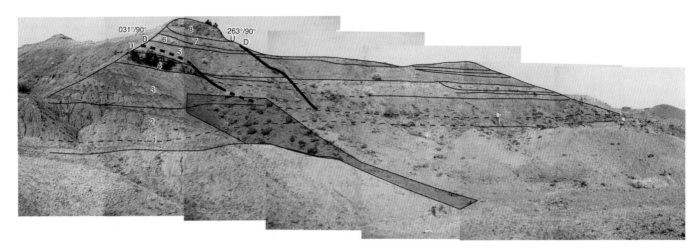

Figure 12. Facies diagram of sediments in the subaerial delta channel–delta plain facies of the Sidi Hakoma Member of the Hadar Formation. The orientations of two faults with small-scale (~1 m) offset are shown. Table 3 contains descriptions of the sedimentological features and interpretations. Dark shaded area indicates the approximate location of the sieving area from which the DIK-1 hominin specimen (3.32 Ma, *Australopithecus afarensis*) was recovered (Alemseged et al., 2006). The remainder of the slope was surveyed by a "group hill crawl."

These sedimentary features suggest subaerially exposed portions of a distributary delta system, in which the anastomosing channels of a braided fluvial system migrated across the low-gradient, and typically flooded delta plains characterized by the poorly developed, dark colored, and seasonally reducing paleosols (Pellic Vertisols; Fig. 11B). The depositional environment provided by the distributary delta system, with a low sediment competence but high sediment capacity, is ideal for the preservation of complete fossils in nearly complete articulation such as the Dikika juvenile hominin ("Selam"; Alemseged et al., 2006). This fossil, in addition to nearly complete articulated specimens of other mammals such as large rhinos and elephants, is preserved in the delta channel sandstones of the Sidi Hakoma Member (Wynn et al., 2006). In general, these sandstones thicken and coarsen to the southwest and generally do not extend into the exposures in the northeastern and eastern areas of the Dikika Research Project area, such as the section at Andalee and lower Meshelu. In the latter areas, laminated lacustrine clays occur in the same stratigraphic intervals.

Flows of the Afar Stratoid Series Basalts (3.3–0.6 Ma) interfinger with the Sidi Hakoma Member deposits in the Dikika Research Project area, having erupted from fissures somewhere to the east or northeast. These flows interfinger with the lacustrine facies of the northeast Dikika Research Project area, where they show hyaloclastic features such as pillow structures and amygdaloidal and slag-like textures, indicating deposition within the Hadar lake. These flows do not extend far into the subaerial facies or further southwest than the most prominent of the northwest-southeast–trending faults exposed in the lower Meshelu drainage.

At least four periods of brief transgression interrupted the dominantly fluvial depositional setting of the middle and upper Hadar Formation. These lacustrine transgressions are represented by: (1) a diatomite a few tens of centimeters above the Sidi Hakoma Tuff (likely the unit previously referred to as the second of a tuff "doublet" for the Sidi Hakoma Tuff; Walter and Aronson, 1993; SH-d in Fig. 7), (2) the sequence between the Sidi Hakoma gastropodite (SH-g in Fig. 7) and the ostracod-bearing upper Sidi Hakoma–lower Denen Dora laminated clays (SH-o in Fig. 7), (3) the Confetti Clay in the Kada Hadar Member (CC in Fig. 7), and (4) the Green Marker Tuff in the uppermost Kada Hadar Member (GMT in Fig. 8). These lacustrine episodes have been thoroughly described in previous work on the Hadar Formation (Aronson and Taieb, 1981; Tiercelin, 1986; Campisano, 2007). They also are extremely laterally extensive and homogeneous, such that they do not differ significantly in their features throughout the Dikika Research Project area, and thus they make excellent stratigraphic markers in the absence of tephras.

The diatom present in SH-d (Fig. 7) is *Aulacoseira granulata* (formerly *Melosira*; Tiercelin, 1986; and L. Vellen, 2004, personal commun.). This taxon is indicative of shallow, turbid freshwater conditions, and waters of relatively low Ca^{2+} concentration (Gómez et al., 1995). The molluscs present in the gastropod bed SH-g (Figs. 6 and 7) include the prosobranch gastropods *Melanoides*, *Bellamya*, and *Cleopatra*, and the unionid bivalve *Chambardia*. These freshwater taxa are indicative of a permanent water body in contact with a well-established fluvial system. *Bellamya* typically inhabit the relatively vegetation-free substrates of lakes, slow-flowing rivers, and streams, and they require hydrographic connections for dispersal (Van Damme, 1984). The larvae of unionid bivalves pass through a parasitic stage in which they are reliant on fish hosts for growth, and their dispersal is therefore restricted to water bodies fish can traverse (Moore, 1969). These mollusc data, combined with other sedimentological evidence, point to a perennial lake connected to a well-established fluvial system, in contrast with those interpreted for the few isolated gastropod-bearing clays of the Busidima Formation discussed next.

Busidima Formation

Depositional environments of the Busidima Formation have also been the subject of previous study (Feibel and Wynn, 1997; Semaw et al., 1997; Quade et al., 2004) and are only summarized here in the context of new information provided by the Dikika Research Project exposures of this formation that have not been studied in detail. The lower Busidima Formation is marked by the onset of stable fluvial sedimentation evident from the stacked sequences of cut-and-fill deposits with highly erosional scours filled with channel clast-supported cobble conglomerates and sandstones, interbedded with silty clay paleosols. This dramatic change in sedimentary facies from the Hadar to Busidima Formation is the likely cause of the change from laterally extensive and generally uniformly thick tephras of the Hadar Formation (typified by the Sidi Hakoma Tuff), to the channelized and laterally discontinuous and erosionally truncated tephras of the Busidima Formation (typified by the Mesele Tuff). The type I conglomerates (sensu Quade et al., 2004) and sandstones of the Busidima Formation contrast with those of the Hadar Formation in that they are often not as tabular and laterally extensive but vary in thickness and extent. These factors have made it difficult to use the coarse clastic beds of the Busidima Formation as stratigraphic markers (Semaw et al., 1997), compared to the laterally extensive nature of the Hadar Formation coarse clastic beds, which do make good markers (Kimbel et al., 1996). Laterally discontinuous fine- to medium-grained tephra deposits also fill many of the lenticular channels of the Busidima Formation and are similarly problematic for lithostratigraphic correlation without using tephra chemistry.

The intervening paleosols of the Busidima Formation are also very different to those of the Hadar Formation because they contain more abundant silt, are generally lighter in color, and have different features identifying their mode of pedogenesis. Many of the Busidima Formation paleosols lack the features of poorly drained Vertisols evident in the Hadar Formation paleosols (Fig. 4F). The Busidima Formation paleosols have surface horizons with fine granular structure, in addition to abundant very fine (<2 mm diameter) rhizocretions and larger root petrifactions (sensu Klappa, 1980), sometimes with carbonized plant remains (Feibel and Wynn, 1997). Many of the surface horizons of the Busidima Formation are typical of Mollisols or Mollic-soils, most notably the paleosol of the capping "Halalalee Bed" (Figs. 4J and 4L). In addition to these fine roots, more abundant in surface horizons, many of the coarser-grained paleosols contain abundant, well-formed root petrifactions, some of which preserve the cellular structure typical of palm wood (Fig. 4H). These observations support a model of more open and well-drained grass-dominated floodplains in the Busidima Formation than the reed bed, marsh, and woody vegetation environments of the delta plains in the Hadar Formation. These sedimentological observations are in agreement with the abundance of bovids of the tribe Reduncini in the Asbole fauna of the Busidima Formation, which suggest an abundance of seasonally flooded grasslands (0.8–0.6 Ma; Geraads et al., 2004; Alemseged et al., 2005; Wynn et al., 2006).

Two models for the change in depositional environments from the Hadar to the Busidima Formation have been presented, both of which have focused on the sedimentological features of the conglomerates that predominate above the Busidima unconformity surface. Yemane (1997) interpreted the conglomerates as an ephemeral braid plain and alluvial fan that extended from the western highlands toward a hypothetical meandering ancestral Awash River, somewhere to the east of Hadar, but curiously not preserved in the sedimentary record at Hadar. In contrast, Quade et al. (2004) noted the similarity of many of the clast-supported conglomerates (type I) to similar cobble bar features observed in the Awash River near Gona, and attributed the type I conglomerates directly to deposition by the ancestral Awash River.

Our observations and our model of the basin structure of the Busidima half-graben preclude the presence of an ancestral Awash River to the east of Dikika. The ancestral Awash must have flowed northward through the basin. However, it seems likely that the axial drainage of this depositional basin (the ancestral Awash River) reflected somewhat different hydrological conditions from the modern Awash River that currently flows through the Dikika Research Project area (Fig. 12C). The modern drainage system in this part of the lower Awash River Valley includes entrenched channels rapidly eroding mostly through fine-grained Pliocene-Pleistocene deposits of the Awash Valley. These sediments constitute the majority of the bed load. Cobbles observed in the Awash River channel today are extremely well rounded and were likely derived directly from secondary erosion of the cobble conglomerates exposed in the Busidima Formation. As Quade et al. (2004) noted, cobble and pebble lag surfaces on channel bars are observed in abandoned channels of the Awash River, especially at the intersection with major ephemeral drainages such as the Kada Gona, Busidima, and Andidi. These second-order tributary streams episodically erode the cobble-dominated highlands and carry a coarse-grained bed load into the mud-dominated and perennially flowing Awash River. Only these high gradient tributary streams have the sediment *competence* to regularly erode and transport such large clasts, due to their steep gradients and ephemerally high flow velocities. In contrast, the Awash River has a high sediment *capacity*, but a much lower gradient, and typically much finer-grained, but higher bed load flow. The cobble- and gravel-dominated channel bars of the modern Awash River extend downstream from these intersections with the ephemeral drainages. However, the typical bed load and bed surface of the Awash River are dominated by much finer-grained sediment, leaving only predominantly clast-supported cobble lags, as observed by Quade et al. (2004), where they are deposited at the intersections with large ephemeral streams. The unique features of the Busidima Formation conglomerates could be accounted for by the intersection of the braided and high-sediment-competence tributary streams, providing coarse clasts from the western highlands, with the high sediment capacity and through-flowing transport of fine-grained deposits in the ancestral Awash River system, which flowed northward through the marginal half-graben. Although it flowed northward through this segment of the Awash Valley, the Awash River was likely later diverted eastward toward terminal basins of the Afar Depression before reaching the obstacle of the Gudda horst (see Fig. 1 for topography of the Awash Valley).

These sedimentological observations, combined with the tectonic model of sediment accumulation in a rift-marginal basin such as the Busidima half-graben, provide a depositional model consistent with both previous sets of observations (Yemane, 1997; Quade et al., 2004). Activity on the As Duma fault, which bounds the Busidima half-graben, is necessary to provide the accommodation space for sediment accumulation, and it is also the primary driving force behind the transport of coarse clastic material of the clast-supported conglomerates. This tectonosedimentary setting is unique to *marginal* half-grabens in which the primary river system is in close proximity to the rift escarpment. This setting contrasts with the majority of the remainder of the course of the modern Awash River, in which the primary river is far removed from clastic input from the high-competence ephemeral rivers of the highlands. For example, further south in the Middle Awash Valley, the Awash River abuts against the inactive flanks of the broad plateaus between the Southern Afar Rift and Awash Valley (Fig. 1). Likewise, north of the Dikika, Gona, and Hadar areas, the Awash River cuts through these plateaus and flows across broad floodplains in the terminal basins. Both of these settings lack the input of coarse clastic material from an active escarpment of the Ethiopian Rift. Such a setting occurs uniquely through the stretch of the Awash River referred to here as the Lower Awash Valley (roughly the Dikika Research Project, Gona, Hadar, and northern Middle Awash areas).

Another unique feature of the Busidima Formation is the well-indurated and calcite-cemented conglomerates. Clasts of the "top conglomerate" are cemented by a well-developed calcic horizon, typical of the subsurface Bk horizon of a paleosol from which the overlying horizons have been removed by erosion. Cementation of this unit has made it more resistant to erosion than other overlying gravels. Thus, it holds up the fossiliferous deposits of the Bironita-Duma Plateau (Geraads et al., 2004). This horizon marks a period of nondeposition, followed by a somewhat changed depositional pattern.

Deposits above the "top conglomerate" are very poorly to noncemented, in contrast to the underlying, relatively consolidated sediments. Also, the overlying sequences of cumulatively developed paleosols are thin, having thin surface horizons with fine granular structure and fine root pedotubules, which are also present but not as common in the underlying mudstones. These paleosols above the "top conglomerate" frequently preserve calcified wood with exquisite details of cellular structure. The paleosols and mudstones above the "top conglomerate" also contain abundant thenardite (also eugsterite, both identified from A.C. Calder, 2005, personal commun.), often leaching out at the contact between the loosely consolidated paleosols and well-lithified conglomerate (Fig. 4K). Thenardite is formed by dehydration of mirabilite above temperatures of 35 °C; mirabilite forms by evaporation of high Mg^{2+}/Ca^{2+} ratio waters following Ca^{2+}-limited calcite precipitation (Eugster and Jones, 1979). These sulfate minerals occur only above the "top conglomerate." Combined with the soil structures present above this marker, these features indicate saline soils in small closed basin depressions with ephemeral water and very high evaporation rates, such as evaporation pans in the Kalahari and similar environments (e.g., Mees, 2003). This episode likely represents the final infilling of the Busidima half-graben and the presence of poorly drained evaporation pans in a highly evaporative environment with a nearby source of Ca^{2+}-deprived, slightly alkaline to neutral and saline floodwaters.

The isolated gastropod-bearing clays above the top conglomerate in Negus Kabri also provide some very distinctive paleoenvironmental evidence to support these sedimentological observations (Fig. 9). These beds include Basommatophora gastropods (including the planorbid genera *Bulinus*, *Gyraulus*, and *Biomphalaria*) as well as *Lymnaea*. These "pulmonates" have both the ability to respire despite stagnant conditions and to survive through intermittent dry periods typical of seasonal closed-basin lakes (Brown, 1980). The combination of these genera is indicative of small, seasonal, and poorly oxygenated water bodies, likely with abundant aquatic macrophyte vegetation such as eelgrass (*Vallisneria aethiopicus*), hornworts (*Ceratophyllum*), reeds, water lilies, and papyrus, which are common habitat for these gastropods (Brown, 1980). On the other hand, the presence of the generally perennial water dwelling *Melanoides tuberculata* (and to some extent *Lymnaea* and *Gyraulus*) in the same beds indicates some source of perennial water, likely to be the through-flowing Awash River. Given the sedimentological setting outlined here, it is likely that seasonal flooding events of the Awash River would have left small ephemeral floodplain lakes in isolated basins, likely to be the habitat of *Bulinus* and *Biomphalaria* (ephemeral floodplain lake depicted in Fig. 11C). This sedimentological and paleoecological setting is in marked contrast to that of the bioclastic gastropod-bearing sands of the Hadar Formation described previously. Conditions similar to those interpreted for the Busidima Formation floodplain lakes are present today at Lake Caddabasa, further south in the Awash Valley (Fig. 1), which is very dissimilar from the interpreted setting of Hadar lake in the Hadar Formation. At Lake Caddabasa, the proximal braid plain of ephemeral rivers from the western highlands merges into a sedimentary basin formed by active tectonics with a through-flowing but sluggish river. Despite its potential as a sedimentological analog for Pliocene-Pleistocene deposits of the Awash Valley (Aronson and Taieb, 1981), the tectonic and sedimentary setting of this area is currently poorly documented.

Since ca. 0.15 Ma, activity appears to have ceased in the marginal half-graben of the Busidima Formation, or at least in the resulting depositional basin. In the current setting of the Lower Awash Valley, the central river system is rapidly eroding through the sediments, rather than accumulating. Sometime after this point, the vast grassland plains represented by the Halalalee Bed paleosol were the final depositional environment to be recorded in the Lower Awash Valley.

TECTONIC AND PALEOENVIRONMENTAL IMPLICATIONS

From the geological framework of stratigraphy and depositional environments presented herein, we can begin to formulate a tectonic history of sedimentation in the Hadar and Busidima Basins and examine the role of these structures in the overall tectonic history of the Afar triple junction region (Fig. 10). Our interpretations of the Hadar Basin, particularly of its lower contact with the Dahla Series Basalts, suggest that this depositional basin formed in response to rapid initiation of extension parallel to the Red Sea Rift system. The extension direction of the Hadar Basin contrasts with that of the Ethiopian Rift system, on the shoulder of which it currently lies. Based on current evidence, it appears that the Hadar Basin was distinct from the basin that contains the older Sagantole Formation and the Bodo deposits (informally designated Formation "W") in the Middle Awash area, the geometry of which is not well understood. The Hadar Formation was deposited on a deeply weathered and syndepositionally faulted surface of the Dahla Series Basalt, which was likely an extensive plain throughout what is now the Lower Awash Valley (Fig. 12). This basaltic basement, which underlies the Hadar Formation, may be contiguous with, or interfinger with, basalts exposed in the western margin of the Gona area (ca. 4.2–4.0 Ma, assigned to the Dahla Series in Semaw et al. [2005] and Adu-Asa and Sagantole Formations in Quade et al. [this volume]).

Following its deposition, much of the Hadar Formation was faulted along structures roughly parallel to the structural grain of the Ethiopian Rift system, and roughly paralleling the As Duma border fault. This shift in the tectonic regime from Red Sea Rift system– to Ethiopian Rift system–parallel extension indicates a migration of the Red Sea Rift system tectonic regime toward the northeast of the Dikika Research Project area. This shift is roughly coincident with migration of the R-R-R triple junction and of the Red Sea Rift system–East African Rift system boundary northward during the same period. This model is consistent with the suggestions of Aronson and Taieb (1981) and Campisano and Feibel (this volume, Chapter 8) and Kalb (1995) about the migration of the Hadar depocenter northeast of the Dikika Research Project area during the transition to what is now attributed to the Busidima Formation. Around the same time, or at least after the deposition of the Hadar Formation, the rift basins of the Southern Afar segment of the Ethiopian Rift system (the Yangudi and Adda-Do) began to extend, and sediment began to accumulate in the rift-axial grabens of the Southern Afar Rift. Meanwhile, faulting on the shoulders of these axial basins uplifted the Hadar Formation strata with respect to the basin centers in the east. This short period of relative uplift and erosion of the Lower Awash Valley (2.9–2.7 Ma) was terminated by rift-flank uplift of the western margin along the As Duma border fault and formation of a marginal half-graben (2.7–0.15 Ma), which captured the northward flow of the Awash River. The Awash River south of the Dikika Research Project area is currently diverted by the volcanic edifice at Mt. Ayelu into this depositional basin, on its course to the terminal depositional basins further east in the Afar Depression. This diversion likely began with the initiation of the Busidima half-graben and the return of sediment accumulation to the Lower Awash Valley.

The Busidima Formation is marked by slow and episodic deposition in a predominantly fluvial environment at the margin of the current western escarpment. This escarpment may have shifted eastward to its present position at the initiation of the As Duma border fault, the age of which we interpret to be 2.7 Ma. The proximity of the uplift of the As Duma border fault to the slowly accumulating fluvial sediments in the half-graben, combined with the relatively arid climate of the Pleistocene, resulted in the unique cobble conglomerate lags that characterize the Busidima Formation. Short-lived, shallow floodplain lakes, similar to Lake Caddabasa, may have persisted during some of this period, as evidenced by thin gastropod-bearing clays in the upper Busidima Formation (containing *Melanoides tuberculata*, *Lymnaea* sp., *Bulinus* sp., *Biomphalaria* sp. [*pfeifferi*], *Gyraulus* sp., and Planorbinae indet.). Meanwhile, sedimentological evidence indicates evaporative pans on the poorly drained floodplains above the "top conglomerate." Sedimentation in the Busidima half-graben terminated with the formation of the Halalalee Bed paleosol, which is currently left stranded well above the active erosion of the entrenched Awash River. At some later stage, the Awash River, which previously flowed northward through the Gona area, was captured by downcutting of drainage systems to the east of the Dikika Research Project area, and the spectacular hominin-bearing badland exposures of the Lower Awash Valley were eroded and exposed along this famous stretch of the Awash River.

Our analysis of the interplay of sedimentation and tectonics, along with similar interpretations of the sedimentary history of the Lower Awash Valley reported in this volume (Quade et al., this volume; Campisano and Feibel, this volume, Chapter 8), highlights the fact that only through a greater understanding of both the tectonic and climatic controls on hominin paleoenvironments will we understand the role of the environment in human evolutionary patterns.

ACKNOWLEDGMENTS

This research was accomplished under a permit from the Authority for Research and Conservation of Cultural Heritage of the Ministry of Tourism and Culture and the Afar Regional State of Ethiopia. We thank the members of the Dikika Research Project, including the many Afars and Issas who have provided endless logistical support, aid, and companionship in the field. The Max Planck Society, the National Geographic Society, the Institute for Human Origins, and the French Center for Ethiopian Studies provided funds and support. Funds for tephrochemical analysis were also provided by the Centre National de la Recherche Scientifique, France, and the Baldwin Memorial Fund, University of Oregon. Chemical analyses were completed at the Facility for Earth and Environmental Analysis at the University of St. Andrews, the Research School of Earth Sciences at the Australian National University, the University of Oregon Center for Advanced Materials Characterization in Oregon (CAMCOR) Microprobe Facility, and the Florida Center for Electron Microscopy.

REFERENCES CITED

Acocella, V., and Korme, T., 2002, Holocene extension direction along the Main Ethiopian Rift: East Africa: Terra Nova, v. 14, p. 191–197, doi: 10.1046/j.1365-3121.2002.00403.x.

Alemseged, Z., and Geraads, D., 2000, A new Middle Pleistocene fauna from the Busidima-Telalak region of the Afar, Ethiopia: Comptes Rendus de l'Académie des Sciences, v. 331, p. 549–556.

Alemseged, Z., Wynn, J.G., Kimbel, W.H., Reed, D., Geraads, D., and Bobe, R., 2005, A new hominin from the Basal Member of the Hadar Formation, Dikika, Ethiopia, and its geological context: Journal of Human Evolution, v. 49, p. 499–514, doi: 10.1016/j.jhevol.2005.06.001.

Alemseged, Z., Bobe, R., Geraads, D., Kimbel, W.H., Reed, D., Spoor, F., and Wynn, J.G., 2006, A juvenile early hominin skeleton from Dikika, Ethiopia: Nature, v. 443, p. 296–301, doi: 10.1038/nature05047.

Aronson, J.L., and Taieb, M., 1981, Geology and paleogeography of the Hadar hominid site, Ethiopia, in Rapp, G.J., and Vondra, C.F., eds., Hominid Sites: Their Geologic Settings: American Association for the Advancement of Science Selected Symposium: Boulder, Colorado, Westview Press, p. 165–196.

Aronson, J.L., Schmitt, T.J., Walter, R.C., Taieb, M., Tiercelin, J.J., Johanson, D.C., Naeser, C.W., and Nairn, A.E.M., 1977, New geochronologic and paleomagnetic data for the hominid-bearing Hadar Formation of Ethiopia: Nature, v. 267, p. 323–327, doi: 10.1038/267323a0.

Audin, L., Quidelleur, X., Coulié, E., Courtillot, V., Gilder, S., Manighetti, I., Gillot, P.Y., Tapponnier, P., and Kidane, T., 2004, Paleomagnetism and K-Ar and $^{40}Ar/^{39}Ar$ ages in the Ali Sabieh area (Republic of Djibouti and Ethiopia): Constraints on the mechanism of Aden ridge propagation into southeastern Afar during the last 10 Myr: Geophysical Journal International, v. 158, p. 327–345, doi: 10.1111/j.1365-246X.2004.02286.x.

Behrensmeyer, A.K., 2008, this volume, Paleoenvironmental context of the Pliocene A.L. 333 "First Family" hominin locality, Hadar Formation, Ethiopia, in Quade, J., and Wynn, J.G., eds., The Geology of Early Humans in the Horn of Africa: Geological Society of America Special Paper 446, doi: 10.1130/2008.2446(09).

Beyene, A., and Abdelsalam, M., 2005, Tectonics of the Afar Depression: A review and synthesis: Journal of African Earth Sciences, v. 41, p. 41–59, doi: 10.1016/j.jafrearsci.2005.03.003.

Bilham, R., Bendick, R., Larson, K., Mohr, P., Braun, P., Tesfaye, S., and Asfaw, L., 1999, Secular and tidal strain across the Main Ethiopian Rift: Geophysical Research Letters, v. 26, p. 2789–2792, doi: 10.1029/1998GL005315.

Brown, D., 1980, Freshwater Snails of Africa and Their Medical Importance: London, Taylor & Francis, 626 p.

Brown, F.H., and Cerling, T.E., 1982, Stratigraphical significance of the Tulu Bor Tuff of the Koobi Fora Formation: Nature, v. 299, p. 212–215, doi: 10.1038/299212a0.

Brown, F.H., Sarna-Wojcicki, A., Meyer, C.E., and Haileab, B., 1992, Correlation of Pliocene and Pleistocene tephra layers between the Turkana Basin of East Africa and the Gulf of Aden: Quaternary International, v. 13/14, p. 55–67, doi: 10.1016/1040-6182(92)90010-Y.

Calais, E., DeMets, C., and Nocquet, J.-M., 2003, Evidence for post-3.16 Ma change in Nubia–Eurasia–North America plate motions: Earth and Planetary Science Letters, v. 216, p. 81–92, doi: 10.1016/S0012-821X(03)00482-5.

Campisano, C., 2007, Tephrostratigraphy and Hominin Paleoenvironments of the Hadar Formation, Afar Depression, Ethiopia [Ph.D. thesis]: New Brunswick, New Jersey, Rutgers University, 616 p.

Campisano, C.J., and Feibel, C.S., 2007, Connecting local environmental sequences to global climate patterns: Evidence from the hominin-bearing Hadar Formation, Ethiopia: Journal of Human Evolution, v. 53, p. 515–527, doi: 10.1016/j.jhevol.2007.05.015.

Campisano, C.J., and Feibel, C.S., 2008, this volume (Chapter 6), Tephrostratigraphy of the Hadar and Busidima Formations at Hadar, Afar Depression, Ethiopia, in Quade, J., and Wynn, J.G., eds., The Geology of Early Humans in the Horn of Africa: Geological Society of America Special Paper 446, doi: 10.1130/2008.2446(06).

Campisano, C.J., and Feibel, C.S., 2008, this volume (Chapter 8), Depositional environments and stratigraphic summary of the Pliocene Hadar Formation at Hadar, Afar Depression, Ethiopia, in Quade, J., and Wynn, J.G., eds., The Geology of Early Humans in the Horn of Africa: Geological Society of America Special Paper 446, doi: 10.1130/2008.2446(08).

Clark, J.D., de Heinzelin, J., Schick, K.D., Hart, W.K., White, T.D., WoldeGabriel, G., Walter, R.C., Suwa, G., Asfaw, B., Vrba, E., and Haile-Selassie, Y., 1994, African Homo erectus: Old radiometric ages and young Oldowan assemblages in the Middle Awash Valley: Ethiopia: Science, v. 264, p. 1907–1910, doi: 10.1126/science.8009220.

Clark, J.D., Beyene, A., WoldeGabriel, G., Hart, W.K., Renne, P., Gilbert, H., Defleur, A., Suwa, G., Katoh, S., Ludwig, K.R., Boisserie, J.-R., Asfaw, B., and White, T., 2003, Stratigraphic, chronological and behavioural contexts of Pleistocene Homo sapiens from Middle Awash, Ethiopia: Nature, v. 423, p. 747–752, doi: 10.1038/nature01670.

Courtillot, V., Achache, J., Landre, F., Bonhomeet, N., Montigny, R., and Féraydm, G., 1984, Episodic spreading and rift propagation; new paleomagnetic and geochronologic data from the Afar nascent passive margin: Journal of Geophysical Research, v. 89, no. B5, p. 3315–3333, doi: 10.1029/JB089iB05p03315.

deMenocal, P.B., and Brown, F.H., 1999, Pliocene tephra correlations between East African hominid localities, the Gulf of Aden, and the Arabian Sea, in Agusti, J., Rook, L., and Andrews, P., eds., Climatic and Environmental Change in the Neogene of Europe: London, Cambridge University Press, p. 23–52.

DiMaggio, E.N., Campisano, C.J., Arrowsmith, J R., Reed, K.E., Swisher, C.C., III, and Lockwood, C.A., 2008, this volume, Correlation and stratigraphy of the BKT-2 volcanic complex in west-central Afar, Ethiopia, in Quade, J., and Wynn, J.G., eds., The Geology of Early Humans in the Horn of Africa: Geological Society of America Special Paper 446, doi: 10.1130/2008.2446(07).

Eagles, G., Gloaguen, R., and Ebinger, C., 2002, Kinematics of the Danakil microplate: Earth and Planetary Science Letters, v. 203, p. 607–620, doi: 10.1016/S0012-821X(02)00916-0.

Ebinger, C.J., 2002, Causes and consequences of lithospheric extension, in Renaut, R.W., and Ashley, G.M., eds., Sedimentation in Continental Rifts: Society for Sedimentary Geology (SEPM) Special Publication 73, p. 11–24.

Ebinger, C.J., Yemane, T., WoldeGabriel, G., and Aronson, J., 1993, Late Eocene–Recent volcanism and faulting in the southern Main Ethiopian Rift system: Journal of the Geological Society of London, v. 150, p. 99–108, doi: 10.1144/gsjgs.150.1.0099.

Eugster, H.P., and Jones, B.F., 1979, Behavior of major solutes during closed-basin brine evolution: American Journal of Science, v. 279, p. 609–631.

Feibel, C.S., and Wynn, J.G., 1997, Preliminary report on the geologic context of the Gona archeological sites: Appendix, in Semaw, S., Late Pliocene Archeology of the Gona River Deposits, Afar, Ethiopia [Ph.D. diss.]: New Brunswick, New Jersey, Rutgers University, p. A257–302.

Geraads, D., 2005, Pliocene Rhinocerotidae (Mammalia) from Hadar and Dikika (Lower Awash, Ethiopia), and a revision of the origin of modern African rhinos: Journal of Vertebrate Paleontology, v. 25, p. 451–461, doi: 10.1671/0272-4634(2005)025[0451:PRMFHA]2.0.CO;2.

Geraads, D., Alemseged, Z., and Bellon, H., 2002, The late Miocene mammalian fauna of Chorora, Awash basin, Ethiopia: Systematics, biochronology and $^{40}K/^{39}Ar$ age of associated volcanics: Tertiary Research, v. 21, p. 113–122.

Geraads, D., Alemseged, Z., Reed, D., Wynn, J.G., and Roman, D., 2004, The Pleistocene fauna (other than primates) from Asbole, Lower Awash Valley, Ethiopia, and its environmental and biochronological implications: Geobios, v. 37, p. 697–718, doi: 10.1016/j.geobios.2003.05.011.

Gómez, N., Riera, J.L., and Sabater, S., 1995, Ecology and morphological variability of Aulacoseira granulata (Bacillariophyceae) in Spanish reservoirs: Journal of Plankton Research, v. 17, p. 1–16, doi: 10.1093/plankt/17.1.1.

Haileab, B., 1994, Melt inclusions in phenocryts of late Neogene tuff from East Africa—Key to correlation of altered tuffs and magma composition: Geological Society of America Abstracts with Programs, v. 26, no. 7, p. 483.

Haileab, B., 1995, Geochronology, Geochemistry and Tephrostratigraphy of the Turkana Basin, Southwestern Ethiopia, Northern Kenya [Ph.D. thesis]: Salt Lake City, University of Utah, 369 p.

Haileab, B., and Brown, F.H., 1992, Turkana Basin–Middle Awash Valley correlations and the age of the Sangatole and Hadar Formations: Journal of Human Evolution, v. 22, p. 453–468, doi: 10.1016/0047-2484(92)90080-S.

Haile-Selassie, Y., Deino, A., Saylor, B., Umer, M., and Latimer, B., 2007, Preliminary geology and paleontology of new hominid-bearing Pliocene localities in the central Afar region of Ethiopia: Anthropological Science, v. 115, p. 215–222, doi: 10.1537/ase.070426.

Hall, C.M., Walter, R.C., Westgate, J.A., and York, D., 1984, Geochronology, stratigraphy and geochemistry of Cindery Tuff in Pliocene hominid-bearing sediments of the Middle Awash, Ethiopia: Nature, v. 308, p. 26–31, doi: 10.1038/308026a0.

Hey, R., 1977, A new class of "pseudofaults" and their bearing on plate tectonics: A propagating rift mode: Earth and Planetary Science Letters, v. 37, p. 321–325, doi: 10.1016/0012-821X(77)90177-7.

Hofmann, C., Courtillot, V., Feraud, G., Rochette, P., Yirgu, G., Ketefo, E., and Pik, R., 1997, Timing of the Ethiopian flood basalt event: Implications for plume birth and global change: Nature, v. 389, p. 838–841, doi: 10.1038/39853.

Johanson, D.C., and Taieb, M., 1976, Plio-Pleistocene hominid discoveries in Hadar, Ethiopia: Nature, v. 260, p. 293–297, doi: 10.1038/260293a0.

Kalb, J., 1993, Refined stratigraphy of the hominid-bearing Awash Group, Middle Awash Valley, Afar Depression, Ethiopia: Newsletters on Stratigraphy, v. 29, p. 21–62.

Kalb, J., 1995, Fossil elephantoids, Awash paleolake basins, and the Afar triple junction, Ethiopia: Palaeogeography, Palaeoclimatology, Palaeoecology, v. 114, p. 357–368, doi: 10.1016/0031-0182(94)00088-P.

Kalb, J., Oswald, E.B., Tebedge, S., Mebrate, A., Tola, E., and Peak, D., 1982a, Geology and stratigraphy of Neogene deposits, Middle Awash Valley, Ethiopia: Nature, v. 298, p. 17–25, doi: 10.1038/298017a0.

Kalb, J.E., Oswald, E.B., Mebrate, A., Tebedge, S., and Jolly, C.J., 1982b, Stratigraphy of the Awash Group, Middle Awash Valley, Afar, Ethiopia: Newsletters on Stratigraphy, v. 11, p. 95–127.

Katoh, S., Nagaoka, S., WoldeGabriel, G., Renne, P., Snow, M.G., Beyene, A., and Suwa, G., 2000, Chronostratigraphy and correlation of the Plio-Pleistocene tephra layers of the Konso Formation, southern Main Ethiopian Rift, Ethiopia: Quaternary Science Reviews, v. 19, no. 13, p. 1305–1317, doi: 10.1016/S0277-3791(99)00099-2.

Kidane, T., Courtillot, V., Manighetti, I., Audin, L., Lahitte, P., Quidelleur, X., Gillot, P.Y., Gallet, Y., Carlut, J., and Haile, T., 2003, New paleomagnetic and geochronologic results from Ethiopian Afar: Block rotations linked to rift overlap and propagation and determination of a 2 Ma reference pole for stable Africa: Journal of Geophysical Research, v. 108, no. B2, p. 2102, doi: 10.1029/2001JB000645.

Kimbel, W.H., Walter, R.C., Johanson, D.C., Reed, K.E., Aronson, J., Assefa, Z., Marean, C.W., Eck, G.G., Bobe, R., Hovers, E., Rak, Y., Vondra, C., Yemane, T., York, D., Chen, Y., Evensen, N.M., and Smith, P.E., 1996, Late Pliocene *Homo* and Oldowan stone tools from the Hadar Formation (Kada Hadar Member), Ethiopia: Journal of Human Evolution, v. 31, p. 549–561, doi: 10.1006/jhev.1996.0079.

Klappa, C.F., 1980, Rhizoliths in terrestrial carbonates: Classification, recognition, genesis and significance: Sedimentology, v. 27, p. 613–629, doi: 10.1111/j.1365-3091.1980.tb01651.x.

Kleinsasser, L.L., Quade, J., McIntosh, W.C., Levin, N.E., Simpson, S.W., and Semaw, S., 2008, this volume, Stratigraphy and geochronology of the late Miocene Adu-Asa Formation at Gona, Ethiopia, *in* Quade, J., and Wynn, J.G., eds., The Geology of Early Humans in the Horn of Africa: Geological Society of America Special Paper 446, doi: 10.1130/2008.2446(02).

Lahitte, P., Gillot, P.-Y., and Courtillot, V., 2003a, Silicic central volcanoes as precursors to rift propagation: The Afar case: Earth and Planetary Science Letters, v. 207, p. 103–116, doi: 10.1016/S0012-821X(02)01130-5.

Lahitte, P., Gillot, P.-Y., Kidane, T., Courtillot, V., and Bekele, A., 2003b, New age constraints on the timing of volcanism in central Afar, in the presence of propagating rifts: Journal of Geophysical Research, v. 108, no. B2, p. 2123, doi: 10.1029/2001JB001689.

Manighetti, I., Tapponnier, P., Courtillot, V., Gruszow, S., and Gillot, P.-Y., 1997, Propagation of rifting along the Arabia-Somalia plate boundary: The Gulf of Aden and Tadjoura: Journal of Geophysical Research, v. 102, p. 2681–2710, doi: 10.1029/96JB01185.

McDougall, I., 1981, ^{40}Ar/^{39}Ar age spectra from the KBS Tuff, Koobi Fora Formation: Nature, v. 294, p. 120–124, doi: 10.1038/294120a0.

McKenzie, D., 1978, Some remarks on the development of sedimentary basins: Earth and Planetary Science Letters, v. 40, p. 25–32, doi: 10.1016/0012-821X(78)90071-7.

Mees, F., 2003, Salt mineral distribution patterns in soils of the Otjomongwa pan, Namibia: CATENA, v. 54, p. 425–437, doi: 10.1016/S0341-8162(03)00135-8.

Menzies, M., Gallagher, K., Yelland, A., and Hurford, A., 1997, Volcanic and non-volcanic rifted margins of the Red Sea and Gulf of Aden: Crustal cooling and margin evolution in Yemen: Geochimica et Cosmochimica Acta, v. 61, p. 2511–2527, doi: 10.1016/S0016-7037(97)00108-7.

Moore, R.C., 1969, Treatise on Invertebrate Paleontology: Bivalvia: New York, Geological Society of America and University of Kansas Press, 952 p.

Oliver Withjack, M., Schlishe, R.W., and Olsen, P.E., 2002, Rift-basin structure and its influence on sedimentary systems, *in* Renaut, R.W., and Ashley, G.M., eds., Sedimentation in Continental Rifts: Society for Sedimentary Geology (SEPM) Special Publication 73, p. 57–82.

Quade, J., Levin, N., Semaw, S., Stout, D., Renne, P., Rogers, M., and Simpson, S., 2004, Paleoenvironments of the earliest stone toolmakers, Gona, Ethiopia: Geological Society of America Bulletin, v. 116, p. 1529–1544, doi: 10.1130/B25358.1.

Quade, J., Levin, N.E., Simpson, S.W., Butler, R., McIntosh, W.C., Semaw, S., Kleinsasser, L., Dupont-Nivet, G., Renne, P., and Dunbar, N., 2008, this volume, The Geology of Gona, *in* Quade, J., and Wynn, J.G., eds., The Geology of Early Humans in the Horn of Africa: Geological Society of America Special Paper 446, doi: 10.1130/2008.2446(01).

Renne, P., WoldeGabriel, G., Hart, W.K., Heiken, G., and White, T., 1999, Chronostratigraphy of the Mio-Pliocene Sagantole Formation, Middle Awash Valley, Ethiopia: Geological Society of America Bulletin, v. 111, p. 869–885, doi: 10.1130/0016-7606(1999)111<0869:COTMPS>2.3.CO;2.

Retallack, G.J., 2001, Soils of the Past: An Introduction to Paleopedology: Oxford, Blackwell, 404 p.

Roche, H., and Tiercelin, J.-J., 1977, Découverte d'une industrie lithique ancienne in situ dans la formation d'Hadar, Afar Central, Ethiopie: Comptes Rendus de l'Académie des Sciences, Série D, v. 284, p. 1871–1874.

Roman, D.C., Campisano, C., Quade, J., DiMaggio, E., Arrowsmith, J R., and Feibel, C., 2008, this volume, Composite tephrostratigraphy of the Dikika, Gona, Hadar, and Ledi-Geraru project areas, northern Awash, Ethiopia, *in* Quade, J., and Wynn, J.G., eds., The Geology of Early Humans in the Horn of Africa: Geological Society of America Special Paper 446, doi: 10.1130/2008.2446(05).

Semaw, S., Renne, P., Harris, J.W.K., Feibel, C.S., Bernor, R.L., Fesseha, N., and Mowbray, K., 1997, 2.5-million-year-old stone tools from Gona, Ethiopia: Nature, v. 385, p. 333–336, doi: 10.1038/385333a0.

Semaw, S., Simpson, S., Quade, J., Renne, P., Butler, R.F., McIntosh, W.C., Levin, N., Dominguez-Rodrigo, M., and Rogers, M., 2005, Early Pliocene hominids from Gona, Ethiopia: Nature, v. 433, p. 301–305, doi: 10.1038/nature03177.

Taieb, M., and Tiercelin, J.-J., 1979, Sédimentation pliocène et paléoenvironnements de rift: Exemple de la formation à Hominidés d'Hadar (Afar Éthiopie): Bulletin de la Societé Géologique de France, v. 7, p. 243–253.

Taieb, M., Coppens, Y., Johanson, D.C., and Kalb, J., 1972, Dépôts sédimentaires et faunes du Plio-Pléistocène de la basse vallée de l'Awash (Afar central, Ethiopie): Comptes Rendus de l'Académie des Sciences, Série D, v. 275, p. 819–882.

Taieb, M., Johanson, D.C., Coppens, Y., and Aronson, J.L., 1976, Geological and palaeontological background of Hadar hominid site, Afar, Ethiopia: Nature, v. 260, p. 289–293, doi: 10.1038/260289a0.

Tefera, M., Chernet, T., and Haro, W., 1996, Geological Map of Ethiopia (2nd ed.): Ethiopian Ministry of Mines and Energy, Ethiopian Geological Survey, scale 1:2,000,000.

Tiercelin, J.-J., 1986, The Pliocene Hadar Formation, Afar Depression of Ethiopia, *in* Frostick, L.E., Renaut, R.W., Reid, I., and Tiercelin, J.-J., eds., Sedimentation in the African Rifts: Geological Society of London Special Publication 25, p. 221–240.

Ukstins, I., Renne, P., Wolfenden, E., Baker, J., and Menzies, M., 2002, Matching conjugate volcanic rifted margins: ^{40}Ar/^{39}Ar chronostratigraphy of pre- and synrift bimodal flood volcanism in Ethiopia and Yemen: Earth and Planetary Science Letters, v. 198, p. 289–306, doi: 10.1016/S0012-821X(02)00525-3.

Van Damme, D., 1984, The Freshwater Mollusca of Northern Africa: Dordrecht, Netherlands, W. Junk Publishers, 164 p.

Vondra, C.F., Yemane, T., Aronson, J.K., and Walter, R.C., 1996, A major disconformity within the Hadar Formation: Geological Society of America Abstracts with Programs, v. 28, no. 6, p. A72.

Walter, R.C., 1981, The Volcanic History of the Hadar Early-Man Site and the Surrounding Afar Region of Ethiopia [Ph.D. thesis]: Cleveland, Case Western Reserve University, 426 p.

Walter, R.C., 1994, Age of Lucy and the First Family: Laser ^{40}Ar/^{39}Ar dating of the Denen Dora Member of the Hadar Formation: Geology, v. 22, p. 6–10, doi: 10.1130/0091-7613(1994)022<0006:AOLATF>2.3.CO;2.

Walter, R.C., and Aronson, J.L., 1982, Revisions of K/Ar ages for the Hadar hominid site, Ethiopia: Nature, v. 296, p. 122–127, doi: 10.1038/296122a0.

Walter, R.C., and Aronson, J.L., 1993, Age and source of the Sidi Hakoma Tuff, Hadar Formation, Ethiopia: Journal of Human Evolution, v. 25, p. 229–240, doi: 10.1006/jhev.1993.1046.

Walter, R.C., Aronson, J.L., Chen, Y., Evensen, N.M., Smith, P.E., York, D., Vondra, C., and Yemane, T., 1996, New radiometric ages for the Hadar Formation above the disconformity: Geological Society of America Abstracts with Programs, v. 28, no. 6, p. A69.

Watchorn, F., Nichols, G., and Bosence, D., 1998, Rift-related sedimentation and stratigraphy, southern Yemen (Gulf of Aden), in Purser, B.H., and Bosence, D.W.J., eds., Sedimentation and Tectonics of Rift Basins in the Red Sea–Gulf of Aden: London, Chapman & Hall, p. 165–189.

White, T.D., Suwa, G., Hart, W.K., Walter, R.C., WoldeGabriel, G., de Heinzelin, J., Clark, J.D., Asfaw, B., and Vrba, E., 1993, New discoveries of *Australopithecus* at Maka in Ethiopia: Nature, v. 366, p. 261–265, doi: 10.1038/366261a0.

White, T., WoldeGabriel, G., Asfaw, B., Ambrose, S., Beyene, Y., Bernor, R.L., Boisserie, J.-R., Currie, B., Gilbert, H., Haile-Selassie, Y., Hart, W.K., Hlusko, L.J., Howell, F.C., Kono, R.T., Lehmann, T., Louchart, A., Lovejoy, C.O., Renne, P., Saegusa, H., Vrba, E., Wesselman, H., and Suwa, G., 2006, Asa Issie, Aramis and the origin of *Australopithecus*: Nature, v. 440, p. 883–889, doi: 10.1038/nature04629.

WoldeGabriel, G., Aronson, J., and Walter, R.C., 1990, Geology, geochronology, and rift basin development in the central sector of the Main Ethiopia Rift: Geological Society of America Bulletin, v. 102, p. 439–458, doi: 10.1130/0016-7606(1990)102<0439:GGARBD>2.3.CO;2.

WoldeGabriel, G., White, T.D., Suwa, G., Renne, P., and de Heinzelin, J., 1994, Ecological and temporal placement of early Pliocene hominids at Aramis, Ethiopia: Nature, v. 371, p. 330–333, doi: 10.1038/371330a0.

WoldeGabriel, G., Heiken, G., White, T., Asfaw, B., Hart, W.K., and Renne, P., 2000, Volcanism, tectonism, sedimentation, and the paleoanthropological record in the Ethiopian Rift system, in McCoy, F.W., and Heiken, G., eds., Volcanic Hazards and Disasters in Human Antiquity: Geological Society of America Special Paper 345, p. 83–99.

WoldeGabriel, G., Haile-Selassie, Y., Renne, P., Hart, W.K., Ambrose, S., Asfaw, B., Heiken, G., and White, T., 2001, Geology and palaeontology of the late Miocene Middle Awash valley, Afar Rift, Ethiopia: Nature, v. 412, p. 175–178, doi: 10.1038/35084058.

Wolfenden, E., Ebinger, C., Yirgu, G., Deino, A., and Ayelew, D., 2004, Evolution of the northern Main Ethiopian Rift: Birth of a triple junction: Earth and Planetary Science Letters, v. 224, p. 213–228, doi: 10.1016/j.epsl.2004.04.022.

Wynn, J.G., Roman, D., and Alemseged, Z., 2004, Geology and stratigraphy of sediments below the Sidi Hakoma Tuff in the Dikika Research area, Lower Awash Valley, Ethiopia: Geological Society of America Abstracts with Programs, v. 36, no. 5, p. A486.

Wynn, J.G., Alemseged, Z., Bobe, R., Geraads, D., Reed, D., and Roman, D., 2006, Geological and palaeontological context of a Pliocene juvenile hominin at Dikika, Ethiopia: Nature, v. 443, p. 332–336, doi: 10.1038/nature05048.

Yemane, T., 1997, Stratigraphy and Sedimentology of the Hadar Formation, Afar, Ethiopia [Ph.D. thesis]: Ames, University of Iowa, 204 p.

MANUSCRIPT ACCEPTED BY THE SOCIETY 17 JUNE 2008

The Geological Society of America
Special Paper 446
2008

Composite tephrostratigraphy of the Dikika, Gona, Hadar, and Ledi-Geraru project areas, northern Awash, Ethiopia

D.C. Roman
Department of Geology, University of South Florida, 4202 E. Fowler Avenue, SCA 528, Tampa, Florida, 33620, USA

C. Campisano
Institute of Human Origins, Arizona State University, P.O. Box 874101, Tempe, Arizona 85287-4101, USA

J. Quade
Department of Geosciences, University of Arizona, Gould-Simpson Building #77, 1040 E. 4th Street, Tucson, Arizona 85721, USA

E. DiMaggio
J R. Arrowsmith
School of Earth and Space Exploration, Arizona State University, Tempe, Arizona 85287, USA

C. Feibel
Department of Anthropology, Rutgers University, 131 George Street, New Brunswick, New Jersey 08901-1414, USA, and Department of Geological Sciences, Rutgers University, 610 Taylor Road, Piscataway, New Jersey 08854-8066, USA

ABSTRACT

Mapping and description of the Hadar and Busidima Formations in the northern Awash valley, Ethiopia, have been greatly aided by the use of tephrostratigraphy and tephra correlation in the Dikika, Gona, Hadar, and Ledi-Geraru paleoanthropological project areas. The Hadar Formation contains at least nine dated tuffs, many of which have been correlated across the northern Awash project areas, and all of which are easily distinguished from each other on the basis of major-element chemistry. The overlying Busidima Formation contains at least 35 distinct tuffs, many of which are firmly or approximately dated. Because of their discontinuous and compositionally similar nature, many of the Busidima Formation tuffs are not correlated across the northern Awash project areas. Trace-element compositional data or detailed stratigraphic information may be necessary for correlation or relative placement of many of the Busidima Formation tuffs. Differences in the frequency, chemistry, and extent of Hadar and Busidima Formation tuffs preserved in the northern Awash valley may ultimately be related to the tectonic evolution of the region throughout the Pliocene-Pleistocene, as well as to basin-scale geological processes. Despite a number of known issues in tephra correlation, the composite tephrostratigraphy assembled for the northern Awash valley demonstrates the effectiveness of this technique, which has played a key role in ongoing efforts to document the geological history of this unique and important region.

Keywords: tephrostratigraphy, East Africa volcanism, volcano-tectonics, rhyolite tuff, Hadar Formation, Busidima Formation.

Roman, D.C., Campisano, C., Quade, J., DiMaggio, E., Arrowsmith, J R., and Feibel, C., 2008, Composite tephrostratigraphy of the Dikika, Gona, Hadar, and Ledi-Geraru project areas, northern Awash, Ethiopia, *in* Quade, J., and Wynn, J.G., eds., The Geology of Early Humans in the Horn of Africa: Geological Society of America Special Paper 446, p. 119–134, doi: 10.1130/2008.2446(05). For permission to copy, contact editing@geosociety.org. ©2008 The Geological Society of America. All rights reserved.

INTRODUCTION

The analysis of tephra layers preserved in the sedimentary record has long been a key tool for establishing the age and context of significant paleoanthropological finds in East Africa (e.g., Feibel, 1999). Tephra, a generic term for all airborne volcanic ejecta, is typically comprised mostly of fragmented volcanic glass shards (volcanic ash) and pumice or lapilli clasts, but it may also contain crystals precipitated from the erupted magma. Potassium-rich minerals contained in a tephra layer (or "tuff") allow direct dating of the tuff through K-Ar or $^{40}Ar/^{39}Ar$ analysis. Even in cases where the tuff contains only glass, it may still be dated indirectly through correlation on the basis of glass chemistry to a tephra from a distant location that represents the same eruption but contains dateable crystals. Tephra analysis is also useful for local or basin-scale reconstruction of stratigraphy, particularly in highly faulted and/or poorly exposed regions of Neogene strata, such as the Afar of Ethiopia. In this context, correlation between individual tephra samples from distinct outcrops on the basis of glass chemistry is sufficient, although dateable material or a correlation to a dated tuff is necessary to place the section in an absolute age frame. Tephra correlation between widespread areas aids not only in interpretations of local geology, but also of basin-scale and regional tectonic history. In addition, the lack of correlation of tephra is a useful indicator of variable volcanological and paleoenvironmental conditions across an area. Furthermore, a complete stratigraphic framework or composite stratigraphy is critical for studies of long-term changes in climate, environment, and fauna, all of which are important for understanding human evolutionary patterns.

The Awash River Valley, located in the Afar region of Ethiopia, has yielded significant paleoanthropological finds for over four decades (e.g., Johanson and Taieb, 1976; Semaw et al., 2005; Alemseged et al., 2006). Four of the active research projects located in the northern Awash valley, the Dikika, Gona, Hadar, and Ledi-Geraru research projects (informally known as the "northern Awash projects"; Fig. 1), have worked to expand geological mapping of this region in recent years, aided strongly by tephra analysis and identification both independently and through collaboration and discussion among the project geologists.

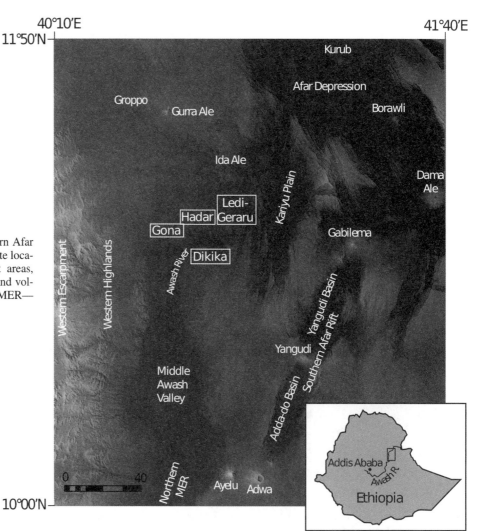

Figure 1. Shaded relief map of the western Afar region of Ethiopia, showing the approximate location of the four northern Awash project areas, along with major physiographic features and volcanic centers (volcano names in italics). MER—Main Ethiopian Rift.

To date, several major tephras (those that are widespread and/or have known extrabasinal correlates) have been identified in one or more of the project areas, and a significant number of tephra correlations have been identified among all four of the project areas. This allows us to construct a preliminary composite tephrostratigraphy for the Pliocene-Pleistocene strata of the northern Awash valley, one of the main aims of this paper.

In this paper, we provide a summary of the tephra data collected by the northern Awash projects through the 2006–2007 field seasons (details of tephra analyses and tephrostratigraphy in individual project areas may be found in: Campisano and Feibel, this volume; DiMaggio et al., this volume; Quade et al., this volume; and Wynn et al., this volume). While many of the mapped tuffs in the northern Awash valley are dated and/or have known stratigraphic positions, some are still unidentified, and we summarize all that is known about them here. We present a composite tephrostratigraphy based on data collected to date by the four northern Awash projects, including age information where known, and we summarize all currently known correlations between the four project areas. We also list any extrabasinal correlations that have been identified. As tephrostratigraphy is based on the assumption that each volcanic event produces tephra with a unique chemical composition, we analyzed the compositional differences between tephras of the Hadar and Busidima Formations and assessed the extent to which they can be uniquely distinguished. Although it is likely that the synthesis presented in this paper will change with time as mapping and analysis progress, we believe that an effort to assess our current knowledge of tephrostratigraphy in the northern Awash valley at this stage is warranted, as it will provide a cohesive framework for future data collection and analysis.

METHODS SUMMARY

Major-element compositions of tephra samples were determined using electron-probe microanalysis (EPMA) by all four of the northern Awash projects. EPMA is a standard technique for the analysis of vitric tephra, and it is well suited to it for a number of reasons. Most importantly, because of the often highly contaminated and heterogeneous nature of tephra samples from the northern Awash valley, a method that allows analysis of individual glass shards is preferred. A summary of analytical conditions and corrections used by each project team is presented next, and a summary of EPMA beam conditions is given in Table 1.

Dikika Research Project

Samples from Dikika prefixed E02- and E03- were analyzed on a Cameca SX-50 electron probe microanalyzer housed in the Department of Geological Sciences at the University of Oregon. Samples prefixed E04- were analyzed on a Cameca SX-50 housed in the Institute for Materials Research at the University of Leeds, England, and samples prefixed E06- were analyzed on a JEOL JXA-8900-R superprobe housed in the Department of Earth Sciences at Florida International University. All Dikika samples were analyzed by D. Roman. The primary considerations in choosing analytical conditions were (1) the potentially high water contents of the glasses, which would result in an elevated sodium migration effect, and (2) a concern that some of the glass shards to be analyzed were extremely thin. Beam settings were 10–12 keV, aimed at limiting the analytical volume to less than 10 μm to facilitate the analysis of fine-grained tephras while producing enough overvoltage to obtain reliable count data for high-energy elements such as iron; a current of 5 nA, chosen to minimize the amount of sodium migration; and a spot size between 5 and 10 μm, again chosen to minimize the amount of sodium migration while still allowing analysis of fine-grained tephra. On-peak count times were 20 s for all analyzed elements. For a subset of early analyses, sodium migration was corrected using a self-correction model, which involved sampling at 4 s intervals and extrapolating backward to time $t = 0$. Silicon and potassium counts were also corrected using this model. However, for consistency with other published tephra analyses, uncorrected analyses were used for tephra correlation, and the corrected data have only been used to assess the magnitude of sodium loss in the tephra analyses.

Gona Research Project

Samples from Gona prefixed BUST-, ASASH-, AST-, GONASH-, and DANASH- were analyzed by J. Quade and students on a Cameca SX-50 housed at the University of Arizona. An original suite of analyses obtained with beam conditions of

TABLE 1. SUMMARY OF ELECTRON-PROBE MICROANALYSIS (EPMA) BEAM CONDITIONS USED TO COLLECT DATA PRESENTED IN THIS STUDY

Project - analysis set	Voltage (keV)	Current (nA)	Spot size (μm)	On-peak count time (s)
Dikika	10–12	5	10	20
Gona - Set 1 (JQ)	15	20	2	20
Gona - Set 2 (JQ)	15	8	10	10
Gona - CF/Utah	15	25	5–25	
Hadar - Rutgers	15	15	12 (rastered)	20
Hadar - AMNH	15	10 (2 for Na, K)	15	
Ledi-Geraru	15	10	15	10

Note: See text for additional details on analysis sets and instrument type and location.

15 keV, 20 nA, and 2 μm spot was partly superseded by a second suite of analyses obtained with a beam set at 15 keV and 8 nA, with a spot size of 10 μm (see Quade et al. [this volume] for details). Count times were 20 s in the original suite of analyses and 10 s in the second suite. No sodium correction model was used. Samples prefixed E93- and E94- were analyzed by C. Feibel (with assistance from Marsha Smith and Bereket Haileab) using a Cameca SX-50 microprobe housed at the University of Utah Department of Geology and Geophysics and beam conditions of 15 keV, 25 nA, and a beam diameter of 5–25 μm. Some of the E93- and E94- Gona samples were reanalyzed at Rutgers University and the American Museum of Natural History, using the instruments and beam conditions described below.

Hadar Research Project

Samples from Hadar were prepared by C. Campisano and analyzed by Godwin Mollel on a JEOL JXA-8600 superprobe housed at Rutgers University. Beam conditions were 15 keV and 15 nA, with the beam rastered over a 12 μm^2 area to minimize sodium loss. Sodium, silicon, and potassium migration were corrected using a self-correction model. On-peak count times were 20 s for all analyzed elements. Some of the Hadar samples were reanalyzed by G. Mollel on a Cameca SX100 microprobe housed at the American Museum of Natural History, using beam conditions of 15 keV and 10 nA, and a spot size of 15 μm for all elements except Na and K, which were analyzed with a beam set at 15 keV and 2 nA to minimize sodium migration. Additional details can be found in Campisano and Feibel (this volume).

Ledi-Geraru Research Project

Samples from Ledi-Geraru were analyzed by E. DiMaggio on a JEOL JXA-8600 superprobe housed at Arizona State University. Beam conditions were 15 keV and 10 nA, with a spot size of 15 μm to minimize alkali loss. No sodium correction model was used. On-peak count times were 10 s for all analyzed elements. Additional details can be found in DiMaggio (2007).

COMPOSITE TEPHROSTRATIGRAPHY

Hadar Formation

The currently known composite tephrostratigraphy of the Pliocene Hadar Formation (Taieb et al., 1976) in the northern Awash valley is summarized in Table 2. Nine distinct Hadar Formation tuffs, plus the Kada Damum Basalt flow, have been mapped in the northern Awash valley to date. Three of the Hadar Formation tephras could be chemically correlated to extrabasinal tephras of known age. These are, in chronological order, the Ikini Tuff, correlated to the Wargolo and VT-3 tuffs of Turkana and the Middle Awash valley, dated to 3.8 Ma by Brown et al. (1992) and deMenocal and Brown (1999); the Arbosh Tuff, correlated to the Maka Sand Tuff of the Middle Awash, dated to ca. 3.42 Ma[1] by White et al. (1993); and the well-known Sidi Hakoma Tuff, correlated to the Tulu Bor β, U-10, and B-β tuffs of the Turkana and Omo regions, and dated to 3.42 Ma by Walter and Aronson (1993). The Sidi Hakoma Tuff has been found in all four of the northern Awash project areas, but the Arbosh and Ikini Tuffs have so far only been identified at Dikika. One new key observation regarding Hadar Formation tephrostratigraphy is that the Arbosh Tuff clearly lies just below the Sidi Hakoma Tuff at Dikika (Wynn et al., this volume), confirming the suggestion by White et al. (1993) that an age of ca. 3.42 Ma is appropriate for this tuff.

Two other Hadar Formation tuffs, the Kada Hadar Tuff and the BKT-2 tuff complex, have also been mapped in all four northern Awash project areas. The Kada Hadar Tuff was dated to 3.20 Ma by Walter (1994) but does not always contain unaltered glass suitable for analysis by EMPA. Where the tuff contains no fresh glass, it is identifiable by the presence of a distinctive green shale bed known as the Confetti Clay, which occurs ~4 m above the Kada Hadar Tuff. The BKT-2 complex has been consistently dated to 2.96 Ma (BKT-2L) to 2.94 Ma (BKT-2U) (Semaw et al., 1997; Campisano, 2007; DiMaggio et al., this volume). This tuff complex is easily distinguishable by the presence of an underlying basaltic tuff known as the Green Marker Bed. The tuff complex is generally described as having at least two additional units, BKT-2L and BKT-2U, which have compositionally identical rhyolitic glasses but chemically distinct feldspar populations, and which also differ by the presence of basaltic scoria in BKT-2U (Campisano and Feibel, this volume). In most places outside of the Ledi-Geraru and eastern Hadar regions, the BKT-2 complex contains only altered (unanalyzable) glass but is still identifiable by its internal stratigraphy and lithology. A detailed description of the BKT-2 tuff complex is provided by Campisano and Feibel (this volume) and DiMaggio et al. (2008, this volume).

The other four northern Awash Hadar Formation tuffs (the Kada Me'e Tuff, Triple Tuff-4, the BKT-1 complex, and an unnamed tuff, HE04-442 from Hadar, in Table 2) preserve no fresh glass but do contain dateable feldspars. Only one of these, the Triple Tuff-4 (TT-4), has been mapped outside of Hadar in the northern Awash valley. TT-4 is easily identifiable in the field by the near-continuous layer of sanidine crystals at its base (Walter, 1994) and its lithologic context within an ostracod-bearing green fissile shale. TT-4 has been dated to 3.24 Ma by Walter (1994) and 3.256 Ma by Campisano (2007). So far, the TT-4 has been positively identified at Hadar, Dikika, and Ledi-Geraru.

Overall, the nine tuffs of the Hadar Formation identified in the northern Awash valley may be easily placed into a composite tephrostratigraphy because many have been directly dated, and stratigraphic relationships between these tuffs have

---

[1]Previously published ^{40}Ar/^{39}Ar ages reported in this study have been increased by approximately 0.65% to reflect the revised age of the monitor mineral used (Fish Canyon Sanidine) from 27.84 to 28.02 Ma (Renne et al., 1998), to be consistent with more recent publications.

TABLE 2. COMPOSITE TEPHROSTRATIGRAPHY OF THE HADAR FORMATION, NORTHERN AWASH REGION

Tuff name	Extrabasinal correlate	Age (Ma)	Age reference	Gona	Dikika	Hadar	Ledi-Geraru
BKT-2 tuff complex		2.94–2.96	Campisano (2007) DiMaggio (this volume) Semaw et al. (1997)	No sample	E05-4033	HE01-151 HE04-457 HE01-182 HE01-184	AM06-1013 LG33E AM06-1017 LG33F
						E01-7355 HE01-185 HE04-456 HE01-186 HE04-458 E01-7357 E01-7183 E01-7358 E01-7356	AM06-1018 LG33G AM06-1019 LG33H AM06-1020 LG22 MLG2002-36a LG15 MLG2002-37c
(Unnamed tuff)		3.01	Campisano (2007)			HE04-442	
BKT-1 tuff complex		3.12*	Campisano (2007)			HE04-431 HE01-180 HE04-432 HE04-459	
Kada Hadar Tuff		3.2	Walter (1994)	GONASH-2 GONASH-10	E05-4024 E03-3041	E01-7341 E02-7413	AM06-1047
Triple Tuffs (TT-4)		3.26	Campisano (2007) Walter (1994)		E02-1074 E03-2013 E02-1075 E03-2030 E02-1089 E05-4023	HE04-427	
Kada Damum Basalt		3.3	Renne et al. (1993)		E03-3044	HE04-460 E01-7343	
Kada Me'e Tuff		3.36	Campisano (2007)			HE01-194 HE04-415	
Sidi Hakoma Tuff	Tulu Bor β U-10 B-β	3.42	Walter and Aronson (1993)	GONASH-6	E02-1067 E02-3012 E02-1096 E02-3017 E02-1051 E02-3018 E02-2003 E02-3019 E02-2006 E02-3020	HE01-207 HE01-208 HE01-140	AM05-101
Arbosh Tuff	Maka Sand Tuff	Ca. 3.4	White et al. (1993)		E02-2005		
Ikini Tuff	Wargolo	3.8†	deMenocal and Brown (1999)		E02-1110		
	VT-3	3.77	White et al. (1993)		E02-2004		

Note: Italicized names/sample numbers indicate tuffs/samples with no fresh (analyzable) glass. Italicized ages were determined by stratigraphic scaling. All other ages were determined by $^{40}Ar/^{39}Ar$ unless otherwise indicated. All $^{40}Ar/^{39}Ar$ ages are reported in reference to a neutron fluence monitor with an age of 28.02 Ma (Renne et al., 1998). Previously published ages not using this monitor age have been recalculated for consistency. Sample numbers correspond to geochemical analyses given in individual project area papers (see text for references).
*Close proximity to base of Kaena subchron.
†Orbitally tuned from Ocean Drilling Program (ODP) marine core.

been extensively documented. Thus, there is currently no ambiguity regarding the relative position of these tuffs, and they are all easily identifiable based on a combination of chemical and field characteristics.

Busidima Formation

Compared to that of the Hadar Formation, the composite tephrostratigraphy of the Pliocene-Pleistocene Busidima Formation (Quade et al., 2004) in the northern Awash valley is significantly less constrained and still under development, because of (1) the discontinuous, cut-and-fill nature of Busidima Formation sedimentation (e.g., Quade et al., this volume), (2) the presence of multiple chemically similar tuffs (discussed in detail in the following sections), (3) a paucity of dateable material in many of the tuffs, and (4) the shorter duration of intensive study of the Busidima Formation (since 2000 versus since ca. 1972 for the Hadar Formation). The current composite tephrostratigraphy of the Busidima Formation is given in Tables 3 and 4 and Figure 2. The relative stratigraphic position of virtually all the tuffs has been established within each project area, and most of the tuffs can be firmly placed in a master stratigraphy encompassing the entire project area (Fig. 2). From these analyses, it is apparent that that the Busidima Formation contains a large number of distinct tuffs.

The absolute ages of 13 of the Busidima Formation tuffs are known either through direct dating or through chemical correlation to a dated extrabasinal tuff, and the approximate ages of 10 of the Busidima Formation tuffs are known from stratigraphic scaling relative to magnetostratigraphic data. Of these, four tuffs have been identified in more than one of the three northern Awash project areas in which the Busidima Formation is exposed (Gona, Hadar, and Dikika). These are the Waidedo Vitric Tuff (WAVT), identified so far at Gona and Dikika and correlated to a tuff of the same name from the Herto area in the Middle Awash valley (Clark et al., 2003); the Bironita Tuff, a correlate of an unnamed tuff from the Bodo-Hargufia-Dawaitoli area in the Middle Awash valley (Clark et al., 1994), identified so far at Gona and Dikika; the AST-3/Fialu Tuff, identified so far at Gona, Dikika, and Hadar; and the Dahuli Tuff, identified so far at Gona and Hadar.

As illustrated in Figure 2, while the tephrostratigraphic sequence within a single project area may be understood, the overall relative stratigraphic positions of the remaining tuffs are known to varying degrees of accuracy. For these tephras, exact stratigraphic placement is dependent upon (1) chemical correlation to a tephra with a known radioisotopic age date and/or (2) knowledge of stratigraphic position with respect to tephras of known age. Fortunately, many of the Busidima Formation tephras contain unaltered glass, and the major-element composition of the glass shards allows them to be grouped and/or correlated to some degree (see following). Seven of the remaining tephras have been identified in more than one of the northern Awash project areas. Several of the currently ambiguous correlations remain to be verified by other means, including trace-element analysis, and are thus not shown in Figure 2.

TEPHRA CHEMISTRY

The majority of the tephras from the northern Awash valley are iron-poor or iron-rich subalkaline rhyolites (Fig. 3). The exceptions are (1) occasional subalkaline basaltic tephras and basalt flows, and (2) rare basaltic andesites and trachydacites. Figure 3 illustrates major-element compositional differences between tephras of the Hadar and Busidima Formations: Hadar Formation rhyolitic tephras have higher SiO_2 contents and are slightly less alkaline on average than Busidima Formation rhyolitic tephras.

Hadar Formation

Tephras of the Hadar Formation preserved in the northern Awash project areas are easily distinguishable on the basis of major-element chemistry. A bivariate plot of CaO versus Fe_2O_3t (total Fe is reported as Fe_2O_3 throughout) in the Hadar Formation tephras (Fig. 4) demonstrates that they may be identified on the basis of these two elements alone. The Kada Hadar Tuff is unique in the Hadar Formation because of its high iron content ($Fe_2O_3t > 5$ wt%), and the other four Hadar Formation tephras containing unaltered glass (with similar Fe_2O_3t contents of ~2–3 wt%) can be identified by their calcium content: the Ikini Tuff has 0.1–0.2 wt% CaO, the Sidi Hakoma Tuff has 0.3–0.4 wt% CaO, the BKT-2 rhyolite tuffs have 0.5–0.6 wt% CaO, and the Arbosh Tuff has a CaO content of 0.7 wt%.

Busidima Formation

Many of the tephras of the Busidima Formation are not easily distinguished on the basis of major-element chemistry, as demonstrated by a bivariate plot of CaO versus Fe_2O_3t (Fig. 5). While several Busidima Formation tephras (the Burahin Dora Tuff, the Bironita Tuff, the Ken Di Tuff, Elamuita Tuff, sub–Ken Di Tuff, Mesele Tuff, Inaalale Tuff, We-Aytu A Tuff, Ridge Tuff, Busidima Tuff, Talata Tuff, and an unnamed tuff-2 [found at both Hadar and Gona]) can be distinguished uniquely on the basis of these two elements and the degree of compositional bimodality, four groups of tephras are indistinguishable on this diagram and on the basis of Fe_2O_3 and CaO compositional data (e.g., Quade et al., this volume; Wynn et al., this volume; Campisano and Feibel, this volume). The first group includes the Waidedo Vitric Tuff, Waterfall, Boolihinan, and Gawis Tuffs. Within this group, the Gawis Tuff is distinguished by its higher aluminum content, and the Waterfall Tuff is distinguished by its high silicon and low potassium contents; however, the Waidedo Vitric Tuff and Boolihinan Tuffs are virtually identical on the basis of major-element composition, despite clear stratigraphic evidence (e.g., Fig. 2) that they represent distinct tuffs. Group two includes the Alakata and Silbo Tuffs, which have slightly different silicon and sodium contents but may also represent a single tuff (pending additional paleomagnetic analysis). Group three includes the We-Aytu B and Biidi Foro Tuffs, which are also indistinguishable and may represent a single tuff (pending trace-element analysis).

TABLE 3. COMPOSITE TEPHROSTRATIGRAPHY OF THE BUSIDIMA FORMATION, NORTHERN AWASH REGION: TUFFS WITH RADIOMETRICALLY DETERMINED AGES

Tuff name	Extrabasinal correlate	Age (Ma)	Age reference	Gona	Dikika	Hadar
Waidedo Vitric Tuff	WAVT[†]	0.16	Clark et al. (2003)	ASASH-3, ASASH-18, ASASH-19, ASASH-21, ASASH-11, ASASH-12, ASASH-14	E06-7013, E06-7050	
Bironita Tuff	Unnamed (MA)	0.64	Clark et al. (1994)	ASASH-6	E02-1151, E02-1126, E02-1111, E06-7002, E02-1127, E03-3065, E06-7008	
~Bironita Tuff[§]		0.64?			E03-3066	
Silbo Tuff	Silbo Tuff	0.75	McDougall and Brown (2006)	BUST-20*, BUST-23		
(Unnamed tuff)		1.64	Quade et al. (2004)	GONASH-16		
Boolihinan Tuff	DEM-4-1	1.65	Quade et al. (this volume)	BSN-12 cr*, BUST-1*, BUST-10		
(Unnamed tuff)		1.96	Quade et al. (this volume)	GONASH-79		
(Unnamed tuff)		2.17	Quade et al. (2004)	GONASH-21		
Salal Me'e Tuff		Ca. 2.2	Campisano (2007)			HE01-103, HE01-202, HE01-107, HE04-448, HE04-449, HE02-7447, HE02-7390
(Unnamed tuff)		2.27	Quade et al. (2004)	GONASH-41		
BKT-3		2.35	Kimbel et al. (1996)			E02-7389, E02-7388, E02-7450, E02-7400, HE01-196, HE02-7387, HE02-7386, HE02-7399, HE01-106, HE02-7385, HE01-176, HE01-175, HE01-108, HE04-447B
~BKT-3[§]		2.35?				HE01-201, HE02-7380, HE01-108
AST-2.75		2.53	Semaw et al. (1997)	No sample		
(Unnamed tuff)		2.53	Quade et al. (2004)	GONASH-14		
(Unnamed tuff)		2.69	Quade et al. (2004)	GONASH-39		

Note: Tuff and sample names in italics indicate no fresh (analyzable) glass. Ages in italics were determined by stratigraphic scaling to magnetostratigraphy. All $^{40}Ar/^{39}Ar$ ages are reported in reference to a neutron fluence monitor with an age of 28.02 Ma (Renne et al., 1998). Sample numbers correspond to geochemical analyses given in individual project area papers (see text for references). Previously published ages not using this monitor age have been recalculated for consistency.
*Samples not plotted in Figures 5 and 6 (these are samples analyzed with 20 nA beam current; see Quade et al., this volume).
[†]WAVT lies directly below the dated Silver Tuff at Konso (Clark et al., 2003).
[§]Not shown as discrete tuffs in Figures 2, 5, and 6 (see Wynn et al. [this volume] and Campisano and Feibel [this volume] for further discussion).

125

TABLE 4. COMPOSITE TEPHROSTRATIGRAPHY OF THE BUSIDIMA FORMATION, NORTHERN AWASH REGION: OTHER TUFFS

Tuff name	Extrabasinal correlate	Age (Ma)	Age reference	Gona	Dikika	Hadar
Butte Tuff		<0.15	Quade et al. (this volume)	GONASH-37* GONASH-40*		
Korina Tuff				ASASH-1* ASASH-16	E06-7022 E06-7023 E06-7049	
Talata Tuff		Ca. 0.45	Quade et al. (this volume)	ASASH-4* ASASH-17		
Gawis Tuff		Ca. 0.5	Quade et al. (this volume)	BUST-21* ASASH-7* ASASH-5* ASASH-8 ASASH-10 ASASH-23 ASASH-22 ASASH-24		
Negus Kabri Tuff					E02-3067 E06-7003	
Odele Tuff				ASASH-2*	E06-7026	
Busidima Tuff		Ca. 0.7	Quade et al. (this volume)	BUST-24		
Dahuli Tuff		0.83	Quade et al. (this volume)	BUST-11* BUST-13* BUST-2 BUST-26 BUST-3* BUST-27		HE01-190 E02-7394 HE01-209 HE01-145
We-Aytu B Tuff					E06-7011	
We-Aytu A Tuff					E06-7010	
(Unnamed tuff-1)					E06-7017 E06-7018	
Upper Hurda Tuff						HE01-164
Ken Di Tuff				GONASH-86	E03-4025 E03-4026	E02-7430
Burahin Dora Tuff				E94-5455 E93-5456		E02-7384 HE01-171 HE01-188 HE01-169 E02-7383
Sub–Ken Di Tuff				GONASH-87		
Alakata Tuff				GONASH-43* GONASH-61* GONASH-25 GONASH-84*		
Ridge Tuff		Ca. 0.9	Quade et al. (this volume)	GONASH-32 GONASH-83		

(continued)

TABLE 4. COMPOSITE TEPHROSTRATIGRAPHY OF THE BUSIDIMA FORMATION, NORTHERN AWASH REGION: OTHER TUFFS (continued)

Tuff name	Extrabasinal correlate	Age (Ma)	Age reference	Gona	Dikika	Hadar
Camp Tuff		1	Quade et al. (this volume)	GONASH-30* GONASH-33 GONASH-31* GONASH-44* GONASH-82		
Elamuita Tuff					E06-7001 E06-7016 E06-7015	
AST-3/Fialu Tuff†		Ca. 1.3	Quade et al. (this volume)	AST-3 GONASH-20 AST3(24) GONASH-68 GONASH-15 GONASH-24 GONASH-18 BUST-18 GONASH-78 AST3(85)	E02-1109 E03-4028 E02-3037 E03-4022 E03-4041 E03-4013 E03-4017	HE01-143 HE01-172 HE01-210 HE01-165 HE01-189
Biido Foro Tuff					E06-7009 E06-7012	
Waterfall Tuff		Ca. 1.7	Quade et al. (this volume)	GONASH-49		HE01-177
Sub-Waterfall Tuff		Ca. 1.7	Quade et al. (this volume)	GONASH-48 BUST-5* GONASH-62*		
Inaalale Tuff				E93-5035	E02-2015 E02-4031 E02-3034	
Mesele Tuff					E04-4034	
Unnamed tuff-2				E93-4973		HE01-168
AST-2.5				No sample		
AST-2				E93-4997 AST2(23)		
AST-1				E93-4971 E93-4992		HE04-443 HE01-167 HE01-142
Unnamed tuff						HE01-441

Note: Tuff and sample names in italics indicate no fresh (analyzable) glass. Ages in italics were determined by stratigraphic scaling to magnetostratigraphy. Sample numbers correspond to geochemical analyses given in individual project area papers (see text for references).
*Samples not plotted in Figures 5 and 6 (these are samples analyzed with 20 nA beam current; see Quade et al., this volume).
†See Quade et al. (this volume) for a discussion of this correlation.

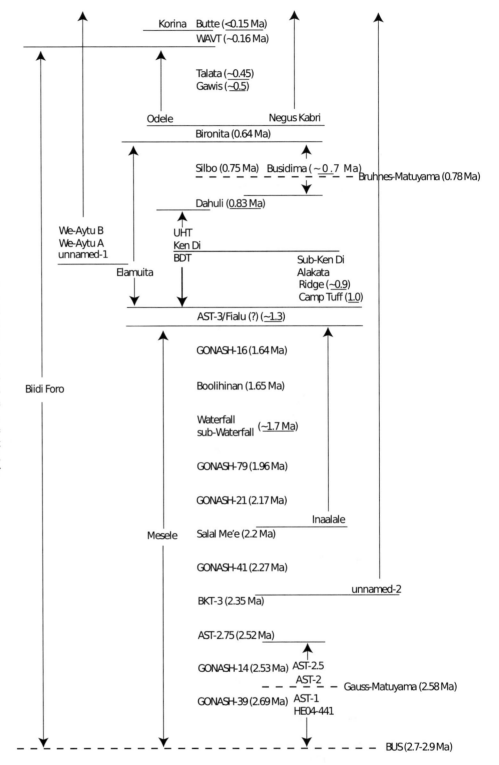

Figure 2. Composite tephrostratigraphy of the Busidima Formation showing absolute or relative chronological order of all analyzed tuffs. Dated tuffs have age in parentheses; see Tables 3 and 4 for dating methods, references, and individual sample numbers (corresponding to geochemical analyses presented in individual project papers [this volume]). Underlined ages were obtained from stratigraphic scaling. BUS—Busidima unconformity surface of Wynn et al. (2006). AST—Artifact Site Tuff, BDT—Bucahin Dora Tuff, BKT—Bouroukie Tuff, UHT—Upper Hurda Tuff, WAVT—Waidedo Vitric Tuff.

Group four includes a large number of tuffs (the Dahuli Tuff, Salal Me'e Tuff [SMT], Negus Kabri Tuff [NKT], sub-Waterfall Tuff, Upper Hurda Tuff [UHT], Camp Tuff, AST-3, BKT-3, an unnamed tuff-1 from Dikika, AST-1, AST-2, and the Odele and Korina Tuffs), several of which (the AST-3, BKT-3, and Dahuli in particular) have wide compositional ranges that overlap with or lie in close proximity to each other and to more compositionally constrained tuffs in this group.

Campisano (2007) demonstrated that the rhyolite tephras of the Busidima Formation from Hadar can be distinguished by major-element composition on the basis of Fe_2O_3t, CaO, and Al_2O_3. We plot the fourth group of Busidima Formation tephras

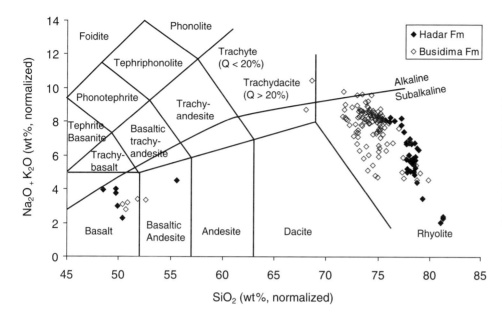

Figure 3. Bivariate plot of total alkalis ($Na_2O + K_2O$) versus SiO_2 (after Le Maitre, 2002) of tephras from the Hadar and Busidima Formations, northern Awash region. Data, consisting of an average analysis for each sample, are plotted on a volatile-free basis (normalized to 100%). The continuous line represents the alkalic-tholeiitic boundary of Irvine and Baragar (1971).

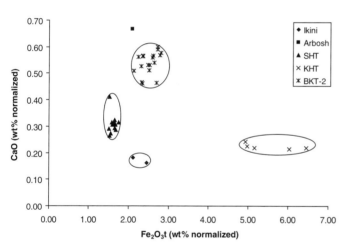

Figure 4. Bivariate plot (CaO versus Fe_2O_3t) of tephra analyses (average analysis for each sample) from the Hadar Formation, northern Awash region. See Table 2 for sample numbers corresponding to geochemical analyses in individual project papers (this volume). BKT—Bouroukie Tuff, KHT—Kada Hadar Tuff, and SHT—Sidi Hakoma Tuff.

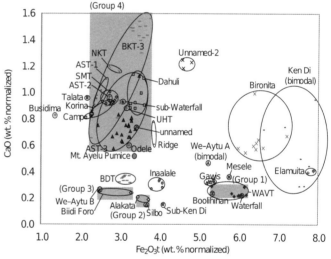

Figure 5. Bivariate plot (CaO versus Fe_2O_3t) of tephra analyses (average analysis for each sample) from the Busidima Formation in the northern Awash region. Four groups of overlapping and/or proximal tephras are shown with shaded regions, and these are discussed in the text. Averages of individual modes for bimodal tuffs are plotted rather than an average analysis for the entire sample. The composition of a pumice sample from the flanks of Mt. Ayelu (E03-3026, Dikika; see Fig. 1 for location) is also shown for comparison. AST—Artifact Site Tuff, BDT—Burahin Dora Tuff, BKT—Bouroukie Tuff, NKT—Negus Kabri Tuff, SMT—Sala Me'e Tuff, UHT—Upper Hurda Tuff, WAVT—Waidedo Vitric Tuff.

(from Fig. 5) in an Fe_2O_3t-CaO-Al_2O_3 ternary diagram in Figure 6, and we find that the group four tuffs again overlap or lie in close proximity to each other in this three-component compositional space. Additional inspection of these tuffs indicates that there is little difference between major-element compositions, despite the fact that these tuffs occupy positions throughout the entire Busidima Formation and thus clearly represent different eruptions. One possibility for the discrepancy between our ternary analysis and the analysis of Campisano (2007) is that we are considering tephras analyzed on several different instruments using a range of beam conditions, while Campisano (2007) considered only analyses from one instrument made under a single set of beam conditions. This is clearly not the only explanation for the compositional overlap of the Busidima Formation tuffs, however, since group one (Fig. 5) represents analyses of two verifiably different tephras from a single instrument and set of beam conditions (the Boolihinan and Waidedo Vitric Tuff; see Quade et al., this volume) and still shows compositional overlap in all major elements (see Quade et al., this volume).

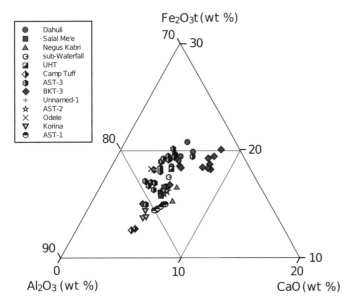

Figure 6. Fe_2O_3t-CaO-Al_2O_3 ternary diagram of "group four" (from Fig. 5) Busidima Formation tephras (average analysis for each sample). Original concentrations of the three oxides were normalized to 100%. AST—Artifact Site Tuff, BKT—Bouroukie Tuff, and UHT—Upper Hurda Tuff.

DISCUSSION

Apparent differences in the character of tuffs from the Hadar and Busidima Formations likely reflect, in large part, changes in sedimentological and tectonic processes affecting the northern Awash region through the Pliocene and Pleistocene. The period represented by the Hadar Formation is characterized by relatively continuous (both lateral and temporal) deposition. This is reflected in the ubiquitous, laterally continuous, and easily traceable nature of the Hadar Formation tuffs, which were deposited in a low-topography and rapidly subsiding sedimentary basin. In contrast, the discontinuous and complex nature of the Busidima Formation tuffs reflects, at least in part, the large number of major erosional events (also indicated by a number of cut-and-fill sequences preserved in the Busidima; e.g., Quade et al., this volume) that took place during the time represented by the Busidima Formation. As discussed in more detail later, additional differences between the Hadar and Busidima Formation tephras may also reflect larger tectonic trends occurring in East Africa during the Pliocene-Pleistocene.

Issues in Tephra Correlation

The utility of interproject tephra correlation in the northern Awash valley has been demonstrated by the work completed by the northern Awash projects. Information gained from intra- and extrabasinal tephra correlation has contributed substantially to the construction of local and regional stratigraphic frameworks, and to dating of key fossils of hominin and other fauna. Information gained from tephra correlation has also provided clues to unmapped or incompletely mapped geology in the four project areas, along with clues about the tectonic and volcanic setting and history of the region. However, questions concerning the northern Awash tephras remain. In particular, key issues to be resolved are the identity of as-yet uncorrelated tephras (e.g., Fig. 2), whether it is possible to distinguish between Busidima Formation tephras of similar or overlapping major-element composition (e.g., Fig. 6) on the basis of trace elements or other information, and whether Busidima Formation tephras with wide compositional ranges (e.g., Fig. 5) represent truly heterogeneous magmas or suffer from analytical errors and inconsistencies.

Several issues inherent in tephra correlation are illustrated by our attempt to synthesize tephra data collected by the four project groups. First and foremost, there is the effect that different analytical methods and instrumental conditions have on analyses of volcanic glasses. The northern Awash projects have carried out analyses of tephra independently of each other and beginning at different times, resulting in a lack of shared methodology among the project groups or a shared reference standard analyzed on each microprobe used to collect the major-element composition data synthesized in this study. Furthermore, tephra analysis has a long history in East Africa, resulting in an extensive database of analyses collected by multiple workers using a range of techniques, instruments, and conditions (Roman and Wynn, 2004). Quantitatively assessing the effect that analytical conditions have on the resulting analysis and factoring this effect into tephra correlation is difficult, if not impossible. However, from a qualitative standpoint, it is clear that the comparison of two tephra analyses obtained using significantly different analytical approaches is often akin to comparing apples to oranges, and thus it is likely to result in meaningless or misleading correlation. There is no simple solution to this problem—a complete treatment of the problem would require agreement on a standard method and conditions (e.g., Froggatt, 1992) followed by reanalysis of hundreds of samples by this method to produce a homogeneous and self-consistent database. A more practical approach is to reanalyze potential correlates under the same instrumental conditions to confirm any hypothesized correlation. This approach has been used effectively, often in combination with analysis of trace-element data obtained by X-ray fluorescence, by several of the northern Awash projects for both intra- and extrabasinal correlations.

To illustrate expected compositional differences due to differences in analytical conditions, we present a list of paired analyses in Table 5, showing the same sample analyzed using two different methods or analytical setups. In general, the elements most likely to be affected by analytical differences are Si, Na, and K (affected by differences in microprobe beam current and/or correction model used, and by analytical method) and Fe (affected by a beam voltage below 8–10 keV). While the northern Awash projects have not agreed upon a standard set of microprobe beam conditions, the range in beam conditions, correction models, and beam apertures used by the individual project groups are shown to have a minor effect on tephra analyses (Table 5). Regardless,

TABLE 5. COMPARISON OF ANALYTICAL CONDITIONS (EACH PAIR SHOWS THE SAME SAMPLE ANALYZED USING TWO DIFFERENT METHODS/MICROPROBE SETTINGS)

Region	BC*	Tuff	Sample	Method†	Na Corr§	No. shards	SiO_2	TiO_2	Al_2O_3	Fe_2O_3	MnO	MgO	CaO	Na_2O	K_2O	Total	Orig. total	Reference
Gona	0.84	Bironita	ASASH-6	EMP (15, 8, 10)	No	20	75.94	0.41	12.34	6.38	0.26	0.01	0.55	1.61	2.50	100.00	98.79	Quade et al. (this volume)
Gona	0.84	Bironita	ASASH-6	EMP (10, 5, 10)	No	10	72.14	0.50	12.29	6.29	0.26	0.01	0.54	4.32	3.64	100.00	91.23	Roman (personal observ.)
Dikika	0.94	Bironita	E02-1127	EMP (10, 5, 10)	No	10	71.77	0.36	12.52	6.36	0.22	0.02	0.59	4.38	3.79	100.00	93.63	Geraads et al. (2004)
Dikika	0.94	Bironita	E02-1127	EMP (10, 5, 10)	Yes	10	(1.14)# 69.83	(0.08) 0.35	(0.17) 12.35	(0.17) 5.99	(0.06) 0.22	(0.02) 0.02	(0.03) 0.58	(0.86) 6.88	(0.32) 3.78	100.00	95.95	Geraads et al. (2004)
Dikika	0.89	Wargolo	E02-1110	EMP (10, 5, 10)	No	11	(1.14) 77.13	(0.08) 0.08	(0.21) 12.20	(0.40) 2.11	(0.05) 0.06	(0.02) 0.03	(0.03) 0.18	(0.87) 3.60	(0.44) 4.61	100.00	93.48	Wynn et al. (2006)
Dikika	0.89	Wargolo	E02-2004	EMP (12, 5, 10)	No	16	(0.55) 77.45	(0.09) 0.11	(0.38) 12.31	(0.13) 2.45	(0.04) 0.06	(0.02) 0.02	(0.04) 0.16	(0.29) 3.57	(0.61) 3.85	100.00	91.99	Wynn et al. (2006)
							(0.88)	(0.04)	(0.26)	(0.13)	(0.05)	(0.04)	(0.03)	(0.65)	(0.42)			
Middle Awash	0.92	Bouri	MA92-01	EMP (15, 15, 10)	No	31	74.98	0.36	9.97	6.71	0.21	0.00	0.26	3.88	3.63	100.00	93.48	Clark et al. (2003)
Middle Awash	0.92	Bouri	MA92-01	Bulk DCP	N/A	N/A	73.22	0.35	9.61	6.35	0.22	0.01	0.26	6.00	3.96	100.00	94.96	Clark et al. (2003)
Middle Awash	0.92	Konso	TA-55	EMP (15, 15, 10)	No	29	75.24	0.36	9.94	6.14	0.22	0.00	0.28	4.43	3.39	100.00	92.82	Clark et al. (2003)
Middle Awash	0.92	Konso	TA-55	Bulk DCP	N/A	N/A	73.60	0.33	9.71	6.15	0.21	0.01	0.25	5.37	4.33	100.00	95.76	Clark et al. (2003)
Middle Awash	0.95	Cindery	ERV-049	EMP (15, .1 mA, 1)	RB	10	77.84	0.24	12.42	3.15	0.08	0.01	1.05	2.78	2.44	100.00	93.14	Brown et al. (1992)
Middle Awash	0.95	Cindery	ERV-049	XRF/NAA	N/A	N/A	76.95	0.23	12.29	3.10	0.07	0.01	1.04	2.72	3.59	100.00	94.21	Haileab and Brown (1992)

*Borchardt coefficient (Borchardt et al., 1972), calculated on basis of all elements (unweighted).
†EMP—electron microprobe (beam conditions: keV, nA, spot size in μm); Bulk DCP—bulk direct coupled plasma spectrometry, XRF—X-ray fluorescence spectrometry, NAA—neutron activation analysis.
§For microprobe analysis, method of correction for sodium migration effect; Yes indicates extrapolation to $t = 0$ (see text), RB indicates beam was rastered across sample during analysis.
#Numbers in parentheses indicate standard deviation (given where data are available).

moderate differences in Si, Fe, and alkali content (alone) between two tephras should not immediately rule out a correlation.

Another key issue in tephra correlation is the way in which a tephra is defined for the purposes of stratigraphic correlation and dating, and under what circumstances to consider a pair or group of discrete tephra samples correlates. In practice, established correlations between tephra samples from the northern Awash valley typically have Borchardt coefficients (Borchardt et al., 1972) above 0.9, although this is not an absolute criterion for correlation. Rather, correlations are generally established based on careful but ultimately subjective consideration of a number of factors, including geochemical similarity, appearance in the field, and available stratigraphic information. For example, northern Awash tephras with bimodal or heterogeneous compositions are not easily analyzed using the Borchardt coefficient approach and must be identified qualitatively or on the basis of quantitative correlation of individual modes. A further question is whether compositionally similar or overlapping tephras (e.g., Figs. 5 and 6) represent distinct eruptions (perhaps from the same volcano or volcanic center) and thus distinct points in time, or whether they represent a combination of compositional heterogeneity in the erupted magma and/or differences in analytical conditions and thus belong to a single eruption. In some cases, trace-element and/or stratigraphic data may be used to resolve this question, but in other cases, it is unclear.

Tephra Source Regions

Most of the tuffs preserved in the northern Awash valley contain evidence of reworking and/or redeposition, making it difficult to locate their source vent(s) through standard techniques of physical volcanology (e.g., Houghton et al., 2000, and references therein). However, several clues about the source regions for these tuffs are contained in their relative frequency, chemistry, and degree of correlation. Exposures of the Hadar Formation in the northern Awash valley date from ca. 3.8 Ma to 2.9 Ma and contain approximately nine mapped tuffs. Exposures of the Busidima Formation date from 2.7 Ma to <0.2 Ma and contain over 35 distinct tuffs. Despite the fact that the time periods represented by these two formations differ (1.1 m.y. versus 2.5 m.y.), and that both formations contain unreported tuffs (e.g., DiMaggio et al., this volume), it is apparent that, in the northern Awash valley, the number of individual tuffs in the Busidima Formation is significantly greater than in the Hadar Formation. Furthermore, there are significant compositional differences on average between the Hadar and Busidima Formation tephras, as is evident in Figure 3 and from a comparison between Figures 4 and 5. In general, Busidima Formation tephras are more calcium- and iron-rich and have lower SiO_2 contents than Hadar Formation tephras, although there are some minor exceptions to this pattern. Finally, many of the Hadar Formation tuffs are widespread across the northern Awash valley (and across East Africa), while the Busidima Formation tuffs are more commonly localized within a project area (reflecting in part the cut-and-fill nature of the Busidima Formation in the northern Awash) and lack definite correlates outside of the Afar.

Our observations are consistent with previous suggestions that, in general, Hadar and Busidima Formation tephras were erupted from different source regions. The lower Hadar Formation tuffs, specifically the Ikini and Sidi Hakoma Tuff, likely represent large eruptions of volcanoes located at some distance away from the northern Awash region, possibly in the Main Ethiopian Rift (e.g., Hart et al., 1992; Walter and Aronson, 1993; Haileab, 1994). This is consistent with our observation that Hadar Formation tephras are generally widespread across the northern Awash region, and they are possibly fewer in number, a possible indication that only the largest East African eruptions that occurred during this time deposited material as far away from the inferred source region as the northern Awash. In contrast, the Busidima Formation tuffs (e.g., AST-1, AST-2; BKT-3) likely represent source regions within the Afar that are closer to the northern Awash (Hart et al., 1992), consistent with the larger number of tuffs in the northern Awash Busidima Formation, representing closer eruptions on a range of scales. Two noteworthy exceptions to this pattern are the 3.9 Ma Cindery Tuff and the 2.94–2.96 Ma BKT-2 tuff complex, which are both thought to have erupted from the Ida Ale volcanic center neighboring Hadar (Hall et al., 1984; Walter et al., 1987; DiMaggio et al., this volume; Fig. 1). Pumice carrying dateable material (i.e., feldspars) is rare in the northern Awash Busidima Formation tuffs, suggesting that the source vents for Busidima Formation tuffs were still at some distance from the northern Awash region. One possible source for some of the Busidima Formation tuffs is Mt. Ayelu (Fig. 1), which has pumices that are similar in composition to tephra group four (Fig. 5). A source model in which the dominant focus of volcanic activity shifts closer to the northern Awash region in the late Pliocene is consistent with a developing model of regional tectonics proposed by Wynn et al. (this volume), in which the local axis of rifting shifted to the southern Afar segment between the time of deposition of the Hadar and Busidima Formations. A shift in the nature of volcanism may have also occurred in conjunction with the hypothesized regional tectonic changes, from large eruptions of highly evolved rhyolite to smaller, more frequent eruptions of less-evolved rhyolites, consistent with the shift in eruption frequency and chemistry observed in the Hadar and Busidima Formation tephras. A more detailed and systematic study (e.g., Hart et al., 1992) of recently obtained geochemical data for the Hadar and Busidima Formation tephras is necessary to test these hypotheses, however.

A significant amount of work remains to be done on the tephrostratigraphy of the northern Awash valley. In addition to ongoing mapping of the project areas, which will surely identify new tephras and new occurrences of known tephras, ongoing analysis of mapped tephras for trace-element composition and to confirm tentative correlations is expected to resolve a number of outstanding questions concerning northern Awash stratigraphy. The key aim of ongoing tephrostratigraphic work in the northern Awash valley is to contribute to the construction of a complete stratigraphic framework for the region and for each project area in order to guide future paleoanthropological, paleontological, and paleoclimate investigations, which tie together in extending our understanding of the environmental context of human evolution. In addition, the data resulting from completed and future tephrostratigraphic analyses in the northern Awash area also have the potential to elucidate the volcanic history of the region, providing significant constraints on the tectonic history of the region and thus furthering our understanding of the processes involved in the development of continental rifts.

CONCLUSIONS

In summary, we conclude that tuffs of the Hadar Formation in the northern Awash valley are well-correlated and easily distinguishable on the basis of major-element chemistry. In contrast, the numerous Busidima Formation tuffs are limited in their lateral extent across the northern Awash valley, less well correlated, and often difficult to distinguish on the basis of major-element chemistry. Correlation of these tuffs may require trace-element chemistry and/or additional stratigraphic information. While a number of factors have the potential to obscure tephra correlations, careful consideration of all available data and reanalysis of key tephras are expected to resolve many of the ambiguities in northern Awash Busidima Formation tephrostratigraphy. Finally, an analysis of the amassed data set of northern Awash tephra analyses may help to test and refine a proposed model of source regions for the tephras of the Hadar and Busidima Formations.

ACKNOWLEDGMENTS

The northern Awash paleoanthropological research projects have been funded by grants from the National Science Foundation, the National Geographic Society, the Leakey Foundation, the Institute of Human Origins (Arizona State University), the Center for Human Evolutionary Studies (Rutgers University), the Max Planck Society, the Baldwin Foundation (University of Oregon), Centre Nationale de la Recherche Scientifique (France), and the Wenner Gren Foundation. For assistance with microprobe analyses, we thank Marsha Smith, Bereket Haileab, Barbara Nash, Ray Lambert, Godwin Mollel, Jerry Delaney, Charles Mandeville, John Donovan, Eric Condliffe, Tom Beasley, and Gordon Moore. For logistical and field support, we would like to thank Zeray Alemseged, Erella Hovers, Don Johanson, Bill Kimbel, Charles Lockwood, Kaye Reed, Sileshi Semaw, and Jonathan Wynn, as well as the Institute of Human Origins, the National Museum of Ethiopia, the Ethiopian Authority for Research and Conservation of Cultural Heritage, and especially our field crews and friends from the Eloaha, Mille, Adaytu, Busidima, and Asaita regions of Ethiopia. Finally, we gratefully acknowledge Shigehiro Katoh and an anonymous reviewer for their exceptionally thorough and constructive reviews of this article.

REFERENCES CITED

Alemseged, Z., Kimbel, W., Spoor, F., Bobe, R., Geraads, D., Reed, D., and Wynn, J.G., 2006, The earliest juvenile hominin skeleton from Dikika, Ethiopia: Nature, v. 443, p. 296–301, doi: 10.1038/nature05047.

Borchardt, G.A., Aruscavage, P.J., and Millard, H.T., 1972, Correlation of the Bishop Ash, a Pleistocene marker bed, using instrumental neutron activation analysis: Journal of Sedimentary Petrology, v. 42, p. 301–306.

Brown, F.H., Sarna-Wojcicki, A.M., Meyer, C.E., and Haileab, B., 1992, Correlation of Pliocene and Pleistocene tephra layers between the Turkana Basin of East Africa and the Gulf of Aden: Quaternary International, v. 13–14, p. 55–67, doi: 10.1016/1040-6182(92)90010-Y.

Campisano, C.J., 2007, Tephrostratigraphy and Hominin Paleoenvironments of the Hadar Formation, Afar Depression, Ethiopia [Ph.D. thesis]: New Brunswick, New Jersey, Rutgers University, 601 p.

Campisano, C.J., and Feibel, C.S., 2008, this volume, Tephrostratigraphy of the Hadar and Busidima Formations at Hadar, Afar Depression, Ethiopia, in Quade, J., and Wynn, J.G., eds., The Geology of Early Humans in the Horn of Africa: Geological Society of America Special Paper 446, doi: 10.1130/2008.2446(06).

Clark, J.D., de Heinzelin, J., Schick, K.D., Hart, W.K., White, T.D., WoldeGabriel, G., Walter, R.C., Suwa, G., Asfaw, B., Vrba, E., and Haile-Selassie, Y., 1994, African *Homo erectus*: Old radiometric ages and young Oldowan assemblages in the Middle Awash valley, Ethiopia: Science, v. 264, p. 1907–1910, doi: 10.1126/science.8009220.

Clark, J.D., Beyene, Y., WoldeGabriel, G., Hart, W., Renne, P., Gilbert, H., Defleur, A., Suwa, G., Katoh, S., Ludwig, K.R., Boisserie, J.-R., Asfaw, B., and White, T.D., 2003, Stratigraphic, chronological and behavioural contexts of Pleistocene *Homo sapiens* from Middle Awash, Ethiopia: Nature, v. 423, p. 747–752, doi: 10.1038/nature01670.

deMenocal, P.B., and Brown, F.H., 1999, Pliocene tephra correlations between East African hominid localities, the Gulf of Aden and the Arabian Sea, in Agusti, J., Rook, L., and Andrews, P., eds., The Evolution of Terrestrial Ecosystems in Europe: Cambridge, UK, Cambridge University Press, p. 23–54.

DiMaggio, E.N., 2007, Volcanic and Stratigraphic Characterization of Pliocene Tephra from the Ledi-Geraru Region of Afar, Ethiopia [M.S. thesis]: Tempe, Arizona State University, 140 p.

DiMaggio, E.N., Campisano, C.J., Arrowsmith, J R., Reed, K.E., Swisher, C.C., III, and Lockwood, C.A., 2008, this volume, Correlation and stratigraphy of the BKT-2 volcanic complex in west-central Afar, Ethiopia, in Quade, J., and Wynn, J.G., eds., The Geology of Early Humans in the Horn of Africa: Geological Society of America Special Paper 446, doi: 10.1130/2008.2446(07).

Feibel, C.S., 1999, Tephrostratigraphy and geological context in paleoanthropology: Evolutionary Anthropology, v. 8, p. 87–100, doi: 10.1002/(SICI)1520-6505(1999)8:3<87::AID-EVAN4>3.0.CO;2-W.

Froggatt, P.C., 1992, Standardization of the chemical analysis of tephra deposits. Report of the ICCT Working Group: Quaternary International, v. 13–14, p. 93–96, doi: 10.1016/1040-6182(92)90014-S.

Geraads, D., Alemseged, Z., Reed, D., Wynn, J., and Roman, D., 2004, The Pleistocene fauna from Asbole, lower Awash Valley, Ethiopia: Geobios, v. 37, p. 697–718, doi: 10.1016/j.geobios.2003.05.011.

Haileab, B., 1994, Geochemistry, Geochronology, and Tephrostratigraphy of Tephra from the Turkana Basin, Southern Ethiopia and Northern Kenya [Ph.D. thesis]: Salt Lake City, University of Utah, 369 p.

Haileab, B., and Brown, F.H., 1992, Turkana Basin–Middle Awash Valley correlations and the age of the Sagantole and Hadar Formations: Journal of Human Evolution, v. 22, p. 453–468, doi: 10.1016/0047-2484(92)90080-S.

Hall, C.M., Walter, R.C., Westgate, J.A., and York, D., 1984, Geochronology, stratigraphy, and geochemistry of Cindery Tuff in Pliocene hominid-bearing sediments of the Middle Awash, Ethiopia: Nature, v. 308, p. 26–31, doi: 10.1038/308026a0.

Hart, W.K., Walter, R.C., and WoldeGabriel, G., 1992, Tephra sources and correlations in Ethiopia: Application of elemental and neodymium isotope data: Quaternary International, v. 13–14, p. 77–86, doi: 10.1016/1040-6182(92)90012-Q.

Houghton, B.F., Wilson, C.J.N., and Pyle, D.M., 2000, Pyroclastic fall deposits, in Sigurdsson, H., ed., Encyclopedia of Volcanoes: London, Academic Press, p. 555–570.

Irvine, T.N., and Baragar, W.P.A., 1971, A guide to the chemical classification of the common volcanic rocks: Canadian Journal of Earth Sciences, v. 8, p. 523–548.

Johanson, D.C., and Taieb, M., 1976, Plio-Pleistocene hominid discoveries in Hadar, Ethiopia: Nature, v. 260, p. 293–297, doi: 10.1038/260293a0.

Kimbel, W.H., Walter, R.C., Johanson, D.C., Reed, K.E., Aronson, J.L., Assefa, Z., Marean, C.W., Eck, G.G., Robe, R., Hovers, E., Rak, Y., Vondra, C., Yemane, T., York, D., Chen, Y., Evensen, N.M., and Smith, P.E., 1996, Late Pliocene *Homo* and Oldowan tools from the Hadar Formation (Kada Hadar Member), Ethiopia: Journal of Human Evolution, v. 31, p. 549–561, doi: 10.1006/jhev.1996.0079.

McDougall, I., and Brown, F.H., 2006, Precise ^{40}Ar/^{39}Ar geochronology for the upper Koobi Fora Formation, Turkana Basin, northern Kenya: Journal of the Geological Society of London, v. 163, p. 205–220, doi: 10.1144/0016-764904-166.

Le Maitre, R.W., ed., 2002, Igneous Rocks: A Classification and Glossary of Terms (2nd ed.): Cambridge, Cambridge University Press, 224 p.

Quade, J., Levin, N., Semaw, S., Stout, D., Renne, R., Rogers, M., and Simpson, S., 2004, Paleoenvironments of the earliest stone toolmakers, Gona, Ethiopia: Geological Society of America Bulletin, v. 116, p. 1529–1544, doi: 10.1130/B25358.1.

Quade, J., Levin, N.E., Simpson, S.W., Butler, R., McIntosh, W.C., Semaw, S., Kleinsasser, L., Dupont-Nivet, G., Renne, P., and Dunbar, N., 2008, this volume, The geology of Gona, Afar, Ethiopia, *in* Quade, J., and Wynn, J.G., eds., The Geology of Early Humans in the Horn of Africa: Geological Society of America Special Paper 446, doi: 10.1130/2008.2446(01).

Renne, P., Walter, R., Verosub, K., Sweitzer, M., and Aronson, J., 1993, New data from Hadar (Ethiopia) support orbitally tuned time-scale to 3.3 Ma: Geophysical Research Letters, v. 20, p. 1067–1070, doi: 10.1029/93GL00733.

Renne, P.R., Swisher, C.C., III, Deino, A.L., Karner, D.B., Owens, T.L., and DePaolo, D.J., 1998, Intercalibration of standards, absolute ages and uncertainties in ^{40}Ar/^{39}Ar dating: Chemical Geology, v. 145, p. 117–152, doi: 10.1016/S0009-2541(97)00159-9.

Roman, D.C., and Wynn, J.G., 2004, Development of an interactive online database of East African tephras: Geological Society of America Abstracts with Programs, v. 36, no. 5, p. 486.

Semaw, S., Renne, P., Harris, J.W.K., Feibel, C.S., Bernor, R.L., Fesseha, N., and Mowbray, K., 1997, 2.5-million-year-old stone tools from Gona, Ethiopia: Nature, v. 385, p. 333–336, doi: 10.1038/385333a0.

Semaw, S., Simpson, S.W., Quade, J., Renne, P.R., Butler, R.F., McIntosh, W.C., Levin, N., Dominguez-Rodrigo, M., and Rogers, M.J., 2005, Early Pliocene hominids from Gona, Ethiopia: Nature, v. 433, p. 301–305, doi: 10.1038/nature03177.

Taieb, M., Johanson, D.C., Coppens, Y., and Aronson, J.L., 1976, Geological and palaeontological background of Hadar hominid site, Afar, Ethiopia: Nature, v. 260, p. 289–293, doi: 10.1038/260289a0.

Walter, R.C., 1994, Age of Lucy and the First Family: Single-crystal ^{40}Ar/^{39}Ar dating of the Denen Dora and lower Kada Hadar Members of the Hadar Formation, Ethiopia: Geology, v. 22, p. 6–10, doi: 10.1130/0091-7613 (1994)022<0006:AOLATF>2.3.CO;2.

Walter, R.C., and Aronson, J.L., 1993, Age and source of the Sidi Hakoma Tuff, Hadar Formation, Ethiopia: Journal of Human Evolution, v. 25, p. 229–240, doi: 10.1006/jhev.1993.1046.

Walter, R.C., Hart, W.K., and Westgate, J.A., 1987, Petrogenesis of a basalt-rhyolite tephra from the west-central Afar, Ethiopia: Contributions to Mineralogy and Petrology, v. 95, p. 462–480, doi: 10.1007/BF00402206.

White, T.D., Suwa, G., Hart, W.K., Walter, R.C., WoldeGabriel, G., Deheinzelin, J., Clark, J.D., Asfaw, B., and Vrba, E., 1993, New discoveries of *Australopithecus* at Maka in Ethiopia: Nature, v. 366, p. 261–265, doi: 10.1038/366261a0.

Wynn, J.G., Alemseged, Z., Bobe, R., Geraads, D., Reed, D., and Roman, D., 2006, Geology and paleontology of the Pliocene hominin locality at Dikika, Ethiopia: Nature, v. 443, p. 332–336, doi: 10.1038/nature05048.

Wynn, J.G., Roman, D.C., Alemseged, Z., Reed, D. Geraads, D., and Munro, S., 2008, this volume, Stratigraphy, depositional environments, and basin structure of the Hadar and Busidima Formations at Dikika, Ethiopia, *in* Quade, J., and Wynn, J.G., eds., The Geology of Early Humans in the Horn of Africa: Geological Society of America Special Paper 446, doi: 10.1130/2008.2446(04).

MANUSCRIPT ACCEPTED BY THE SOCIETY 17 JUNE 2008

The Geological Society of America
Special Paper 446
2008

Tephrostratigraphy of the Hadar and Busidima Formations at Hadar, Afar Depression, Ethiopia

Christopher J. Campisano*
Institute of Human Origins, Arizona State University, P.O. Box 874101, Tempe, Arizona 85287-4101, USA

Craig S. Feibel*
Department of Anthropology, Rutgers University, 131 George Street, New Brunswick, New Jersey 08901-1414, USA, and Department of Geological Sciences, Rutgers University, 610 Taylor Road, Piscataway, New Jersey 08854-8066, USA

ABSTRACT

This paper documents the lithology and geochemistry of vitric tephra deposits from the Pliocene-Pleistocene Hadar and Busidima Formations from the early hominin site of Hadar in Ethiopia. Vitric tephras of the Hadar Formation (ca. 3.45–2.9 Ma) are limited to certain facies of the Sidi Hakoma Tuff, the Kada Hadar Tuff, and the Bouroukie Tuff 2 (BKT-2) Complex, the latter of which is discussed in detail in this study. In contrast, this systematic study identified at least 12 distinct vitric tephras preserved in the Busidima Formation at Hadar (ca. 2.7–0.81 Ma), which are represented by no less than 20 chemical modes. These analyses are used to construct the first tephrostratigraphic-based sequence for the highly complex and discontinuous Busidima Formation deposits preserved at Hadar. Busidima Formation correlations have also been established between Hadar and neighboring project areas, specifically Dikika and Gona. Artifact Site Tuff 3 (AST-3), the Inaalale Tuff, and the Ken Di Tuff are correlated between Hadar and Dikika. AST-1, AST-3, the Ken Di Tuff, the Dahuli Tuff, and several localized tuffs of the Busidima Formation are correlated between Hadar and Gona. However, tuffs associated with the earliest archaeology in the two regions, namely AST-2 from Gona and BKT-3 from Hadar, were not identified outside their respective project areas. Nonetheless, the sequence of tephra provides important information for the placement and relationship of archaeological and paleontological sites both within Hadar and between Hadar and adjacent project areas.

Keywords: tephra, Pliocene, chronostratigraphy, geochemistry, hominin.

INTRODUCTION

The Hadar and Busidima Formations exposed within the Hadar Research Project area in the Afar region of the Ethiopian Rift are composed of ~200 m of fluviolacustrine sediments (Figs. 1 and 2) (Taieb et al., 1976; Johanson et al., 1978; Taieb and Tiercelin, 1979; Aronson and Taieb, 1981; Tiercelin, 1986; Campisano and Feibel, this volume). The middle Pliocene Hadar Formation makes up the majority of this sequence (~155 m) and extends from ca. 3.45 Ma to the Busidima unconformity surface (BUS), a regional angular unconformity at ca. 2.9 Ma that separates the Hadar Formation from the overlying Busidima Formation (Quade et al., 2004, this volume; Wynn et al.,

*Campisano: campisano@asu.edu; Feibel: feibel@rci.rutgers.edu.

Campisano, C.J., and Feibel, C.S., 2008, Tephrostratigraphy of the Hadar and Busidima Formations at Hadar, Afar Depression, Ethiopia, *in* Quade, J., and Wynn, J.G., eds., The Geology of Early Humans in the Horn of Africa: Geological Society of America Special Paper 446, p. 135–162, doi: 10.1130/2008.2446(06). For permission to copy, contact editing@geosociety.org. ©2008 The Geological Society of America. All rights reserved.

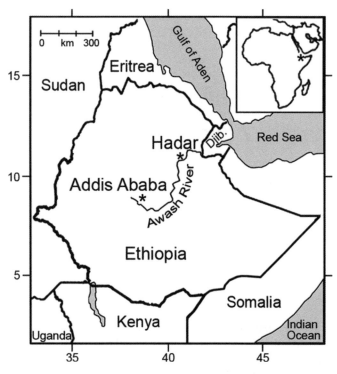

Figure 1. Location map of the Hadar paleoanthropological site in Ethiopia.

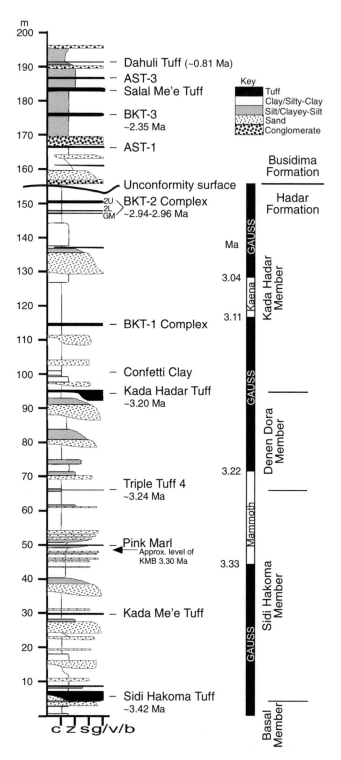

Figure 2. Composite stratigraphic section of the Hadar and Busidima Formations at Hadar. Tuffs and major marker beds are labeled alongside the section. ^{40}Ar/^{39}Ar dates and paleomagnetic transitions are from Schmitt and Nairn (1984), Walter and Aronson (1993), Renne et al. (1993), Walter (1994), Kimbel et al. (1994, 1996, 2004), Semaw et al. (1997), and Campisano (2007). Previously published ^{40}Ar/^{39}Ar dates have been recalculated to reflect the updated age of the Fish Canyon sanidine standard (see text footnote 2). C—clay; Z—silt; S—sand; G/V/B—gravel/volcaniclastics/bioclastics.

2006, this volume). At Hadar, the base of the Busidima Formation is marked by a conglomerate (the KH-5 conglomerate) that displays up to 10 m of erosional relief and contains abundant reworked carbonate nodules at its base. Sediments of the Busidima Formation at Hadar are coeval with those previously and informally referred to as the "upper" Kada Hadar Member of the Hadar Formation (e.g., Kimbel et al., 1996; Vondra et al., 1996; Yemane, 1997; Campisano, 2007). Hadar Formation deposits are laterally extensive, relatively uninterrupted by faulting or areas of discontinuous exposure, and have yielded abundant vertebrate fossils, including the early hominin *Australopithecus afarensis* (Johanson and Taieb, 1976; Johanson et al., 1978, 1982). Busidima Formation sediments at Hadar range in age from younger than 2.7 to ca. 0.81 Ma and have yielded material attributed to *Homo* aff. *H. habilis* and Oldowan stone tools (Kimbel et al., 1996, 1997). In contrast to the pre-unconformity sequence of the Hadar Formation, the postunconformity strata of the Busidima Formation are poorly preserved and exceptionally discontinuous at Hadar, where there are numerous erosional surfaces marked by conglomerates and cut-and-fill cycles.

Lithostratigraphic correlations within the continuous Hadar Formation deposits are relatively uncomplicated. The majority of the Hadar Formation volcanic marker beds do not preserve a vitric component but are correlatable by their often unique outcrop appearance, lithology, and stratigraphic context, if not by direct tracing of the beds (e.g., the Kada Me'e Tuff, TT-4, BKT-1). Conversely, the stratigraphically short interval of the Busidima Formation at Hadar preserves numerous vitric tephra deposits. However, these tephras are typically fluvially reworked, lithologically indistinct, and either preserved in highly localized patches of sediment or as discontinuous lenses among more continuous strata.

This study provides the first systematic geochemical analysis of vitric tephra from the Pliocene-Pleistocene strata of the Hadar Research Project area of Ethiopia. The characterization of Hadar Formation tephras provides a comparative data set for identifying correlates outside of the Hadar project area. The characterization of Busidima Formation tephras not only provides the first tephrostratigraphic-based stratigraphic sequence for the discontinuous postunconformity deposits within Hadar, but it also establishes confirmed tephrostratigraphic links to nearby project areas, such as Gona and Dikika. Finally, detailed lithologic, geographic, and contextual documentation of these tephra deposits, particularly for the complex Busidima Formation units, is included to serve as a reference for current and future geologists at Hadar and in the surrounding regions.

BACKGROUND TO THE STUDY

Tephrostratigraphy, the characterization and correlation of volcanic tephra, has become an indispensable tool for the stratigraphic correlation of local and regional sequences at East African paleontological sites (e.g., Brown, 1982; Feibel et al., 1989; Brown et al., 1992, 2006). The general theory behind tephrostratigraphy is that the tephra distributed across a landscape during a volcanic eruption provides a widespread isochronous marker over a large area that can be preserved in the geologic record. Assuming it is unaltered, the volcanic glass component of the tephra preserves the composition of the magma at the time of eruption. As the melt phase that produces the glass concentrates the incompatible elements, which are not limited in composition by crystal structure considerations, the composition of glass from one eruption to another tends to be geochemically distinct (Feibel, 1999; Sarna-Wojcicki, 2000). It is this unique geochemical signature that is used to distinguish tephra deposits from different eruptions, even those from the same volcano.

The first geochemical analyses of select Hadar tephra were presented by Walter (1981). Unlike this study, Walter's focus was less on intraformational correlations and more on identifying the volcanic sources for the Hadar tephra. Using many of the same samples, the geochemical source of tephras from the Afar region was further investigated by Hart and colleagues (1992). Despite some of the claims made, early attempts to correlate tephras between the Kada Hadar and Kada Gona regions were tentative at best and made primarily on lithostratigraphic similarity (e.g., Roche and Tiercelin, 1977; Walter, 1981; Tiercelin, 1986) and limited geochemical evidence (Walter, 1981).

A detailed study of the Busidima Formation deposits in Gona by Feibel and Wynn appeared as an appendix in S. Semaw's (1997) dissertation on the archaeology of the Gona artifacts and did much to correct and clarify the tephrostratigraphy in the eastern part of the Gona project area (see also Semaw et al., 1997). The comprehensive tephrostratigraphic analysis conducted by Feibel identified at least 17 different chemical compositions among the Gona tephra deposits. These analyses served as the original comparative data set for the study of the Busidima Formation tephrostratigraphy at Hadar (Campisano, 2007), the relevant details of which are discussed in this paper. Complementary papers in this volume present the first tephrostratigraphic framework for the Busidima Formation in the Dikika project area (Wynn et al., this volume) and an expanded study of the Gona project area tephra (Quade et al., this volume). Collectively, these data greatly enhance our understanding of the stratigraphy and regional variation of the Busidima Formation (see also Roman et al., this volume).

MATERIALS AND METHODS

Tephra samples were collected by the authors at Hadar from 2001 to 2004 and by one of us (CSF) at Gona in 1993 and 1994 (Figs. 3 and 4). Results from this study were also compared against tephra data from the Gona and Dikika project areas (Quade et al., this volume; Wynn et al., this volume). Only Gona and Dikika tephras that were found to have correlates to the Hadar region or are of important stratigraphic significance are included in this study.

Samples were prepared in a standard technique for tephra as outlined by Feibel (1999). Samples were sieved in brass screens to concentrate the 60–120 mesh (250–125 µm = fine sand) fraction. To remove any calcium carbonate from the sample, the material was bathed in a 10% HNO_3 solution until any reaction had run to completion. To remove clays adhering to the glass shards, the material was bathed in a 5% HF solution in an ultrasonic bath for 2 min intervals and then repeatedly rinsed and decanted with hot water. Finally, the samples were rinsed and bathed in distilled water for 12–24 h and then dried in an oven. In most cases, samples were sufficiently cleaned to be composed of 85%–98% glass shards. For a few samples, such as those with less than 50% glass shards, the material was separated using a Franz isodynamic magnetic separator to concentrate the nonmagnetic glass shards. Bouroukie Tuff 2 (BKT-2) feldspars and magnetites were prepared in the same manner but were handpicked from the original bulk samples. Samples were then mounted in epoxy discs, polished, and carbon coated for electron microprobe analysis.

Geochemical analyses in this study were conducted by electron microprobe analysis (EMPA). All analyzed elements except Ba, Cl, F, and Zr were converted to and reported by their oxide percentage, with Fe converted to Fe_2O_3. Unnormalized analytical totals are reported in this study. In general, one point per glass shard and 15 shards per sample were analyzed. Virtually all Hadar samples were analyzed on the Rutgers University Microanalysis JEOL JXA-8600 superprobe operating at 15 kV and 15 nA with a rastering electron beam over a 12 µm^2 area. A 20 s on-peak count time was used for each element, with a 10 s off-peak count time for background. A combination of natural and synthetic feldspars and pyroxene standards was used for calibration. A volatile correction routine (Donovan, 2000) was applied to minimize the effect

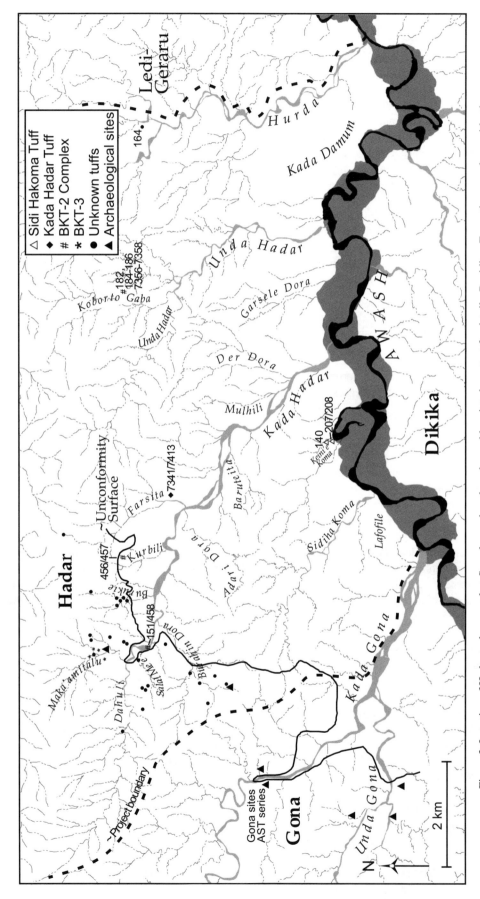

Figure 3. Location of Hadar tephra samples. Sample numbers have been abbreviated, see text for details. Busidima Formation samples are identified in Figure 4. Position of archaeological sites and unconformity surface in the Gona region were adapted from Quade et al. (2004).

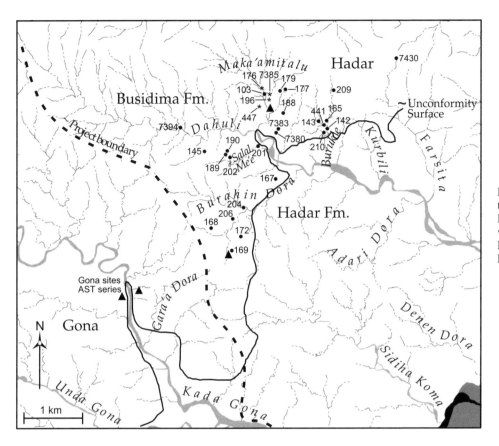

Figure 4. Location of Busidima Formation tephra samples from Hadar. For locations where more than one sample was collected, only one sample is listed. Legend and figure details are as in Figure 3.

of Na, K, and Si loss during analysis. All quantitative analysis was performed by wavelength-dispersive spectrometry (WDS). Additional information on sample preparation and the analytical methods and conditions of the Rutgers University microprobe is given in McHenry (2004, 2005) and Campisano (2007).

Most analyses of samples collected by C. Feibel from the Gona region were analyzed at the University of Utah on a Cameca SX50 electron microprobe operating at 15 kV and 25 nA with an electron beam diameter from 5 to 25 μm and a 20 s count time. Additional information on the analytical methods and conditions of the University of Utah microprobe is described in Nash (1992) and Brown et al. (2006). Additional analyses of both Hadar and Gona vitric samples and BKT-2 feldspars and magnetites were conducted at the American Museum of Natural History (AMNH) Electron Microprobe Facility. Samples were analyzed on a Cameca SX100 electron microprobe operating at 15 kV and 10 nA with a beam diameter of 15 μm for all elements except Na and K, which used a 2 nA current to minimize volatilization. Count times for Na, K, and Cl were 40 s, 60 s for F, and 30 s for all other elements.

DATA REDUCTION AND ANALYSIS

No single value characterizes the geochemistry of a tuff. Even replicate analyses of the same sample will not result in an exact match. Most of the error within a sample analysis, or between multiple analyses of the same sample from different runs, is related to intershard variability within an individual sample (either natural or from alteration) and the analytical precision of the microprobe as opposed to instrumentation problems or the lack of reproducibility (Sarna-Wojcicki and Davis, 1991). As noted by others (e.g., Feibel, 1999; Sarna-Wojcicki, 2000), the best evidence for correlation comes not from a single quantitative or statistical measure, but from consistencies in physical, chemical, stratigraphic, and chronological data from a large comparative database.

Analytical totals of individual volcanic glass shards rarely equal 100%. In addition to the precision of the microprobe, this is primarily due to the presence of unanalyzed water in the sample (both original and from secondary hydration) and the diagenetic and analytical mobility of alkalis (Nielsen and Sigurdsson, 1981; Cerling et al., 1985). Although the water content can be estimated by the direct analysis of oxygen (Nash, 1992) and can account for up to 10% in samples from East Africa (e.g., Brown et al., 2006), the Rutgers and AMNH microprobe facilities were not set up to analyze for oxygen. Individual glass shard analyses that had totals less than 85% or more than 102% were rejected (the exception being sample HE04-441, discussed later). This lower limit is a few percent lower than typically encountered for East African tephra. Select individual analyses with totals less than 88% (~10% of all shards analyzed) were included in calculating sample averages if it was demonstrated that they likely represented the natural variation of the sample, and were not simply a poor analysis. This was determined by inspecting the full range of a sample's individual shard totals and noting whether the removal/inclusion of an analysis affected only the sample's standard deviation of an oxide, or had a notable impact on the oxide mean as well. Unnormalized analytical totals are reported in this study. The disadvantages associated with simply recalculating oxide compositions to 100%, particularly when oxygen is not measured, have been previously outlined (e.g., Brown et al., 1992; Hunt and Hill, 1993). These include inflating silica and alumina to unrealistic values and errors in recalculations when the original concentrations of alkalis are unknown and their measured values are highly variable from sample to sample.

For each sample analyzed, if more than one obvious chemical mode was recognized (e.g., basaltic versus rhyolitic glass), it was divided into separate populations. A mean and standard deviation were calculated for each population, and replicate samples that were analyzed on different runs were treated independently during data analysis. Any analysis that was more than three standard deviations from the sample mean in any detectable element or two standard deviations from the sample mean in any three elements was excluded from the population, and the mean was recalculated. If multiple "excluded" shards were similar in composition, they were grouped into a separate population, or mode, and considered to be a characteristic of that sample with a separate mean and standard deviation. Averages of homogeneous populations are presented in the data tables as the sample number with the suffix .AV (e.g., HE01-142.AV). If a sample has two or more distinct populations, then they are distinguished in the data tables by .AV1 and .AV2 suffixes (e.g., HE01-188.AV1 and HE01-188.AV2). If a single shard failed to match any chemical mode of a tephra, it is listed in the data tables as an isolated shard with its analysis number as its suffix (e.g., HE01-140.04). The summary table (Table 1) only displays sample populations/modes and select single-shard analyses that either form distinct modes (e.g., the "high Fe" BKT-3 mode) or are correlated to other sample modes. The results of all microprobe analyses are provided in Campisano (2007) and the GSA Data Repository (Table DR1).[1]

The primary method for establishing correlations in this study was by the sorting of data in a spread sheet and identifying potential correlates by pattern matching. Most tephras are compositionally distinct, and many of the differences in the geochemistry become evident when sorting various elements/oxides without having to resort to statistical tests, particularly as no statistical technique can automatically identify correlates, and the criteria for correlation are more a matter of interpretation. Initial sorting was done principally on the concentrations of iron, calcium, and aluminum. Subsequent discriminations were made on the basis of magnesium, manganese, titanium, and, to an extent, chlorine. The concentrations of sodium, potassium, and silica can vary widely as these elements are easily leached during early weathering (depending on the depositional environment) and may be poorly analyzed by EMP (Nielsen and Sigurdsson, 1981; Cerling et al., 1985; McHenry, 2004; WoldeGabriel et al., 2005), but they can still be useful in the sorting process. Barium, fluorine, and zirconium were not consistently analyzed in all sample runs, so they were of only limited use in aiding correlations in this study.

With an emphasis on Fe_2O_3, Al_2O_3, CaO, MgO, MnO, and TiO_2, tephra samples that had elemental concentrations within one standard deviation of each other were considered to be chemical correlates. Single-shard analyses that were excluded from sample averages were sorted and evaluated in a similar manner and then compared to both tephra averages and other single-shard analyses. Before making formal tephra correlations, chemical correlates were evaluated within the context of the stratigraphic sequences in which they were found and the homotaxial relationships between exposures. Preliminary comparisons between correlations using normalized versus unnormalized data sets showed no differences.

As an exploratory technique to identify potentially correlatable tephra samples, this study utilized a modified version of Euclidean distance (termed EDm here) as outlined by Perkins and colleagues (1995), which takes into account estimates of the analytical precision of the instrument:

$$\text{ED}m = \sqrt{\sum_{k=1}^{n}\left[\frac{(x_{k1} - x_{k2})^2}{2\sigma_k^2}\right]}, \quad (1)$$

where x_{k1} = the concentration of element x_k in tephra sample 1; x_{k2} = the concentration of element x_k in tephra sample 2; n = the number of elements used in the comparison; and σ_k = the standard deviation for analytical precision of element x_k. The analytical precision estimate (σ_k) was calculated by averaging the differences of individual oxides produced from multiple analytical runs of 10 Hadar tephra samples on the Rutgers University microprobe resulting in the following values: Al_2O_3 (0.1), Fe_2O_3 (0.045), CaO (0.01), MgO (0.003), MnO (0.004), TiO_2 (0.005), and Cl (0.025) (Campisano, 2007).

Euclidean distance (ED) is simply the linear distance between two points in multidimensional space (Ludwig and Reynolds, 1988). Although there is no way to statistically test the certainty of a classification, the graphical representation of the data provides a convenient way to sort through a large amount of data to identify clusters of potentially correlatable samples. In order to create a graphical representation of the data, a dendrogram was constructed (Fig. 5) using the cluster analysis program in MSVP v.3.13 (Kovach Computing) and the modified Euclidean distance equation of Perkins and colleagues (1995) as the distance measurement and UPGMA as the clustering method.

To verify that the "known" samples discussed in this study were, in fact, definitive correlates of previously named and analyzed tuffs, the chemical compositions from Walter (1981) and Hart et al. (1992) for the Sidi Hakoma Tuff (SHT), Kada Hadar Tuff (KHT), BKT-3, Artifact Site Tuff 1 (AST-1), AST-2, and

---
[1]GSA Data Repository item 2008195, individual microprobe analyses of vitric tephras from the Hadar and Busidima Formations, is available at www.geosociety.org/pubs/ft2008.htm, or on request from editing@geosociety.org, Documents Secretary, GSA, P.O. Box 9140, Boulder, CO 80301-9140, USA.

AST-3 were compared to the type section samples analyzed in this study (Table 2). As the data from Hart et al. (1992) were presented only as values normalized to 100%, both the data from Walter (1981) and this study were also normalized for comparison. The only differences between the results are slightly lower Fe_2O_3 concentrations and slightly higher Al_2O_3 concentrations reported by Walter (1981) and Hart et al. (1992) compared to this study. A comparison between the unnormalized values from Walter (1981) and this study shows that the minor range of variation can be accounted for by the standard deviation in the various oxides/elements. As such, it is concluded that there are no sampling errors in identifying previously named and analyzed tuffs.

STRATIGRAPHIC SEQUENCE OF THE HADAR VITRIC TEPHRA

The vast majority of the Hadar tephra is rhyolitic and subalkaline in composition, whereas the rest is typically basaltic. The results of the tephrostratigraphic analysis are presented in stratigraphic sequence of the tephra deposits starting at the base of the Hadar Formation exposed at Hadar (Table 1). In instances where unnamed tephras at Hadar correlate to tephras previously named and analyzed from Gona or Dikika, they will be subsumed under the name provided for the Gona/Dikika tephra. For previously unnamed tephras in this study that do not correlate to Gona or Dikika, they will either be formally named or identified by sample number, depending on the significance and/or lateral extent of the newly described tephra.

Analysis of samples included in this discussion, or at least potential correlates within and between project areas, in the same laboratory and/or under the same analytical conditions would clearly provide more definitive evidence for correlation. Unfortunately, this ideal approach is not always feasible in many circumstances (see Roman et al., this volume). The results and correlations proposed in this study and displayed in Table 1 and Figure 6 are a current working model based on the data available and are subject to change depending on future research and analyses.

Sidi Hakoma Tuff (SHT)

The Sidi Hakoma Tuff marks the base of the Sidi Hakoma Member and is the stratigraphically lowest tephra unit preserved in the Hadar project area. It is typically preserved in the Hadar region as a widespread white bentonite, ~20 cm thick, but vitric channel-fill and crevasse-splay deposits are preserved on the flanks of Keini'e Koma and have been dated to ca. 3.42 Ma[2] (Walter and Aronson, 1993). Samples HE01-207 and -208 were collected from the basal and upper portion of the 3-m-thick channel-fill tephra deposit on the east side of Keini'e Koma

(Fig. 3), respectively, where the basal component is finer grained and lighter in color compared to the upper component. Sample HE01-140 was collected from the meter-thick vitric floodplain/crevasse-splay deposit on the north side of Keini'e Koma. Glasses from all three samples of the Sidi Hakoma Tuff form a tight chemical group and are distinct from all other Hadar tephras in having a significantly low Fe_2O_3 concentration and a relatively low CaO concentration (Table 1). As noted in previous studies (Brown, 1982; Brown et al., 1992; Walter and Aronson, 1993), the chemical composition of the Sidi Hakoma Tuff matches that of the Tulu Bor Tuff (= Tuff B) from the Turkana Basin, particularly the Tulu Bor-β. A possible correlation may exist for the single outlier shard HE01-140.04, which is distinct from all the other Sidi Hakoma Tuff shards analyzed but very similar to the composition of Tuff B-γ (Haileab, 1995) (Table 3). However, Tuff B-γ has yet to be identified outside the Shungura region, and HE01-140.04 could represent a non–Sidi Hakoma Tuff shard that coincidentally has a similar chemical composition to Tuff B-γ. Correlates to the Sidi Hakoma Tuff have also been identified in the Middle Awash region (White et al., 1993) and in ocean drill cores from the Gulf of Aden (Sarna-Wojcicki et al., 1985) and the Arabian Sea (deMenocal and Brown, 1999).

Kada Hadar Tuff (KHT)

The Kada Hadar Tuff marks the base of the Kada Hadar Member and is most commonly preserved as a beige, silty bentonite, 20–80 cm thick. In the Farsita Wadi, where samples E01-7341 and E02-7413 were collected, the Kada Hadar Tuff is preserved as a series of four vitric channel-fill deposits (alternating cutbank surfaces), 4–6 m thick over a distance of ~200 m (Fig. 3). Sanidines collected from the Kada Hadar Tuff between the Gona and Sidiha Koma Wadis have been dated to ca. 3.20 Ma (Walter, 1994). Glass from the Kada Hadar Tuff is compositionally distinct from virtually all other Hadar tephras by having high Fe_2O_3 concentrations combined with the lowest Al_2O_3 and CaO concentrations (Table 1). The only other sample composition remotely similar to the Kada Hadar Tuff is from E02-7430.AV1 (Fig. 5), but the latter contains higher concentrations of Fe_2O_3 and CaO and is located stratigraphically more than 75 m above the Kada Hadar Tuff. A small channel-fill tephra deposit preserved in the western Ledi-Geraru region northeast of Hadar has been geochemically correlated to the Kada Hadar Tuff (DiMaggio et al., 2006), and Yemanc (1997) also reported a 5–6-m-thick and 30–40-m-wide Kada Hadar Tuff channel-fill deposit at Merketa Negeria Koma (west of the Gona Wadi).

Bouroukie Tuff 2 (BKT-2 Complex)

Special attention has been paid to the Bouroukie Tuff 2 (BKT-2), the stratigraphically highest tephra unit of the Hadar Formation. It is a primary air-fall tephra with feldspar crystals suitable for dating and is laterally extensive throughout Hadar and adjacent regions, including the Ledi-Geraru, Dikika, and

---

[2]Previously published $^{40}Ar/^{39}Ar$ ages reported in this study have been increased by approximately 0.65% to reflect the revised age of the monitor mineral used (Fish Canyon sanidine) from 27.84 to 28.02 Ma (Renne et al., 1998); this correction is supported by recent geochronological analysis of select Hadar tephra (Campisano, 2007).

TABLE 1. SUMMARY OF MICROPROBE ANALYSES OF VITRIC TEPHRA FROM THE HADAR AND BUSIDIMA FORMATIONS

Sample/Shard	n	SiO_2	Al_2O_3	Fe_2O_3	CaO	K_2O	Na_2O	MgO	MnO	TiO_2	Cl	Totals	Lab	Location	Notes
Sidi Hakoma Tuff															
=Tulu Bor β															
HE01-207.AV	7	73.25 ± 0.28	11.67 ± 0.07	1.55 ± 0.07	0.28 ± 0.03	2.08 ± 0.25	3.79 ± 0.54	0.04 ± 0.01	0.06 ± 0.03	0.15 ± 0.03	0.10 ± 0.01	93.00 ± 0.87	AM	Keini'e Koma	Channel
HE01-208.AV	13	73.83 ± 0.29	11.79 ± 0.10	1.55 ± 0.03	0.28 ± 0.01	1.96 ± 0.15	4.39 ± 0.13	0.06 ± 0.01	0.04 ± 0.01	0.13 ± 0.03	0.10 ± 0.02	94.23 ± 0.45	AM	Keini'e Koma	Channel
HE01-140.AV	6	73.05 ± 0.19	11.66 ± 0.06	1.56 ± 0.10	0.27 ± 0.03	2.05 ± 0.17	3.92 ± 0.32	0.05 ± 0.01	0.04 ± 0.02	0.16 ± 0.04	0.09 ± 0.01	92.86 ± 0.31	AM	Keini'e Koma	Crevasse
≈Tulu Bor γ															
HE01-140.04	1	68.87	12.75	2.96	0.56	1.80	3.56	0.13	0.04	0.33	0.07	91.17	AM	Keini'e Koma	Crevasse
Kada Hadar Tuff															
E01-7341.AV	13	69.48 ± 1.36	8.78 ± 0.16	5.62 ± 0.32	0.19 ± 0.03	1.37 ± 0.16	1.57 ± 0.37	0.01 ± 0.01	0.19 ± 0.06	0.27 ± 0.03	0.23 ± 0.04	87.71 ± 1.29	RU	Farsita	
E02-7413.AV	14	70.79 ± 2.30	9.01 ± 0.32	5.37 ± 0.36	0.20 ± 0.03	1.60 ± 0.16	2.47 ± 0.43	0.02 ± 0.01	0.20 ± 0.04	0.27 ± 0.04	0.19 ± 0.04	90.10 ± 2.83	RU	Farsita	
Green Marker															
E01-7355.AV	25	44.71 ± 3.03	12.27 ± 0.45	15.00 ± 0.32	8.70 ± 0.52	0.79 ± 0.11	2.85 ± 0.17	5.04 ± 0.24	0.21 ± 0.06	2.59 ± 0.12	0.04 ± 0.03	92.19 ± 3.36	RU	Koborto Gaba	
E01-7182.AV	2	50.11 ± 1.97	14.58 ± 1.82	15.65 ± 1.58	8.49 ± 2.42	0.41 ± 0.58	1.87 ± 2.56	5.60 ± 2.86	0.25 ± 0.09	2.55 ± 0.33	0.02 ± 0.03	99.54 ± 3.13	RU	Koborto Gaba	
BKT-2L & -2U															
Rhyolitic mode(s)															
E01-7356.AV	11	67.24 ± 1.37	11.39 ± 0.23	2.50 ± 0.18	0.51 ± 0.06	3.13 ± 0.90	4.13 ± 0.73	0.02 ± 0.03	0.06 ± 0.04	0.15 ± 0.02	0.13 ± 0.03	89.26 ± 2.00	RU	Koborto Gaba	2L
HE01-184.AV	9	71.53 ± 0.59	11.90 ± 0.09	2.63 ± 0.06	0.54 ± 0.01	2.75 ± 0.43	4.93 ± 0.25	0.01 ± 0.01	0.06 ± 0.01	0.15 ± 0.05	0.13 ± 0.01	94.83 ± 0.39	AM	Koborto Gaba	2U(a)
HE01-184.27	1	73.06	11.47	2.24	0.41	3.14	4.86	0.01	0.06	0.16	0.13	95.76	AM	Koborto Gaba	2U(a)
E01-7357.1518	1	70.02	11.25	2.26	0.40	2.95	4.24	0.03	0.09	0.12	0.17	91.54	RU	Koborto Gaba	2U(a)
HE01-186.AV	3	71.98 ± 0.96	11.65 ± 0.20	2.37 ± 0.18	0.48 ± 0.08	3.14 ± 0.52	4.68 ± 0.34	0.02 ± 0.01	0.06 ± 0.02	0.16 ± 0.03	0.12 ± 0.03	94.84 ± 0.78	AM	Koborto Gaba	2U(c)
E01-7358.AV	10	68.49 ± 1.39	11.25 ± 0.27	2.42 ± 0.20	0.42 ± 0.05	3.11 ± 0.36	4.09 ± 0.74	0.01 ± 0.01	0.05 ± 0.05	0.14 ± 0.02	0.16 ± 0.02	90.13 ± 1.29	RU	Koborto Gaba	2U(b)
Basaltic shards															
E01-7357.1514	1	49.44	14.36	12.77	7.55	0.31	3.97	4.05	0.17	3.11	0.06	95.78	RU	Koborto Gaba	2U(a)
E01-7358.1396	1	48.57	12.71	17.35	8.39	0.99	2.56	4.22	0.28	3.05	0.02	98.14	RU	Koborto Gaba	2U(b)
HE04-441															
HE04-441.AV	12	53.17 ± 0.88	13.83 ± 0.16	15.34 ± 0.34	9.30 ± 0.26	0.83 ± 0.06	2.46 ± 0.34	4.87 ± 0.19	0.23 ± 0.08	3.13 ± 0.17	0.03 ± 0.02	103.21 ± 1.07	RU	Burukie	
AST-1															
Rhyolitic mode															
E93-4992.AV	9	71.11 ± 0.49	12.22 ± 0.13	2.27 ± 0.21	0.92 ± 0.07	3.66 ± 0.19	3.98 ± 0.34	0.20 ± 0.04	0.07 ± 0.02	0.34 ± 0.07	0.07 ± 0.01	94.93 ± 0.44	AM	Gona	Type
E93-4992.AV	28	72.46 ± 0.77	12.77 ± 0.14	2.42 ± 0.23	0.98 ± 0.09	3.82 ± 0.15	3.35 ± 0.16	0.21 ± 0.04	0.08 ± 0.03	0.36 ± 0.06	0.07 ± 0.01	96.64 ± 0.53	UU	Gona	Type
HE01-167.AV	7	65.35 ± 3.81	12.32 ± 0.30	2.35 ± 0.30	0.96 ± 0.11	3.76 ± 0.26	3.67 ± 0.22	0.21 ± 0.04	0.08 ± 0.07	0.39 ± 0.10	0.09 ± 0.03	89.17 ± 3.95	RU	Burahin Dora	
HE01-167.AV	8	71.06 ± 1.01	12.20 ± 0.21	2.19 ± 0.31	0.85 ± 0.11	3.72 ± 0.13	3.47 ± 1.13	0.19 ± 0.06	0.08 ± 0.01	0.29 ± 0.09	0.06 ± 0.01	94.20 ± 0.94	AM	Burahin Dora	
HE04-443.AV	14	70.95 ± 2.72	12.25 ± 0.20	2.20 ± 0.18	0.87 ± 0.08	3.73 ± 0.15	3.43 ± 0.70	0.17 ± 0.03	0.05 ± 0.06	0.31 ± 0.07	0.09 ± 0.02	94.04 ± 2.41	RU	Burahin Dora	
E93-4971.AV	7	70.34 ± 0.61	12.18 ± 0.15	2.22 ± 0.17	0.89 ± 0.11	3.84 ± 0.15	3.77 ± 0.34	0.18 ± 0.04	0.09 ± 0.02	0.30 ± 0.07	0.07 ± 0.01	93.96 ± 0.63	UU	Burahin Dora	
HE01-142.AV	14	65.96 ± 2.95	12.30 ± 0.15	2.24 ± 0.20	0.89 ± 0.07	3.70 ± 0.12	3.46 ± 0.53	0.18 ± 0.03	0.07 ± 0.04	0.36 ± 0.06	0.07 ± 0.02	89.23 ± 3.01	RU	Burukie	
Basaltic shards															
E93-4971.15	1	51.51	13.09	13.81	8.07	1.15	2.18	4.04	0.28	2.89	0.04	97.18	UU	Burahin Dora	
E93-4992.24	1	50.88	13.30	14.56	8.61	0.98	2.31	4.54	0.23	2.85	0.03	98.34	UU	Gona	Type
AST-2															
E93-4997.AV	31	72.49 ± 0.38	12.70 ± 0.09	2.63 ± 0.06	0.93 ± 0.03	3.74 ± 0.11	3.56 ± 0.18	0.06 ± 0.01	0.08 ± 0.02	0.18 ± 0.04	0.07 ± 0.01	96.57 ± 0.50	UU	Gona	Type

(continued)

143

TABLE 1. SUMMARY OF MICROPROBE ANALYSES OF VITRIC TEPHRA FROM THE HADAR AND BUSIDIMA FORMATIONS (continued)

Sample/Shard	n	SiO$_2$	Al$_2$O$_3$	Fe$_2$O$_3$	CaO	K$_2$O	Na$_2$O	MgO	MnO	TiO$_2$	Cl	Totals	Lab	Location	Notes
E02-7380															
Rhyolitic mode															
E02-7380.AV1	10	70.14 ± 1.40	11.29 ± 0.13	2.45 ± 0.34	0.83 ± 0.11	3.50 ± 0.19	3.82 ± 0.62	0.01 ± 0.01	0.07 ± 0.04	0.20 ± 0.03	0.09 ± 0.03	92.38 ± 2.10	RU	Kada Hadar	
E02-7380.AV	7	70.47 ± 1.54	11.37 ± 0.17	2.76 ± 0.59	0.88 ± 0.14	3.33 ± 0.33	4.07 ± 0.24	0.02 ± 0.01	0.09 ± 0.06	0.23 ± 0.04	0.09 ± 0.01	93.32 ± 1.61	RU	Kada Hadar	
HE01-108.AV	4	70.74 ± 1.02	11.53 ± 0.12	2.76 ± 0.24	0.76 ± 0.09	4.06 ± 0.25	3.67 ± 0.24	0.01 ± 0.00	0.07 ± 0.02	0.23 ± 0.01	0.10 ± 0.02	94.02 ± 0.94	AM	Maka'amitalu	
HE01-201.AV	6	70.68 ± 0.47	11.54 ± 0.14	2.83 ± 0.22	0.86 ± 0.11	3.91 ± 0.13	3.71 ± 0.10	0.02 ± 0.01	0.07 ± 0.02	0.22 ± 0.04	0.09 ± 0.01	94.02 ± 0.62	AM	Salal Me'e	
E93-4986.17	1	69.60	11.78	2.63	0.66	2.97	2.13	0.02	0.08	0.24	0.13	90.24	UU	Maka'amitalu	BKT-3
E93-4986.16	1	69.44	11.88	3.06	0.77	2.69	1.73	0.02	0.09	0.25	0.10	90.02	UU	Maka'amitalu	BKT-3
Basaltic mode															
E02-7380.AV2	4	48.95 ± 1.38	12.37 ± 0.60	15.92 ± 0.96	8.36 ± 0.40	0.80 ± 0.10	1.88 ± 0.08	4.39 ± 0.28	0.29 ± 0.02	3.59 ± 0.31	0.01 ± 0.01	96.57 ± 0.99	UU	Kada Hadar	
BKT-3															
Dominant mode															
HE04-447B.AV	12	67.84 ± 4.01	12.65 ± 0.38	3.24 ± 0.26	1.39 ± 0.13	3.35 ± 0.24	3.48 ± 1.02	0.37 ± 0.05	0.12 ± 0.05	0.45 ± 0.09	0.08 ± 0.03	92.98 ± 4.52	RU	Maka'amitalu	Topotype
HE01-106.AV	6	68.72 ± 0.54	12.31 ± 0.15	3.11 ± 0.06	1.39 ± 0.11	3.33 ± 0.05	3.98 ± 0.19	0.35 ± 0.03	0.10 ± 0.03	0.44 ± 0.05	0.06 ± 0.01	93.89 ± 0.53	AM	Maka'amitalu	
HE01-108.04	1	68.80	12.22	3.22	1.44	3.27	3.87	0.38	0.21	0.49	0.07	94.07	AM	Maka'amitalu	
HE01-176.AV	9	64.28 ± 1.18	12.42 ± 0.15	3.11 ± 0.17	1.38 ± 0.17	3.51 ± 0.14	3.87 ± 0.23	0.36 ± 0.03	0.12 ± 0.02	0.45 ± 0.07	0.08 ± 0.02	89.59 ± 1.64	RU	Maka'amitalu	
HE01-176.AV	10	63.62 ± 1.24	11.82 ± 0.15	3.29 ± 0.17	1.39 ± 0.09	3.38 ± 0.06	3.64 ± 0.40	0.39 ± 0.02	0.12 ± 0.05	0.44 ± 0.05	0.07 ± 0.04	88.14 ± 1.23	AM	Maka'amitalu	
HE01-196.AV	3	68.73 ± 0.45	12.23 ± 0.21	3.05 ± 0.19	1.40 ± 0.17	3.33 ± 0.08	3.92 ± 0.12	0.34 ± 0.03	0.11 ± 0.02	0.47 ± 0.06	0.07 ± 0.01	93.73 ± 0.14	AM	Maka'amitalu	
E02-7385.AV	10	68.08 ± 0.76	12.51 ± 0.13	3.29 ± 0.54	1.30 ± 0.18	3.46 ± 0.21	3.91 ± 0.35	0.37 ± 0.03	0.13 ± 0.07	0.45 ± 0.05	0.07 ± 0.01	93.56 ± 0.85	RU	Maka'amitalu	
E02-7386.AV	11	65.05 ± 1.01	12.39 ± 0.58	3.06 ± 0.21	1.26 ± 0.10	3.20 ± 0.14	3.18 ± 0.34	0.36 ± 0.05	0.14 ± 0.04	0.46 ± 0.04	0.07 ± 0.02	89.17 ± 1.48	RU	Maka'amitalu	
E02-7387.AV	10	65.67 ± 0.93	12.17 ± 0.25	3.27 ± 0.19	1.35 ± 0.15	3.11 ± 0.13	3.55 ± 0.35	0.38 ± 0.06	0.11 ± 0.05	0.51 ± 0.07	0.07 ± 0.02	90.18 ± 1.11	RU	Maka'amitalu	
E02-7388.AV1	7	65.77 ± 0.54	12.56 ± 0.49	3.26 ± 0.13	1.29 ± 0.13	3.23 ± 0.13	3.90 ± 0.24	0.37 ± 0.03	0.15 ± 0.05	0.48 ± 0.08	0.06 ± 0.02	91.07 ± 0.43	RU	Maka'amitalu	
E02-7389.AV1	6	64.74 ± 0.68	11.77 ± 0.28	3.23 ± 0.86	1.26 ± 0.20	2.71 ± 0.34	3.02 ± 0.73	0.36 ± 0.09	0.09 ± 0.04	0.52 ± 0.14	0.06 ± 0.02	87.76 ± 1.89	RU	Salal Me'e	
E02-7399.AV	11	67.29 ± 0.62	12.44 ± 0.12	3.27 ± 0.30	1.20 ± 0.17	3.36 ± 0.13	3.89 ± 0.63	0.39 ± 0.06	0.12 ± 0.06	0.44 ± 0.06	0.07 ± 0.02	92.47 ± 0.99	RU	Salal Me'e	
E02-7400.AV	15	66.88 ± 2.36	12.33 ± 0.27	3.26 ± 0.26	1.29 ± 0.12	3.35 ± 0.18	3.50 ± 0.36	0.37 ± 0.05	0.12 ± 0.05	0.43 ± 0.07	0.08 ± 0.03	91.62 ± 2.86	RU	Maka'amitalu	
E02-7450.AV	2	66.66 ± 0.22	11.81 ± 0.40	2.95 ± 0.05	1.30 ± 0.04	2.99 ± 0.03	2.29 ± 0.95	0.33 ± 0.01	0.08 ± 0.01	0.43 ± 0.02	0.06 ± 0.00	88.91 ± 1.13	RU	Maka'amitalu	
E93-4986.AV1	11	67.90 ± 0.91	12.66 ± 0.18	3.38 ± 0.30	1.51 ± 0.25	3.00 ± 0.34	2.14 ± 0.54	0.40 ± 0.11	0.13 ± 0.02	0.49 ± 0.11	0.07 ± 0.01	91.75 ± 1.50	UU	Maka'amitalu	
"Low-Fe" mode															
HE01-175.AV	2	69.65 ± 0.73	10.66 ± 0.14	1.98 ± 0.03	0.57 ± 0.04	1.34 ± 0.01	2.64 ± 0.04	0.08 ± 0.01	0.05 ± 0.07	0.24 ± 0.04	0.03 ± 0.01	87.24 ± 0.80	RU	Maka'amitalu	Channel
HE01-106.02	1	74.13	10.75	1.62	0.38	4.62	2.63	0.11	0.01	0.30	0.01	94.55	AM	Maka'amitalu	
HE01-107.441	1	68.64	11.51	1.77	0.48	4.10	3.96	0.07	0.04	0.24	0.09	90.90	RU	Maka'amitalu	Channel
HE01-196.04	1	72.97	12.07	1.71	0.47	6.29	2.50	0.10	0.03	0.24	0.01	96.42	AM	Maka'amitalu	
HE01-196.06	1	72.74	10.87	2.04	0.42	5.77	2.77	0.07	0.04	0.34	0.05	95.18	AM	Maka'amitalu	
E02-7385.517	1	73.87	11.12	1.95	0.48	5.40	2.82	0.09	0.11	0.26	0.08	96.16	RU	Maka'amitalu	
E02-7430.542	1	73.02	11.94	1.98	0.57	4.96	3.52	0.03	0.09	0.27	0.02	96.40	RU	Farsita	
"High-Fe" mode															
HE04-447B.490	1	62.03	12.74	4.83	2.21	3.22	4.09	0.71	0.15	0.87	0.05	90.90	RU	Maka'amitalu	Topotype
HE01-176.545	1	63.68	12.43	4.23	2.23	3.15	3.51	0.66	0.13	0.69	0.09	90.82	RU	Maka'amitalu	
E02-7387.319	1	62.76	12.99	4.99	2.34	2.58	3.56	0.79	0.13	1.07	0.08	91.30	RU	Maka'amitalu	
E02-7389.352	1	63.92	11.82	4.59	2.18	2.64	2.84	0.69	0.08	0.77	0.07	89.60	RU	Maka'amitalu	Channel
E02-7390.372	1	64.54	14.18	4.23	1.98	3.32	3.56	0.55	0.18	0.73	0.06	93.34	RU	Maka'amitalu	
E02-7450.386	1	66.59	13.29	4.51	2.25	2.87	3.53	0.67	0.22	0.98	0.02	94.92	RU	Maka'amitalu	
E93-4986.2	1	65.99	13.02	4.37	2.21	2.89	2.35	0.68	0.19	0.80	0.06	92.55	UU	Maka'amitalu	
Basaltic mode(s)															
E02-7388.AV2	4	49.21 ± 0.94	13.61 ± 0.12	13.89 ± 0.54	8.57 ± 0.25	0.84 ± 0.13	2.26 ± 0.19	4.79 ± 0.19	0.22 ± 0.02	3.38 ± 0.07	0.03 ± 0.01	96.79 ± 1.29	RU	Maka'amitalu	Channel
E02-7390.AV2	7	50.94 ± 1.92	13.41 ± 0.55	14.11 ± 1.01	7.88 ± 1.03	0.93 ± 0.31	2.31 ± 0.15	4.12 ± 0.55	0.25 ± 0.02	2.88 ± 0.17	0.02 ± 0.01	96.83 ± 1.49	RU	Maka'amitalu	
E93-4986.AV2	5	49.99 ± 0.87	13.06 ± 0.10	13.91 ± 0.59	8.40 ± 0.50	1.19 ± 0.14	2.07 ± 0.15	4.51 ± 0.31	0.24 ± 0.02	3.11 ± 0.24	0.04 ± 0.01	96.55 ± 0.76	UU	Maka'amitalu	Channel
E02-7386.309	1	48.12	12.99	14.68	8.84	0.78	2.19	4.97	0.17	3.62	0.03	96.37	RU	Maka'amitalu	
E02-7389.347	1	46.72	12.33	12.80	8.48	0.69	2.50	4.92	0.17	3.11	0.00	91.74	RU	Maka'amitalu	
E02-7450.374	1	46.62	11.09	13.89	7.68	0.76	1.66	3.83	0.23	3.01	0.01	88.79	RU	Maka'amitalu	
HE01-176.546	1	50.72	13.96	14.71	9.61	0.89	2.38	5.19	0.17	3.37	0.00	101.00	RU	Maka'amitalu	

(continued)

TABLE 1. SUMMARY OF MICROPROBE ANALYSES OF VITRIC TEPHRA FROM THE HADAR AND BUSIDIMA FORMATIONS (continued)

Sample/Shard	n	SiO$_2$	Al$_2$O$_3$	Fe$_2$O$_3$	CaO	K$_2$O	Na$_2$O	MgO	MnO	TiO$_2$	Cl	Totals	Lab	Location	Notes
HE01-168															
HE01-168.AV	9	68.94 ± 0.81	12.52 ± 0.16	4.13 ± 0.18	1.10 ± 0.04	3.48 ± 0.40	3.96 ± 0.80	0.03 ± 0.01	0.14 ± 0.04	0.31 ± 0.03	0.15 ± 0.02	90.10 ± 1.61	RU	Burahin Dora	
HE01-188.AV2	2	63.43 ± 0.09	12.57 ± 0.09	4.60 ± 0.45	0.94 ± 0.15	4.47 ± 0.12	4.56 ± 0.38	0.00 ± 0.00	0.15 ± 0.01	0.31 ± 0.01	0.13 ± 0.04	91.16 ± 0.08	RU	Lip/Road	
E93-4973.AV	15	67.97 ± 0.53	13.10 ± 0.20	4.42 ± 0.13	1.13 ± 0.07	2.66 ± 0.43	2.05 ± 0.62	0.03 ± 0.01	0.16 ± 0.01	0.35 ± 0.03	0.13 ± 0.01	92.05 ± 1.39	UU	Burahin Dora	
E02-7383.249	1	59.29	12.46	5.06	0.96	4.23	5.48	0.00	0.19	0.33	0.12	88.14	UU	Kada Hadar	
Salal Me'e Tuff															
Rhyolitic mode															
HE01-103.AV	7	69.99 ± 0.64	12.31 ± 0.17	2.40 ± 0.07	0.88 ± 0.02	4.04 ± 0.11	3.88 ± 0.31	0.16 ± 0.01	0.06 ± 0.02	0.30 ± 0.04	0.09 ± 0.01	94.18 ± 0.76	AM	Maka'amitalu	Channel
HE01-107.AV	10	66.58 ± 0.83	12.51 ± 0.12	2.50 ± 0.16	0.92 ± 0.07	4.04 ± 0.15	3.90 ± 0.18	0.18 ± 0.02	0.09 ± 0.05	0.30 ± 0.02	0.09 ± 0.02	91.10 ± 1.06	RU	Maka'amitalu	Channel
HE01-202.AV	14	63.43 ± 1.18	12.43 ± 0.17	2.54 ± 0.41	0.83 ± 0.09	4.12 ± 0.22	4.54 ± 0.35	0.19 ± 0.03	0.11 ± 0.04	0.31 ± 0.03	0.08 ± 0.03	88.59 ± 1.41	RU	Salal Me'e	Channel
HE01-202.AV	13	68.28 ± 1.43	12.61 ± 0.22	2.70 ± 0.30	0.89 ± 0.07	3.85 ± 0.31	3.72 ± 0.31	0.20 ± 0.03	0.09 ± 0.04	0.34 ± 0.04	0.10 ± 0.02	92.78 ± 1.95	RU	Salal Me'e	Channel
HE04-448.AV	5	69.78 ± 0.58	12.60 ± 0.18	2.57 ± 0.19	0.89 ± 0.07	3.84 ± 0.08	4.25 ± 0.11	0.18 ± 0.04	0.08 ± 0.02	0.31 ± 0.03	0.08 ± 0.00	94.66 ± 0.56	AM	Maka'amitalu	Air-fall
HE04-449.AV	11	70.28 ± 0.51	12.56 ± 0.10	2.51 ± 0.11	0.87 ± 0.06	3.96 ± 0.09	4.20 ± 0.08	0.18 ± 0.02	0.07 ± 0.02	0.29 ± 0.05	0.08 ± 0.01	95.11 ± 0.49	AM	Maka'amitalu	Air-fall
E02-7390.AV1	6	67.00 ± 1.88	12.50 ± 0.65	2.53 ± 0.31	0.83 ± 0.11	3.57 ± 0.27	2.98 ± 0.32	0.17 ± 0.03	0.08 ± 0.04	0.35 ± 0.04	0.10 ± 0.04	90.10 ± 3.13	RU	Maka'amitalu	Channel
E02-7447.AV1	11	68.82 ± 1.51	12.44 ± 0.32	2.64 ± 0.16	0.86 ± 0.11	3.72 ± 0.52	3.86 ± 0.28	0.18 ± 0.03	0.09 ± 0.06	0.29 ± 0.05	0.11 ± 0.02	93.02 ± 2.36	RU	Maka'amitalu	Air-fall
Basaltic mode(s)															
HE01-202.343	1	48.79	13.91	13.32	4.44	1.84	3.03	4.30	0.30	2.45	0.07	92.44	RU	Salal Me'e	Channel
E02-7447.AV2	3	50.14 ± 0.83	13.54 ± 0.40	15.54 ± 0.35	8.91 ± 0.23	0.59 ± 0.05	2.51 ± 0.43	5.20 ± 0.10	0.20 ± 0.06	2.81 ± 0.04	0.03 ± 0.01	99.48 ± 1.67	RU	Maka'amitalu	Air-fall
Inaalale Tuff															
HE01-177.AV	6	65.24 ± 0.75	11.53 ± 0.16	3.55 ± 0.06	0.30 ± 0.04	3.32 ± 0.26	4.44 ± 0.77	0.08 ± 0.01	0.14 ± 0.04	0.23 ± 0.02	0.35 ± 0.02	89.16 ± 0.95	RU	Maka'amitalu	
E93-5035.AV	17	69.31 ± 0.57	11.63 ± 0.19	3.74 ± 0.07	0.31 ± 0.02	3.33 ± 0.68	4.47 ± 0.71	0.08 ± 0.02	0.17 ± 0.02	0.27 ± 0.03	0.30 ± 0.03	93.80 ± 1.19	UU	Gara Dora	
AST-3															
E94-5451.AV	15	67.23 ± 1.61	11.76 ± 0.08	2.79 ± 0.20	0.62 ± 0.02	4.09 ± 0.35	4.75 ± 0.38	0.00 ± 0.01	0.08 ± 0.04	0.23 ± 0.02	0.15 ± 0.02	91.71 ± 1.81	RU	Gona	Type
E94-5451.AV	15	67.96 ± 0.74	11.73 ± 0.15	2.96 ± 0.12	0.69 ± 0.04	4.09 ± 0.34	3.88 ± 0.23	0.01 ± 0.01	0.08 ± 0.04	0.21 ± 0.02	0.15 ± 0.03	91.76 ± 1.12	RU	Gona	Type
E94-5452.AV	12	67.78 ± 0.60	11.80 ± 0.16	2.95 ± 0.12	0.68 ± 0.04	4.15 ± 0.31	4.01 ± 0.31	0.00 ± 0.01	0.06 ± 0.04	0.21 ± 0.02	0.14 ± 0.02	91.80 ± 0.68	RU	Gona	Type
E94-5451.AV	14	66.45 ± 1.11	11.40 ± 0.16	3.05 ± 0.12	0.65 ± 0.04	4.16 ± 0.45	3.80 ± 0.21	0.01 ± 0.02	0.07 ± 0.03	0.22 ± 0.02	0.14 ± 0.02	89.95 ± 1.23	RU	Gona	Type
E94-5452.AV	8	67.87 ± 1.49	11.84 ± 0.11	2.97 ± 0.37	0.61 ± 0.03	4.05 ± 0.15	4.43 ± 0.31	0.01 ± 0.02	0.07 ± 0.05	0.24 ± 0.03	0.12 ± 0.05	92.18 ± 1.68	RU	Gona	Type
HE01-143.AV	10	67.16 ± 0.89	11.41 ± 0.16	2.97 ± 0.13	0.71 ± 0.04	3.78 ± 0.28	3.72 ± 0.14	0.01 ± 0.01	0.10 ± 0.05	0.23 ± 0.03	0.10 ± 0.03	90.18 ± 1.20	RU	Burukie	
HE01-165.AV	10	67.47 ± 1.87	11.47 ± 0.19	2.97 ± 0.08	0.72 ± 0.05	3.68 ± 0.44	3.69 ± 0.29	0.01 ± 0.01	0.07 ± 0.05	0.26 ± 0.04	0.09 ± 0.02	90.43 ± 2.07	RU	Burukie	
HE01-172.AV	7	66.34 ± 2.06	11.54 ± 0.24	2.96 ± 0.12	0.73 ± 0.03	3.83 ± 0.43	3.83 ± 0.39	0.01 ± 0.01	0.11 ± 0.06	0.25 ± 0.03	0.10 ± 0.03	89.70 ± 2.77	RU	Burahin Dora	
HE01-189.AV	15	64.85 ± 1.45	11.75 ± 0.10	2.89 ± 0.29	0.59 ± 0.03	4.10 ± 0.27	4.42 ± 0.37	0.00 ± 0.00	0.08 ± 0.04	0.22 ± 0.03	0.14 ± 0.03	89.06 ± 1.74	RU	Salal Me'e	
HE01-189.AV	13	69.03 ± 0.75	11.87 ± 0.14	3.03 ± 0.29	0.66 ± 0.03	3.96 ± 0.44	4.02 ± 0.24	0.01 ± 0.01	0.09 ± 0.04	0.22 ± 0.03	0.14 ± 0.02	93.03 ± 1.10	RU	Salal Me'e	
HE01-210.AV	14	70.63 ± 3.23	11.52 ± 0.24	2.84 ± 0.38	0.60 ± 0.10	3.78 ± 0.25	3.70 ± 0.48	0.01 ± 0.01	0.09 ± 0.04	0.25 ± 0.07	0.11 ± 0.02	93.55 ± 3.21	RU	Burukie	
Burahin Dora Tuff															
Dominant mode															
HE01-169.AV	12	67.68 ± 0.86	11.63 ± 0.18	2.93 ± 0.08	0.33 ± 0.02	3.18 ± 0.63	4.20 ± 0.58	0.00 ± 0.00	0.11 ± 0.04	0.17 ± 0.03	0.17 ± 0.02	90.40 ± 1.07	RU	Burahin Dora	
HE01-171.AV	4	70.38 ± 0.22	11.50 ± 0.07	2.90 ± 0.06	0.31 ± 0.04	3.93 ± 0.06	4.30 ± 0.26	0.00 ± 0.00	0.06 ± 0.04	0.17 ± 0.03	0.15 ± 0.00	93.71 ± 0.11	AM	Burahin Dora	
HE01-188.AV	15	71.29 ± 1.75	11.93 ± 0.21	2.90 ± 0.10	0.33 ± 0.05	3.45 ± 0.44	4.48 ± 0.72	0.00 ± 0.01	0.09 ± 0.05	0.16 ± 0.04	0.16 ± 0.04	94.78 ± 2.07	RU	Lip/Road	
HE01-188.AV1	7	66.63 ± 1.79	11.69 ± 0.16	2.79 ± 0.08	0.32 ± 0.03	3.42 ± 0.61	4.32 ± 0.37	0.00 ± 0.00	0.10 ± 0.05	0.17 ± 0.03	0.18 ± 0.02	89.63 ± 1.44	RU	Kada Hadar	
E02-7383.AV	3	69.42 ± 0.67	11.67 ± 0.22	2.80 ± 0.22	0.30 ± 0.04	3.70 ± 0.50	5.24 ± 0.26	0.00 ± 0.00	0.10 ± 0.03	0.16 ± 0.03	0.17 ± 0.03	93.56 ± 1.44	RU	Kada Hadar	
E02-7383.AV1	11	66.35 ± 2.70	11.63 ± 0.17	2.86 ± 0.19	0.29 ± 0.01	3.69 ± 0.46	4.60 ± 0.98	0.00 ± 0.00	0.08 ± 0.06	0.14 ± 0.03	0.15 ± 0.03	89.80 ± 2.45	RU	Burukie	
E02-7384.AV	12	68.49 ± 1.64	11.58 ± 0.27	2.80 ± 0.30	0.32 ± 0.02	2.94 ± 0.74	4.35 ± 0.88	0.00 ± 0.00	0.08 ± 0.04	0.17 ± 0.03	0.17 ± 0.03	90.89 ± 2.92	RU	Burukie	
E02-7384.AV	9	66.43 ± 1.97	11.58 ± 0.27	2.90 ± 0.33	0.28 ± 0.02	3.22 ± 0.68	4.10 ± 1.34	0.00 ± 0.00	0.09 ± 0.05	0.15 ± 0.03	0.18 ± 0.04	88.93 ± 1.89	RU	Divide	
E94-5455.AV	13	67.64 ± 1.62	11.63 ± 0.17	2.84 ± 0.10	0.33 ± 0.03	3.44 ± 0.70	4.09 ± 0.63	0.00 ± 0.01	0.09 ± 0.06	0.17 ± 0.03	0.17 ± 0.02	90.39 ± 2.15	RU	Divide	
E94-5456.AV	10	68.27 ± 0.72	11.58 ± 0.12	2.84 ± 0.10	0.30 ± 0.03	3.76 ± 0.47	4.08 ± 0.81	0.00 ± 0.00	0.09 ± 0.04	0.16 ± 0.02	0.17 ± 0.02	91.25 ± 0.85	RU		
"Low-Fe" shards															
HE01-171.01	1	73.56	11.53	1.16	0.34	4.88	3.37	0.05	0.04	0.19	0.07	95.20	AM	Burahin Dora	
HE01-188.577	1	68.67	11.60	1.09	0.33	2.97	3.56	0.05	0.04	0.19	0.07	88.56	RU	Lip/Road	

(continued)

TABLE 1. SUMMARY OF MICROPROBE ANALYSES OF VITRIC TEPHRA FROM THE HADAR AND BUSIDIMA FORMATIONS (continued)

Sample/Shard	n	SiO$_2$	Al$_2$O$_3$	Fe$_2$O$_3$	CaO	K$_2$O	Na$_2$O	MgO	MnO	TiO$_2$	Cl	Totals	Lab	Location
Ken Di Tuff														
E02-7430.AV1	5	67.85 ± 0.90	8.85 ± 0.21	6.90 ± 0.46	0.26 ± 0.02	1.85 ± 1.12	3.97 ± 1.14	0.02 ± 0.01	0.26 ± 0.06	0.34 ± 0.03	0.17 ± 0.02	90.48 ± 1.58	RU	Farsita
E02-7430.AV2	2	63.14 ± 0.36	11.72 ± 0.19	6.09 ± 0.21	0.66 ± 0.07	4.03 ± 0.22	5.60 ± 0.30	0.05 ± 0.00	0.26 ± 0.10	0.51 ± 0.04	0.11 ± 0.02	92.18 ± 0.81	RU	Farsita
E02-7430.AV3	2	62.50 ± 1.44	12.22 ± 0.28	7.26 ± 0.02	0.87 ± 0.03	2.84 ± 1.55	5.17 ± 1.19	0.08 ± 0.00	0.35 ± 0.09	0.57 ± 0.01	0.09 ± 0.01	91.95 ± 4.41	RU	Farsita
Upper Hurda Tuff														
HE01-164.AV	16	69.01 ± 2.75	12.49 ± 0.18	2.98 ± 0.15	0.83 ± 0.06	3.96 ± 0.21	4.06 ± 0.21	0.22 ± 0.02	0.09 ± 0.04	0.46 ± 0.06	0.11 ± 0.02	94.21 ± 2.80	RU	Hurda
Dahuli Tuff														
HE01-145.AV	3	66.22 ± 4.29	12.25 ± 0.14	3.29 ± 0.61	1.01 ± 0.22	4.19 ± 0.20	3.92 ± 0.38	0.04 ± 0.03	0.11 ± 0.03	0.31 ± 0.02	0.00 ± 0.00	91.43 ± 4.02	RU	Dahuli
HE01-190.AV	9	69.01 ± 0.22	12.48 ± 0.09	3.15 ± 0.08	1.07 ± 0.05	3.74 ± 0.09	4.42 ± 0.17	0.17 ± 0.06	0.10 ± 0.01	0.27 ± 0.07	0.10 ± 0.02	94.65 ± 0.28	AM	Salal Me'e
HE01-190.388	1	69.91	12.34	3.23	0.81	3.80	3.54	0.04	0.11	0.26	0.10	89.42 ± 6.28	RU	Salal Me'e
HE01-209.AV	17	69.34 ± 2.85	12.31 ± 0.18	3.18 ± 0.21	0.87 ± 0.07	3.64 ± 0.24	4.39 ± 0.17	0.01 ± 0.01	0.11 ± 0.05	0.22 ± 0.04	0.13 ± 0.04	94.20 ± 2.83	RU	Burukie/Kurbili
E02-7394.AV	14	67.47 ± 2.46	11.81 ± 0.22	3.22 ± 0.14	0.90 ± 0.11	3.67 ± 0.43	3.51 ± 0.37	0.04 ± 0.01	0.09 ± 0.04	0.26 ± 0.05	0.10 ± 0.03	91.06 ± 2.89	RU	Dahuli
E93-4975.AV	18	69.19 ± 0.46	12.17 ± 0.11	3.24 ± 0.12	0.93 ± 0.05	3.73 ± 0.14	4.07 ± 0.39	0.01 ± 0.01	0.13 ± 0.02	0.23 ± 0.04	0.13 ± 0.01	93.92 ± 0.65	UU	Cusrali

Note: Tephra are listed in proposed stratigraphic order. All values reported as wt%. AV results listed as mean and one standard deviation. Fluorine values are not reported as the element was not consistently analyzed. RU—Rutgers University; AM—American Museum of Natural History (AMNH); UU—University of Utah.

Gona project areas. Walter (1981) divided the BKT-2 into three different volcanic components, which are collectively referred to as the BKT-2 Complex in this study. Investigations by Campisano and colleagues in the Ledi-Geraru region, just east of Hadar, have identified as many as seven distinct volcanic units associated with the BKT-2 Complex encased in a laminated diatomite (DiMaggio et al., this volume). The following lithologic descriptions of the BKT-2 Complex are the same as those provided in the accompanying article by Campisano and Feibel (this volume).

Green Marker Bed (GMB)

The lowermost horizon of the BKT-2 Complex, BKT-2L$_1$, is also known as the Green Marker Tuff (GMT) or, perhaps more appropriately, the Green Marker Bed (GMB), as it does not always show evidence of being tuffaceous. Samples E01-7355 and HE01-182 are from the Green Marker Bed in the Koborto Gaba region (Fig. 3), where it is a 5-cm-thick, laminated dark green unit composed primarily of dark brown (basaltic) vitric tephra and augite crystals. It is encased in a brown, massive claystone and underlies BKT-2L by ~50 cm. Although Walter (1981) described the unit as grading westward into a thin green tuffaceous sand as far as the Gona Wadi, it is more often preserved as a laminated olive-green sandy claystone grading into alternating laminated green and brown clays. The Green Marker Bed is commonly 10–20 cm thick but reaches a thickness of 85 cm in portions of the Burahin Dora and upper Kada Hadar Wadis. Sample HE01-151 is from the base of the Green Marker Bed exposed in a large cutbank in the upper Kada Hadar Wadi (Fig. 3), where it contains macroscopic phenocrysts.

Analysis of sample E01-7355 demonstrates the unimodal basaltic character of the Green Marker Bed (Table 1). One analysis from HE01-182 falls within this basaltic glass range, but all other analyses represent minerals, particularly titanomagnetite and a green to dark green augite. Although a few basaltic glass shards were noticed before preparing sample HE01-151, only magnetite grains remained after preparing the sample as the rare, thin basaltic shards were dissolved by excessive HF during sample cleaning. The octahedral titanomagnetite grains analyzed from samples HE01-151 and HE01-182 are generally similar in chemical composition (Table 4), but without a more extensive comparative data set, it is difficult to determine the degree of similarity. Regardless, the correlation between the Green Marker Bed deposits in the Kada Hadar and Koborto Gaba regions proposed by Walter (1981) is still believed to be valid.

BKT-2L

The BKT-2L in the Koborto Gaba is a 10-cm-thick, powdery, white to yellowish white, crystal-lithic–rich unit with a minor amount of rhyolitic vitric tephra. The powdery nature of BKT-2L is possibly the result of alteration of the vitric material as well as the alteration of some of the feldspars as witnessed by crystal morphs that disintegrate to powder under light pressure. BKT-2L deposits in the central Hadar region (e.g., in the Farsita, Kurbili, Burukie, and Kada Hadar Wadis) are similar in

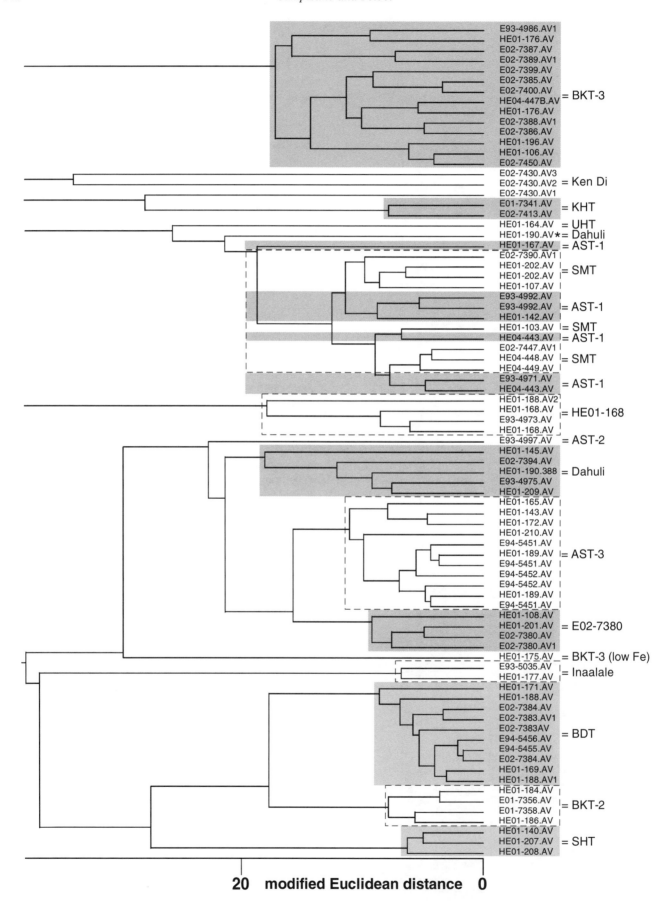

Figure 5. Detail of the cluster analysis dendrogram of rhyolitic vitric tephras from the Hadar and Busidima Formations based on the normalized sample mean composition of Al_2O_3, Fe_2O_3, CaO, MgO, MnO, TiO_2, and Cl. The similarity index is a modified version of the Euclidean distance as discussed in the text. Compared to the results presented in Table 1, HE01-190 (asterisked) is not grouped with the other Dahuli Tuff samples, and there is poor discrimination between AST-1 and the Salal Me'e Tuff (SMT), which are known to be stratigraphically distinct. AST—Artifact Site Tuff; BDT—Burahin Dora Tuff; BKT—Bouroukie Tuff; KHT—Kada Hadar Tuff; SHT—Sidi Hakoma Tuff; SMT—Salal Me'e Tuff.

◀─────────────────────────────

appearance to deposits in the Koborto Gaba, but they are slightly thinner (5–7 cm). BKT-2L deposits in the central region also lack a vitric component, and feldspars are sometimes completely altered. Anorthoclase feldspars from BKT-2L from Gona, Hadar, and Ledi-Geraru have been dated to ca. 2.96 Ma (Semaw et al., 1997; Campisano, 2007; DiMaggio et al., this volume). Glass analysis of sample E01-7356 from the Koborto Gaba produced a single rhyolitic mode (Table 1). It should be noted, however, that most of the totals in the microprobe analysis for the sample are relatively low, averaging only ~89%.

BKT-2U

Depending on the location, BKT-2U (= BKT-2u) is separated from BKT-2L by 0.4–2.2 m of brown clays that typically preserve linked pentagonal and hexagonal desiccation cracks infilled with BKT-2U material. Anorthoclase feldspars from BKT-2U from Hadar and Ledi-Geraru have been dated to ca. 2.94 Ma (Campisano, 2007; DiMaggio et al., this volume; Figure 1 of Walter, 1994). BKT-2U in the Koborto Gaba is 45 cm thick and composed of three subunits, informally labeled here as BKT-2Ua, -b, and -c from bottom to top. The lowest unit is 15–20 cm thick and very similar in lithology and appearance to BKT-2L. The middle unit is also 15–20 cm thick and consists dominantly of crystal-lithic material and dark brown vesicular/scoriaceous basaltic glass, but it also includes rhyolitic glass and a minor quantity of small, light-colored pumices. The capping 2–4 cm is very similar in lithology to the basal BKT-2U unit, but it is exceptionally calcareous. Whereas the Green Marker Bed and BKT-2L appear to be continuously preserved across Hadar, BKT-2U is not consistently found in association with them. Where preserved outside of the Koborto Gaba, the thickness varies significantly, from 15 cm in the Farsita to 75 cm in the upper Kurbili. In this area, BKT-2U is crystal-rich and yellowish white in appearance, like BKT-2L, but is thicker, coarser grained, and contains reworked brown clay fragments, small tuff clasts, and occasional gastropod shells. Unlike in the Koborto Gaba, BKT-2U exposures in central and western Hadar preserve no vitric component, and the small tuff clasts have been altered in situ to fine clay while preserving the original pumiceous fabric.

Glass analysis of sample HE01-184 from BKT-2Ua in the Koborto Gaba produced a very tight rhyolitic mode with only one minor outlier, HE01-184.27 (Table 1). Analysis of sample E01-7357, collected from the same location, produced only three glass data points. The first is unlike any other Hadar tephra. The second,

TABLE 2. COMPARATIVE GEOCHEMISTRY OF KNOWN HADAR AND BUSIDIMA FORMATION TUFFS

Tuff	Sample	SiO_2	Al_2O_3	Fe_2O_3	CaO	K_2O	Na_2O	MgO	MnO	TiO_2	Total	Reference
AST-1		75.76	13.61	2.02	0.93	3.87	3.16	0.29		0.35	100.00	Walter (1981)
AST-1		75.19	13.50	2.01	0.93	3.84	3.99	0.19		0.35	100.00	Hart et al. (1992)
AST-1	E93-4992.AV	74.91	12.87	2.39	0.97	3.86	4.19	0.21	0.07	0.36	100.00	This study*
AST-1	E93-4992.AV	74.97	13.22	2.50	1.01	3.96	3.46	0.22	0.09	0.37	100.00	Feibel and Wynn (1997)
AST-2		77.79	13.93	2.61	1.02	3.26	1.13	0.05		0.22	100.00	Walter (1981)
AST-2		74.63	13.39	2.54	0.95	3.40	4.76	0.08	0.04	0.20	99.99	Hart et al. (1992)
AST-2	E93-4997.AV	75.07	13.15	2.73	0.97	3.88	3.69	0.06	0.08	0.18	100.00	Feibel and Wynn (1997)
AST-3		77.35	13.63	3.00	0.76	3.88	1.20	0.02		0.21	100.00	Walter (1981)
AST-3		74.45	13.17	2.90	0.74	3.75	4.79			0.20	100.00	Hart et al. (1992)
AST-3	E94-5451.AV	73.31	12.83	3.05	0.68	4.46	5.17	0.00	0.08	0.25	100.00	This study*
AST-3	E94-5451.AV	74.06	12.78	3.22	0.75	4.46	4.23	0.01	0.09	0.23	100.00	This study*
AST-3	E94-5451.AV	73.88	12.68	3.39	0.72	4.62	4.23	0.02	0.07	0.24	100.00	This study*
AST-3	E94-5452.AV	73.84	12.85	3.22	0.74	4.52	4.37	0.00	0.07	0.23	100.00	This study*
AST-3	E94-5452.AV	73.62	12.84	3.22	0.66	4.39	4.80	0.01	0.07	0.26	100.00	This study*
BKT-3		74.21	13.99	2.92	1.45	3.34	3.25	0.41		0.44	100.00	Walter (1981)
BKT-3		73.51	13.87	2.89	1.44	3.31	4.14	0.41		0.44	100.01	Hart et al. (1992)
BKT-3	HE04-447B.AV	79.37	13.61	3.48	1.50	3.61	3.74	0.39	0.13	0.49	100.00	This study
KHT	KHT-25	78.87	10.30	5.48	0.26	1.64	2.66	0.02		0.27	99.99	Walter (1981)
KHT	KHT-29	79.28	10.47	5.11	0.22	1.71	3.25	0.01		0.36	100.00	Walter (1981)
KHT	KHT-25	76.99	8.75	5.70	0.17	1.75	4.08			0.27	100.00	Hart et al. (1992)
KHT	KHT-29	79.10	10.55	5.77	0.19	1.62	4.25	0.02	0.21	0.32	100.01	Hart et al. (1992)
KHT	E01-7341.AV	78.60	10.03	6.44	0.22	1.57	1.84	0.01	0.22	0.31	100.00	This study
KHT	E02-7413.AV	79.37	10.01	6.02	0.21	1.74	2.68	0.02	0.22	0.29	100.00	This study
SHT	SHT-41	78.55	13.39	1.44	0.31	2.03	4.06	0.04		0.18	100.00	Walter (1981)
SHT	SHT-41	77.23	12.85	1.84	0.27	2.15	5.41	0.05	0.03	0.18	100.01	Hart et al. (1992)[†]
SHT	HE01-207.AV	78.77	12.55	1.67	0.31	2.24	4.08	0.05	0.06	0.16	100.00	This study
SHT	HE01-208.AV	78.36	12.51	1.65	0.29	2.08	4.66	0.06	0.05	0.13	100.00	This study

Note: All values are reported as wt% and normalized since no raw totals were provided in Hart et al. (1992).
*Collected as part of Feibel and Wynn's (1997) study but reanalyzed in this study.
[†]X-ray fluorescence (XRF) analysis.

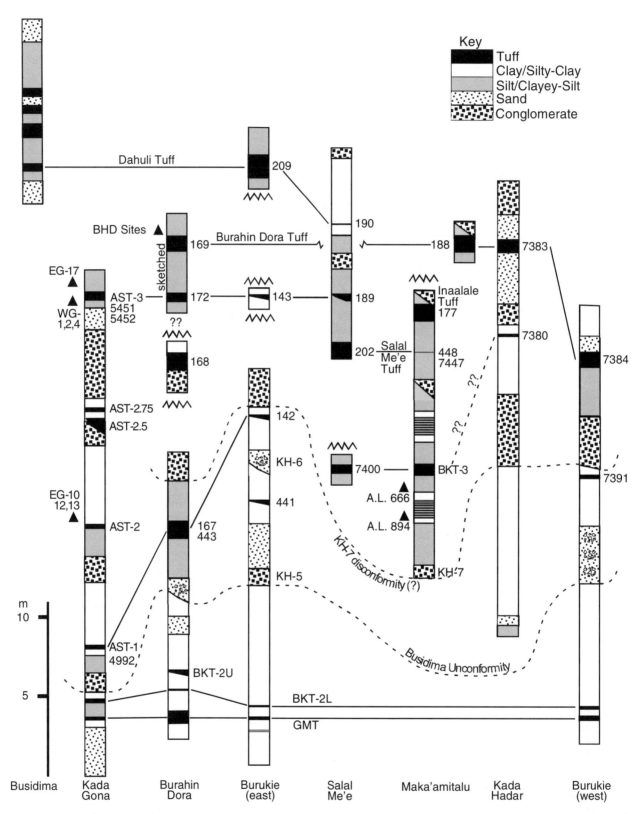

Figure 6. Stratigraphic relationships between select tuffs of the Busidima Formation. Sample numbers have been abbreviated. With the exception of Burukie (east), most sections progress left to right in a general eastward direction. The stratigraphic position of HE01-168 to HE01-169 and -172 is not definitive. The Ken Di and Upper Hurda Tuffs are not depicted. Sections from the Kada Gona and Busidima were adapted from Semaw et al. (1997) and Quade et al. (2004). A.L.—Afar Locality; AST—Artifact Site Tuff; BHD—Burahin Dora; EG—East Gona; GMT—Green Marker Tuff; KH—Kada Hadar; WG—West Gona.

TABLE 3. COMPARATIVE GEOCHEMISTRY OF THE SIDI HAKOMA (HADAR) AND TULU BOR (TURKANA BASIN) TUFFS

Sample	Tuff	n	SiO_2	Al_2O_3	Fe_2O_3	CaO	K_2O	Na_2O	MgO	MnO	TiO_2	Ba	Cl	Total
83-1720[†]	Tulu Bor β	19	74.00	11.80	1.57	0.30	3.30	3.27	0.06	0.05	0.15	0.02	0.10	94.60
K93-5050.AV*	Tulu Bor β		73.90	11.80	1.53	0.29	1.49	3.40	0.06	0.04	0.11	0.02	0.10	99.04
K93-5064.AV*	Tulu Bor β		74.25	11.86	1.48	0.28	1.52	3.35	0.06	0.04	0.12	0.02	0.10	99.67
K92-4858.AV*	Tulu Bor β	18	74.26	12.19	1.58	0.30	1.45	2.08	0.06	0.05	0.12		0.11	98.80
HE01-207.AV	Sidi Hakoma	7	73.25	11.67	1.55	0.28	2.08	3.79	0.04	0.06	0.15	0.02	0.10	93.00
HE01-208.AV	Sidi Hakoma	13	73.83	11.79	1.55	0.28	1.96	4.39	0.06	0.04	0.13		0.10	94.23
HE01-140.AV	Sidi Hakoma	6	73.05	11.66	1.56	0.27	2.05	3.92	0.05	0.04	0.16	0.02	0.09	92.86
ETH128, M2[†]	B-γ (=Tulu Bor γ)	5	69.80	13.20	3.00	0.54	2.75	1.56	0.12	0.11	0.14	0.10	0.07	91.60
K92-4858.4*	Tulu Bor γ	1	70.15	13.10	2.87	0.56	0.61	1.10	0.11	0.09	0.34		0.10	97.67
K92-4858.5*	Tulu Bor γ	1	70.60	13.25	2.97	0.54	1.22	1.53	0.12	0.10	0.28		0.10	98.74
K93-5064.4*	Tulu Bor γ	1	69.02	12.99	2.84	0.54	0.98	1.87	0.12	0.07	0.28	0.10	0.07	97.98
K93-5050.14*	Tulu Bor γ	1	69.39	13.17	3.29	0.64	1.03	1.83	0.16	0.07	0.37	0.18	0.07	98.96
HE01-140.04	Sidi Hakoma	1	68.87	12.75	2.96	0.56	1.80	3.56	0.13	0.04	0.33	0.10	0.07	91.17

*Feibel (unpublished).
[†]Haileab (1995).

E01-7357.1518, matches the outlier from HE01-184, suggesting that these two points may represent variation within the larger BKT-2U mode as opposed to a distinct mode. The third point, E01-7357.1514, is the only basaltic shard analyzed from BKT-2Ua.

Glass analysis of sample E01-7358 from BKT-2Ub in the Koborto Gaba produced a dominant rhyolitic mode and a single basaltic glass shard (E01-7358.1396) (Table 1). Analyses of basaltic material were disproportionately low compared to its presence in the tuff outcrop. This was primarily influenced by the selective analysis of easily identifiable rhyolitic glass shards compared to the thin-walled and vesicular basaltic scoria. A more detailed analysis of BKT-2U deposits from the Ledi-Geraru region has identified both rhyolitic and basaltic compositions from a sample chemically correlated to E01-7358 (DiMaggio et al., this volume).

Glass analysis of sample HE01-186 from BKT-2Uc in the Koborto Gaba involved only three rhyolitic glass shards, the rest being feldspars (Table 1). This is not surprising considering the overall rarity of glass in the unit.

Similarities and Differences of BKT-2 Subunits

The unimodal basaltic composition of the Green Marker Bed clearly distinguishes this tuff from the other components of the BKT-2 Complex. Although tephra analyses of BKT-2L and -2U components from the Koborto Gaba are distinguishable from non-BKT-2 tephra, they are surprisingly similar to each other (Fig. 5). While the averages of the tephra modes for each unit are slightly different, there is overlap in the standard deviations of virtually all oxides/elements between any two samples of BKT-2L and -2U (Table 1). The only possible exception to this observation is sample HE01-184, from the base of the BKT-2U sequence in the Koborto Gaba, with mean concentrations of Al_2O_3, Fe_2O_3, and K_2O slightly higher than in other BKT-2 samples. However, data plots of individual shards demonstrate that other BKT-2 samples have individual shards similar to that of HE01-184. Thus, while BKT-2L and -2U glasses from the Koborto Gaba are distinguishable from non-BKT-2 tephra, there are little to no data to suggest that BKT-2L and any subunits of BKT-2U can be chemically distinguished from each other based on major-element analysis. It is possible that trace-element analysis would define differences between the two tuffs that cannot be done using major elements alone.

Unlike the glass chemistry analyses, BKT-2L and -2U from both the Kada Hadar and Koborto Gaba regions may be distinguished by their feldspar compositions. Whereas BKT-2L anorthoclase feldspar compositions overlap those of BKT-2U in most major oxides/elements, there is a distinct separation in K_2O and CaO concentrations, which are inversely proportional to each other in the BKT-2 feldspars (Table 4). This separation between BKT-2L and -2U feldspars is clearly evident in a bivariate K_2O-CaO scatter plot of mean sample values, where BKT-2L is more calcic and less potassic than BKT-2U (Fig. 7A). In the bivariate plot of all individual BKT-2 feldspar analyses, there is a minor degree of overlap between the two tuffs, specifically between BKT-2L and the lowest component of BKT-2U, BKT-2Ua. Some of this overlap, however, may be the result of analytical differences between the two different microprobes used for analysis. Feldspar chemistry for samples E01-7356, -7357, and -7358 was recorded while analyzing for glass on the Rutgers University probe. The other samples presented were analyzed specifically for feldspar geochemistry on the AMNH probe, which allows for K and Na to be analyzed at different conditions to avoid elemental loss (larger beam and smaller current). In the K_2O-CaO scatter plot of individual BKT-2 feldspar using only the AMNH results, only a single BKT-2U analysis overlaps with the spread of BKT-2L samples (Fig. 7B). No distinction was noted between feldspar compositions of the individual BKT-2U subunits (i.e., a, b, and c) from the Koborto Gaba.

A similar distinction between BKT-2L and -2U feldspar K/Ca ratios was observed during $^{40}Ar/^{39}Ar$ analysis of the tuffs[3] (Campisano, 2007). K/Ca ratios average 0.67 ± 0.46 (mean ± 1σ) for BKT-2L ($n = 47$) and 2.11 ± 0.58 for BKT-2U ($n = 49$), with only minor overlap between individual feldspar ratios (Fig. 7C). An explanation for the lithologic stratification visible

---
[3]As calculated from the measurements of argon isotopes produced from K and Ca during the irradiation of the samples in analyses conducted at the Noble Gas Laboratory, Rutgers University.

TABLE 4. SUMMARY OF ELECTRON MICROPROBE ANALYSES OF BKT-2 COMPLEX FELDSPARS AND MAGNETITES

Sample	n	SiO_2	Al_2O_3	Fe_2O_3	CaO	K_2O	Na_2O	MgO	MnO	TiO_2	Ba	Cl	Totals	Lab	ID	Location
HE01-151(M) AV	5	0.07	0.93	70.90	0.00	0.02	0.00	0.76	0.43	26.02	0.09	0.01	99.23	AM	GMB	Kada Hadar
HE01-182(M).AV	5	0.14	1.64	75.42	0.00	0.01	0.02	0.88	0.39	21.81	0.08	0.01	100.38	AM	GMB	Korborto Gaba
E01-7356(F).AV	14	64.51	20.18	0.27	1.97	1.54	8.95	0.00	0.01	0.03		0.01	97.47	RU	BKT-2L	Koborto Gaba
HE01-183(F).AV	6	64.07	20.37	0.25	2.16	1.47	9.18	0.00	0.01	0.01	0.22	0.00	97.73	AM	BKT-2L	Koborto Gaba
HE04-456(F).AV	6	64.83	20.81	0.25	2.51	1.30	9.24	0.00	0.01	0.02	0.21	0.00	99.17	AM	BKT-2L	Kurbili
HE04-458(F).AV	7	64.63	21.41	0.31	3.17	1.10	9.11	0.00	0.01	0.01	0.15	0.00	99.90	AM	BKT-2L	Kada Hadar
HE04-457(F).AV	7	66.69	20.04	0.24	1.37	2.35	9.13	0.00	0.01	0.01	0.28	0.00	100.12	AM	BKT-2U	Kurbili
E01-7357(F).AV	22	64.15	19.79	0.35	1.64	1.82	9.08	0.00	0.01	0.03		0.01	96.89	RU	BKT-2U(a)	Koborto Gaba
HE01-184(F).AV	8	66.17	20.09	0.26	1.49	2.11	9.07	0.00	0.01	0.03	0.25	0.00	99.49	AM	BKT-2U(a)	Koborto Gaba
HE01-184(F).AV	2	66.74	20.55	0.33	1.69	1.84	9.48	0.00	0.01	0.00		0.00	100.65	AM	BKT-2U(a)	Koborto Gaba
E01-7358(F).AV	8	64.34	19.34	0.31	1.17	2.40	9.29	0.00	0.01	0.03		0.01	96.91	RU	BKT-2U(b)	Koborto Gaba
HE01-185(F).AV	8	66.49	20.06	0.23	1.31	2.38	8.99	0.00	0.01	0.01	0.29	0.01	99.79	AM	BKT-2U(b)	Koborto Gaba
HE01-186(F).AV	7	66.44	20.04	0.23	1.37	2.25	9.17	0.01	0.01	0.02	0.32	0.01	99.86	AM	BKT-2U(c)	Koborto Gaba
HE01-186(F).AV	7	66.93	20.28	0.28	1.42	2.07	9.39	0.00	0.00	0.01		0.00	100.38	AM	BKT-2U(c)	Koborto Gaba

Note: (M)—magnetite, (F)—anorthoclase feldspars. All values are reported as wt%. RU—Rutgers University; AM—American Museum of Natural History (AMNH).

in BKT-2U is still being sought, particularly in the Ledi-Geraru region where the deposit reaches a meter in thickness and has been sampled and analyzed at a finer level of resolution, but it may be the result of a mixed-magma eruption (e.g., Walter, 1981; DiMaggio et al., this volume).

HE04-441

This unnamed tuff is the stratigraphically lowest Busidima Formation tephra unit yet identified at Hadar. It is located between the first conglomerate of the Busidima Formation (the KH-5 conglomerate) and a cemented sandstone/pebble-sand conglomerate (the KH-6 sandstone) (Fig. 6). This discontinuous tephra deposit ranges from 20 to 40 cm in thickness and is only exposed across a few tens of meters along the western bank of the eastern branch of the Burukie Wadi (Fig. 4). Much of the unit is devitrified and preserved as a fine-grained silt, but where glass exists, it is of basaltic composition. Although the individual shard analysis totals for the sample range from 100% to 105% (i.e., greater than the 102% cutoff), the data are included here as they still provide some geochemical characterization of the tephra. Similarly, even with this caveat, no potential correlates of HE04-441 were identified, indicating that it represents a unique tephra.

Artifact Site Tuff 1 (AST-1)

Two tephra deposits sampled from Hadar are proposed as Artifact Site Tuff 1 (AST-1) correlates based on comparison to the AST-1 stratotype E93-4992 collected from Gona. The first correlate is sample HE01-142 from the eastern bank of the eastern branch of the Burukie Wadi (Fig. 4). The sample is from a silty, fine-grained tephra, ~20–30 cm thick, located just below a 4-m-thick conglomerate (likely the KH-7 conglomerate) that caps the exposure and often erodes to below the level of the tephra (Fig. 6). Duplicate samples, HE01-167 and HE04-443, from the west bank of the main branch of the Burahin Dora Wadi are also correlated to AST-1 and sample HE01-142 (Fig. 4). These samples were collected from an almost white-colored, very fine-grained, but vitric, tephra, ~1.2 m thick, just a few meters below the KH-7 conglomerate that caps the section (Fig. 6). The description of AST-1 from the Kada Gona region, 1–1.3 m thick and largely altered to white claystone, also matches the description of the Hadar correlates. Although the University of Utah analysis of E93-4992 has slightly higher SiO_2, Al_2O_3, and Fe_2O_3 values than the Hadar correlates analyzed at Rutgers University, it also has a higher total sum value. When E93-4992 was reanalyzed at Rutgers in the same analytical run as HE04-443, the results were indistinguishable, not only from each other, but also from other proposed Hadar AST-1 correlates from other runs (Table 1).

The dominant chemical mode and a single basaltic shard reported from sample E93-4971, collected from the Burahin Dora Wadi (but not from the same exposure as HE01-167), also groups with the dominant rhyolitic mode of AST-1 and the single basaltic shard reported from the stratotype E93-4992 (Table 1).

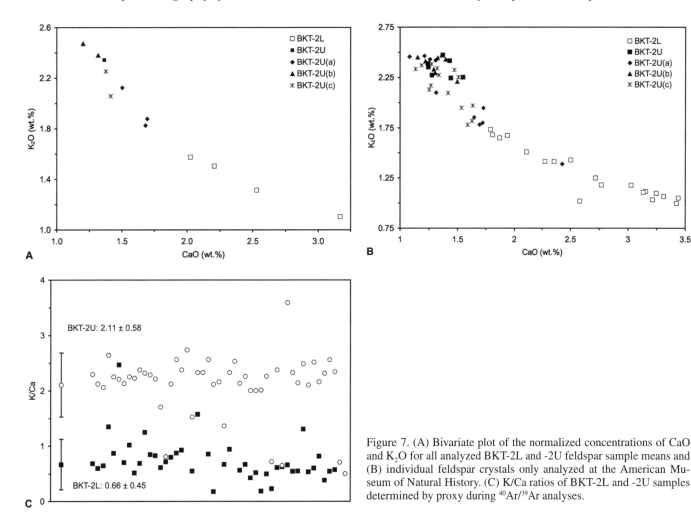

Figure 7. (A) Bivariate plot of the normalized concentrations of CaO and K_2O for all analyzed BKT-2L and -2U feldspar sample means and (B) individual feldspar crystals only analyzed at the American Museum of Natural History. (C) K/Ca ratios of BKT-2L and -2U samples determined by proxy during $^{40}Ar/^{39}Ar$ analyses.

Sample E02-7391, from just below the KH-7 conglomerate in the western Burukie region (Figs. 4 and 6), is at an appropriate stratigraphic level to be an AST-1 candidate, but analysis of this glass-poor sample produced only mineral values and no glass results to confirm a correlation.

Artifact Site Tuff 2 (AST-2)

No tephras yet analyzed from Hadar match the chemical composition of the AST-2 stratotype sample E93-4997 collected at Gona. This is an important note as AST-2 in Gona immediately underlies the oldest archaeological sites there. Also, AST-2 has previously been proposed as a correlate to tephra at Hadar, specifically BKT-2 (Walter, 1981; Tiercelin, 1986), but such correlation is clearly negated by the data presented here.

E02-7380

Samples HE01-201 and E02-7380, from the Salal Me'e and upper Kada Hadar Wadis, respectively, lie at a stratigraphic level above the KH-7 conglomerate similar to BKT-3, and were identified by the authors as potential BKT-3 correlates in the field. Sample HE01-201 was collected from a vitric tephra up to a meter thick located just above the KH-7 conglomerate, not far from the Salal Me'e's confluence with the Kada Hadar Wadi (Fig. 4). Sample E02-7380 was collected from a 10-cm-thick, yellow-brown vitric tephra ~3 m above the KH-7 conglomerate in a steep ravine leading up from a cutbank on the north/east bank of the upper Kada Hadar Wadi opposite the Salal Me'e Wadi (Figs. 4 and 6).

Despite the stratigraphic positions of HE01-201 and E02-7380, their chemical compositions do not appear to support a correlation to BKT-3, as they are distinctly lower in CaO, MgO, and TiO_2 concentrations than the dominant BKT-3 mode, discussed next (Table 1). However, these two tephras may have some relation to BKT-3, as several single-shard analyses from BKT-3 samples (E93-4986.16 and .17 and possibly E02-7389.354) and the dominant mode in HE01-108 ($n = 4$) have a similar composition (Table 1). It may be possible that these individual shards and the four correlated shards from sample HE01-108 represent a reworked component of a distinct tephra, represented by samples HE01-202 and E02-7380, into the BKT-3 deposit. However, it is unusual that so few other BKT-3 samples would preserve such a composition, particularly E02-7387 and -7388, which were collected at the same location as HE01-202. An unlikely possibility is that samples HE01-201 and E02-7380 are part of the BKT-3 complex but represent a mode that is frequent in these two samples and rare or absent in all other BKT-3 samples. Sample E02-7380 also preserves a minor basaltic mode ($n = 4$) that is similar to the basaltic mode of BKT-3, but it has higher Fe_2O_3 and lower Al_2O_3 concentrations and may be evidence against a correlation between the two samples. While this problem may be resolved with future analyses, it presents only a minor issue in the overall story of stratigraphic correlation for the Busidima Formation. As only sample E02-7380 was placed into a stratigraphic section, both it and HE01-201 will be grouped together and referred to as tuff E02-7380 (Fig. 6; Table 1) until the details or significances of these units are better understood.

Bouroukie Tuff 3 (BKT-3)

The main tuff of the archaeology-bearing Maka'amitalu Basin is the BKT-3, which lies ~80 cm above the archaeological site A.L. (Afar Locality) 666 and 3 m above site A.L. 894. BKT-3 has been dated to ca. 2.35 Ma (Kimbel et al., 1996), making the stone tools from A.L. 666 and 894 and the early *Homo* maxilla from A.L. 666 among the oldest yet discovered (Kimbel et al., 1996, 1997; Hovers, 2003). Virtually all sediments in the Maka'amitalu lie above the KH-7 conglomerate, which essentially forms the floor of the basin in a large part of the area. BKT-3 is noticeable as a distinct unit about halfway up the exposures that ring the basin, and it is present in most knolls and small peninsulas of exposures within the basin. Exposures of BKT-3 appear to continue south along the western bank of the Maka'amitalu toward, and possibly into, the Dahuli and Salal Me'e Wadis. The thickness of BKT-3 ranges between 35 and 80 cm, and it is tan to gray in color. The size fraction and percentage of glass to silt in the sampling locations of BKT-3 vary both laterally and vertically. BKT-3 is interpreted to represent a floodplain deposit, and it is possible that the lithologic variation recorded could be due to hydraulic sorting during deposition of the tephra, or different phases of deposition.

Twelve different BKT-3 samples collected from the Maka'amitalu Basin are included in this study[4] (Fig. 4). Among them, sample HE04-447(A&B) was collected from the area defined as the type locality of BKT-3 by Walter (1981) "ca. 100 m west of A.L. 666" (Kimbel et al., 1996, p. 551). This deposit of BKT-3 is distinctly coarser grained and has a higher glass component than other BKT-3 sampling locations, but it still includes a silt and sand component.

BKT-3 is not only one of the most important of Busidima Formation tuffs at Hadar, but it is also the most geochemically complex. The results from the BKT-3 samples display one dominant chemical mode that represents the original BKT-3 eruptive chemistry and three minor modes (two silicic, one basaltic) that were most likely incorporated during eruption or fluvial reworking prior to deposition (Fig. 8; Table 1). The two minor silicic modes of BKT-3 only become apparent when individual outlier shard compositions from the various BKT-3 samples are grouped together. In other words, with one exception (HE01-175), no analyzed BKT-3 sample contained more than one shard from a specific silicic minor mode. It is also worth noting that all BKT-3 samples analyzed preserved evidence of at least one minor mode. The dominant mode of BKT-3 is present in all 12 of the BKT-3 samples from the Maka'amitalu Basin and represents ~80% of all BKT-3 shards analyzed. This mode is characterized by a combination of relative high concentrations of Fe_2O_3, CaO, and MgO and often a high range of intra-sample variation in Al_2O_3 and sometimes in Fe_2O_3 (Table 1).

The first minor mode present in BKT-3 is easily distinguished from the dominant mode of BKT-3 by having lower concentrations of virtually all oxides, particularly Fe_2O_3, CaO, and MgO, and higher concentrations of SiO_2 and K_2O. This "low-Fe" mode is present in BKT-3 samples HE01-106, E02-7385 (1 shard each), HE01-196 (2 shards), and is also present as single shards in two samples of the Salal Me'e Tuff (HE01-107 and -175), discussed later (Figs. 8 and 9; Table 1). In contrast to the first minor mode, the second minor mode of BKT-3 has distinctly higher concentrations of Fe_2O_3 and CaO compared to the dominant mode, as well as the highest MgO and TiO_2 concentrations of any nonbasaltic Hadar tephra. This "high-Fe" minor mode is present only as single shards in five BKT-3 samples: HE01-176, HE04-447B, E93-4986, E02-7387, E02-7450, and possibly E02-7389 (Table 1). Similar to the "low-Fe" mode, this "high-Fe" mode is also present as a single shard in a sample from the Salal Me'e Tuff (E02-7390.372) (Fig. 9). Other chemical compositions are recorded in the analyses of individual BKT-3 shards (Campisano, 2007) but do not appear to form any additional coherent chemical modes and most likely represent reworked detrital shards.

The basaltic mode of BKT-3 is present as a single shard in four BKT-3 samples: HE01-176, E02-7386, E02-7389, and E02-7450, and as a multishard mode in two BKT-3 samples: E93-4986 ($n = 5$) and E02-7388 ($n = 4$) (Table 1). It is also present as a multishard mode in a sample from the Salal Me'e Tuff (E02-7390). It is possible that this basaltic mode is not a reworked/incorporated component of BKT-3 but is associated with the original volcanic event that produced the dominant, silicic mode. This interpretation remains questionable as the mode is present in such low proportion in only half of the BKT-3 samples analyzed and is also present in samples of the Salal Me'e Tuff.

Sample HE01-108 presents an interesting problem. Although the sample was collected at the same location as E02-7387 and -7388, only one of the five glass shards analyzed (HE01-108.04) correlates with the other two samples that represent the dominant BKT-3 mode (Table 1). As discussed earlier, the remaining four shards represent a mode that appears to correlate to samples HE01-201 and E02-7380 (Fig. 9) but has not yet been detected in the analyses of E02-7387 and -7388.

BKT-3 is present in the Salal Me'e Wadi, as evident by the correlation to samples E02-7399 and -7400 sampled near the con-

---

[4]The exact sampling locations of BKT-3 in the Maka'amitalu Basin are presented in Campisano (2007).

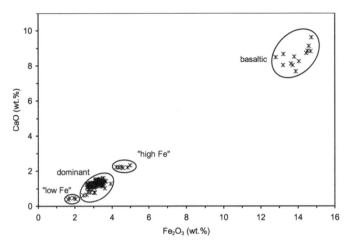

Figure 8. Bivariate plot of the unnormalized concentrations of Fe_2O_3 and CaO for individual BKT-3 glass shards. The four chemical modes of BKT-3 are identified.

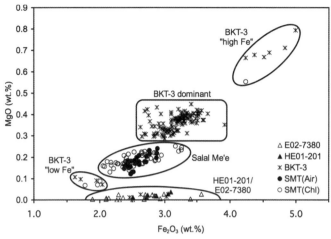

Figure 9. Bivariate plot of the unnormalized concentrations of Fe_2O_3 and MgO for individual silicic glass shards from BKT-3, the Salal Me'e Tuff, and E02-7380/HE01-201. The legend refers to the tephra from which a shard was collected, whereas the labeled groupings refer to the tuff and/or mode the shard is chemically correlated to. SMT(Air) = Salal Me'e Tuff air-fall deposit; SMT(Chl) = Salal Me'e Tuff channel deposit. Note the presence of both "low-Fe" ($n = 3$) and "high-Fe" ($n = 1$) BKT-3 correlates in the Salal Me'e Tuff channel, but not air-fall, deposit, as well as the dominant mode of BKT-3 sample HE01-108 ($n = 4$) grouped with tuff E02-7380.

fluence of the Salal Me'e and the Kada Hadar Wadis, just south of the Maka'amitalu Basin. Thus, BKT-3 does extend outside the Maka'amitalu Basin, but it does not do so by much. BKT-3 has yet to be identified in the Gona region, negating the proposed lithostratigraphic correlation by Tiercelin (1986).

HE01-168

This unnamed tephra is located along a cutbank of an east-west–trending tributary of the southern branch of the Burahin Dora Wadi (Fig. 4). The road along the Gona-Hadar divide runs just to the west of the sampled area. The sample is from a gray, coarse-grained vitric tephra, ~30 cm thick, resting atop a 1.7 m conglomerate and overlain by about a meter of clays (Fig. 6). A similar chemical composition to HE01-168 is observed in sample E93-4973, collected from a tephra in the same general region of the Burahin Dora above the AST-1 correlate E93-4971. Most oxide values between HE01-168 and E93-4973 are indistinguishable, but whereas SiO_2 and Fe_2O_3 concentrations overlap within a single standard deviation, Al_2O_3 concentrations between the two are only within two standard deviations of each other (Table 1). Although a correlation between these two tephra units is not definitive, they are tentatively correlated here based on their similarity in most major oxides/elements, relatively high Fe_2O_3 concentrations, stratigraphic position, and geographic location. As both HE01-168 and the underlying conglomerate are discontinuous and only exposed along a 10–15-m-long cutbank in the drainage, the stratigraphic position of HE01-168 is by no means definitive. Although these units could not be traced laterally to other tephra samples in the Burahin Dora Wadi, it clearly lies above AST-1 in the Burahin Dora sequence and most likely lies below the Burahin Dora Tuff (discussed later), as samples HE01-188 and E02-7383 appear to contain a few reworked shards of tuff HE01-168 within them (Table 1).

Salal Me'e Tuff (SMT)

The Salal Me'e Tuff is a new name proposed for three tephra deposits correlated between the Salal Me'e Wadi and Maka'amitalu Basin. The stratotype for this tuff is sample HE01-202 from a coarse-grained, mixed-composition tephra approximately a meter thick, discontinuously exposed in the western reaches of the Salal Me'e Wadi (Fig. 4). As analyses focused primarily on rhyolitic tephra, only one basaltic glass shard was detected in HE01-202. The Salal Me'e Tuff does show a degree of chemical variation as two different runs of the same sample resulted in not only slightly different mean values, but also displayed relatively large standard deviations for certain oxides, such as Fe_2O_3. The Salal Me'e Tuff is broadly similar in composition to AST-1 (Table 1; Fig. 5), but, on average, it has higher Al_2O_3 and Fe_2O_3 concentrations and is lithologically and stratigraphically distinct from AST-1. The Salal Me'e Tuff is also relatively similar to AST-2 but can be distinguished on the basis of MgO and TiO_2, as well as on stratigraphic position.

Samples HE01-103 and -107 from a localized (<10 m long) channel-fill tephra deposit in the Maka'amitalu Basin north of A.L. 894 (Fig. 4) not only match the chemical composition of HE01-202, but the deposits are virtually identical in outcrop appearance. A thin (2–4 cm), dark, tephra preserved ~7 m above BKT-3 in the northeast portion of the Maka'amitalu Basin (Figs. 4 and 6) represented by samples HE04-448, -449, and E02-7447 is also proposed as a correlate to the Salal Me'e Tuff. The tephra contains mixed-composition glass shards, as well as a crystal-lithic component, but it is homogenized into the surrounding soil outside of the sampling area and likely represents an air-fall tephra. Samples HE04-448 and -449 underwent magnetic separation during sample preparation to produce a more pure glass separate for analysis, but it also resulted in the removal of the basaltic component from the analyzed sample. Sample E02-7447, however, was not separated during preparation, and three shards comprise the basaltic mode of E02-7447 (AV2). The single basaltic shard HE01-202.343 displays only a general similarity to the basaltic mode of E02-7447, but it is difficult to discount the correlation on the basis of a single shard.

Although samples HE04-448 and -449 contain plagioclase feldspars, attempts to obtain a reliable age for the Salal Me'e Tuff are still in progress. Single- and multigrain $^{40}Ar/^{39}Ar$ analyses of these samples clearly identify the presence of Miocene-age xenocrysts, but may also indicate a juvenile population with an age of ca. 2.2 Ma (Campisano, 2007). While this date remains questionable, it may correlate with an age of 1.9–2.0 Ma reported for an unnamed tephra above BKT-3 reported by Walter et al. (1996).

It is likely that the coarse-grained vitric channel deposits in the Salal Me'e Wadi and Maka'amitalu Basin represent the fluvially reworked component of the primary air-fall tephra present in the Maka'amitalu Basin. In the Maka'amitalu Basin, the isolated channel fill is not laterally continuous with any other sediments adjacent to it, but the top of the exposure is topographically less than 50 cm below the level of BKT-3 on a nearby exposure. However, the air-fall tephra is topographically 7 m above the channel fill in the Maka'amitalu Basin and stratigraphically above BKT-3. This can be explained either by the fluvial channel deposit having eroded into underlying sediment to below the level of BKT-3, or having filled in a natural channel or topographic low that already existed.

As mentioned in the discussion of the BKT-3, the compositions of a few individual "outlier" shards of the Salal Me'e Tuff also correlate with the minor modes of the BKT-3. Specifically, two analyses from HE01-175 and one analysis from HE01-107 match the "low-Fe" BKT-3 mode, one analysis from E02-7390 matches the "high-Fe" BKT-3 mode, and a basaltic mode of E02-7390 ($n = 7$) appears to match the basaltic mode of BKT-3 (Fig. 9; Table 1). The occurrence of infrequent, but nonetheless correlated glass shards between the minor modes of the BKT-3 and Salal Me'e Tuffs suggests that both tuffs were deposited in the Maka'amitalu Basin by a fluvial system that drained watersheds that preserved BKT-3 and Salal Me'e tephras as well as tephras from previous eruptive events. When rain waters flushed the watershed system(s) following BKT-3 and Salal Me'e eruptions, both events could have incorporated tephra left on the landscape from previous eruptions and redeposited the reworked material in the Maka'amitalu Basin along with the BKT-3, and later, the Salal Me'e Tuff.

This scenario is preferred over possible alternative mechanisms because it is only the reworked channel deposit of the Salal Me'e Tuff, not the air-fall unit, that contains the shards shared with BKT-3. If the shared chemical compositions between the Salal Me'e and BKT-3 Tuffs were the result of incorporating material during the original eruptive event, then we would expect to find shards of the shared composition in both the channel and air-fall units, not just the channel deposit. Additionally, if the shared compositions between the two tuffs were the result of the Salal Me'e Tuff incorporating reworked components of the BKT-3, then the dominant mode of BKT-3 would also be expected to be preserved in Salal Me'e Tuff deposits, not just the exceedingly rare minor modes.

Inaalale Tuff

The Inaalale Tuff is a newly described, widespread vitric tephra from the Dikika project area that is proposed as a correlate to a previously unnamed tephra sample from Hadar, HE01-177 (Wynn et al., this volume). Sample HE01-177 was collected from a light gray vitric tephra less than 2 m above the Salal Me'e Tuff air-fall unit (HE04-448) in the northeast portion of the Maka'amitalu Basin (Figs. 4 and 6). The tuff is over a meter thick in some places, but it is very fine-grained, only exposed in a small area around the sampling site, and is often eroded into by an overlying carbonate-cemented pebble-sand conglomerate. Although every analysis but one resulted in a total of just under 90%, there are no Hadar tephras that appear to correlate with HE01-177. Sample HE01-177 is distinct in having a relatively high Fe_2O_3 concentration but very low CaO concentration compared to other nonbasaltic high-Fe tephra (Table 1). This combination is also observed in sample E93-5035 from the top of a section in the Gara'a Dora, a tributary of the Kada Gona, not far from the Hadar-Gona divide. Most oxide values between HE01-177 and E93-5035 are indistinguishable, but whereas Al_2O_3 concentrations overlap within a single standard deviation, SiO_2 and Fe_2O_3 concentrations between the two samples are similar, but only within two standard deviations of each other. The unnormalized data from the Dikika Inaalale Tuff samples display higher Al_2O_3 concentrations and much larger standard deviations in Fe_2O_3 concentrations compared to both HE01-177 and E93-5035 (Table 5). The normalized data for these tephra presented in Wynn et al. (this volume) essentially eliminate the differences observed between HE01-177 and E93-5035 in the unnormalized SiO_2 and Fe_2O_3 concentrations but slightly increase the difference in their Al_2O_3 concentrations. This gap in Al_2O_3 concentrations, however, is subsequently filled in with the normalized values from the Dikika Inaalale Tuff samples.

The stratigraphic position of both HE01-177 and E93-5035 in relation to AST-3 was uncertain at both Hadar and Gona, as they are found at the same general level, but not in the same section or immediate vicinity to be certain. The tuffs were tentatively placed below AST-3, a position confirmed by their correlation to the Inaalale Tuff at Dikika, where it is stratigraphically below AST-3. Although a correlation of the Inaalale Tuff among the three project areas is not definitive, they are tentatively correlated here based on their unique Fe_2O_3/CaO ratio, similarity in most major elements, and similar stratigraphic position.

TABLE 5. COMPARATIVE GEOCHEMISTRY OF SELECT HADAR, GONA, AND DIKIKA TEPHRAS

Sample	Tuff	n	SiO_2	Al_2O_3	Fe_2O_3	CaO	K_2O	Na_2O	MgO	MnO	TiO_2	Cl	F	Total
Inaalale Tuff														
E93-5035.AV	Gona	17	69.31	11.63	3.74	0.31	3.33	4.47	0.08	0.17	0.27	0.30		93.80
HE01-177.AV	Hadar	6	65.24	11.53	3.55	0.30	3.32	4.44	0.08	0.14	0.23	0.35		89.16
E02-2015.AV*	Dikika	20	68.08	12.08	3.72	0.28	3.13	2.87	0.08	0.15	0.29			90.68
E03-3034.AV*	Dikika	10	70.48	12.00	3.54	0.25	3.21	3.03	0.08	0.12	0.25			92.95
E05-4031.AV*	Dikika	10	71.01	11.91	3.87	0.26	3.19	3.49	0.07	0.38	0.22			94.41
Ken Di Tuff														
Mode 1														
E02-7430.AV1	Hadar	5	67.85	8.85	6.90	0.26	1.85	3.97	0.02	0.26	0.34	0.17		90.48
E05-4025.AV1*	Dikika	9	70.86	9.08	7.02	0.27	2.01	2.36	0.02	0.52	0.29			92.46
E05-4026.AV1*	Dikika	10	70.84	9.10	7.44	0.26	2.10	2.29	0.02	0.48	0.35			92.94
Gonash-86.AV†	Gona	12	71.07	9.44	7.15	0.28	1.49	2.67	0.02	0.28	0.33	0.13	0.08	92.40
Mode 2(?)														
E02-7430.AV2	Hadar	2	63.14	11.72	6.09	0.66	4.03	5.60	0.05	0.26	0.51	0.11		92.18
E05-4026.AV2*	Dikika	3	66.81	11.84	7.05	0.67	3.24	3.47	0.05	0.51	0.39			94.12
Mode 3(?)														
E02-7430.AV3	Hadar	2	62.50	12.22	7.26	0.87	2.84	5.17	0.08	0.35	0.57	0.09		91.95
E05-4026.AV3*	Dikika	2	66.09	12.99	6.53	0.83	3.14	2.80	0.08	0.54	0.51			93.57
Mode 2+3														
E02-7430.AV2+3	Hadar	4	62.82	11.97	6.68	0.77	3.44	5.39	0.07	0.30	0.54	0.10		92.07
E05-4026.AV2+3*	Dikika	5	66.52	12.30	6.84	0.73	3.20	3.20	0.06	0.52	0.44			93.90
E05-4025.AV2*	Dikika	6	66.19	12.32	6.85	0.73	1.99	2.37	0.07	0.50	0.45			91.52
Dahuli Tuff														
E93-4975.AV	Gona	18	69.19	12.17	3.24	0.93	3.73	4.07	0.01	0.13	0.23	0.13		93.92
HE01-190.AV	Hadar	9	69.01	12.48	3.15	1.07	3.74	4.42	0.17	0.10	0.27	0.10	0.13	94.65
HE01-190.388	Hadar	1	69.61	12.34	3.23	0.81	3.80	3.54	0.04	0.11	0.26	0.10		89.42
HE01-209.AV	Hadar	17	69.34	12.31	3.18	0.87	3.64	4.39	0.01	0.11	0.22	0.13		94.20
E02-7394.AV	Hadar	14	67.47	11.81	3.22	0.90	3.67	3.51	0.04	0.09	0.26	0.10		91.06
HE01-145.AV	Hadar	3	66.22	12.25	3.29	1.01	4.19	3.92	0.04	0.11	0.31	0.08		91.43
BUST-2.AV†	Gona	20	72.99	12.75	3.21	0.90	3.56	4.32	0.00	0.11	0.24		0.08	98.02
BUST-2.AV§	Gona	17	72.43	12.67	3.25	0.88	2.61	1.54	0.00	0.11	0.25		0.15	93.77
BUST-3.AV§	Gona	21	72.78	12.77	3.28	0.89	2.71	1.60	0.00	0.12	0.22		0.14	94.39
BUST-11.AV§	Gona	21	73.32	12.83	3.28	0.90	2.86	1.66	0.00	0.10	0.22		0.15	95.19
BUST-13.AV§	Gona	19	73.72	12.89	3.19	0.87	2.86	1.78	0.00	0.10	0.23		0.14	95.64
BUST-26.AV†	Gona	12	71.85	12.88	3.28	0.85	3.52	4.17	0.00	0.11	0.23	0.10	0.06	96.86
BUST-27.AV†	Gona	16	72.06	13.11	3.17	0.86	3.50	4.21	0.00	0.11	0.21	0.08	0.07	97.27

Note: Hadar samples were run at 12 μm, 15 nA, 15 kV.
*Wynn et al. (this volume); samples run at 5–10 μm, 5 nA, 10–12 kV.
†Quade et al. (this volume); samples run at 10 μm, 8 nA, 15 kV.
§Quade et al. (this volume); samples run at 2 μm, 20 nA, 15 kV.

Artifact Site Tuff 3 (AST-3)

Five tephra deposits sampled from the Hadar region are proposed as AST-3 correlates based on the recent analysis of the AST-3 stratotypes E94-5451 and -5452 upsection from the Gona archaeological sites of EG12 and EG13. Multiple analyses from these type-section samples demonstrate that AST-3 has a broad geochemical signature in several oxides (Fig. 10; Table 1). If it wasn't for these multiple runs, the actual chemical variation of AST-3 may have gone unnoticed, and several correlates proposed here may have been considered distinct tuffs from AST-3. The chemical composition of AST-3 is similar to the Burahin Dora Tuff (discussed later), but it is distinguishable by higher CaO and TiO concentrations. Compared to the Upper Hurda Tuff (discussed later), AST-3 is distinguishable by lower CaO, MgO, and TiO_2 concentrations. Although the full details are not provided, based on its possible correlation to the Fialu Tuff, which overlies a tuff dated to ca. 1.64 Ma, Quade et al. (this volume) estimate the age of AST-3 at ca. 1.3 Ma. AST-3 has also been recorded from the Dikika project area (Wynn et al., this volume).

Samples HE01-143, -165, and -210 from three separate locations within the Burukie Wadi are correlated with AST-3 (Fig. 4). Sample HE01-143 was collected from a light-colored, fine-grained tephra within the western branch of the Burukie Wadi. The localized patch of tephra lies at the base of a small cutbank in the stream bed and reaches a maximum thickness of 65 cm, but it is cut out by an overlying conglomerate. Sample HE01-165 was collected from a collapsing peninsula of exposure in the northern portion of the eastern branch of the Burukie Wadi, north of both HE01-142 (AST-1) and HE01-210. Sample HE01-210 was collected from the western bank of the eastern branch of the Burukie Wadi, immediately north of where the Burukie Wadi splits into two branches. Despite being at approximately the same level as, and located just south of and across the drainage from, AST-1 sample HE01-142, the two tephras clearly do not correlate with each other. Thus, it appears that both AST-1 and AST-3 are

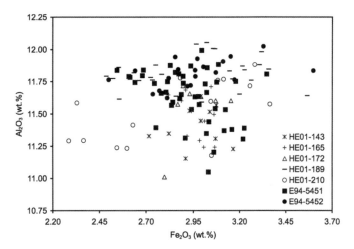

Figure 10. Bivariate plot of the unnormalized concentrations of Fe_2O_3 and Al_2O_3 for individual AST-3 glass shards. Solid squares and solid circles represent AST-3 type section samples E94-5451 and -5452, respectively, which demonstrate the compositional variability of the tuff.

present in a short stratigraphic interval in the eastern branch of the Burukie Wadi above the BKT-2 Complex.

Sample HE01-172, collected from a discontinuous, beige, fine-grained vitric tephra, ~25 cm thick, in a tributary of the southern branch of the Burahin Dora Wadi (Fig. 4), is also placed within the AST-3 group. Finally, sample HE01-189 from the western reaches of the Salal Me'e Wadi is also correlated to AST-3. This sample was collected from a virtually pure vitric tephra lens up to 40 cm thick, ~3 m above Salal Me'e Tuff sample HE01-202 and 4 m below Dahuli Tuff sample HE01-190 (discussed later) (Figs. 4 and 6). Replicate analyses of sample HE01-189 also demonstrate the geochemical variation of AST-3 within any single sample and may, in fact, be characteristic of AST-3 (Fig. 10).

Burahin Dora Tuff (BDT)

The Burahin Dora Tuff is a new name proposed for six tephra samples correlated both within Hadar and between Hadar and Gona. The stratotype for this tuff is sample HE01-169, collected from a gray, coarse-grained vitric tephra that ranges from 30 cm to 1 m in thickness in a tributary of the southern branch of Burahin Dora Wadi (Fig. 4). The Burahin Dora Tuff is well exposed, although not continuously, along both sides of the tributary, but only for a distance of a few tens of meters, and it may represent a channel-fill deposit. The Burahin Dora Tuff is significant because stone tools and mammal fossils were discovered on the exposed top surface of the tuff itself, as well as on the surface of the sediments immediately overlying the tuff. The chemical composition of the Burahin Dora Tuff is similar to AST-3 but can be distinguished from it by lower CaO and TiO_2 concentrations (Table 1). Sample HE01-171 correlates with HE01-169 and was sampled a few tens of meters lateral to it. Minor differences between the two samples, particularly in Al_2O_3 and MnO, help to display the geochemical variation inherent in the tuff.

Samples E02-7383 and -7384, from two different lenticular channel-fill tephra deposits in the upper Kada Hadar region, are also correlated with the Burahin Dora Tuff. Sample E02-7383 was collected from a 70-cm-thick vitric tephra located ~5.5 m above tuff E02-7390 and separated from it by a conglomerate (Figs. 4 and 6). Sample E02-7384 was collected from a lens of tephra up to a meter thick located ~3 m above the KH-7 conglomerate in exposures between the Burukie Wadi and the Maka'amitalu Basin (Fig. 6). The final Burahin Dora Tuff correlate is the dominant mode of sample HE01-188 (AV1). Located not far from E02-7383 and -7384, HE01-188 was collected from a 30–60-cm-thick tephra just outside the Maka'amitalu Basin, adjacent to the dirt track to Eloaha just before it turns to drop down into the basin. As with the stratotype samples, the duplicate analysis of all three of the tephra samples just discussed helps to display the geochemical variation of the Burahin Dora Tuff (Fig. 11; Table 1). A second mode composed of two shards is also observed in HE01-188 (AV2). This mode is distinct from the dominant mode in having higher concentrations of virtually all elements analyzed except SiO_2 and Cl. A single shard from E02-7383 (.249) may correlate with this minor mode and would strengthen the argument for a correlation between the two tephras. Additionally, a possible single-shard correlation between HE01-188.577 and HE01-171.01, which have the lowest Fe_2O_3 concentrations reported in this study, also supports the correlation of HE01-188 to the Burahin Dora Tuff (Fig. 11; Table 1).

The three high-Fe shards from HE01-188 (AV2) and E02-7383, noted already, appear to correlate with samples HE01-168 and E93-4973 from the Burahin Dora Wadi (Fig. 11; Table 1). Assuming that their presence in the Burahin Dora Tuff represents postdepositional reworking of tuff HE01-168, it implies that HE01-168 should be stratigraphically lower than the Burahin Dora Tuff. Thus, the tuff represented by sample HE01-168 would be stratigraphically above AST-1 but below both HE01-169 (BDT) as well as HE01-172 (AST-3), which is stratigraphically below HE01-169 in an uninterrupted section of the drainage (Fig. 6).

Two tephra samples collected from the Kada Gona region, E94-5455 and -5456, also correlate with the Burahin Dora Tuff. These gray, vitric samples were collected in the relatively flat-lying areas not far from the Burahin Dora Wadi but on the Gona side of the Hadar-Gona divide. Considering the proximity of these tephras to HE01-169, such a correlation is not surprising.

Ken Di Tuff

The Ken Di Tuff is a recently named tuff from the Dikika project area (Wynn et al., this volume) that is proposed as a correlate to a previously unnamed tephra sample from Hadar, E02-7430 (Campisano, 2007), and Gona tephra sample, Gonash-86 (Quade et al., this volume). Sample E02-7430 was collected from a fine-grained vitric tephra approximately a meter thick in the very upper reaches of the Farsita Wadi, just below the level of the plateau (Fig. 4). Analysis of the sample demonstrated

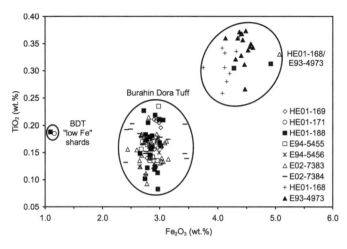
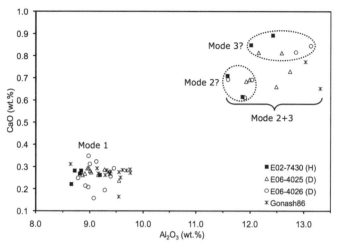

Figure 11. Bivariate plot of the unnormalized concentrations of Fe_2O_3 and TiO_2 for individual glass shards from the Burahin Dora Tuff (BDT) and tuff HE01-168. Note the apparent presence of reworked shards of tuff HE01-168 in BDT samples HE01-188 ($n = 2$) and E02-7383 ($n = 1$).

Figure 12. Bivariate plot of the unnormalized concentrations of Al_2O_3 and CaO for individual glass shards from the Ken Di Tuff at Hadar, Dikika, and Gona. The dashed circles represent the two minor modes identified in Hadar sample E02-7430 (AV2 and AV3) and possibly in Dikika sample E06-4026, but not E06-4025. Individual Dikika and Gona shard data were provided by D. Roman and J. Quade, respectively. See Table 5 for details.

three chemical modes for this tuff, all of which are the most Fe-rich nonbasaltic tephra analyzed from Hadar (Fig. 12; Table 1). The first mode, E02-7430.AV1 ($n = 5$), is easily distinguished from E02-7430.AV2 and .AV3 by its lower Al_2O_3 and CaO concentrations. Modes E02-7430.AV2 and .AV3 appear to be distinguishable in several oxides, including Al_2O_3, Fe_2O_3, and CaO. However, as both modes are represented by only two shards each, it may be likely that they are actually part of a single, but highly variable mode. This possibility may be supported by analyses of Ken Di Tuff samples from Dikika (Wynn et al., this volume) and Gona (Quade et al., this volume).

The major mode of the Dikika Ken Di Tuff samples E04-4025 and -4026 (here, AV1) matches E02-7430.AV1 rather well, although there is some difference in the Fe_2O_3 concentrations between the two Dikika samples, and both have much higher MnO concentrations compared to E02-7430 (Fig. 12; Table 5). Similarly, the major mode of Gona sample Gonash-86 is very similar to the composition of the Dikika and Hadar samples. The Gona tephra has a higher average Al_2O_3 concentration, particularly when compared to E02-7430.AV1 (Table 5), but it exhibits intershard variability that clearly overlaps with both the Hadar and Dikika values (Fig. 12). The MnO concentration of Gonash-86 is comparable to that of E02-7430.AV1 rather than the Dikika samples.

Dikika sample E04-4026 possibly contains two minor modes similar to E02-7430.AV2 and .AV3, with a distinction in the Al_2O_3 and CaO concentrations. In contrast, Dikika sample E04-4025 contains a second mode with a large range in several oxides, particularly Al_2O_3, Fe_2O_3, and CaO, that overlap with the two potential minor modes found in E02-7430 (Hadar) and possibly E04-4026 (Dikika), but it cannot be differentiated into two distinct modes (Fig. 12; Table 5). Thus, sample E04-4025 may indicate that the chemical variation observed in the minor mode(s) of E02-7430 and E04-4026 represents only a single, highly variable second mode. Gona sample Gonash-86 may provide some support for this interpretation. Two shards of Gonash-86 clearly represent at least a second mode of the sample. Although they exhibit a slight difference in their CaO values (Fig. 12), both are within the range of E04-4025, and there is no clear distinction in Al_2O_3 or Fe_2O_3 concentrations as observed in E02-7430.AV2 and .AV3. A comparison of the second mode of E04-4025 (AV2) to an undifferentiated second mode of E04-4026 and a combined second mode of E02-7430 (AV2+3) shows strong similarities in most major oxides (Table 5). Future analyses will hopefully verify whether this similarity accurately reflects a single chemical population or if it is an artifact of averaging two distinct populations. Nonetheless, we accept the correlation of the Ken Di Tuff among Hadar, Gona, and Dikika based on its distinctive bi- or multimodal iron-rich composition, as evident in these four samples.

Upper Hurda Tuff (UHT)

Represented by sample HE01-164, the Upper Hurda Tuff is a new name proposed for a coarse-grained, vitric channel-fill tephra located in the northern reaches of the Hurda, just below the level of the plateau in that area. The 0.25–1.35-m-thick tephra overlies a conglomerate that contains pieces of fossilized wood, and numerous Acheulean stone tools have been found lying on the surface of the tuff. The chemical composition of the Upper Hurda Tuff can be distinguished from other Hadar tephras by its combination of relatively high MgO and TiO_2 concentrations, which are similar to AST-1, but with a higher Fe_2O_3 concentration (Table 1). Although the Upper Hurda Tuff has not been identified outside its sampling area, it was given a formal name because numerous Acheulean stone tools are at least surficially associated with the tuff. As its exact stratigraphic position is debatable, the Upper Hurda Tuff could be stratigraphically lower than the Ken Di Tuff.

The Dahuli Tuff (= Kuseralee Tuff)

The term Dahuli Tuff was originally applied to a tephra in the Dahuli Wadi by Yemane (1997) that was believed to be the stratigraphically highest tephra unit in the Hadar sequence. Although no geochemical analyses, lithologic description, or specific geographic location was included in the description of the Dahuli Tuff, the name will be applied here to samples HE01-145 and E02-7394, collected in separate locations within the Dahuli Wadi (Fig. 4). Both samples are exceptionally fine-grained with a high silt component. Sample HE01-145 was collected in a basin of exposures on the south bank of the wadi, where the tuff is discontinuously exposed in the area, but it is more commonly homogenized into the powder-like silts that characterize this part of the Dahuli. Analysis of sample HE01-145 produced only three useable shard values, one of which displays the highest Fe_2O_3 and CaO concentrations for any Dahuli Tuff sample but is otherwise comparable in other oxides/elements (Fig. 13). Sample E02-7394 was collected several hundred meters northwest of HE01-145 on the north bank of the wadi. At present, the Dahuli Tuff is the only tephra known from the wadi and has relatively high Fe_2O_3 and CaO concentrations, similar to BKT-3, but with much lower MgO and TiO_2 concentrations (Table 1).

Two other samples at Hadar correlate with the Dahuli Tuff. Sample HE01-190, located a few meters above HE01-189 (AST-3) in the western reaches of the Salal Me'e Wadi (Figs. 4 and 6), not only matches the chemical composition of HE01-145, but it is geographically close, at a similar stratigraphic level, and lithologically similar. Sample HE01-190 was analyzed at both Rutgers University and the AMNH. The Rutgers analysis produced only one useable shard value (HE01-190.388), but it is similar to other proposed Dahuli Tuff samples. The AMNH analysis ($n = 9$) is similar to HE01-190.388 and other proposed correlates in most oxide values, but notably higher in CaO and MgO concentrations (Fig. 13; Table 1). It is uncertain whether the difference between the two analyses of sample HE01-190 is the due to geochemical variation of the Dahuli Tuff, particularly in CaO and MgO, or the analytical differences between the two laboratories. The second proposed Hadar correlate of the Dahuli Tuff is sample HE01-209 from the northern region of the divide between the Burukie and Kurbili Wadis just below the plateau surface (Fig. 4). Sample HE01-209 was collected from a tan tuffaceous silt layer, ~1 m thick, that is not consistently tuffaceous but can be traced eastward to between the Kurbili and Farsita Wadis and westward to the exposures stratigraphically above the Maka'amitalu Basin that cross the dirt track leading to/from the town of Eloaha.

The Dahuli Tuff at Hadar is proposed as a correlate to a previously identified tuff from the Gona project area based on comparisons to sample E93-4975 from the "Cusrali" drainage of the Gona region collected in 1993, and a tephra formerly referred to

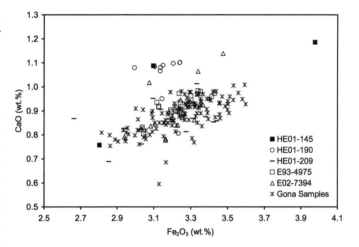

Figure 13. Bivariate plot of the unnormalized concentrations of Fe_2O_3 and CaO for individual glass shards from the Dahuli Tuff at Hadar and Gona. Individual Gona shard data were provided by J. Quade. See Table 5 for details.

as the Kuseralee Tuff (Quade et al., this volume). The Kuseralee Tuff was first noted by Quade et al. (2004, their Fig. 3), but they did not provide any geochemical or geochronological information. In light of the recognition of this tuff by Yemane (1997), the name Dahuli (in place of Kuseralee) has been adopted by both the Hadar and Gona projects (see also Quade et al., this volume). Paleomagnetic work conducted by the Gona team has identified the Brunhes-Matuyama boundary (ca. 0.78 Ma) less than a meter above the Dahuli Tuff, providing an approximate age of 0.81 Ma for the tuff (Quade et al., this volume).

The geochemical comparisons of Dahuli Tuff samples between Hadar and Gona show a similarity in most oxides/elements (Fig. 13; Table 5). SiO_2 and Al_2O_3 concentrations are higher in the Gona analyses of Quade et al. (this volume), but so are the total sum values. With the exception of HE01-190. AV, MgO concentrations are low but measurable in the Hadar samples, whereas none was detected in the Gona samples. The difference in analytical conditions could easily account for the minor differences observed, particularly because E93-4975, collected in the vicinity of the other Gona Dahuli samples (J. Quade, 2007, personal commun.), is indistinguishable from the Hadar samples. While we believe that the available evidence indicates a correlation between the Hadar and Gona samples, analysis of these samples in the same laboratory under the same analytical conditions would provide the definitive evidence.

DISCUSSION

Since the identity and sequence of Hadar Formation tephras have already been well established, this discussion section focuses on the Busidima Formation tephras at Hadar above the level of the Busidima unconformity surface marked by the KH-5 conglomerate. In contrast to providing a simple stratigraphic summary of the Busidima Formation at Hadar, this tephrostratigraphic study instead testifies to the complex nature of the deposits. For the most part, exposures of Busidima Formation tephras are laterally limited within any particular drainage and are often confined to that drainage alone. Construction of a composite sequence is complicated by the fact that while any single tephra in an exposure or section may correlate to another, rarely does more than one tephra correlate between the same two sections/exposures. For example, while AST-3 is present in the both the Burukie and the Salal Me'e, and the Salal Me'e Tuff is present in both the Salal Me'e and the Maka'amitalu, there are no direct tephrostratigraphic links between the Burukie and the Maka'amitalu (Table 6; Fig. 6). Thus, the feasibility of constructing a single comprehensive composite section for the Busidima Formation at Hadar is limited. Similarly, even ascertaining the exact stratigraphic relationship among all the individual tuffs is a difficult task, and the order of some tuffs presented in this study remains tentative.

Overall, the large-scale clusters formed in the normalized EDm dendrograms match the proposed pattern-matching tephra correlations rather well, with only a few notable discrepancies (Fig. 5). The AMNH analysis of HE01-190 was separated from the single-shard Rutgers analysis HE01-190.388 and from the Dahuli Tuff cluster, likely due to its higher CaO and MgO concentrations. Additionally, no clear distinction was identified between AST-1 and the Salal Me'e Tuff. While the EDm dendrograms are a helpful guide, particularly when working with a large data set, such discrepancies illustrate why numerical or statistical methods alone should not be relied upon for conclusive tephrostratigraphic correlations. It is apparent from this study that the actual linkage distances between samples or groups of samples are not a clear indication of correlation. For example, the linkage distance for two different analyses of sample E02-7380 is virtually the same distance as between BKT-2L and -2U (Fig. 5). It should be noted, however, that linkage distances are also a product of the overall quality of the data, including the analytical precision values estimated earlier.

Tuff HE04-441 is currently the only tuff at Hadar preserved between the KH-5 and KH-6 conglomerates, but it occurs only in a small area of the Burukie Wadi. Between the KH-6 and the KH-7 conglomerates, there is at least one tuff, AST-1, preserved in the Burukie and Burahin Dora Wadis, and possibly a second tuff, if E02-7391 does not represent an AST-1 correlate.

BKT-3 and HE01-201/E02-7380 appear to be at a stratigraphically similar level above the KH-7 conglomerate, but their exact relationship to each other, both stratigraphically and geochemically, is not fully understood. BKT-3 is present in both the Salal Me'e Wadi and Maka'amitalu Basin above the KH-7 conglomerate.

The localized tuff HE01-168 is suggested as the next highest tuff in the sequence, but it could easily be moved to anywhere in the sequence between AST-1 and AST-3. It is stratigraphically above AST-1 in the Burahin Dora, and the presence of potentially reworked shards of HE01-168 in the Burahin Dora Tuff suggests that it would be stratigraphically below the Burahin Dora Tuff. By

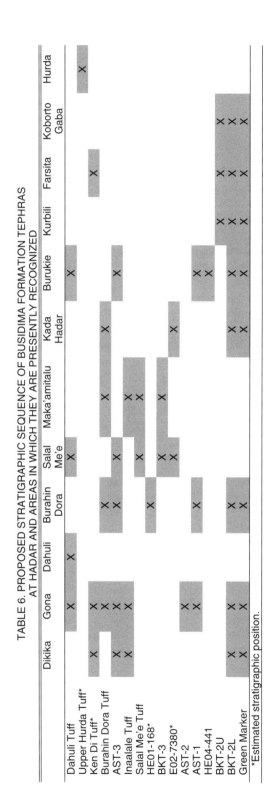

TABLE 6. PROPOSED STRATIGRAPHIC SEQUENCE OF BUSIDIMA FORMATION TEPHRAS AT HADAR AND AREAS IN WHICH THEY ARE PRESENTLY RECOGNIZED

extension, as AST-3 is found downsection from the Burahin Dora Tuff, HE01-168 would presumably be stratigraphically below AST-3 as well. The only reason why HE01-168 is placed stratigraphically below the Salal Me'e Tuff in this discussion is because the tuff has not been identified in the sequence between the Salal Me'e Tuff and AST-3 in the Salal Me'e section (Fig. 6). Similarly, HE01-168 could be placed stratigraphically below BKT-3 (still above KH-7) but is placed above it for the time being because at least one conglomerate is known to exist between BKT-3 and the Salal Me'e Tuff, suggesting the HE01-168 tuff could have been removed by an erosional event if it did exist in this position.

The Salal Me'e Tuff is the next tuff in the sequence and is preserved as both an air-fall and fluvial channel-fill deposit in the Maka'amitalu Basin and as channel fill in the Salal Me'e Wadi. The Inaalale Tuff and AST-3 are positioned at a similar distance above the Salal Me'e Tuff in the Maka'amitalu Basin and Salal Me'e Wadi, respectively, but the Inaalale Tuff at Dikika has been demonstrated to lie below AST-3.

AST-3 is present in the Burukie, Salal Me'e, and Burahin Dora Wadis at Hadar (Table 6; Fig. 6). Its identification in the Burukie Wadi was somewhat surprising, as one sample (HE01-210) was collected at what was believed to be the same stratigraphic level as HE01-142, an AST-1 correlate. The presence of both AST-1 and AST-3 in such a short stratigraphic interval in the Burukie Wadi illustrates both the importance of tephrostratigraphic analyses for establishing true correlations as well as the complex nature of the Busidima Formation conglomerates and their questionable reliability when used for even local stratigraphic correlations. The Burahin Dora Tuff is placed above AST-3 based on the observed sequence in the Burahin Dora Wadi (Fig. 6), the apparent high stratigraphic position of samples E02-7383 and -7384 in the upper Kada Hadar and western Burukie, respectively, and the position of proposed correlates E94-5455 and -5456 from the Gona side of the Hadar-Gona divide.

The Ken Di Tuff, the Upper Hurda Tuff, and the Dahuli Tuff are believed to be the stratigraphically highest three tuffs at Hadar, but their exact stratigraphic relationship to each other is still tentative and potentially interchangeable. The Ken Di Tuff is placed as the lowest of the three tuffs because it was collected just below a plateau surface that Dahuli Tuff sample HE01-209 is believed to be above. The Upper Hurda Tuff was collected from the northeastern corner of the project area, far from the other analyzed tephras, and it is the most stratigraphically uncertain of the tuffs. Its position above the Ken Di Tuff is simply based on the fact that it too is located near a plateau surface, but it was not identified in the stratigraphic sequence between BKT-2 and the Ken Di Tuff in the Farsita Wadi, and it is associated with Acheulean artifacts that are not present in the Farsita. The position of the Dahuli Tuff as the stratigraphically highest tuff preserved at Hadar is tentative as well. Its position is based on the young date of ca. 0.81 Ma assigned to it, as well as the position of Dahuli samples E02-7394 and HE01-209, which can be geographically and/or stratigraphically placed above all other tephras analyzed from the Busidima Formation at Hadar with the exception the Upper Hurda Tuff.

The pattern of tephra preservation and the absence of correlatable tephras may be equally informative as those that do correlate. For example, the three easternmost exposed Busidima Formation tephras from Dikika (AST-3, Inaalale, and Ken Di) are the only three that are currently correlated from Dikika to Hadar (Wynn et al., this volume). Similarly, as evident in this study and that of Quade et al. (this volume), most of the Gona-Hadar tephra correlations are restricted to the eastern Gona area. Deposits much younger than the Dahuli Tuff (ca. 0.81 Ma) appear to be absent from Hadar. While we presumed this to be the case based on the stratigraphic and geographic position of the Dahuli Tuff, the apparent absence of the Bironita Tuff (ca. 0.64 Ma) at Hadar, a widespread marker identified at Dikika, Gona, and the Middle Awash (Wynn et al., this volume; Quade et al., this volume; Clark et al., 1994), may provide supportive evidence for this inference. As our understanding of the regional stratigraphy continues to improve, the pattern(s) of tephra preservation in the region will likely provide additional insights into the development of the Hadar Basin and Busidima half-graben as well as their depositional and postdepositional histories (e.g., Quade et al., this volume; Wynn et al., this volume).

CONCLUSIONS

Vitric tephra deposits are relatively rare in the Hadar Formation; they are essentially limited to the channel deposits of the Sidi Hakoma and Kada Hadar Tuffs and the eastern exposures of the BKT-2 Complex. The Sidi Hakoma Tuff has already been shown to correlate to the both the Tulu Bor (= Tuff B) α and β, and evidence in this study suggest that the Tuff B-γ may be present as well.

BKT-2 has traditionally been treated and identified as a single unit. Several lines of evidence suggest that these tuffs do, in fact, represent different volcanic events and should be treated individually within the context of the BKT-2 Complex. The Green Marker Bed is distinct from BKT-2L and -2U in both glass geochemistry and lithology. BKT-2L and -2U are similar in glass geochemistry and in general lithology but can be differentiated in aspects of their feldspar geochemistry and lithologic detail in addition to their separate stratigraphic positions. The various lines of evidence indicate that while BKT-2L and -2U are similar in several aspects, they do represent the result of two separate, but closely related eruptions from the same volcanic center, most likely the nearby Ida Ale volcanic complex as suggested by Walter (1981).

At least 12 distinct vitric tephras are preserved in the Busidima Formation at Hadar and are represented by no less than 20 chemical modes. It is likely that several more individual tephras are represented by bentonites observed in the field as well as by tephras that remain to be discovered or analyzed. Despite this abundance, only three Busidima Formation tuffs are currently correlated between Hadar and Dikika (the Inaalale Tuff, AST-3, and the Ken Di Tuff) and six are currently correlated between Hadar and Gona (AST-1, the Inaalale Tuff, AST-3, the Burahin Dora Tuff, the Ken Di Tuff, and the Dahuli Tuff). Although they are not discussed in this study, at least seven distinct tephras from Busidima Formation deposits were identified in a reanalysis of Feibel's unpublished tephra data from Gona. Additionally, Quade et al. (2004, this volume) reported several other tephras between BKT-2 and the Dahuli Tuff in the eastern Gona region that are not accounted for in this study or Feibel's unpublished data. Tuffs directly associated with the earliest archaeological sites at Gona and Hadar do not have correlates between the two regions, namely AST-2 and BKT-3,

The correlation established between the Dahuli Tuff at Hadar and Gona provides not only an upper age estimate for the sediments at Hadar of ca. 0.81 Ma, but it also demonstrates the dramatic differences in sedimentation rates and sediment preservation between the Hadar and Busidima Formations. Whereas roughly 500,000 yr of sediment is preserved in the 155 m thickness of the Hadar Formation at Hadar, perhaps as much as 2 m.y. of sediment is preserved in only 40 m of the Busidima Formation at Hadar.

The geological studies at Hadar, Gona, and Dikika have demonstrated the complexity of the Busidima Formation stratigraphy and the necessity of a strong tephrostratigraphic framework for understanding the relationships both within and between project areas. As evident by discussions presented elsewhere in this volume (e.g., Quade et al., this volume; Wynn et al., this volume), the utility of these correlations goes beyond creating stratigraphic sequences—they are also useful in understanding local and regional tectonic and paleogeographic issues.

ACKNOWLEDGMENTS

This project was funded by grants from the National Science Foundation, the National Geographic Society, the Institute of Human Origins (Arizona State University), and the Center for Human Evolutionary Studies (Rutgers University) to both the authors and directors of the Hadar Research Project. For laboratory access and assistance in the analysis of samples at the University of Utah, Rutgers University, and the American Museum of National History, we thank Marsha Smith, Bereket Haileab, Barbara Nash, Ray Lambert, Jerry Delaney, Charles Mandeville, Carl Swisher, Brent Turrin, and especially Godwin Mollel. We would also like to thank Jay Quade, Jonathan Wynn, Diana Roman, and Erin DiMaggio for data sharing and discussions on regional tephrostratigraphy. For logistical and field support, we would like to thank Bill Kimbel, Don Johanson, Erella Hovers, Gerry Eck, the Institute of Human Origins, the National Museum of Ethiopia, the Ethiopian Authority for Research and Conservation of Cultural Heritage (A.R.C.C.H.), and especially our field crew and friends from the Eloaha region. We thank Jay Quade, Barbara Nash, and John Westgate for providing thoughtful and constructive reviews of this manuscript.

REFERENCES CITED

Aronson, J.L., and Taieb, M., 1981, Geology and paleogeography of the Hadar hominid site, Ethiopia, in Rapp, G., and Vondra, C.F., eds., Hominid Sites: Their Geologic Settings: American Association for the Advancement of Science Selected Symposium: Boulder, Colorado, Westview Press, p. 165–195.

Brown, F.H., 1982, Tulu Bor Tuff at Koobi Fora correlated with the Sidi Hakoma Tuff at Hadar: Nature, v. 300, p. 631–635, doi: 10.1038/300631a0.

Brown, F.H., Sarna-Wojcicki, A.M., Meyer, C.E., and Haileab, B., 1992, Correlation of Pliocene and Pleistocene tephra layers between the Turkana Basin of East Africa and the Gulf of Aden: Quaternary International, v. 13–14, p. 55–67, doi: 10.1016/1040-6182(92)90010-Y.

Brown, F.H., Haileab, B., and McDougall, I., 2006, Sequence of tuffs between the KBS Tuff and the Chari Tuff in the Turkana Basin, Kenya and Ethiopia: Journal of the Geological Society of London, v. 163, p. 185–204, doi: 10.1144/0016-764904-165.

Campisano, C.J., 2007, Tephrostratigraphy and Hominin Paleoenvironments of the Hadar Formation, Afar Depression, Ethiopia [Ph.D. thesis]: New Brunswick, New Jersey, Rutgers University, 600 p.

Campisano, C.J., and Feibel, C.S., 2008, this volume, Depositional environments and stratigraphic summary of the Pliocene Hadar Formation at Hadar, Afar Depression, Ethiopia, in Quade, J., and Wynn, J.G., eds., The Geology of Early Humans in the Horn of Africa: Geological Society of America Special Paper 446, doi: 10.1130/2008.2446(08).

Cerling, T.E., Brown, F.H., and Bowman, J.R., 1985, Low-temperature alteration of volcanic glass: Hydration, Na, K, ^{18}O and Ar mobility: Chemical Geology, v. 52, p. 281–293.

Clark, J.D., de Heinzelin, J., Schick, K.D., Hart, W.K., White, T.D., WoldeGabriel, G., Walter, R.C., Suwa, G., Asfaw, B., Vrba, E., and Haile-Selassie, Y., 1994, African *Homo erectus*: Old radiometric ages and young Oldowan assemblages in the Middle Awash Valley, Ethiopia: Science, v. 264, p. 1907–1910, doi: 10.1126/science.8009220.

deMenocal, P.B., and Brown, F.H., 1999, Pliocene tephra correlations between East African hominid localities, the Gulf of Aden and the Arabian Sea, in Agusti, J., Rook, L., and Andrews, P., eds., The Evolution of Terrestrial Ecosystems in Europe: Cambridge, UK, Cambridge University Press, p. 23–54.

DiMaggio, E.N., Campisano, C.J., Arrowsmith, J.R., Reed, K.E., and Lockwood, C.A., 2006, Geochemistry and tephrochronology of the BKT-2 and Kada Hadar tephras in the middle Ledi-Geraru region of Afar, Ethiopia: Eos (Transactions, American Geophysical Union), v. 87, p. 52.

DiMaggio, E.N., Campisano, C.J., Arrowsmith, J.R., Reed, K.E., Swisher, C.C., III, and Lockwood, C.A., 2008, this volume, Correlation and stratigraphy of the BKT-2 volcanic complex in west-central Afar, Ethiopia, in Quade, J., and Wynn, J.G., eds., The Geology of Early Humans in the Horn of Africa: Geological Society of America Special Paper 446, doi: 10.1130/2008.2446(07).

Donovan, J., 2000, Probe for Windows: Analysis and Automation for EPMA, Version 5.11: Vienna, Ohio, Advanced Microbeam.

Feibel, C.S., 1999, Tephrostratigraphy and geological context in paleoanthropology: Evolutionary Anthropology, v. 8, p. 87–100, doi: 10.1002/(SICI)1520-6505(1999)8:3<87::AID-EVAN4>3.0.CO;2-W.

Feibel, C.S., Brown, F.H., and McDougall, I., 1989, Stratigraphic context of fossil hominids from the Omo Group deposits: Northern Turkana Basin, Kenya and Ethiopia: American Journal of Physical Anthropology, v. 78, p. 595–622, doi: 10.1002/ajpa.1330780412.

Haileab, B., 1995, Geochemistry, Geochronology and Tephrostratigraphy of Tephra from the Turkana Basin, Southern Ethiopia and Northern Kenya [Ph.D. thesis]: Salt Lake City, University of Utah, 369 p.

Hart, W.K., Walter, R.C., and WoldeGabriel, G., 1992, Tephra sources and correlations in Ethiopia: Application of elemental and neodymium isotope data: Quaternary International, v. 13–14, p. 77–86, doi: 10.1016/1040-6182(92)90012-Q.

Hovers, E., 2003, Treading carefully; site formation processes and Pliocene lithic technology, in Martinez-Moreno, J.M., Mora, R., and de la Torre, I., eds., Oldowan: Rather More Than Smashing Stones: Barcelona, Universitat Autonoma de Barcelona, p. 145–164.

Hunt, J.B., and Hill, P.G., 1993, Tephra geochemistry: A discussion of some persistent analytical problems: The Holocene, v. 3, p. 271–278, doi: 10.1177/095968369300300310.

Johanson, D.C., and Taieb, M., 1976, Plio-Pleistocene hominid discoveries in Hadar, Ethiopia: Nature, v. 260, p. 293–297, doi: 10.1038/260293a0.

Johanson, D.C., Taieb, M., Gray, B.T., and Coppens, Y., 1978, Geological framework of the Pliocene Hadar Formation (Afar, Ethiopia) with notes on paleontology including hominids, in Bishop, W.W., ed., Geological Background to Fossil Man: Edinburgh, Scotland, Scottish Academic Press, p. 549–564.

Johanson, D.C., Taieb, M., and Coppens, Y., 1982, Pliocene hominids from the Hadar Formation, Ethiopia (1973–1977): Stratigraphic, chronological, and paleoenvironmental contexts, with notes on hominid morphology and systematics: American Journal of Physical Anthropology, v. 57, p. 373–402, doi: 10.1002/ajpa.1330570402.

Kimbel, W.H., Johanson, D.C., and Rak, Y., 1994, The first skull and other new discoveries of *Australopithecus afarensis* at Hadar, Ethiopia: Nature, v. 368, p. 449–451, doi: 10.1038/368449a0.

Kimbel, W.H., Walter, R.C., Johanson, D.C., Reed, K.E., Aronson, J.L., Assefa, Z., Marean, C.W., Eck, G.G., Robe, R., Hovers, E., Rak, Y., Vondra, C., Yemane, T., York, D., Chen, Y., Evensen, N.M., and Smith, P.E., 1996, Late Pliocene *Homo* and Oldowan tools from the Hadar Formation (Kada Hadar Member), Ethiopia: Journal of Human Evolution, v. 31, p. 549–561, doi: 10.1006/jhev.1996.0079.

Kimbel, W.H., Johanson, D.C., and Rak, Y., 1997, Systematic assessment of a maxilla of *Homo* from Hadar, Ethiopia: American Journal of Physical Anthropology, v. 103, p. 235–262, doi: 10.1002/(SICI)1096-8644(199706)103:2<235::AID-AJPA8>3.0.CO;2-S.

Kimbel, W.H., Rak, Y., and Johanson, D.C., 2004, The Skull of *Australopithecus afarensis*: New York, Oxford University Press, 254 p.

Ludwig, A.J., and Reynolds, J.F., 1988, Statistical Ecology: A Primer on Methods and Computing: New York, Wiley Press, 337 p.

McHenry, L.J., 2004, Characterization and Correlation of Altered Plio-Pleistocene Tephra Using a "Multiple Technique" Approach: Case Study at Olduvai Gorge, Tanzania [Ph.D. thesis]: New Brunswick, New Jersey, Rutgers University, 382 p.

McHenry, L.J., 2005, Phenocryst composition as a tool for correlating fresh and altered tephra, Bed I, Olduvai Gorge, Tanzania: Stratigraphy, v. 2, p. 101–115.

Nash, W.P., 1992, Analysis of oxygen with the electron-microprobe: Applications to hydrated glass and minerals: The American Mineralogist, v. 77, p. 453–456.

Nielsen, C.H., and Sigurdsson, H., 1981, Quantitative methods for electron microprobe analysis of sodium in natural and synthetic glasses: The American Mineralogist, v. 66, p. 547–552.

Perkins, M.E., Nash, W.P., Brown, F.H., and Fleck, R.J., 1995, Fallout tuffs of Trapper-Creek, Idaho: A record of Miocene explosive volcanism in the Snake River Plain volcanic province: Geological Society of America Bulletin, v. 107, p. 1484–1506, doi: 10.1130/0016-7606(1995)107<1484:FTOTCI>2.3.CO;2.

Quade, J., Levin, N., Semaw, S., Stout, D., Renne, R., Rogers, M., and Simpson, S., 2004, Paleoenvironments of the earliest stone toolmakers, Gona, Ethiopia: Geological Society of America Bulletin, v. 116, p. 1529–1544, doi: 10.1130/B25358.1.

Quade, J., Levin, N.E., Simpson, S.W., Butler, R., McIntosh, W.C., Semaw, S., Kleinsasser, L., Dupont-Nivet, G., Renne, P., and Dunbar, N., 2008, this volume, The geology of Gona, Afar, Ethiopia, *in* Quade, J., and Wynn, J.G., eds., The Geology of Early Humans in the Horn of Africa: Geological Society of America Special Paper 446, doi: 10.1130/2008.2446(01).

Renne, P., Walter, R., Verosub, K., Sweitzer, M., and Aronson, J., 1993, New data from Hadar (Ethiopia) support orbitally tuned time-scale to 3.3 Ma: Geophysical Research Letters, v. 20, p. 1067–1070, doi: 10.1029/93GL00733.

Renne, P.R., Swisher, C.C., Deino, A.L., Karner, D.B., Owens, T.L., and DePaolo, D.J., 1998, Intercalibration of standards, absolute ages and uncertainties in $^{40}Ar/^{39}Ar$ dating: Chemical Geology, v. 145, p. 117–152, doi: 10.1016/S0009-2541(97)00159-9.

Roche, H., and Tiercelin, J.J., 1977, Découverte d'une industrie lithique ancienne in situ dans la formation d'Hadar, Afar Central, Éthiopie: Comptes Rendus de l'Académie des Sciences, v. 284D, p. 1871–1874.

Roman, D.C., Campisano, C., Quade, J., DiMaggio, E., Arrowsmith, R., and Feibel, C., 2008, this volume, Composite tephrostratigraphy of the Dikika, Gona, Hadar, and Ledi-Geraru project areas, northern Awash, Ethiopia, *in* Quade, J., and Wynn, J.G., eds., The Geology of Early Humans in the Horn of Africa: Geological Society of America Special Paper 446, doi: 10.1130/2008.2446(05).

Sarna-Wojcicki, A., 2000, Tephrochronology, *in* Stratton Noller, J., Sowers, J.M., and Lettis, W.R., eds., Quaternary Geochronology: Methods and Applications: Washington, D.C., American Geophysical Union, p. 357–377.

Sarna-Wojcicki, A., and Davis, J.O., 1991, Quaternary tephrochronology, *in* Morrison, R.B., ed., Quaternary Non-Glacial Geology; Conterminous United States: Boulder, Colorado, Geological Society of America, Geology of North America, v. K-2, p. 93–116.

Sarna-Wojcicki, A.M., Meyer, C.E., Roth, P.H., and Brown, F.H., 1985, Ages of tuff beds at East African early hominid sites and sediments in the Gulf of Aden: Nature, v. 313, p. 306–308, doi: 10.1038/313306a0.

Schmitt, T.J., and Nairn, A.E.M., 1984, Interpretations of the magnetostratigraphy of the Hadar hominid site, Ethiopia: Nature, v. 309, p. 704–706, doi: 10.1038/309704a0.

Semaw, S., 1997, Late Pliocene Archaeology of the Gona River Deposits, Afar, Ethiopia [Ph.D. thesis]: New Brunswick, New Jersey, Rutgers University, 302 p.

Semaw, S., Renne, P., Harris, J.W.K., Feibel, C.S., Bernor, R.L., Fesseha, N., and Mowbray, K., 1997, 2.5-million-year-old stone tools from Gona, Ethiopia: Nature, v. 385, p. 333–336, doi: 10.1038/385333a0.

Taieb, M., and Tiercelin, J.J., 1979, Sédimentation Pliocène et paléoenvironments de rift: Exemple de la formation à Hominidés d'Hadar (Afar, Éthiopie): Bulletin de la Société Géologique de France, v. 21, p. 243–253.

Taieb, M., Johanson, D.C., Coppens, Y., and Aronson, J.L., 1976, Geological and palaeontological background of Hadar hominid site, Afar, Ethiopia: Nature, v. 260, p. 289–293, doi: 10.1038/260289a0.

Tiercelin, J.J., 1986, The Pliocene Hadar Formation, Afar Depression of Ethiopia, *in* Frostick, L.E., Renaut, R.W., Reid, I., and Tiercelin, J.J., eds., Sedimentation in the African Rifts: Oxford, UK, Blackwell Scientific, p. 221–240.

Vondra, C.F., Yemane, T., Aronson, J.L., and Walter, R.C., 1996, A major disconformity within the Hadar Formation: Geological Society of America Abstracts with Programs, v. 28, no. 6, p. 69.

Walter, R.C., 1981, The Volcanic History of the Hadar Early-Man Site and the Surrounding Afar Region of Ethiopia [Ph.D. thesis]: Cleveland, Case Western Reserve University, 426 p.

Walter, R.C., 1994, Age of Lucy and the First Family: Single-crystal $^{40}Ar/^{39}Ar$ dating of the Denen Dora and lower Kada Hadar Members of the Hadar Formation, Ethiopia: Geology, v. 22, p. 6–10, doi: 10.1130/0091-7613(1994)022<0006:AOLATF>2.3.CO;2.

Walter, R.C., and Aronson, J.L., 1993, Age and source of the Sidi Hakoma Tuff, Hadar Formation, Ethiopia: Journal of Human Evolution, v. 25, p. 229–240, doi: 10.1006/jhev.1993.1046.

Walter, R.C., Aronson, J.L., Chen, Y., Evensen, N., Smith, P.E., York, D., Vondra, C.F., and Yemane, T., 1996, New radiometric ages for the Hadar Formation above the disconformity: Geological Society of America Abstracts with Programs, v. 28, no. 6, p. 69.

White, T.D., Suwa, G., Hart, W.K., Walter, R.C., WoldeGabriel, G., de Heinzelin, J., Clark, J.D., Asfaw, B., and Vrba, E., 1993, New discoveries of *Australopithecus* at Maka in Ethiopia: Nature, v. 366, p. 261–265, doi: 10.1038/366261a0.

WoldeGabriel, G., Hart, W.K., Katoh, S., Beyene, Y., and Suwa, G., 2005, Correlation of Plio-Pleistocene tephra in Ethiopian and Kenyan rift basins: Temporal calibration of geological features and hominid fossil records: Journal of Volcanology and Geothermal Research, v. 147, p. 81–108, doi: 10.1016/j.jvolgeores.2005.03.008.

Wynn, J.G., Alemseged, Z., Bobe, R., Geraads, D., Reed, D., and Roman, D.C., 2006, Geological and palaeontological context of a Pliocene juvenile hominin at Dikika, Ethiopia: Nature, v. 443, p. 332–336, doi: 10.1038/nature05048.

Wynn, J.G., Roman, D.C., Alemseged, Z., Reed, D., and Geraads, D., 2008, this volume, Stratigraphy, depositional environments, and basin structure of the Hadar and Busidima Formations at Dikika, Ethiopia, *in* Quade, J., and Wynn, J.G., eds., The Geology of Early Humans in the Horn of Africa: Geological Society of America Special Paper 446, doi: 10.1130/2008.2446(04).

Yemane, T., 1997, Stratigraphy and Sedimentology of the Hadar Formation [Ph.D. thesis]: Ames, Iowa State University, 182 p.

MANUSCRIPT ACCEPTED BY THE SOCIETY 17 JUNE 2008

Correlation and stratigraphy of the BKT-2 volcanic complex in west-central Afar, Ethiopia

Erin N. DiMaggio*
School of Earth and Space Exploration, Arizona State University, Tempe, Arizona 85287-1404, USA

Christopher J. Campisano
Institute of Human Origins, Arizona State University, P.O. Box 874101, Tempe, Arizona 85287-4101, USA

J Ramón Arrowsmith
School of Earth and Space Exploration, Arizona State University, Tempe, Arizona 85287-1404, USA

Kaye E. Reed
Institute of Human Origins, Arizona State University, P.O. Box 874101, Tempe, Arizona 85287-4101, USA

Carl C. Swisher III
Department of Earth and Planetary Sciences, Rutgers University, 610 Taylor Road, Piscataway, New Jersey 08854, USA

Charles A. Lockwood
Department of Anthropology, University College London, Gower Street, London WC1E 6BT, UK

ABSTRACT

Located adjacent to the paleoanthropological site of Hadar in Afar, Ethiopia, the Ledi-Geraru project area preserves multiple tephra deposits within the Pliocene sediments of the hominin-bearing Hadar Formation. Tephra deposits of the Bouroukie Tuff 2 volcanic complex (BKT-2) are important regional markers, and here we provide correlations between the Hadar and Ledi-Geraru project areas using major-element glass chemistry, stratigraphic relationships, outcrop characteristics, and $^{40}Ar/^{39}Ar$ dates. These correlations greatly expand existing temporal and spatial resolution, aid in interpretations of regional depositional environments, and increase the documented extent of BKT-2 to ~600 km². BKT-2 exposures at Ledi-Geraru are the thickest and most complete yet observed. There, the BKT-2 complex is preserved as two air-fall lapilli layers, BKT-2U (<97 cm thick) and BKT-2L (<9 cm thick), separated by <2.5 m of silts and clays or diatomite that overlie the Green Marker Bed (GMB), a laminated ash tuff. Measured sections were evaluated to create a stratigraphy-based model of paleolandscape variations using BKT-2 tephra as laterally extensive isochronous surfaces.

---

*erin.dimaggio@asu.edu

DiMaggio, E.N., Campisano, C.J., Arrowsmith, J R., Reed, K.E., Swisher, C.C., III, and Lockwood, C.A., 2008, Correlation and stratigraphy of the BKT-2 volcanic complex in west-central Afar, Ethiopia, *in* Quade, J., and Wynn, J.G., eds., The Geology of Early Humans in the Horn of Africa: Geological Society of America Special Paper 446, p. 163–177, doi: 10.1130/2008.2446(07). For permission to copy, contact editing@geosociety.org. ©2008 The Geological Society of America. All rights reserved.

BKT-2 was mainly erupted into lacustrine and nearshore environments. The eastern Ledi-Geraru region was likely located at the depocenter of an expansive fluviolacustrine network. Representing the last major lacustrine phase of the Hadar Formation, lateral facies variations show the westward expansion of a lacustrine setting ca. 2.96 Ma followed by eastward regression initiated sometime prior to the eruption of BKT-2U ca. 2.94 Ma. High-resolution, well-correlated, and temporally constrained stratigraphic records are key to the interpretation of paleoenvironmental variation in East Africa.

Keywords: tephrostratigraphy, tephrochronology, Hadar Formation, paleoenvironment, Pliocene, Afar Depression.

INTRODUCTION

Geological and geochronological investigations at fossiliferous sites in west-central Afar have contributed paleoenvironmental models of the northern Awash Basin and provided insight into the paleoenvironmental history of hominid habitats (e.g., Taieb and Tiercelin, 1979; Aronson and Taieb, 1981; Tiercelin, 1986; Quade et al., 2004; Campisano and Feibel, this volume, Chapter 8; Quade et al., this volume; Wynn et al., this volume). Additionally, the landscape variation captured in detailed sedimentological records provides a spatial and temporal context for the rich paleontological and archaeological record preserved in the region (e.g., Johanson and Taieb, 1976; White et al., 1994; Kimbel et al., 1996; Alemseged et al., 2006). Analyses of tephra —a collective term for volcanic ejecta—are essential to these studies because they establish geochemical and temporal links within and between basins. This study contributes a high-resolution stratigraphy-based model of paleoenvironmental variations ca. 2.95 Ma. We use the Bouroukie Tuff 2 volcanic complex (BKT-2)—a series of three discrete laterally extensive tephra deposits—to correlate sections from the Ledi-Geraru project area into the regional stratigraphic framework of the Pliocene Hadar Formation.

The Ledi-Geraru project area is bounded in the south and southwest by the Ledi, Hurda, and Awash Rivers, by the Awash River in the east, and by the Bati-Mille road in the north (Fig. 1). Originally mapped by Taieb et al. (1976), the Ledi-Geraru region has not been well described in the literature, even though it exposes sediments coeval with well-studied sites such as Hadar, adjacent to the south-

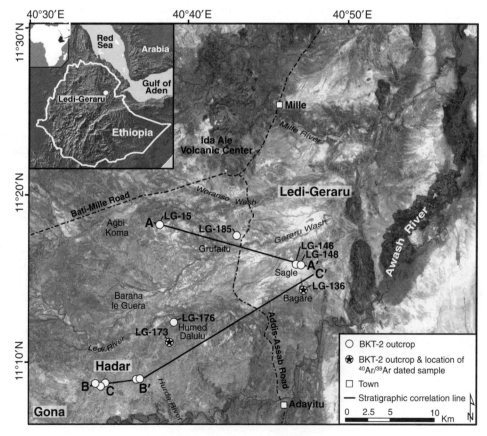

Figure 1. Location of the Ledi-Geraru field site in west-central Afar, Ethiopia. The Landsat image (bands 4–3–2) of the Ledi-Geraru, Hadar, and Gona research areas is overlain by Bouroukie Tuff 2 volcanic complex (BKT-2) sites, region names, and stratigraphic correlation lines. Vegetation and basalts are dark gray, and white to light gray shades are sediment, gravels, and rhyolite at the Ida Ale volcanic center. Region names for BKT-2 sites in Hadar from west to east are: Kada Hadar, Kurbili, Farsita, Unda Hadar, and Koborto Gaba.

west. In addition to the recovery of two hominid molars, investigations from three field seasons (2002, 2004, 2005) have created a basic stratigraphic framework for the entirety of the Hadar Formation, identified new tephras, and documented fossil-rich strata for faunal analysis (Arrowsmith et al., 2004; Reed et al., 2003). However, more analyses are necessary to firmly link these important sediments into the regional stratigraphy. To address these questions, we described and sampled tephra deposits and measured detailed stratigraphic sections during our 2006 field season.

BKT-2 deposits are prominent stratigraphic markers in the northern Awash region and the most widely recognized tephras in Ledi-Geraru (e.g., Figs. 2 and 3). As defined at Hadar and Gona, BKT-2 lies within the Kada Hadar Member of the Hadar Formation (Taieb et al., 1976; Aronson et al., 1977), which is significant as a series of deposits stratigraphically above fauna associated with and including *Australopithecus afarensis*. Dated to ca. 2.96 and ca. 2.94 Ma (BKT-2L and BKT-2U, respectively) (Semaw et al., 1997; Campisano, 2007; this chapter), BKT-2 tephras were deposited during a critical time period of climate variability and faunal evolution between 2.6 and 3.0 Ma (deMenocal, 1995, 2004). The identification of BKT-2 in Ledi-Geraru increases our understanding of the regional geography and paleoenvironment in several ways. First, it enables field documentation and geochemical characterization of thicker and better-preserved BKT-2 deposits—a pattern consistent with the eastward thickening of the tephras noted at Hadar (Walter, 1981). Second, this correlation links a portion of the Ledi-Geraru sedimentary sequence into a regional stratigraphic

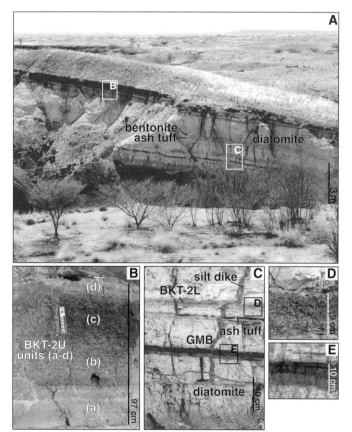

Figure 2. (A) BKT-2 outcrop in laminated diatomite silts at site LG-136. (B) BKT-2U deposit showing the physical expression of tephra units (a–d): (a) crystal-rich lapilli tephra, (b–c) graded basaltic and rhyolitic lapilli tephra, (d) rhyolitic pumice ash. (C) GMB tuff and BKT-2L crosscut by a silty dike. (D) BKT-2L encased in laminated diatomite. (E) GMB overlying a silty sill encased in diatomite. Note that additional thin tuffs and a bentonite layer are present in this section.

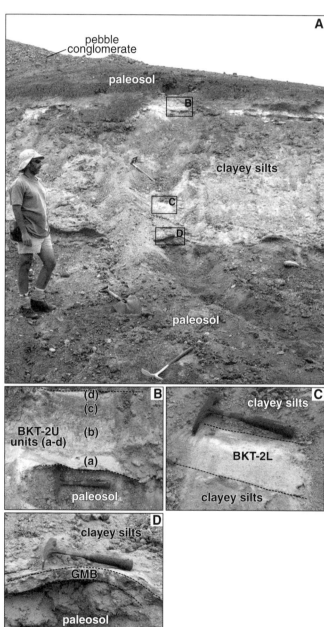

Figure 3. (A) BKT-2 outcrop in paleosols and clayey silts at site LG-173. (B) BKT-2U deposit showing the physical expression of tephra units (a–d). (C) BKT-2L encased in clayey silts. (D) Undulating GMB tuff mantling a paleosol and capped by clayey silts and a pebble conglomerate. For scale, person is 1.8 m tall, and hammer is 28 cm long.

framework spanning multiple project areas, thereby increasing the spatial breadth over which we can interpret paleolandscape variability. As a result, a more complete assessment of basin-scale variations can be made during this critical time period.

In this study, we correlate BKT-2 exposures between the Hadar and Ledi-Geraru project areas to interpret paleodepositional variations. These correlations are substantiated by detailed stratigraphy, major-element glass composition, and ^{40}Ar/^{39}Ar dates. Our interpretations of lithofacies organization and variation across this region indicate a paleolake depocenter located in eastern Ledi-Geraru and show a westward expansion of the lacustrine system ca. 2.96 Ma followed by eastward regression, which initiated sometime prior to ca. 2.94 Ma.

GEOLOGIC SETTING AND STRATIGRAPHY

The Ledi-Geraru project area lies within the west-central Afar sedimentary basin (Fig. 1; Taieb et al., 1976). This basin is situated in the Afar Depression of Ethiopia, south of the Tendaho-Goba'ad discontinuity, which separates the Main Ethiopian Rift from the Red Sea and Gulf of Aden Rift segments (Tesfaye et al., 2003). The structural setting of the Afar Depression is unique, characterized as the transition from purely continental rifting in the south (Main Ethiopian Rift) to ocean spreading centers in the Red Sea and Gulf of Aden (e.g., Wolfenden et al., 2004; Beyene and Abdelsalam, 2005). This extension has produced basin subsidence and subsequent sedimentation. Sediments in the Ledi-Geraru expose almost the entirety of the thick (~150–300 m), fossiliferous Pliocene Hadar Formation, which is classically divided by marker tephra beds into the Basal, Sidi Hakoma, Denen Dora, and Kada Hadar Members (Taieb et al., 1976). The Hadar Formation is part of the Miocene to Pleistocene Awash Group that is exposed along the tributaries of the Awash River (Taieb et al., 1976; Kalb et al., 1982). Sediments preserved are dominantly lacustrine, lake-margin, and floodplain clays and silts and fluvial-deltaic sands (Tiercelin, 1986; Campisano, 2007; Campisano and Feibel, this volume, Chapter 8). They overlie and are interbedded within the 1–4 Ma Afar Stratoid Series volcanic rocks, including primary and reworked tephra, and altered bentonite-clay units (Lahitte et al., 2003). Structurally, the formation dips north 2°–4° and is overprinted by NNE-striking faults with offsets of less than 40 m (Taieb et al., 1976; Arrowsmith et al., 2004).

Sediments of the Kada Hadar Member are poorly exposed in Ledi-Geraru, but they record a stronger lacustrine signature than coeval sediments in Hadar (Campisano and Feibel, this volume, Chapter 8; Dupont-Nivet et al., this volume). The lower portion of the Kada Hadar Member is dominated by laminated silty clays that contain leaf impressions, fish scales, and gastropod shells with lesser pedogenically modified clays and channel sands. Overlying this sequence, well-developed clay Vertisols with abundant carbonate nodules and root traces are cut by thick channel sands and associated silts. In addition to capping the BKT-2 complex, the uppermost 5–10 m of this section records a strong lacustrine signal indicated by the presence of undisturbed laminated diatomite, pedogenically modified diatomaceous silts, and laminated clays. In contrast to the diatomites preserved at Ledi-Geraru region, the upper Kada Hadar Member lacustrine sequence at Hadar consists of lacustrine mudstone with variable evidence of postdepositional pedogenic modification (Tiercelin, 1986; Campisano, 2007; Campisano and Feibel, this volume, Chapter 8). A major unconformity that defines the base of the Busidima Formation (Quade et al., 2004; Wynn et al., this volume) is recognized in the stratigraphy at Ledi-Geraru as a carbonate-rich pebble conglomerate that forms an erosive surface above underlying clays and silts of the Hadar Formation.

FIELD AND LABORATORY METHODS

Section Measurements and Sampling Techniques

Field investigations were conducted in Ledi-Geraru in January 2006. Stratigraphic sections were measured and described at seven BKT-2 localities (Fig. 1). Multiple characteristics were recorded at decimeter scale, including sediment type and structures, depositional modification, color, and pedogenic modifications. Geologic samples from most tephra deposits were sampled for geochemical analysis and dating.

Glass Composition

BKT-2 pumice samples were prepared using ~200 g of a dry sieved, >0.5 mm sample split. Samples were rinsed in distilled water, agitated for <1 min in a 5% HF solution, and then re-rinsed repeatedly in distilled water. This process was repeated if adhering clays were not adequately removed. Approximately 40–80 grains of clear, fresh-looking, unaltered glass fragments were handpicked from the bulk sample and gently crushed. A dense glass-epoxy mixture was created for each sample. Thick sections were polished using grit paper, and 0.25 µm diamond paste was used to obtain a high polish. All electron microprobe (EMP) samples were coated with a 150 ± 25 Å carbon film before analysis (Kerrick et al., 1973).

Major-element glass compositions were acquired using a JEOL JXA-8600 superprobe at Arizona State University operating at 15 kV and 10 nA with a defocused electron beam (15 µm diameter) to minimize alkali loss (Froggatt, 1992; Hunt and Hill, 2001). A 10 s counting time was used to analyze major elements, including Si, Ti, Al, Mn, Fe, Mg, Ca, Na, and K. Common standards were analyzed to calibrate each element, including Los Posos Rhyolite, San Carlos Olivine, and synthetic fayalite. Using secondary and backscattered electron imagery, individual glass shards were examined for compositional homogeneity and adequate thickness and width and selected for analysis. Typically, 20–30 different glass shards were analyzed for each sample. Data were corrected using atomic number (Z), absorption (A), and fluorescence (F) corrections or ZAF corrections. Glass shard data are reported as unnormalized means with associated standard deviation for each sample population (Table 1).

TABLE 1. MAJOR-ELEMENT CHEMISTRY OF BKT-2 IN LEDI-GERARU–ELECTRON MICROPROBE ANALYSES

Region	Site	Tephra	Sample ID	n		SiO$_2$	TiO$_2$	Al$_2$O$_3$	MnO	Fe$_2$O$_3$	MgO	CaO	Na$_2$O	K$_2$O	Total
BKT-2 Mode 1: Rhyolitic															
Bagare	LG-136	BKT-2L	AM06-1013	23	avg	72.88	0.15	12.21	0.06	2.20	0.04	0.53	2.55	2.76	93.39
					±1σ	0.65	0.03	0.10	0.04	0.14	0.06	0.04	0.40	0.35	1.16
Bagare	LG-136	BKT-2U (a)	AM06-1017	22	avg	73.14	0.17	12.36	0.08	2.56	0.05	0.56	2.03	2.74	93.68
					±1σ	0.46	0.04	0.16	0.03	0.19	0.07	0.05	0.57	0.38	0.91
Bagare	LG-136	BKT-2U (b)	AM06-1018	22	avg	73.69	0.16	12.13	0.07	2.34	0.03	0.50	2.46	2.77	94.16
					±1σ	0.91	0.03	0.35	0.03	0.27	0.04	0.07	0.50	0.37	0.96
Bagare	LG-136	BKT-2U (c)	AM06-1019	23	avg	73.58	0.16	12.06	0.07	2.37	0.04	0.50	2.43	2.78	93.99
					±1σ	0.83	0.04	0.31	0.04	0.25	0.05	0.06	0.45	0.32	1.03
Bagare	LG-136	BKT-2U (d)	AM06-1020	25	avg	72.41	0.15	12.19	0.07	2.42	0.06	0.52	2.53	2.84	93.19
					±1σ	1.10	0.02	0.25	0.03	0.25	0.08	0.08	0.53	0.33	1.42
Bagare	LG-136	BKT-2U (d)	LG33E	5	avg	71.53	0.16	12.46	0.06	2.40	0.01	0.52	2.28	2.55	91.97
					±1σ	1.11	0.03	0.40	0.05	0.18	0.02	0.10	0.76	0.47	1.13
Bagare	LG-136	BKT-2U (b-c)	LG33F	13	avg	72.56	0.16	12.21	0.05	2.15	0.01	0.49	2.53	2.78	92.94
					±1σ	1.24	0.03	0.31	0.05	0.69	0.02	0.06	0.52	0.27	1.50
Bagare	LG-136	BKT-2U (a)	LG33G	11	avg	70.91	0.16	12.40	0.07	2.49	0.03	0.54	2.19	2.38	91.16
					±1σ	1.78	0.02	0.48	0.05	0.25	0.03	0.05	0.64	0.61	2.19
Bagare	LG-136	BKT-2L	LG33H	18	avg	73.09	0.15	12.53	0.06	2.20	0.02	0.53	2.25	2.30	93.13
					±1σ	1.48	0.03	0.46	0.05	0.20	0.02	0.05	0.60	0.63	1.75
Sagle	LG-146	BKT-2	LG22	13	avg	72.46	0.16	12.09	0.07	2.17	0.01	0.43	2.52	2.53	92.44
					±1σ	0.88	0.03	0.12	0.06	0.19	0.01	0.04	0.35	0.35	1.18
E. Agbi Koma	LG-15	BKT-2	LG15	9	avg	72.94	0.16	12.58	0.06	2.09	0.01	0.52	2.53	2.12	93.02
					±1σ	1.30	0.04	0.17	0.03	0.49	0.02	0.07	0.47	0.48	1.37
Humed Dalulu	LG-173	BKT-2U (a)	MLG2002-36	2	avg	73.49	0.16	11.88	0.07	2.48	0.05	0.51	2.82	2.87	94.35
					±1σ	0.51	0.03	0.35	0.01	0.30	0.07	0.10	0.27	0.12	0.67
Humed Dalulu	LG-173	BKT-2U (a)	MLG2002-36	1	N.A.	73.07	0.15	12.29	0.05	1.98	0.12	0.47	2.37	2.53	93.04
Humed Dalulu	LG-173	BKT-2U (c)	MLG2002-37	8	avg	74.68	0.15	12.03	0.05	2.22	0.03	0.44	2.83	2.59	95.03
					±1σ	0.99	0.03	0.40	0.04	0.23	0.03	0.04	0.50	0.34	1.77
Koborto Gaba*		BKT-2U (a)	HE01-184	12	avg	72.84	0.17	12.40	0.06	2.51	0.02	0.55	2.38	2.41	93.34
					±1σ	1.19	0.03	0.18	0.03	0.14	0.04	0.04	0.23	0.27	1.08
BKT-2 Mode 2: Basaltic															
Bagare	LG-136	BKT-2U (a)	AM06-1017	11	avg	49.52	2.87	13.07	0.28	15.03	3.88	8.43	3.07	1.00	97.14
					±1σ	1.32	0.50	0.34	0.08	1.24	0.27	0.39	0.15	0.17	0.85
Bagare	LG-136	BKT-2U (b)	AM06-1018	10	avg	49.45	2.59	12.98	0.25	14.67	3.75	8.15	3.21	1.05	96.10
					±1σ	1.20	0.29	0.40	0.07	0.79	0.27	0.31	0.19	0.09	1.01
Bagare	LG-136	BKT-2U (c)	AM06-1019	10	avg	48.53	3.20	12.55	0.25	15.10	3.95	8.52	2.94	1.00	96.05
					±1σ	2.08	0.80	0.46	0.07	1.16	0.44	0.63	0.20	0.16	1.05
Humed Dalulu	LG-173	BKT-2U (c)	MLG2002-37	3	avg	50.05	3.42	13.66	0.27	15.82	4.50	9.17	2.78	0.82	100.49
					±1σ	0.57	0.09	0.18	0.03	0.11	0.17	0.26	0.52	0.01	1.11
Humed Dalulu	LG-173	BKT-2U (c)	MLG2002-37	1		55.93	2.40	12.19	0.22	13.94	3.69	7.59	3.43	1.13	100.52

Notes: See text for electron microprobe (EMP) running conditions and analysis location; n is the number of analyses; avg is the average value associated with ±1σ standard deviation.
*Hadar BKT-2U comparison sample; sample split was provided by C. Campisano. See Figure 1 for sample locations.

^{40}Ar/^{39}Ar Analyses

Anorthoclase from four samples of the Ledi-Geraru BKT-2 sampled at two sites (LG-136 and LG-173; Figs. 1–3) was prepared at Arizona State University for ^{40}Ar/^{39}Ar dating. Bulk tephra samples were oven dried, sieved, and a coarse fraction (>0.25 mm) was retained. Approximately 30–50 of the least-altered, euhedral anorthoclase crystals between 0.25 mm and 0.5 mm were handpicked from each sample. These crystals exhibited the least amount of alteration (given the sample population) and were generally euhedral with minor fracturing, clear to slightly cloudy, and inclusion free. To remove any adhering materials, crystals were gently agitated for ~10–30 s in a 5% HF solution and rinsed 2–3 times in distilled water.

Anorthoclase crystals were dated by ^{40}Ar/^{39}Ar laser incremental-heating techniques at Rutgers University following procedures recently outlined in Carr et al. (2007). All ^{40}Ar/^{39}Ar ages were calculated relative to Fish Canyon Tuff (FCT) sanidine using an adopted age of 28.02 Ma (Renne et al., 1998). Data and incremental-heating spectra are provided in Table 2 and Figures 4 and 5.

ANALYSIS AND CORRELATION OF BKT-2

Physical Description

To clarify terminology, we use the term "tephra" to refer to any deposit composed of pyroclasts, or fragments released in a volcanic eruption. Unconsolidated or loosely consolidated tephra are lapilli or ash deposits, while the lithified or mainly consolidated equivalent is a tuff (Cas and Wright, 1988). The BKT-2 complex in the Ledi-Geraru area was identified at seven sites that span ~18 km from Bagare and Sagle in the east, Humed Dalulu in the south near Hadar, to just east of Agbi Koma in the north (Fig. 1). From top to bottom, the BKT-2 complex at Ledi-Geraru contains three primary air-fall beds, the upper (BKT-2U) and lower (BKT-2L) tephra layers, which are underlain by 1–2 m of sediment, and an ash tuff called the Green Marker Bed (GMB = BKT-2L$_1$; after Walter, 1981) (Figs. 2 and 3). We also consistently identified an ash tuff and a bentonite (altered volcanic ash) layer preserved between BKT-2L and -2U, and a second ash tuff underlying BKT-2L. Aside from thicknesses and degree of alteration, this basic tephra sequence is similar at all Ledi-Geraru sites.

The Green Marker Bed (GMB) is a 3–6-cm-thick ash tuff that is black or dark green and white and is observed ~0.5–1 m below BKT-2L (Figs. 2C, 2E, and 3D). Upper and lower contacts are sharp with 1–2 mm dark red-orange alteration staining. Typically, the GMB is finely laminated, compact, and breaks along weaker planes developed in a slightly coarser dark ash fraction.

BKT-2L is a loosely consolidated, 2–9-cm-thick, crystal-lithic lapilli tephra (Figs. 2C, 2D, and 3C). Most pumice clasts are altered to bentonite, but minor amounts of fresh glass are preserved. Upper and lower contacts are sharp. The phenocryst assemblage is dominated by plagioclase feldspar (oligoclase) with orange-colored alteration. Regionally, BKT-2L varies in degree of glass alteration and crystal preservation.

BKT-2U is a 30–97-cm-thick primary fallout eruptive sequence that preserves a record of explosive volcanism involving both rhyolitic and basaltic magmas (Figs. 2B and 3B). It was sampled and described in four informal units (a, b, c, and d; Fig. 6) to capture distinct physical variability within a single air-fall sequence. The basal unit (a) comprises approximately one-third of the total tephra thickness (depending on the completeness of the deposit) and is a crystal-lithic lapilli tephra with rhyolitic pumice. The phenocryst assemblage is dominated by subhedral anorthoclase mantled by rhyolitic vesicle wall glass, slender prismatic iron-rich clinopyroxene, and minor hornblende and oxides. Units b and c (<50 cm total thickness) sharply overlie unit a and contain compositionally bimodal and physically graded juvenile glass. Rhyolitic pumice makes up ~45% of the total components of b. Unit b grades into c, which contains 75% dark brown basaltic scoria. The uppermost unit (d) is a thinner, less well-preserved, tan-colored ash with pumice lapilli. It is unconsolidated, slope forming, and generally slightly reworked into overlying sediments or completely eroded.

At seven sites in the Ledi-Geraru, BKT-2L and BKT-2U show little variation in physical expression and are equivalent to descriptions of the same at Hadar (Walter, 1981; Walter and Aronson, 1982; Campisano and Feibel, this volume, Chapter 6). However, the physical expression of the GMB is more variable between research areas. The presence and consistent physical grading of juvenile rhyolitic pumice and basaltic scoria in BKT-2U (see following discussion) make this deposit a highly diagnostic marker bed.

Glass Composition

Electron microprobe analyses of BKT-2U and -2L rhyolitic glasses have consistent major-element weight percent concentrations (Table 1). Mode 1 is characterized as a high-silica (~73%) glass with moderate Al_2O_3 (~12%) and CaO (~0.5%). Rhyolite glass composition does not distinguish BKT-2U from BKT-2L, although Campisano and Feibel (this volume, Chapter 6) show clear distinctions in feldspar composition. Unlike BKT-2L, BKT-2U glasses display two distinct compositional modes. Analyses of scoria clasts from BKT-2U define mode 2 as largely basaltic (SiO_2 ~48%–50%) with characteristically high Fe_2O_3, TiO_2, MgO, and CaO values (Table 1). The bimodal nature of this tephra is not a reflection of posteruption contamination. Rather, it is a primary feature indicating the eruption of a heterogeneous magma, likely from the Ida Ale volcanic center (Fig. 1; Walter, 1981; DiMaggio, 2007). The Ida Ale volcanic center is accepted as the source of BKT-2 tephra because of the similar major-element, trace-element, and isotopic composition of rhyolitic glasses and feldspars. Additionally, the Ida Ale volcanic center is close to BKT-2 deposits (10–20 km; Fig. 1), and

TABLE 2. ^{40}Ar/^{39}Ar ANALYTICAL DATA FOR BKT-2 FROM THE LEDI-GERARU REGION

Sample / Lab ID# Increment	Laser (W)	Ca/K	^{36}Ar/^{39}Ar	^{40}Ar*/^{39}Ar	Mol ^{39}Ar	%^{39}Ar	%^{40}Ar*	Age (Ma) (±1σ)
AM05-122 ($J = 2.699 \times 10^{-4} \pm 1.868 \times 10^{-7}$)								
30411-01A	2	0.478	0.47012	10.329	0.029	0.6	6.9	5.023 ± 0.986
30411-01B	4	0.824	0.01274	7.697	0.212	4.7	67.3	3.744 ± 0.088
30411-01C	6	0.906	0.00245	7.033	0.548	12.2	91.1	3.421 ± 0.035
30411-01D	8	0.980	0.00124	6.245	0.741	16.4	95.0	3.039 ± 0.026
•30411-01E	10	1.035	0.00103	6.100	0.876	19.4	95.9	2.968 ± 0.022
•30411-01F	12	1.037	0.00266	6.022	0.769	17.1	89.0	2.930 ± 0.026
•30411-01G	14	0.984	0.00057	6.187	0.458	10.2	98.0	3.010 ± 0.038
•30411-01H	17	0.951	0.00404	6.034	0.315	7.0	83.9	2.936 ± 0.057
•30411-01I	21	0.977	0.00268	6.011	0.176	3.9	88.8	2.925 ± 0.096
•30411-01J	26	0.928	0.00345	6.124	0.163	3.6	86.2	2.980 ± 0.105
•30411-01K	32	0.961	0.01505	5.876	0.074	1.6	57.1	2.859 ± 0.232
•30411-01L	40	0.918	0.00926	6.449	0.147	3.3	70.5	3.138 ± 0.119
%^{39}Ar on plateau = 66.1; plateau age = 2.961 ± 0.014 Ma								
AM05-125 ($J = 2.694 \times 10^{-4} \pm 1.812 \times 10^{-7}$)								
30413-01A	2	1.371	0.96960	4.773	0.020	0.5	1.6	2.319 ± 1.758
•30413-01B	4	1.222	0.02479	6.495	0.209	5.4	47.1	3.155 ± 0.097
•30413-01C	6	1.231	0.00455	6.190	0.552	14.3	82.7	3.007 ± 0.036
•30413-01D	8	1.324	0.00186	6.196	0.796	20.6	92.6	3.009 ± 0.025
•30413-01E	10	1.281	0.00128	6.170	0.816	21.2	94.9	2.997 ± 0.026
•30413-01F	12	1.263	0.00103	6.098	0.517	13.4	96.0	2.962 ± 0.034
•30413-01G	13	1.330	0.00079	6.191	0.327	8.5	97.1	3.007 ± 0.055
•30413-01H	17	1.416	0.00170	6.133	0.232	6.0	93.2	2.979 ± 0.073
•30413-01I	21	1.724	0.00198	6.019	0.182	4.7	92.1	2.923 ± 0.093
•30413-01J	26	1.591	0.00192	5.851	0.110	2.8	92.1	2.842 ± 0.152
•30413-01K	32	2.001	0.00495	6.126	0.015	0.4	81.6	2.975 ± 1.072
•30413-01L	40	1.458	0.00276	6.279	0.081	2.1	89.2	3.050 ± 0.199
%^{39}Ar on plateau = 99.1; plateau age = 2.996 ± 0.013 Ma								
AM05-139 ($J = 2.695 \times 10^{-4} \pm 1.835 \times 10^{-7}$)								
30412-01A	2	1.317	0.31592	6.532	0.037	1.4	6.5	3.174 ± 0.683
30412-01B	4	1.226	0.00437	6.389	0.197	7.1	83.7	3.105 ± 0.089
•30412-01C	6	1.243	0.00154	6.038	0.373	13.5	93.7	2.934 ± 0.047
•30412-01D	8	1.276	0.00105	6.183	0.452	16.3	96.0	3.005 ± 0.040
•30412-01E	10	1.338	0.00578	6.152	0.627	22.7	78.8	2.989 ± 0.034
•30412-01F	12	1.367	0.00309	5.986	0.429	15.5	87.4	2.909 ± 0.043
•30412-01G	14	1.580	0.00343	6.244	0.222	8.0	86.8	3.034 ± 0.078
•30412-01H	17	1.675	0.00482	6.054	0.217	7.8	81.6	2.942 ± 0.081
•30412-01I	21	1.613	0.00710	5.766	0.098	3.5	73.9	2.802 ± 0.178
•30412-01J	26	1.560	0.02102	4.708	0.035	1.3	43.3	2.288 ± 0.478
30412-01K	32	1.407	0.01593	6.464	0.029	1.1	58.1	3.141 ± 0.576
30412-01L	40	1.505	0.01634	5.945	0.050	1.8	55.5	2.889 ± 0.335
%^{39}Ar on plateau = 88.7; plateau age = 2.965 ± 0.019 Ma								
AM05-141 ($J = 2.696 \times 10^{-4} \pm 1.793 \times 10^{-7}$)								
30414-01A	2	0.374	0.51882	14.627	0.012	0.3	8.7	7.101 ± 1.716
30414-01B	4	0.622	0.06439	8.074	0.095	2.1	29.8	3.924 ± 0.217
30414-01C	6	0.643	0.01704	6.800	0.250	5.4	57.6	3.305 ± 0.079
30414-01D	8	0.648	0.01378	6.277	0.454	9.8	60.8	3.051 ± 0.049
•30414-01E	10	0.645	0.00852	6.012	0.799	17.3	70.7	2.923 ± 0.031
•30414-01F	12	0.655	0.00334	6.039	0.944	20.4	86.3	2.936 ± 0.023
•30414-01G	14	0.590	0.00181	6.072	1.018	22.1	92.2	2.952 ± 0.021
•30414-01H	17	0.572	0.00487	6.036	0.366	7.9	81.0	2.934 ± 0.051
•30414-01I	21	0.678	0.00806	6.260	0.294	6.4	72.7	3.043 ± 0.064
•30414-01J	26	0.599	0.01877	6.255	0.141	3.1	53.1	3.041 ± 0.130
•30414-01K	32	1.115	0.03168	5.832	0.114	2.5	38.5	2.835 ± 0.163
30414-01L	40	0.546	0.03934	5.309	0.129	2.8	31.4	2.581 ± 0.155
%^{39}Ar on plateau = 79.6; plateau age = 2.944 ± 0.013 Ma								

Note: These samples were also analyzed for major-element glass chemistry (Table 1): AM05-122=AM06-1013; AM05-125=AM06-1017. Enclosed analyses were used in calculating plateau ages.

170 DiMaggio et al.

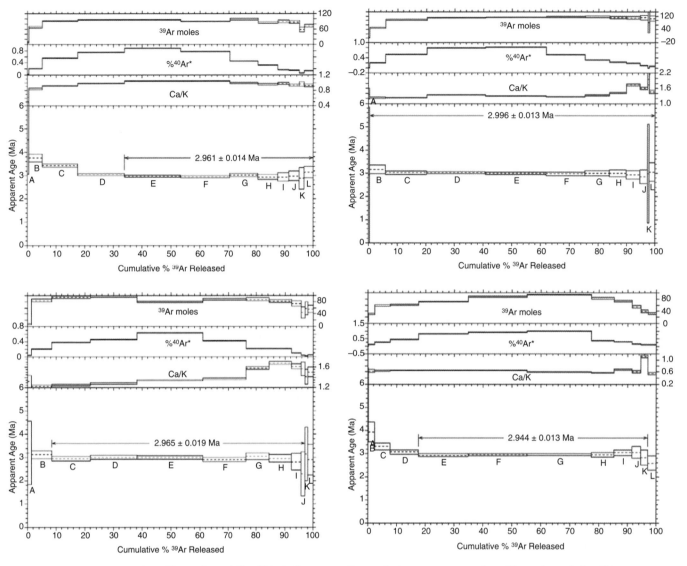

Figure 4. Age spectrum and auxiliary plots of $^{40}Ar/^{39}Ar$ bulk step-heating results for BKT-2L samples AM05-122 (top) and AM05-139 (bottom). See Table 2.

Figure 5. Age spectrum and auxiliary plots of $^{40}Ar/^{39}Ar$ bulk step-heating results for BKT-2U samples AM05-125 (top) and AM05-141 (bottom). See Table 2.

the timing of its explosive activity is coeval with sedimentation in that region (Walter, 1981).

Glass compositions from this study and Campisano and Feibel (this volume, Chapter 6) support the correlation of BKT-2 between Ledi-Geraru and Hadar. To assess interlaboratory reproducibility and to further test the correlation, a sample split of BKT-2U (HE01-184) from the Koborto Gaba in Hadar was reanalyzed at Arizona State University using the same preparation techniques and instrumental conditions that were used for Ledi-Geraru samples. Geochemical results of sample HE01-184 are consistent with its previously established composition (Campisano, 2007), and all analyzed elements are within one standard deviation of Ledi-Geraru samples (Table 1), confirming this correlation.

$^{40}Ar/^{39}Ar$ Analyses

The $^{40}Ar/^{39}Ar$ dating of the Ledi-Geraru BKT-2 samples provides further support for correlation with Hadar BKT-2. All dated anorthoclase samples yielded well-behaved release spectra (Figs. 4 and 5) and well-defined plateaus that incorporate greater than 65% of the total ^{39}Ar released. BKT-2L samples AM05-122 from LG-136 and AM05-139 from LG-176 yielded plateau ages of 2.961 ± 0.014 Ma and 2.965 ± 0.019 Ma, respectively. These ages are indistinguishable from single-crystal $^{40}Ar/^{39}Ar$ total fusion ages reported for BKT-2L from Hadar of 2.96 ± 0.04 Ma (Campisano, 2007), and from Gona of 2.960 ± 0.006 Ma (Semaw et al., 1997, recalculated using 28.02 Ma FCT). BKT-2U samples AM05-125 from LG-136 and AM05-141 from

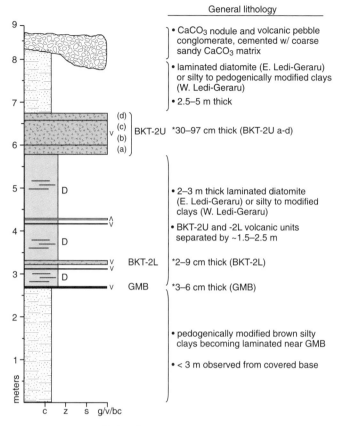

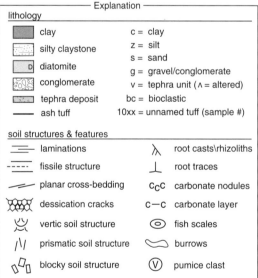

Figure 6. (Top) General stratigraphy and variation of BKT-2 deposits in Ledi-Geraru. The tephra-encasing lithology varies from laminated diatomite to clayey silts and carbonate-rich paleosols. See Figures 7–9 for detailed stratigraphic sections containing BKT-2. (Bottom) Stratigraphic explanation of symbols used in Figures 6–9.

LG-176 yielded plateau ages of 2.996 ± 0.013 Ma and 2.944 ± 0.013 Ma, respectively. Sample AM05-141 is indistinguishable from BKT-2U total fusion ages from Hadar of 2.94 ± 0.02 Ma (Campisano, 2007) and 2.94 Ma (Fig. 1 of Walter, 1994; recalculated using 28.02 Ma FCT). Feldspars from sample AM05-125 appear slightly more fractured and altered compared to those from sample AM05-141, which may account for the slightly anomalous age of 2.996 Ma. A similar situation was also encountered in an equivalent subsample of BKT-2U feldspars from Hadar, which yielded a similarly anomalous age of ca. 3.01 Ma (Campisano, 2007).

BKT-2 STRATIGRAPHY AND VARIATION

Sediments were initially mapped at Ledi-Geraru by Arrowsmith et al. (2004) with revisions in 2005–2006. We confirmed the presence of BKT-2 in regions where the Kada Hadar Member was identified. The very shallow north dips (<4°) and low topographic relief produce broad exposures of BKT-2 in this region. Stratigraphic sections containing BKT-2 deposits were measured and described at seven locations (Fig. 1). The resulting sections (Figs. 7–9) span ~20 km across Ledi-Geraru and record in detail ~2–12 m of upper Kada Hadar Member strata that encase the BKT-2 complex.

Sites LG-136, LG-146, and LG-148

Site LG-136 in the Bagare region (Figs. 2, 7, and 9) preserves the most complete exposure of BKT-2, including the GMB, BKT-2L, and BKT-2U. Two thin bentonite layers (<2 cm thick) and two scoriaceous ash tuffs (~2 cm thick) are also present. Here, these units are encased in a 6 m section of predominately tan to white diatomite crosscut by medium-brown silty dikes that overlie an ~2.5 m calcareous Vertisol and massive clays. A scoriaceous tuff and the GMB are interbedded in the diatomite near the base of the section. The GMB is finely laminated with white and black-green–colored ash. Overlying the bed by ~0.5 m, BKT-2L is preserved as a 9-cm-thick tan to orange crystal lapilli tephra with altered glass. It is separated from BKT-2U by 2.5 m of diatomite, a scoriaceous tuff, and a pink bentonite layer 4–10 cm above the tuff. The upper lapilli deposit is 97 cm thick and forms a sharp contact against underlying diatomite. It is the thickest and best-preserved lapilli deposit identified at Ledi-Geraru. The upper unit (d) is slightly reworked into overlying laminated diatomite. This section is capped by gravels that may represent the base of the Busidima Formation (Quade et al., 2004).

Sites LG-146 and LG-148 in the Sagle region (Figs. 7 and 9) preserve a 70–80-cm-thick BKT-2 deposit contained within laminated diatomite. A more complete section was measured at LG-148, where multiple thin tuffs, a pink bentonite, and BKT-2L and -2U are encased in diatomite with a slightly higher clay/silt content than site LG-136. Underlying the GMB, carbonate-rich sands and conglomerates are interbedded within thick vertic clays and silts. These sediments overlie a vesicular basalt flow at LG-148.

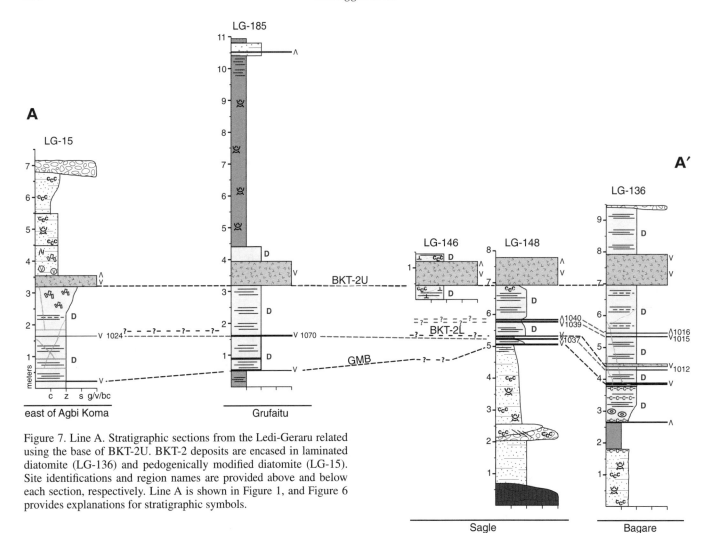

Figure 7. Line A. Stratigraphic sections from the Ledi-Geraru related using the base of BKT-2U. BKT-2 deposits are encased in laminated diatomite (LG-136) and pedogenically modified diatomite (LG-15). Site identifications and region names are provided above and below each section, respectively. Line A is shown in Figure 1, and Figure 6 provides explanations for stratigraphic symbols.

Sites LG-15 and LG-185

Site LG-15 is located ~5 km east of Agbi Koma, south of the Bati-Mille road (Fig. 1). At this site, two scoriaceous tuffs (the lowest likely the GMB) are preserved in laminated silty diatomite that becomes pedogenically modified ~0.5 m below BKT-2U (Fig. 7). A 3-m-thick, carbonate-rich, silty-clay paleosol overlies the BKT-2 complex and is disconformably overlain by a capping pebble conglomerate. Site LG-185 is located less than 1 km west of the Addis-Assab road in Grufaitu (Fig. 1). Here, BKT-2 is encased in massive to laminated silty white diatomite (Fig. 7). A thick clay Vertisol overlies the diatomite and tephra deposits, and gravels are absent in the stratigraphy. BKT-2L was not observed at either LG-15 or LG-185.

Sites LG-176 and LG-173

Located in southwest Ledi-Geraru, the stratigraphic sections measured at sites LG-176 and LG-173 are the closest sections to Hadar (Fig. 1). They are dominated by massive, laminated, and pedogenically modified brown and green clays and clayey silts (Figs. 3 and 9). BKT-2U at site LG-173 is 30–48 cm thick and is encased in a sequence of carbonate-rich Vertisols and dark brown blocky paleosols (Fig. 3B). Approximately 2 m above BKT-2U, a pebble unit, rich in volcanic and carbonate clasts, erodes into the blocky paleosol. A similar pebble conglomerate erodes into the ash (unit d) of BKT-2U at site LG-176. At both sites, we identified a thin bentonite layer, a scoriaceous tuff, BKT-2L (Fig. 3C), and the GMB (Fig. 3D).

Stratigraphic Correlation

Stratigraphic correlations of BKT-2 and associated sediments capture discrete facies changes that permit a regional (Hadar and Ledi-Geraru) assessment of paleoenvironmental variation across space and time (Figs. 7–9). Sections along line A (Fig. 7) run northwest to southeast ~18 km across central Ledi-Geraru. Five stratigraphic sections can be correlated based on sharp basal contacts of BKT-2U. At each site, BKT-2U overlies or is encased within laminated diatomaceous silts (2–4 m thick)

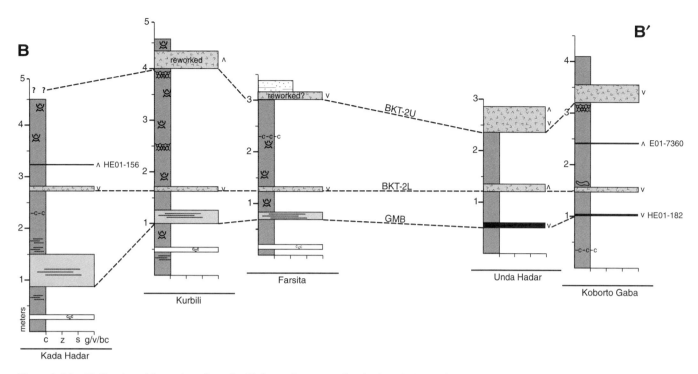

Figure 8. Line B. Stratigraphic sections from the Hadar project area related using the base of BKT-2L. BKT-2 deposits are encased in pedogenically modified and laminated clays and silts. Sections were redrawn from Walter (1981) and Campisano (2007), and regional names are provided below the section. Line B is shown in Figure 1, and Figure 6 provides explanations for stratigraphic symbols.

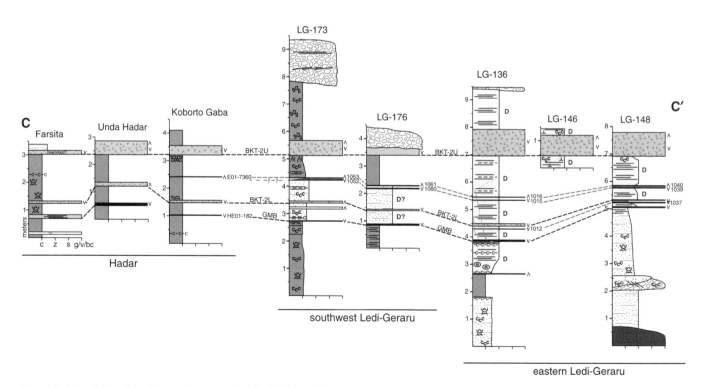

Figure 9. Line C. Stratigraphic sections spanning the Hadar and Ledi-Geraru project areas. BKT-2U is predominately encased in pedogenically modified and laminated silts in Hadar and western Ledi-Geraru, and it is encased in diatomaceous silts in eastern Ledi-Geraru. Site identifications are provided above each section. Line C is shown in Figure 1, and Figure 6 provides explanations for stratigraphic symbols.

deposited on clay to silty clays and calcareous Vertisols. The GMB, BKT-2L, and additional thin bentonite layers and tuffs are interbedded within the diatomite and are correlated based on field descriptions and relative stratigraphic position. In the northern exposures (LG-15 and LG-185), massive calcareous silty clays and Vertisols (>3 m thick) dominate the sediments overlying BKT-2U, whereas laminated and massive diatomite encases BKT-2U in the southeast (LG-136 and LG-146). Pebble conglomerates likely belonging to the younger Busidima Formation (Quade et al., 2004) have abundant carbonate nodules and disconformably overlie sediment at sites LG-15 and LG-136.

Line B (Fig. 8) is a compilation of five stratigraphic sections drawn from descriptions in Walter (1981) and Campisano (2007). These sections are 3–5 m thick, span ~7 km west to east across the Hadar project area, and were correlated using BKT-2L. Unlike at Ledi-Geraru, sediments encasing the BKT-2 complex at Hadar are primarily massive brown to light brown clays. Unaltered volcanic glass is only preserved in the Unda Hadar and Koborto Gaba regions. Additionally, BKT-2U is not consistently preserved at all BKT-2 exposures, and where preserved, it is typically underlain by a network of desiccation cracks infilled with BKT-2U material (Campisano and Feibel, this volume, Chapter 6).

Line C (Fig. 9) encompasses the Hadar and Ledi-Geraru project areas. At most sites in the Ledi-Geraru, the GMB overlies dark brown Vertisols containing abundant carbonate nodules that grade into massive or laminated, olive green to light brown silts and silty clays. The GMB overlies massive clays at Hadar. At site LG-136 in eastern Ledi-Geraru, laminated diatomaceous silts encase the BKT-2 complex, whereas massive silts and massive to laminated clayey silts with carbonate nodules and minor pedogenic modification comprise sediments at site LG-173 in southwestern Ledi-Geraru and sites in Hadar at this time. Slightly diatomaceous silts overlie the GMB and BKT-2L at site LG-173.

DEPOSITIONAL ENVIRONMENTS

Large-scale paleolandscape changes inferred from lithofacies variations in the Hadar Formation were originally constructed by Tiercelin (1986). Recently, new geological and paleontological work has improved these original interpretations of Hadar Formation depositional environments (Wynn et al., 2006; Campisano and Feibel, this volume, Chapter 8; Campisano and Reed, 2007). The new data presented in this study provide a high-resolution depositional history from the time of the BKT-2 complex in the Ledi-Geraru and Hadar regions. The following is an interpretation of depositional environments ca. 2.9 Ma in this region based on spatial and temporal facies variations inferred from Figures 7–9.

The base of the sequence begins several meters stratigraphically below the GMB. The presence of well-developed, carbonate-rich clay and silty clay Vertisols, coarse pebble conglomerates, and sand facies suggests that subaerially exposed floodplain environments dominated the Ledi-Geraru at that time. Below the GMB at site LG-136, an ~1-m-thick laminated diatomite that contains carbonate layers and ostracods overlying massive clays likely indicates that this region was at or near the lacustrine depocenter. Across Ledi-Geraru and Hadar, clays immediately underlying and associated with the GMB are laminated or massive, consistent with lacustrine to lake-marginal settings. Stratigraphic sections from the Kada Gona region document the preservation of the GMB in lacustrine sediments (Semaw et al., 1997; see sections 5 and 6 in Data Repository item 2008216 from Quade et al., this volume). Collectively, these records clearly show the expansion of lacustrine conditions across this region, which likely extended to the lower slopes of the western margin of the Afar Depression. BKT-2L also appears to have been deposited during lake transgression maximum (ca. 2.96 Ma), as it is encased in massive to laminated mudstone and silts in the west (Hadar and Gona), grading to laminated diatomite in the east (LG-136 at Ledi-Geraru). During this time (Fig. 10, bottom), lacustrine and lake-margin settings extended across much of central Ledi-Geraru and may have engulfed the Ida Ale volcanic center (Walter, 1981).

Sometime prior to the eruption of BKT-2U (ca. 2.94 Ma), lithofacies changes at Hadar and sites LG-15, LG-173, and LG-176 mark the onset of an eastward lacustrine regression (Fig. 10, top). This regression is indicated by sub–BKT-2U pedogenic modification at Hadar and western Ledi-Geraru exposures, as well as the decrease in diatomaceous sediment in western Ledi-Geraru. Subsequent pedogenic modifications of delta-plain or nearshore clays and silt deposits above BKT-2U attest to the continued regression of the lake after 2.94 Ma at all but the eastern Ledi-Geraru sites. To the east, such as at LG-136, laminated diatomite facies remain unchanged following the deposition of BKT-2U, confirming the sustained presence of the lake in this region. Based upon this evidence, depositional environments to the west and north of the paleodepocenter appear to have been controlled by fluctuation between an ephemeral shallow lake and exposed floodplain/lake-margin deposits with subsequent subaerial exposure and pedogenesis.

Sedimentary records from multiple basins in East Africa coeval with the upper Kada Hadar Member do not indicate the presence of a pronounced lake phase ca. 2.9 Ma (Trauth et al., 2005, and references therein). Although these records are not temporally well-constrained at that time, dry climatic conditions are indicated in Kenya and Ethiopia (e.g., Deino et al., 2006; Sturchio et al., 1993; Brown and Feibel, 1991). Lacustrine deposits suggesting high East African lake levels are not recorded in those regions until ca. 2.7 Ma. Thus, the lake phase in west-central Afar at Ledi-Geraru at 2.9 Ma probably represents a basin-scale condition rather than a prolonged high lake phase throughout East Africa. The wet phase at Ledi-Geraru was thus controlled by a regional climate shift, or varying fault activity and loci along the eastern margin of the basin.

CONCLUSIONS

New tephra descriptions, $^{40}Ar/^{39}Ar$ dates, and major-element glass compositions permit us to make chronostratigraphic correlations among intrabasin research areas to greatly extend our

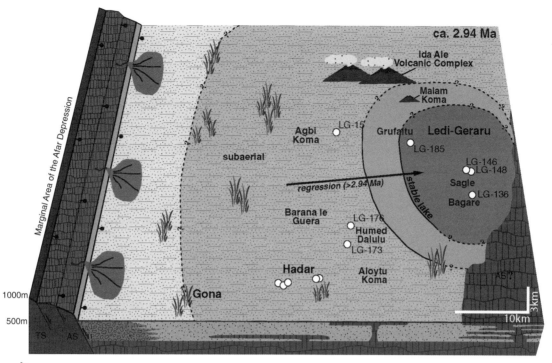
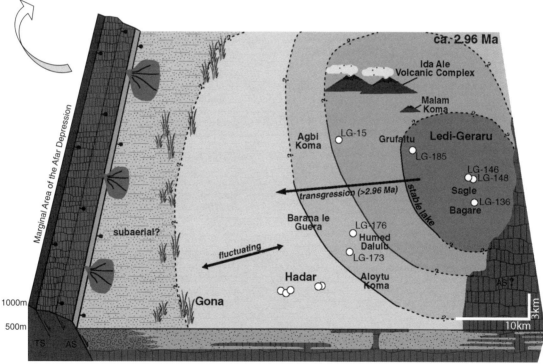

Figure 10. Depositional environment reconstruction of the Ledi-Geraru, Hadar, and Gona regions ca. 2.96 Ma (bottom) and ca. 2.94 Ma (top) based on interpretation of stratigraphic correlations shown in Figures 7–9 and sections GON99-5 and -6 (see Data Repository item 2008216[1]). Relative depth and stability of the lake are depicted by shades of gray, where dark indicates greater stability. Solid black lines indicate boundaries of the lake based on stratigraphic evidence. The extent of various lake stages is unknown and denoted by a dashed line and question marks. BKT-2 stratigraphy sites are denoted as white circles in Ledi-Geraru and Hadar. Volcanic rocks: TS—Trap Series; AS—Afar Stratoid Series. Gray, textured polygons along the edges of the figure represent basalt and basalt flows; however, their extent and exposure during this time are uncertain. Only the marginal region of the much larger Ethiopian western escarpment is depicted.

[1]GSA Data Repository item 2008216, stratigraphic sections and dating results, is available at www.geosociety.org/pubs/ft2008.htm, or on request from editing@geosociety.org, Documents Secretary, GSA, P.O. Box 9140, Boulder, CO 80301-9140, USA.

spatial resolution for interpreting regional depositional environments ca. 2.96–2.94 Ma. Detailed stratigraphic sections provide a high-resolution record during the deposition of the last major lacustrine phase of the Hadar Formation associated with the BKT-2 complex. Our interpretation of lithofacies organization and variation across the Ledi-Geraru region indicates westward expansion of a paleolake ca. 2.96 Ma followed by eastward regression prior to the eruption of BKT-2U ca. 2.94 Ma. The paleolake depocenter at 2.96 Ma was located in the region of Sagle, east of the modern Addis-Assab road in eastern Ledi-Geraru (Fig. 10). Detailed assessments such as these of lacustrine deposits contribute high-fidelity records that are essential for disentangling basin-scale environment signals in hominin-bearing strata.

ACKNOWLEDGMENTS

We thank our Ethiopian colleagues in the Afar and in Addis Ababa for help in many aspects of this work, and the Authority for Research and Conservation of Cultural Heritage (ARCCH) for permission to carry out the research. Our excellent 2006 field crew was directed by Mesfin Mekonen. Solomon Kebede was our ARCCH representative. We gratefully acknowledge all of their help. We would like to thank Gordon Moore for providing skillful assistance on the electron microprobe and the valuable guidance from Amanda Clarke and Kim Genareau at Arizona State University. This project was funded in part by the Leakey Foundation, the Wenner Gren Foundation for Anthropological Research, and the Institute of Human Origins and School of Human Evolution and Social Change at Arizona State University.

REFERENCES CITED

Alemseged, Z., Spoor, F., Kimbel, W.H., Bobe, R., Geraads, D., Reed, D., and Wynn, J.G., 2006, A juvenile early hominin skeleton from Dikika, Ethiopia: Nature, v. 443, p. 296–301, doi: 10.1038/nature05047.

Aronson, J.L., and Taieb, M., 1981, Geology and paleogeography of the Hadar hominid site, Ethiopia, *in* Rapp, G., and Vondra, C.F., eds., Hominid Sites: Their Geologic Setting: American Association for the Advancement of Science Selected Symposium: Boulder, Colorado, Westview Press, p. 165–195.

Aronson, J.L., Schmitt, T.J., Walter, R.C., Taieb, M., Tiercelin, J.J., Johanson, D.C., Naeser, C.W., and Nairn, A.E.M., 1977, New geochronologic and palaeomagnetic data for the hominid-bearing Hadar Formation of Ethiopia: Nature, v. 267, p. 323–327, doi: 10.1038/267323a0.

Arrowsmith, J R., Reed, K., Lockwood, C., and Jones, K., 2004, Geological mapping and tephrostratigraphy of the Hadar Formation near 11.25°N and 40.75°E (Afar region, Ethiopia): Geological Society of America Abstracts with Programs, v. 36, no. 5, p. 487.

Beyene, A., and Abdelsalam, M.G., 2005, Tectonics of the Afar Depression: A review and synthesis: Journal of African Earth Sciences, v. 41, p. 41–59, doi: 10.1016/j.jafrearsci.2005.03.003.

Brown, F.H., and Feibel, C.S., 1991, Stratigraphy, depositional environments and palaeogeography of the Koobi Fora Formation, *in* Harris, J.M., ed., Koobi Fora Research Project, Volume 3: Oxford, Clarendon Press, p. 1–30.

Campisano, C.J., 2007, Tephrostratigraphy and Hominin Paleoenvironments of the Hadar Formation, Afar Depression, Ethiopia [Ph.D. thesis]: New Brunswick, New Jersey, Rutgers University, 600 p.

Campisano, C.J., and Feibel, C.S., 2008, this volume (Chapter 6), Tephrostratigraphy of the Hadar and Busidima Formations at Hadar, Afar Depression, Ethiopia, *in* Quade, J., and Wynn, J.G., eds., The Geology of Early Humans in the Horn of Africa: Geological Society of America Special Paper 446, doi: 10.1130/2008.2446(06).

Campisano, C.J., and Feibel, C.S., 2008, this volume (Chapter 8), Depositional environments and stratigraphic summary of the Pliocene Hadar Formation at Hadar, Afar Depression, Ethiopia, *in* Quade, J., and Wynn, J.G., eds., The Geology of Early Humans in the Horn of Africa: Geological Society of America Special Paper 446, doi: 10.1130/2008.2446(08).

Campisano, C.J., and Reed, K.E., 2007, Spatial and temporal patterns of *Australopithecus afarensis* habitats at Hadar, Ethiopia: Philadelphia, Pennsylvania, Paleoanthropology Society, Annual Meeting Abstracts, p. A6.

Carr, M.J., Saginor, I., Alvarado, G.E., Bolge, L.L., Lindsay, F.N., Milidakis, K., Turrin, B.D., Feigenson, M.D., and Swisher, C.C., III, 2007, Element fluxes from the volcanic front of Nicaragua and Costa Rica: Geochemistry, Geophysics, Geosystems, v. 8, p. Q06001, doi: 10.1029/2006GC001396.

Cas, R.A.F., and Wright, J.V., 1988, Volcanic Successions: Modern and Ancient: London, Chapman & Hall, 528 p.

Deino, A.L., Kingston, J.D., Glen, J.M., Edgar, R.K., and Hill, A., 2006, Precessional forcing of lacustrine sedimentation in the late Cenozoic Chemeron Basin, Central Kenya Rift, and calibration of the Gauss/Matuyama boundary: Earth and Planetary Science Letters, v. 247, p. 41–60, doi: 10.1016/j.epsl.2006.04.009.

deMenocal, P.B., 1995, Plio-Pleistocene African climate: Science, v. 270, p. 53–59, doi: 10.1126/science.270.5233.53.

deMenocal, P.B., 2004, African climate change and faunal evolution during the Pliocene-Pleistocene: Earth and Planetary Science Letters, v. 220, p. 3–24, doi: 10.1016/S0012-821X(04)00003-2.

DiMaggio, E.N., 2007, Volcanic and Stratigraphic Characterization of Pliocene Tephra from the Ledi-Geraru Region of Afar, Ethiopia [M.S. thesis]: Tempe, Arizona State University, 120 p.

Dupont-Nivet, G., Sier, M., Campisano, C.J., Arrowsmith, J R., DiMaggio, E., Reed, K., Lockwood, C., Franke, C., and Hüsing, S., 2008, this volume, Magnetostratigraphy of the eastern Hadar Basin (Ledi-Geraru research area, Ethiopia) and implications for hominin paleoenvironments, *in* Quade, J., and Wynn, J.G., eds., The Geology of Early Humans in the Horn of Africa: Geological Society of America Special Paper 446, doi: 10.1130/2008.2446(03).

Froggatt, P.C., 1992, Standardization of the chemical analysis of tephra deposits: Report of the ICCT working group: Quaternary International, v. 13–14, p. 93–96, doi: 10.1016/1040-6182(92)90014-S.

Hunt, J.B., and Hill, P.G., 2001, Tephrological implications of beam size–sample size effects in electron microprobe analysis of glass shards: Journal of Quaternary Science, v. 16, p. 105–117, doi: 10.1002/jqs.571.

Johanson, D.C., and Taieb, M., 1976, Plio-Pleistocene hominid discoveries in Hadar, Ethiopia: Nature, v. 260, p. 293–297, doi: 10.1038/260293a0.

Kalb, J.E., Oswald, E.B., Tebedge, S., Mebrate, A., Tola, E., and Peak, D., 1982, Geology and stratigraphy of Neogene deposits, Middle Awash valley, Ethiopia: Nature, v. 298, p. 17–25, doi: 10.1038/298017a0.

Kerrick, D.M., Eminhizer, L.B., and Villaume, J.F., 1973, The role of carbon film thickness in electron microprobe analysis: The American Mineralogist, v. 58, p. 920–925.

Kimbel, W.H., Walter, R.C., Johanson, D.C., Reed, K.E., Aronson, J.L., Assefa, Z., Marean, C.W., Eck, G.G., Bobe, R., Hovers, E., Rak, Y., Vondra, C., Yemane, T., York, D., Chen, Y., Evensen, N.M., and Smith, P.E., 1996, Late Pliocene *Homo* and Oldowan tools from the Hadar Formation (Kada Hadar Member), Ethiopia: Journal of Human Evolution, v. 31, p. 549–561, doi: 10.1006/jhev.1996.0079.

Lahitte, P., Gillot, P.Y., Kidane, T., Courtillot, V., and Bekele, A., 2003, New age constraints on the timing of volcanism in central Afar, in the presence of propagating rifts: Journal of Geophysical Research, v. 108, no. B2, p. 2123, doi: 10.1029/2001JB001689.

Quade, J., Levin, N., Semaw, S., Strout, D., Renne, P., Rogers, M., and Simpson, S., 2004, Paleoenvironments of the earliest stone toolmakers, Gona, Ethiopia: Geological Society of America Bulletin, v. 116, p. 1529–1544, doi: 10.1130/B25358.1.

Quade, J., Levin, N.E., Simpson, S.W., Butler, R., McIntosh, W.C., Semaw, S., Kleinsasser, L., Dupont-Nivet, G., Renne, P., and Dunbar, N., 2008, this volume, The geology of Gona, Afar, Ethiopia, *in* Quade, J., and Wynn, J.G., eds., The Geology of Early Humans in the Horn of Africa: Geological Society of America Special Paper 446, doi: 10.1130/2008.2446(01).

Reed, K.E., Lockwood, C.A., and Arrowsmith, J.R., 2003, Faunal comparison between the Middle Ledi and Hadar hominin sites, Ethiopia: Time, landscape, and depositional environment: American Journal of Physical Anthropology, p. 176–177.

Renne, P.R., Swisher, C.C., III, Deino, A.L., Karner, D.B., Owens, T.L., and DePaolo, D.J., 1998, Intercalibration of standards, absolute ages and uncertainties in ^{40}Ar/^{39}Ar dating: Chemical Geology, v. 145, p. 117–152, doi: 10.1016/S0009-2541(97)00159-9.

Semaw, S., Renne, P., Harris, J.W.K., Feibel, C.S., Bernor, R.L., Fesseha, N., and Mowbray, K., 1997, 2.5-million-year-old stone tools from Gona Ethiopia: Nature, v. 385, p. 333–336, doi: 10.1038/385333a0.

Sturchio, N.C., Dunkley, P.N., and Smith, M., 1993, Climate-driven variations in geothermal activity in the northern Kenya Rift valley: Nature, v. 362, p. 233–234, doi: 10.1038/362233a0.

Taieb, M., and Tiercelin, J.J., 1979, Sédimentation Pliocène et paléoenvironnements de rift: Exemple de la formation à Hominidés d'Hadar (Afar, Éthiopie): Bulletin de la Société Géologique de France, v. 21, p. 243–253.

Taieb, M., Johanson, D.C., Coppens, Y., and Aronson, J.L., 1976, Geological and palaeontological background of Hadar hominid site, Afar, Ethiopia: Nature, v. 260, p. 289–293, doi: 10.1038/260289a0.

Tesfaye, S., Harding, D.J., and Kusky, T.M., 2003, Early continental breakup boundary and migration of the Afar triple junction, Ethiopia: Geological Society of America Bulletin, v. 115, p. 1053–1067, doi: 10.1130/B25149.1.

Tiercelin, J.J., 1986. The Pliocene Hadar Formation, Afar depression of Ethiopia, in Frostick, L.E., Renaut, R.W., Reid, I., and Tiercelin, J.J., eds., Sedimentation in the African Rifts: Oxford, UK, Blackwell Scientific, p. 221–240.

Trauth, M.H., Maslin, M.A., Deino, A., and Strecker, M.R., 2005, Late Cenozoic moisture history of East Africa: Science, v. 309, no. 5743, p. 2051–2053, doi: 10.1126/science.1112964.

Walter, R.C., 1981, Volcanic History of the Hadar Early-Man Site and Surrounding Afar Region of Ethiopia [Ph.D. thesis]: Cleveland, Ohio, Case Western Reserve, 426 p.

Walter, R.C., 1994, Age of Lucy and the First Family: Single-crystal ^{40}Ar/^{39}Ar dating of the Denen Dora and lower Kada Hadar Members of the Hadar Formation, Ethiopia: Geology, v. 22, p. 6–10, doi: 10.1130/0091-7613 (1994)022<0006:AOLATF>2.3.CO;2.

Walter, R.C., and Aronson, J.L., 1982, Revisions of K/Ar ages for the Hadar hominid site, Ethiopia: Nature, v. 296, p. 122–127, doi: 10.1038/296122a0.

White, T.D., Suwa, G., and Asfaw, B., 1994, *Australopithecus ramidus*, a new species of early hominid from Aramis, Ethiopia: Nature, v. 371, p. 306–312, doi: 10.1038/371306a0.

Wolfenden, E., Ebinger, C., Yirgu, G., Deino, A., and Ayalew, D., 2004, Evolution of the northern Main Ethiopian rift: Birth of a triple junction: Earth and Planetary Science Letters, v. 224, p. 213–228, doi: 10.1016/j.epsl.2004.04.022.

Wynn, J.G., Alemseged, Z., Bobe, R., Geraads, D., Reed, D., and Roman, D.C., 2006, Geological and palaeontological context of a Pliocene juvenile hominin at Dikika, Ethiopia: Nature, v. 443, p. 332–336, doi: 10.1038/nature05048.

Wynn, J.G., Roman, D.C., Alemseged, Z., Reed, D. Geraads, D., and Munro, S., 2008, this volume, Stratigraphy, depositional environments, and basin structure of the Hadar and Busidima Formations at Dikika, Ethiopia, in Quade, J., and Wynn, J.G., eds., The Geology of Early Humans in the Horn of Africa: Geological Society of America Special Paper 446, doi: 10.1130/2008.2446(04).

MANUSCRIPT ACCEPTED BY THE SOCIETY 17 JUNE 2008

Depositional environments and stratigraphic summary of the Pliocene Hadar Formation at Hadar, Afar Depression, Ethiopia

Christopher J. Campisano*
Institute of Human Origins, Arizona State University, P.O. Box 874101, Tempe, Arizona 85287-4101, USA

Craig S. Feibel*
Department of Anthropology, Rutgers University, 131 George Street, New Brunswick, New Jersey 08901-1414, USA, and
Department of Geological Sciences, Rutgers University, 610 Taylor Road, Piscataway, New Jersey 08854-8066, USA

ABSTRACT

The Pliocene Hadar Formation (Ethiopia) preserves a rich geological and paleontological record germane to our understanding of early hominin evolution. At the Hadar Research Project area, ~155 m of Hadar Formation strata span the interval from ca. 3.45 to 2.90 Ma and consist of floodplain paleosols (dominantly Vertisols), fluvial and deltaic sands, and both pedogenically modified and unmodified lacustrine clays and silts. Clays and silts constitute the majority of the Hadar sediments. In the absence of clear lacustrine indicators, most of these fine-grained sediments are interpreted as fluvial floodplain or delta-plain deposits that exhibit varying degrees of pedogenic modification. Lacustrine and lake-margin deposits are represented by laminated mudstones, gastropod coquinas, limestones, and certain pedogenically modified and unmodified strata preserving gastropods, ostracods, and aquatic vertebrate remains. Most sands can be attributed to channel and point-bar deposits of a large-scale meandering river system or associated crevasse-splay and distributary-channel deposits.

Fluvial-deltaic deposition predominated at Hadar. The lacustrine depocenter was located east and northeast of Hadar, but lacustrine transgressions into the region were a regular occurrence. Evidence presented here suggests that during lacustrine-dominated intervals, lake water depths at Hadar were most likely relatively shallow and included repeated regression events across a low-gradient shoreline. Vertebrate remains at Hadar are disproportionately recovered from fluvial and deltaic sands and silts. This is most likely a taphonomic effect related to the low preservation potential of bones in Vertisols, which are common at Hadar, as opposed to their original distribution across the paleolandscape.

Keywords: *Australopithecus afarensis*, hominin, paleoenvironment, sedimentology, rift valley.

---

*Campisano, corresponding author: campisano@asu.edu; Feibel: feibel@rci.rutgers.edu.

INTRODUCTION

The Pliocene site of Hadar, Ethiopia, is best known for its abundant remains of fossil mammals, particularly the early hominin *Australopithecus afarensis* (e.g., Johanson and Taieb, 1976; Johanson et al., 1978, 1982; Gray, 1980; Kimbel et al., 1994). Geological investigations at Hadar and the surrounding area have continued hand-in-hand with paleoanthropological and paleontological ventures for more than three decades and have provided a wealth of information pertinent to early hominin evolution. Whereas more recent publications on the Hadar geology have focused on updating the region's chronological framework (e.g., Walter and Aronson, 1993; Walter, 1994; Kimbel et al., 1996), more general geologic overviews (e.g., Aronson and Taieb, 1981; Tiercelin, 1986) have not kept pace. Building upon previously published information, this study presents an updated discussion of depositional environments for Hadar (ca. 3.45–2.9 Ma) across space and through time.

Although only representing about a half-million years, the fluviolacustrine deposits of the Hadar Formation at Hadar preserve a detailed geologic record. The rate of sediment deposition at Hadar is roughly twice that of other East African hominin localities of comparable age and is largely responsible for the rich paleontological record preserved. For example, during the 3.45–2.90 Ma interval, the average sedimentation rate at Hadar was 30 cm/k.y. (Campisano and Feibel, 2007), whereas rates for the Shungura, Nachukui, and Koobi Fora Formations in the Turkana Basin ranged from 13 to 18 cm/k.y. (de Heinzelin and Haesaerts, 1983; Harris et al., 1988; Feibel, 1988). This high-resolution record provides a strong framework for understanding the paleoenvironmental context of *A. afarensis*. Additionally, recent investigations have been successful in integrating the lateral and temporal variation observed in depositional environments with their paleontological assemblages to provide a more detailed understanding of *A. afarensis* habitats (Campisano, 2007; Campisano and Reed, 2007).

Geologic Setting and Stratigraphic Framework

The Hadar Research Project area is located along the Awash River in the Afar Depression of the Ethiopian Rift, ~300 km northeast of Addis Ababa in the Afar Regional State of Ethiopia (~11°06′N, 40°35′E) (Fig. 1). The area's name is derived from "Kada Hadar," the local Afar name for the large ephemeral tributary of the Awash that cuts through the central exposures in the area. The current Hadar Research Project area is ~100 km², slightly smaller than its pre-1990s size, as some regions have since been allocated to separate research projects in adjoining areas. At present, the Hadar Research Project area extends east to the Hurda Wadi (boundary with the Ledi-Geraru Research Project), south to the Awash River (boundary with the Dikika Research Project), west to the topographic divide with the Gona

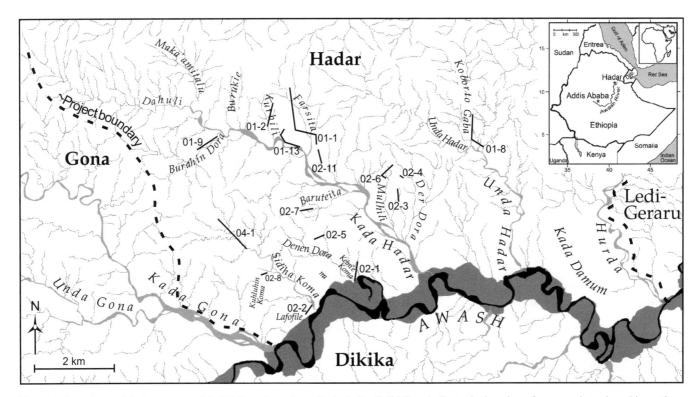

Figure 1. Location and drainage map of the Hadar paleoanthropological site. Solid lines indicate the location of measured stratigraphic sections used in this study (section numbers have been abbreviated). Dashed lines indicate approximate boundaries between Hadar and adjacent project areas. Base map is courtesy of G. Eck.

Wadi (boundary with the Gona Paleoanthropological Research Project), and north to the Ledi Wadi (Fig. 1).

Hadar is situated within a unique geomorphological setting that is commonly referred to as the Afar Depression (a.k.a. Afar Triangle or Afar triple junction), which was formed by extension of the Nubian, Arabian, and Somalian plates (McKenzie et al., 1970). This triple rift (RRR) junction marks the intersection of the spreading ridges forming the east-trending Gulf of Aden Rift, the northwest-trending Red Sea Rift, and the northeast-trending Ethiopian Rift. The Afar Depression is roughly 200,000 km² in area and is bounded to the west by the Ethiopian Escarpment and Plateau, to the southeast by the Somalian Escarpment and Plateau, and to the northeast and east by the relatively low-lying Danakil block/horst (a microplate) and Ali-Sabieh block, respectively (Mohr, 1975; Beyene and Abdelsalam, 2005). Hadar is located southwest of the Tendaho-Goba'ad discontinuity, which separates the Ethiopian Rift system from the Red Sea Rift system (Tesfaye et al., 2003; Wolfenden et al., 2004; Beyene and Abdelsalam, 2005). Most of the Afar Depression lies between 100 and 800 m in elevation, and elevations decrease toward the northeast; Hadar lies at ~500 m above sea level.

Sediments of the Hadar Formation (Taieb et al., 1976) are a component of an ill-defined sedimentary basin, ~60 km wide and 150 km long, that lies parallel and adjacent to the Ethiopian Escarpment, its western boundary, and extends eastward to the Guda, Magenta, and Galalu horsts. This basin is referred to as the west-central Afar sedimentary basin (Taieb et al., 1976), and the sediments form part of the Awash Group (Kalb et al., 1982), which subsumed the Central Afar Group of Taieb et al. (1972).

As exposed in the Hadar project area, the Pliocene Hadar Formation is composed of close to 155 m of fluviolacustrine sediments (Fig. 2) that preserve a high-resolution record of environmental conditions and change from ca. 3.45 to ca. 2.9 Ma. A regional angular unconformity separates the Hadar Formation from the overlying Busidima Formation (Quade et al., 2004; Wynn et al., this volume). The Pliocene-Pleistocene Busidima Formation is poorly expressed at Hadar (Campisano and Feibel, this volume), but it contains some of the earliest evidence of the genus *Homo* and Oldowan stone tools (e.g., Afar Locality [A.L.] 666 and 894) (Kimbel et al., 1996, 1997; Hovers, 2003).

The first discussion of the Hadar Formation was by Taieb et al. (1972), who divided their study interval of 80+ m of sediment into seven units, labeled A–G. This stratigraphic framework was replaced when the Hadar Formation was formally defined by Taieb et al. (1976), and their lithostratigraphic framework has essentially remained unchanged since. Using laterally extensive tephra horizons, the Hadar Formation is divided into four members: the Basal Member, the Sidi Hakoma Member, the Denen Dora Member, and the Kada Hadar Member. These members have been further subdivided into submembers (e.g., DD-1, -2, -3) on the basis of other lithostratigraphic markers, primarily sand bodies, and some submembers have been divided into specific units (e.g., DD-3s) (Fig. 2). Due to the lateral variation of the deposits, certain submembers or units are not consistently preserved

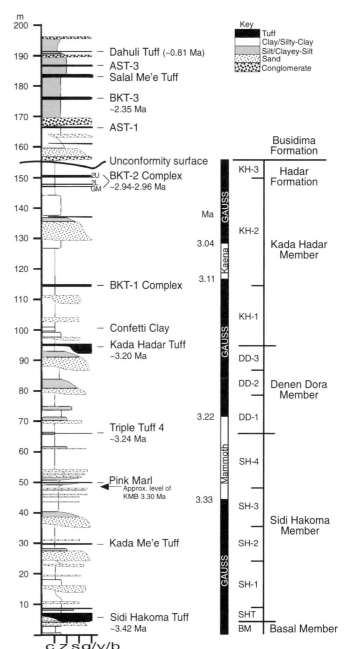

Figure 2. Composite stratigraphic section of the Hadar and Busidima Formations at Hadar. Tuffs and major marker beds are labeled alongside the section. ^{40}Ar/^{39}Ar dates and paleomagnetic transitions are from Schmitt and Nairn (1984); Walter and Aronson (1993); Renne et al. (1993); Walter (1994); Kimbel et al. (1994, 1996, 2004); Semaw et al. (1997); and Campisano (2007). Previously published ^{40}Ar/^{39}Ar dates have been recalculated to reflect the updated age of the Fish Canyon sanidine standard (see text footnote 2).

and/or identifiable throughout the entire region. Although they do not always correspond to sedimentary cycles, and to an extent may even be time transgressive, these submembers serve as the stratigraphic units to which paleontological specimens are assigned and the framework in which paleoenvironmental information has been analyzed (e.g., Johanson et al., 1978; Gray, 1980; Campisano, 2007; Reed, 2008).

The sediments at Hadar are essentially flat-lying, with a general dip of only 2° to the north-northwest in the central region of the project area, but up to 6° in the east near Kada Damum. As a result of this dip and headward erosion, the oldest of the Hadar sediments are found close to the Awash River, and progressively younger sediments are exposed toward the north. Strata are typically exposed on knolls and along ridges between 25 and 45 m in height and are relatively uninterrupted by faulting. With the exception of three north-south–striking normal faults with 5–30 m offset (down-dropped to the east), most faults in the area are minor, with less than 2 m offset. Most of the descriptions and interpretations presented in this paper are consistent with, but expand upon, previous publications and research, specifically Taieb and Tiercelin (1979, 1980), Aronson and Taieb (1981), Tiercelin, (1986), and Yemane (1997), the latter of which also includes sedimentological and petrographic analyses of many of these deposits.

DEPOSITIONAL ENVIRONMENTS OF THE HADAR FORMATION

The first conceptual model of depositional environments at Hadar was a three-phase model developed by Taieb and Tiercelin (1979, 1980; see also Johanson et al., 1978, 1982). This model was based on high-energy streams flowing eastward from the Ethiopian Plateau toward a lacustrine depocenter (Lake Hadar), where small streams bifurcated across low-relief floodplains into marshy zones and larger streams created distributary networks that built small deltas into the lake. The differences between these phases were primarily based upon the extent of the lake, which was relatively restricted in phase I (Basal Member to SH-3), much more expansive but with fluctuations in phase II (SH-4 to the Confetti Clay), and gave way to terrestrial/deltaic deposits that ultimately filled in the basin (above the Confetti Clay). Subsequent work by Aronson and Taieb (1981) focused the environmental interpretations less on the lake, and more on the fluvial deposits of the Hadar Formation. Their model consisted primarily of braided streams flowing across a delta plain and into the lake, with three lacustrine transgressions (late Basal Member, late Sidi Hakoma Member, and early Kada Hadar Member times). Details of the depositional environments were expanded by Tiercelin (1986), which more accurately documented the complexity and variation of depositional environments through time.

Yemane's (1997) much-needed update on the depositional environments of the Hadar Formation utilized facies associations based on lithofacies codes developed by Miall (1977, 1978). Yemane (1997) described these depositional environments in terms of four facies associations that represent: (1) alluvial fan, braided stream, floodplain, and braid delta deposits, (2) meandering stream and associated floodplain deposits, (3) marginal and shallow lacustrine deposits, and (4) delta-plain deposits, including distributary channel, marsh, and floodplain deposits.

Although a characterization of all the Hadar sediments into a few general categories is useful for some endeavors, doing so may mask or overgeneralize potentially important differences between similar lithologic packages or lateral variation that would otherwise be treated as a single facies. In contrast to previous studies of depositional environments that were organized via facies descriptions (e.g., Tiercelin, 1986; Yemane, 1997), the descriptions and discussion in this study progress chronologically through the formation and pay particular attention to the lateral variation within and between depositional environments.

Basal Member

The Basal Member (BM) is defined as those sediments overlying Dahla Series basalts to the base of the Sidi Hakoma Tuff (Taieb et al., 1978; Wynn et al., this volume). Although the Basal Member is more than 60 m thick (Tiercelin, 1986; Wynn et al., 2006), the majority of the member is exposed south of the Awash River, presently under study by the Dikika Research Project, and extends to older than 3.80 Ma (Wynn et al., 2006). Up to 65 m of Basal Member deposits have also been reported from the As Aela area of Gona (Quade et al., this volume). North of the Awash River in the Hadar region, the upper ~10 m of the Basal Member are exposed almost exclusively on the flanks of Keini'e Koma[1] and Lafofile Koma just above the modern Awash floodplain, preserving two distinct depositional environments (Fig. 3).

The base of this exposure is dominantly massive or laminated brown clays and silty clays preserving fish bone fragments that represent primary lacustrine deposition, consistent with interpretations from the Dikika region (Wynn et al., 2006). Two gastropod limestone units (5–10 cm thick each), ostracods, and fish nest structures preserved toward the top of this lacustrine sequence suggest a shallow-lake or marginal/shoreline environment and a period of regression. The virtually monospecific ostracod fauna (*Cyprideis* sp. SH1) from this lacustrine phase indicates an aquatic environment of variable salinity, ranging as high as mesohaline, and probably with seasonal variation (Peypouquet et al., 1983; Tiercelin, 1986).

The upper part of the Basal Member at Keini'e Koma is dominated by a large, well-sorted, trough cross-bedded sand body, the Basal Member sand (BM-2s). On the south side of Keini'e Koma, the BM-2s reaches a thickness of ~4.6 m where it has eroded down to 1.5 m below the level of the limestones. The BM-2s itself is subsequently channeled into by the overlying Sidi Hakoma Tuff channel, which marks the base of the Sidi Hakoma Member (Figs. 3 and 4A). This erosional sand represents a major channel occupying the area and marks a dramatic shift from the lacustrine-dominated units below. The Basal Member sands were originally attributed to a meandering fluvial system, probably narrow-channel meandering streams based on

---
[1]This study uses the American English approximation of Afar geographic terminology compiled by Dr. G. Eck and several Afar field assistants and translators during the 1990–2001 field seasons (Appendix 1 *in* Campisano, 2007).

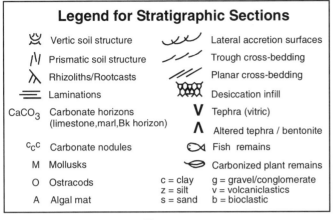

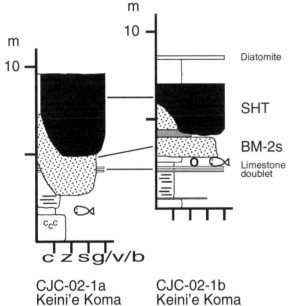

Figure 3. Stratigraphic sections of the Basal Member from Hadar. Solid lines indicate known correlations. See Figure 2 for the geographic locations of sections. SHT—Sidi Hakoma Tuff.

their geometry (e.g., width to thickness ratio) (Tiercelin, 1986; Yemane, 1997). However, recent work by the Dikika Research Project has suggested that the Basal Member sands preserved directly south of Hadar are part of a delta-channel system on a delta plain (Alemseged et al., 2005; Wynn et al., 2006). Unfortunately, determination of the difference between the different types of sand channels at Hadar is difficult in such a limited exposure. Based on its thickness, sedimentary structure, and erosional nature, if the BM-2s was part of a delta-channel system, it likely represented a large channel on an upper delta plain.

Sidi Hakoma Member

The Sidi Hakoma Member is defined as those sediments from the base of the Sidi Hakoma Tuff (SHT) to the base of Triple Tuff 4 (TT-4), and it has been divided into four submembers (Fig. 2). The thickness of the Sidi Hakoma Member increases from west to east. Taieb and Tiercelin (1980) reported a thickness of 45 m in the Kabara Wadi (south of the Awash across from the Sidiha Koma Wadi), 60 m in the central part of Hadar (exactly matching measurements made in this study), 90 m in the Unda Hadar Wadi, and up to 130 m in the Hurda Wadi. The majority of this thickness difference occurs in the lower half of the Sidi Hakoma Member below the Kada Damum Basalt and has been attributed to a hypothetical syndepositional growth fault (Aronson and Taieb, 1981). Similar syndepositional growth faults are also recorded from Basal Member exposures at Dikika (Wynn et al., this volume). The section described here is based on the well-exposed area of outcrops between Keini'e Koma, Lafofile, and the Denen Dora Ridge (Fig. 5).

SH-1 Submember

The SH-1 submember is defined as those sediments from the base of the Sidi Hakoma Tuff (SHT) to the base of the SH-2 sand (Fig. 2), and it preserves both lacustrine and fluvial-deltaic depositional environments.

At Keini'e Koma, the Sidi Hakoma Tuff is preserved as a vitric channel fill, ~3 m thick, and it consists of two geochemically identical tephra units dated to ca. 3.42 Ma[2] (Walter and Aronson, 1993). The basal unit is a fine-grained, light-gray tephra with sets of climbing ripples and cross-bedding. The upper unit, which cuts down into the lower, is a more coarse-grained, darker gray tephra with trough and planar cross-bedding and soft-sediment deformation structures. Immediately lateral to and at a stratigraphic level near the top of the channel edge, there is a thick (>1 m) vitric Sidi Hakoma Tuff deposit that spreads out northward from Keini'e Koma, but it is still rather narrowly constrained, and it contains several alternating units of coarse- and fine-grained tephra. In most areas, however, the Sidi Hakoma Tuff is preserved as a thin (~20 cm), but laterally extensive pinkish white bentonite.

The widespread occurrence of the Sidi Hakoma Tuff allows the depositional history of this easily correlatable deposit to be reconstructed across the greater Hadar region. The Sidi Hakoma Tuff channel-fill deposit at Keini'e Koma occupies the same geographic location as the BM-2s channel (Figs. 3 and 4A), suggesting that these nested channel deposits may be related and/or took advantage of preexisting channel forms. With the exposures preserved at Hadar, it is not possible to determine whether the Basal Member and Sidi Hakoma Tuff channel systems represent a continuum of the same system at different times or different channel episodes altogether. Regional mapping of depositional environments at Dikika by Wynn et al. (2006) has demonstrated that the Sidi Hakoma Tuff is also associated with shallow lacustrine, offshore lacustrine delta, subaqueous delta channel, and upper delta-plain fluvial channel environments.

---

[2]Previously published ^{40}Ar/^{39}Ar ages reported in this study have been increased by approximately 0.65% to reflect the revised age of the monitor mineral used (Fish Canyon sanidine) from 27.84 to 28.02 Ma (Renne et al., 1998); this correction is supported by recent geochronological analysis of select Hadar tephra (Campisano, 2007).

Figure 4. (A) Relationship between the Sidi Hakoma Tuff and the BM-2 sand, south side of Keini'e Koma. (B) Calcite-sand crystal, SH-2 sand, Keini'e Koma. (C) Flamingo nest constructed from gastropod coquina, Denen Dora Ridge. Length of pen is ~13.5 cm. (D) Kada Hadar Tuff cutbank surfaces (outlined), Farsita Wadi. Channel width is ~25 m. (E) Foreset (FS) and topset (TS) beds of the Gilbert-style delta, Kuhluhin Koma. (F) BKT-2U (with subunits indicated) overlying desiccation-cracked mudstone, Koborto Gaba Wadi. Length of trowel is ~25.5 cm.

The two tephra units within the channel suggest two different phases of deposition, perhaps representing different seasonal or yearly flooding events that washed tephra from the catchment area into the fluvial system. The location of the channel edge preserved at Keini'e Koma, combined with the shape of the channel observed in plan view, suggests that it represents the northern cutbank edge of a meander. Similarly, based on the location and shape of the northward-spreading vitric deposit, we interpret it as a crevasse-splay deposit where floodwaters breached the channel at the position of the cutbank. The coarse- and fine-grained units in this deposit may represent a series of distinguishable minor and major flooding/breaching events, while the extensive pinkish white bentonite is interpreted as more typical floodplain deposition from the channel.

Sediments between the Sidi Hakoma Tuff and the SH-1 sand are dominantly silty clays and clays with distinctive colors that are traceable between the Keini'e Koma and Lafofile regions. The Sidi Hakoma Tuff bentonite is overlain by a sequence of olive-green silty clays and an ~30-cm-thick diatomaceous shale preserved at Keini'e Koma and Lafofile (Fig. 5) as well as in the Dikika region (Alemseged et al., 2005; Wynn et al., 2006). The diatomaceous shale, located up to a meter above the Sidi Hakoma Tuff bentonite, has previously been reported as a second bentonite unit (e.g., Gray, 1980; Walter, 1981; Johanson et al., 1982; many published stratigraphic sections). However, Tiercelin (1986, p. 225–226) mentions a "laterally extensive whitish gray shale 30 cm thick" half a meter above the Sidi Hakoma Tuff containing the diatom *Aulacoseira* (formerly *Melosira*) *granulata*, which we have confirmed. Overlying the diatomite, this sequence transitions upward from laminated olive-green and olive-gray clay and sandy clay to blue-gray then brown Vertisol mudstones.

The post–Sidi Hakoma Tuff deposits are interpreted as either a shallow lacustrine or ephemeral floodplain lake interval, where *Aulacoseira granulata* is an indicator of turbid, but fresh, waters (Tiercelin, 1986). Sediments between the diatomite and SH-1s appear to record the regression of the lake to more terrestrial deposition as noted by the upsection increase in Vertisol development as well as the change from reduced clays (olive-green and blue-gray) to oxidized brown floodplain clays. The lack of clear beach deposits (sand and gastropods) in this lacustrine interval at Hadar suggests that the post–Sidi Hakoma Tuff lake had a relatively flat shore consisting primarily of mud.

The SH-1 sand is not a laterally continuous unit, and it is typically preserved as small channel bodies, 1–1.5 m thick.

Between the SH-1s and the SH-2s are primarily brown clays and silts with varying degrees of Vertisol development and a few discontinuous thin sand lenses (Fig. 5). As with the Basal Member sands, Yemane (1997) classified the SH-1s as a narrow to sheetlike sand body from a meandering stream, whereas proposed correlative units in the Dikika region are interpreted as part of a subaerial distributary delta system (Wynn et al., this volume). Although both interpretations are generally similar, the pattern of SH-1s channel distributions at Hadar appear to be more consistent with distributary channels across an exposed delta plain as suggested by Wynn et al. (2006, this volume). The calcareous silts and clays between the SH-1s and the SH-2s represent floodplain deposits. These deposits are slightly coarser at Lafofile (silts and sandy silts versus silty clays) and contain abundant carbonate nodules and rhizoliths compared to Keini'e Koma (Fig. 5). This may indicate that there was a more proximal floodplain environment around Lafofile, perhaps with more vegetation, whereas the region around Keini'e Koma was a distal floodplain environment.

SH-2 Submember

The SH-2 submember is defined as those sediments from the base of the SH-2 sand to the base of the SH-3 sand (Fig. 2), and it preserves both fluvial and palustrine/lacustrine depositional environments.

The base of the SH-2 submember is marked by the SH-2 sand, a coarse-grained and upward-fining unit that is more than 3 m thick with well-preserved cross-bedding and lateral accretion surfaces. It appears to be the most laterally continuous of the Sidi Hakoma sands, it preserves abundant fossilized faunal remains, and it also contains distinctive calcite-sand megacrystals produced by calcite crystals incorporating and cementing sand during growth (C. Vondra, 2001, personal commun.; Fig. 4B). Yemane (1997) classified the SH-2s into the same category as the SH-1s (narrow to sheetlike sand body from a meandering stream), but based on its thickness, lateral extent, and well-preserved lateral accretion surfaces, we interpret the SH-2s to represent channel and point-bar deposits of a much larger and more energetic meandering fluvial system compared to the SH-1s system. The SH-2s fines upward into fine sands, silts, and Vertisol mudstones representative of floodplain deposits.

Approximately a meter above the SH-2s, there is the ca. 3.36 Ma Kada Me'e Tuff (KMT) (Campisano, 2007) (Fig. 5B). The Kada Me'e Tuff complex is ~45 cm thick and consists primarily of olive-gray laminated claystone interbedded with up to three thin (1–3 cm) pink and white bentonites, a crystal-lithic bentonite, and a thin (2 cm) sand with abundant unionid bivalves (B. Van Bocxlaer, 2008, personal commun.). Taieb and Tiercelin (1980) and Tiercelin (1986) described a 20–40-cm-thick coal/lignite layer in lacustrine sediments between the SH-2s and SH-3s in the Hurda Wadi as well as a correlative 20 cm unit of green shale interbedded with thin coal layers north of Keini'e Koma, which correlates to the thin darker units within the Kada Me'e Tuff complex. In the Denen Dora region, a distinct, very

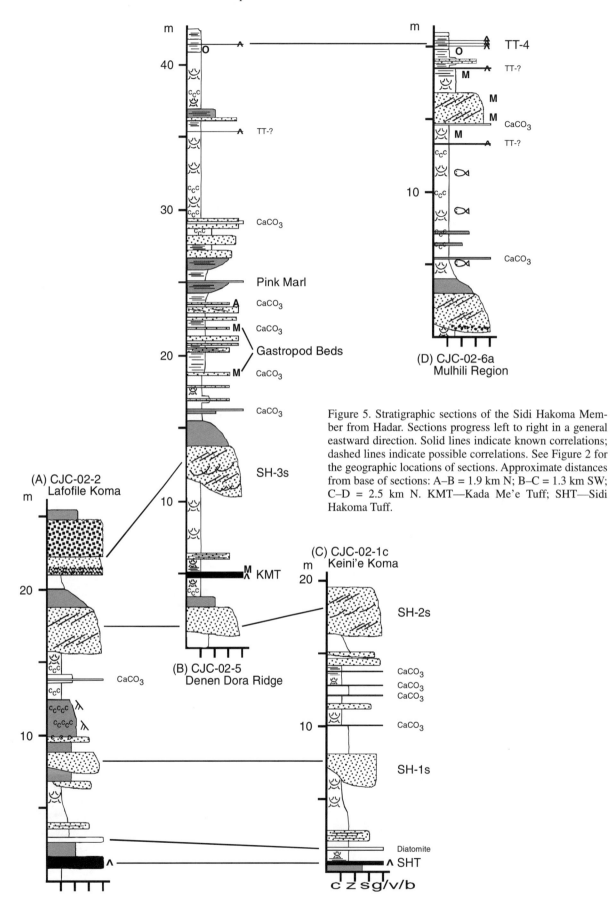

Figure 5. Stratigraphic sections of the Sidi Hakoma Member from Hadar. Sections progress left to right in a general eastward direction. Solid lines indicate known correlations; dashed lines indicate possible correlations. See Figure 2 for the geographic locations of sections. Approximate distances from base of sections: A–B = 1.9 km N; B–C = 1.3 km SW; C–D = 2.5 km N. KMT—Kada Me'e Tuff; SHT—Sidi Hakoma Tuff.

dark gray (almost black) Vertisol clay, 3–4.5 m thick, is preserved between the Kada Me'e Tuff and the SH-3s (Fig. 5B).

Sediments associated with the Kada Me'e Tuff mark a lacustrine transgression following the SH-2 fluvial system. Evidence that the depocenter of Lake Hadar was located to the east is supported by the thickness of the lignite layer and associated palustrine and lacustrine deposits there, as described by Tiercelin (1986), which include an abundant gastropod fauna and ostracods. Ostracods below the lignite in the Hurda Wadi consists of two species, *Cyprideis* sp. SH1 and *Cyprideis* sp. KH1, the morphology of which (tubercles) implies oligohaline waters (Peypouquet et al., 1983). Peypouquet et al. (1983) suggested that the sublignite depositional environment in the Hurda was a paludal environment characterized by a high input of river water and dissolved organic matter. Natrojarosite in the lignite layer is consistent with formation in an anoxic, reducing, and acidic environment without clastic input (Tiercelin, 1986). Following lignite deposition in the Hurda region, the abundance of *Cyprideis* sp. KH1 (without tubercles) increases, suggesting a transition from marsh to lacustrine environments and salinity reaching mesohaline levels (Peypouquet et al., 1983). A portion of this paludal or perilacustrine environment was present at least as far west as the present Denen Dora Wadi, if not further. The very dark gray Vertisol between the Kada Me'e Tuff and the SH-3s in the Denen Dora region may correlate with the more continuously lacustrine strata to the east. However, the thick and strongly developed Vertisol indicates a high rate of pedogenesis (e.g., Kraus, 1999), and possibly a relatively substantial period of subaerial exposure.

Since the Kada Me'e Tuff–lignite complex and overlying dark gray clays are not present in all exposures between the SH-2s and SH-3s (such as at Lafofile), this may suggest a more localized ponding or embayment of the paludal/lacustrine environment in the Denen Dora region. The thickness between the SH-2s and the SH-3s varies significantly between Keini'e Koma, Lafofile, and Denen Dora (Fig. 5), as does the relative position of the Kada Me'e Tuff between the SH-2s and SH-3s. Assuming that the Kada Me'e Tuff is an isochronous unit deposited on relatively flat topography, this may reflect the time-transgressive nature of the SH-2s as it meandered across the Hadar landscape as well as the erosional nature of SH-3s.

SH-3 and SH-4 Submembers

The SH-3 submember includes sediments from the base of the SH-3 sand to the upper gastropod layer below the Pink Marl and the Kada Damum Basalt (Fig. 2), and it preserves both fluvial and beach/nearshore depositional environments. The SH-4 submember includes sediments from the upper gastropod layer to the base of TT-4, and it preserves beach/nearshore and lacustrine depositional environments. A clear boundary between the SH-3 and SH-4 was never formalized. Johanson et al. (1978) included the gastropod units at the base of the SH-4 above a thin, but discontinuous sand (the SH-4 sand), whereas Gray (1980) drew the division at the top of the gastropod horizons. Although we adopt the usage of Gray (1980) for ease in correlating, the lower half of the SH-4 submember is essentially a continuation of upper SH-3 environments, and they will be discussed together.

In the Denen Dora region, the SH-3 sand ranges from 2 to 3.5 m thick, fines upward from very coarse-grained to pebbly sand, and displays cross-bedding and lateral accretion surfaces (Fig. 5B). At Lafofile, the general thickness of the SH-3s is the same, but the sequence contains conglomerate units up to 2 m thick with cobbles up to 10 cm in diameter either interstratified within or capping coarse-grained sand (Fig. 5A). At Lafofile, unlike the exposures in the Denen Dora region, the SH-3s is separated from the SH-2s by only a few meters of overbank deposits.

As with the SH-2 sand, we interpret the laterally continuous SH-3 sand as representing channel and point-bar deposits of a meandering river. Based upon the geometry and lateral extent of the SH-3s (from west of the Gona Wadi to the Unda Hadar), Yemane (1997) classified it as a sheetlike sand body from a meandering stream. The conglomeratic facies of the SH-3s could be related to a change in topography or slope of the river channel that transported and concentrated a cobble bed load in this location. However, as a similar situation is repeated in the DD-3s and lower Kada Hadar Member, it more likely represents a change in localized sediment source. Braided streams transporting alluvial-fan material from the nearby western margin could have served as tributaries to the meandering axial system. The intersection of these two systems west of Hadar would have introduced cobble material into the SH-3s fluvial system, while subsequent eastward reworking of this material would have led to conglomerate deposition at Hadar. Similar to the SH-2 system, the SH-3 sand is overlain by pedogenically modified mudstones representative of overbank floodplain deposits.

In the region where the SH-3 and SH-4 submembers are best exposed, between the Denen Dora Ridge and Lafofile, SH-3s overbank deposits are overlain by a thick package (~10–12 m) of clays and silty clays alternating with thin silt and sand units. This interval also includes two to three gastropod coquinas composed of *Melanoides tuberculata* and *Bellamya unicolor* (commonly referred to as the "Gastropod Beds" at the SH-3/SH-4 boundary), an algal encrusted carbonate mat, and an extensive marl marker bed known as the "Pink Marl" (Fig. 5B). Numerous flamingo nest structures constructed of the gastropod packed sandstone were also found at the level of the Gastropod Beds (Fig. 4C).

This post–SH-3s sequence marks the beginning of the Hadar Formation's most significant lacustrine transgression. The laminated silty clays and silts reflect primary lacustrine deposition, whereas the thin, fine-grained, laminated sands most likely represent clastic input into the lake, perhaps during periods of high rainfall and runoff. Cross-bedded sands, such as those incorporating reworked and fragmented gastropods, represent beach deposits. Shoreline/beach deposits are also indicated by flamingo nest structures and crocodile remains. Oxygen isotope analyses of shells from the Gastropod Beds indicate the least evaporated phase of Lake Hadar, probably reflecting the initial stage of the lacustrine transgression (Hailemichael et al., 2002). Subsequent carbonate horizons such as the algal mat and Pink Marl indicate a transition from shallow/marginal lacustrine to mud flats with desiccation cracks from subaerial exposure.

In the Unda Hadar, a SH-3/SH-4 submember planar crossbedded sand with gastropods at its base may represent a lateral equivalent of the beach deposits further west, but at an earlier period of lake transgression. Also in the Unda Hadar region, the Kada Damum Basalt (= Kadada Moumou) flow erupted into a wet environment (floodplain or marsh) during SH-4 time, ca. 3.30 Ma (Renne et al., 1993), as evident by contact features, including a basal (2–3 cm) chilled zone. The basalt baked sediments to a depth of 2–3 m, and contact metamorphism of an underlying lignite layer suggests a close proximity of the eruption fissure (Walter and Aronson, 1982; Tiercelin, 1986).

The uppermost part of the Sidi Hakoma Member is primarily olive-gray and grayish brown claystone that is laminated or massive with varying degrees of Vertisol development. These claystones contain patches of carbonate nodules, concentrations of gastropods and ostracods, occasional fish bones and scales, and a horizon of septarian concretions. Several thin 2–3 cm bentonites representing the "Triple Tuffs" also occur in these clay horizons including Triple Tuff 4 (TT-4), a ubiquitous white bentonite that marks the base of the Denen Dora Member. Approximately 3 m below TT-4 in the Mulhili/Der Dora region, a planar crossbedded sand body up to 2 m thick contains gastropods (*Melanoides tuberculata*, *Cleopatra bulimoides*, and *Bellamya unicolor*), pelycopods, and in some areas, mammal bones (Fig. 5D).

Despite lacustrine indicators such as gastropods, ostracods, and fish remains, much of this sequence shows clear evidence of subsequent subaerial exposure. For example, in the Denen Dora region, several packages of thick (3–6 m) mudstones with well-developed vertic structure and occasional carbonate nodules are interstratified with thinner packages of laminated, nonpedogenic mudstone (Fig. 5B). These alternating packages of laminated to paleosol mudstones indicate fluctuations from periods of stable lacustrine conditions to periods of lacustrine regression and subaerial exposure that may be climatic in nature. The evidence for pedogenesis of primary lacustrine sediments is given by the periodic preservation of relict horizontal lamination, septarian concretions, and gastropod shells or molds within the Vertisols.

The presence of septarian concretions (5 cm in diameter) within a moderately developed Vertisol in the Denen Dora and Baruteita regions ~2 m below TT-4 is of particular interest. Septarian concretions occur primarily in fine-grained marine deposits and occasionally in some lacustrine mud rocks and fluvial floodplain deposits (Pratt, 2001). The process of crack formation in these nodules is a complex issue that is still debated (Hounslow, 1997; Raiswell and Fisher, 2000; Pratt, 2001). Although the timing of crack formation most likely postdates the pedogenic processes of the Vertisol in which the nodules are found, the physical and chemical properties of these septarian concretions suggest that the primary origin of the uncracked nodules is consistent with a lacustrine environment.

Possible climatically controlled cyclicity may be evident in the upper Sidi Hakoma lacustrine deposits (Campisano and Feibel, 2007). The ~60,000 yr section of lacustrine deposits between the approximate level of the Kada Damum Basalt and TT-4 (3.30–3.24 Ma) in the Mulhili region (Fig. 5D) records three packages of sediment, the lower two of which are identical pair sets. Each of these sets is composed of an identical series of three colored clay sequences that transition upward from dark grayish brown to very dark gray to dark olive-gray. Although unit thicknesses within each sequence are not equal, the thicknesses of corresponding units between the two sequences do match. Pedological analyses have not yet been conducted to identify any other differences among the clay sequences, but it is possible that the color variation represents differences in water-table level related to differences in lake level. The oxidized, brown clays may represent shallower, more oxygenated conditions that transition to the reduced dark gray and olive-gray clays, which may indicate deeper-water conditions. If correct, this section may not only suggest that precessional-scale cycles are recorded at Hadar, but potentially subprecessional-scale cycles as well.

The uppermost Sidi Hakoma sediment package in the Mulhili region (Fig. 5D) preserves nearshore and beach deposits. Beach deposits are indicated by planar-bedded sands containing reworked pelecypods and gastropods. Only a few hundred meters east of these beach deposits, the gastropod assemblage is associated with unconsolidated fine-grained sand and more completely preserved gastropod shells that indicate a slightly more offshore depositional environment. Oxygen isotope analysis of this gastropod assemblage (or its lateral equivalent) indicates a large cyclic variation in $\delta^{18}O$ within individual shells that represents strong seasonal variation (Hailemichael et al., 2002). It has been suggested that during the latter stages of this lacustrine interval, a shallow, partially isolated embayment may have existed at Hadar, which underwent rapid expansions and contractions (Hailemichael et al., 2002).

Denen Dora Member

The Denen Dora Member is defined as those sediments between the base of TT-4 and the base of the Kada Hadar Tuff (KHT), and it has been divided into three submembers (Fig. 2). The thickness of the member is relatively consistent across the Hadar region, with most measurements near 30 m, making it the thinnest of the members (Fig. 6). The member appears thickest in the Denen Dora and Der Dora regions with minor thinning to both the west and east, a pattern also recorded by Tiercelin (1986). The base of the Denen Dora Member was originally defined as the base of the first of three thin bentonitic tuffs referred to as the Triple Tuffs (TT-1, TT-2, TT-3) (Taieb et al., 1976). Subsequent study has demonstrated the presence of at least five, and possibly up to eight, thin tuffs, although their preservation varies significantly across the region (Tiercelin, 1986; Yemane, 1997). As TT-4 (the original TT-2) is the most laterally extensive and distinctive of the tuffs, as well as being the only one that has been dated, it has assumed the position of the base of the Denen Dora Member.

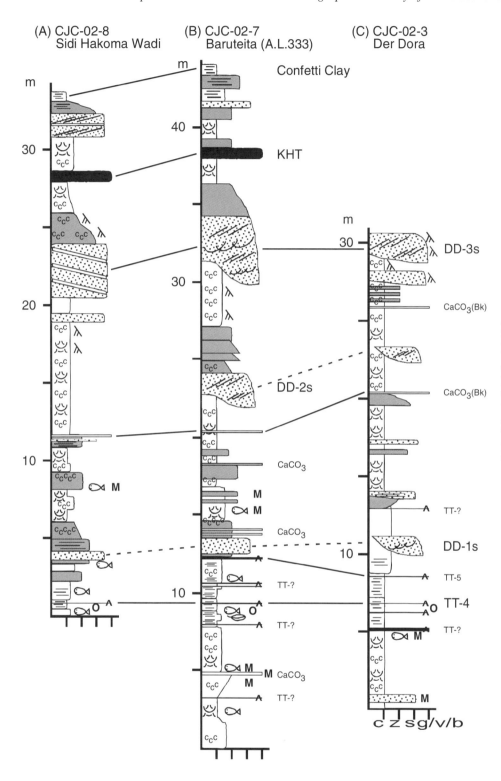

Figure 6. Stratigraphic sections of the Denen Dora Member from Hadar. Sections progress left to right in a general eastward direction. Solid lines indicate known correlations; dashed lines indicate possible correlations. See Figure 2 for the geographic locations of sections. Approximate distances from base of sections: A–B = 1.8 km NE; B–C = 2.3 km E. KHT—Kada Hadar Tuff.

DD-1 Submember

The DD-1 submember was originally proposed for those sediments from the base of TT-4 to the base of the first sand unit (Fig. 2). As noted by Gray (1980), depending on the location, the first sand unit encountered above TT-4 can be either the DD-1 or DD-2 sand. Thus, no formal boundary was given, and there has been a minor degree of overlap between the DD-1 and DD-2 submembers in previous publications (e.g., Gray, 1980; Johanson et al., 1978). Because of this inconsistency, we use the base of the DD-2 sand (or its lateral equivalent) as a more reliable upper boundary for the DD-1 submember in this discussion.

TT-4 is a ubiquitous ~5-cm-thick white bentonite that occasionally preserves sanidine crystals at its base and has been dated to ca. 3.24 Ma (Walter, 1994) and ca. 3.256 (Campisano, 2007). The bentonite is encased in a laminated olive-gray claystone ~1 m thick that weathers in a fissile manner. Ostracods (*Cyprideis* sp. KH1) are also present in the claystone (Peypouquet et al., 1983), with a particularly high concentration 30–50 cm below TT-4 (Fig. 6). This ostracod-bearing shale variably preserves nodules of radiating acicular barite (1–3 cm in diameter), carbonized plant material, fish remains, and beige ovate concretions (2–3 cm in diameter), which have been described as apatite nodules with a nucleus of fish debris (Tiercelin, 1986).

The bottom few meters of the Denen Dora Member above TT-4 are essentially a continuation of the upper Sidi Hakoma Member deposits, and they are often treated as a single unit (SH-4/DD-1) in discussions of paleontological assemblages and depositional environments. These sediments consist of laminated and massive claystone or silty claystone, thin bentonites commonly 1–3 cm thick, as well as a thicker (>5 cm) and relatively laterally extensive pinkish white bentonite referred to as TT-5 (Fig. 5). The clays are sometimes pedogenically modified in areas and occasionally preserve fish remains and gastropods as well as carbonate and septarian nodules.

This basal sequence of the Denen Dora Member records a continuation of lacustrine deposition that commenced during SH-4 time, including a widespread transgression recorded by the ostracod-bearing shale associated with TT-4. The morphology of the *Cyprideis* sp. KH1 ostracod assemblage from the TT-4 shale implies a variable salinity, including mesohaline levels. However, observed dissolution of carapaces indicates that Lake Hadar at TT-4 time was less alkaline and had a higher concentration of ions compared to earlier lacustrine/palustrine phases (Peypouquet et al., 1983). Additionally, the relatively larger size of the *Cyprideis* suggests adequate dissolved nutrients in the waters, particularly phosphate and silica (Peypouquet et al., 1983), which could also have been provided by volcanic activity associated with the Triple Tuffs. Tiercelin (1986) also noted a poorly preserved diatom flora (?*Thalassiosira* sp.) in the eastern Der Dora region that also suggests highly saline waters (citing Hecky and Kilham, 1973). Apatite nodules are also consistently associated with the TT-4 ostracod-bearing shale. Although the fish remains within these concretions are often unidentifiable in hand samples, one of us (C.J.C.) found a very large nodule in the Middle Ledi region east of Hadar, which formed around a partial catfish cranium approximate 6 cm in length.

Whereas the numerous, thin, bentonite tuff deposits preserved in the lacustrine sequence may represent an increase in volcanic activity during upper SH-4 and DD-1 times, they may equally be a preservation effect of deposition in a lacustrine environment. If these thin (1–5 cm) air-fall units were deposited under subaerial conditions, it is likely that they would have been reworked by wind and/or water or incorporated into soils during pedogenesis and gone unrecognized in the field. These air-fall deposits are more likely to be preserved in the quiet lacustrine settings in which the majority of the Hadar Formation tuffs are found. The alteration of these tuffs to bentonites is the result of interaction with the saline and/or alkaline waters that existed during their deposition in the Pliocene or postdepositionally with saline, alkaline rift groundwater.

The lithology of the Denen Dora Member changes slightly above TT-5 and includes a decrease in laminated sediments and in increase in clastic material and pedogenically modified mudstones. Where present, the discontinuous DD-1 sand is a fine-grained sand body up to a meter thick and occasionally fossiliferous. There is enough regional variation in preservation and structure to suggest that although sands are repeated laterally at the same stratigraphic level (Fig. 6), they do not represent a single, consistent sand body.

Sediments between the DD-1s and DD-2s are dominantly laminated or massive silts and clays with varying degrees of Vertisol development that include carbonate nodules as well as the occasional preservation of gastropod shells, shell molds, and fish remains (Fig. 6). Consistently preserved below the DD-2s in most regions, there is a several-meter-thick sequence of maroon to grayish brown clay underlain by an olive-gray clay. These clays have moderate to well-developed vertic structure, contain carbonate nodules, and include a notable carbonate-rich soil horizon (Bk) between them, which is preserved as either concentrated nodules (2–4 cm in diameter) or a continuous bed. This sequence of Vertisols separated by a carbonate horizon can be used as a stratigraphic marker to indicate the approximate level of the DD-2s when it is not preserved.

The sedimentology and discontinuous nature of the DD-1 sand(s) fit well as a system of distributary channels on a delta plain following the regression of the lower Denen Dora lacustrine phase. The clays and silty clays between the DD-1 and DD-2 sands preserved throughout the Hadar region are consistent with a flat delta plain or wet floodplain with possible minor lacustrine transgressions. The dominance of fine-grained sediments as well as the occasional presence of fish remains, gastropods, and a limestone/marl imply primary deposition under low-energy and wet conditions. However, most of the mudstones are preserved as Vertisols with pedogenic carbonate nodules, indicating stable periods of soil development as well.

DD-2 Submember

The DD-2 submember is defined as those sediments from the base of the DD-2 sand (or its lateral equivalent) to the base of the DD-3 sand (Fig. 2), and it preserves primarily delta-plain depositional environments.

Similar to the DD-1s, the DD-2s is also laterally discontinuous but can reach up to 2.5 m in thickness (Fig. 6). Although in some sections the DD-2s is clearly preserved in channel form as a fining-upward sand body with cross-bedding, more often it is present as a package of sandy silts and silts. These silts grade into overlying brown silty clays and clays with strong Vertisol development, abundant carbonate nodules, and rhizoliths. Investigations into the taphonomy and depositional environment of the A.L. 333

hominin assemblage by Behrensmeyer (this volume) have added significantly to the understanding of the DD-2 submember.

As with the DD-1s, the discontinuous nature of the DD-2s suggests a system of fluvial or fluvial-deltaic channels on an exposed delta plain. Studies of the DD-2 sand and silts between the Denen Dora Ridge and the eastern Farsita/Ado Golo region indicate a roughly north-northeast–directed main channel ~40 m wide and up to 3 m deep. Smaller, shallower tributaries and runnels appear to have fed into this channel from the south, and it possibly diverged into multiple channels to the north (Behrensmeyer et al., 2003; Behrensmeyer, this volume). The size and scale of rhizoliths in the silts lateral to the DD-2 channel as well as in the upper channel fill are the appropriate size for grasses, shrubs, and bushes, but probably not large trees (Behrensmeyer, this volume). As in DD-1 time, the consistent stratigraphic position of the sub–DD-2 Bk horizon and DD-2s suggests that the topography of the region was relatively flat and featureless during DD-2 time. A thick clay Vertisol with abundant carbonate nodules and rhizoliths overlies the DD-2 channels in the Denen Dora and Baruteita regions, particular in the vicinity of A.L. 333 (Fig. 6B). Studies of this paleosol indicate similarities to modern soils of wooded-grasslands in semiarid regions (Radosevich et al., 1992).

The lithology and lateral facies changes of the DD-1 and DD-2 sands from west to east have also been used as evidence of a meandering stream giving way to deltaic deposition, including distributary channels and subaqueous delta-front deposits (Yemane, 1997). Denen Dora Member sediments in the Unda Hadar Wadi up to the level of DD-3 fit well with this interpretation of longer-lasting lacustrine and delta-front environments in the eastern regions of Hadar.

DD-3 Submember

The DD-3 submember is defined as those sediments from the base of the DD-3 sand to the base of the Kada Hadar Tuff (KHT) (Fig. 2), and it preserves primarily fluvial depositional environments. The fluvial sands and overbank deposits of the DD-3 submember are the most fossiliferous of the entire Hadar Formation.

The DD-3 sand is the most laterally continuous and extensive sand body in the Hadar Formation, but it displays distinct lateral variation in thickness and lithology. In general, the DD-3s is a fining-upward sand averaging 2.5–5 m in thickness with an erosional base and well-developed cross-bedding and lateral accretion surfaces. The sedimentology and structure of the DD-3 sand, particularly its lateral accretion surfaces, and its ubiquitous preservation are more consistent with a large sheetlike sand body from a meandering river system as discussed by Yemane (1997) as opposed to a deltaic distributary network (Johanson et al., 1982) or braided river system (Tiercelin, 1986).

Exposures of the DD-3s at the western edge of Hadar are preserved as a multistoried sand body >5 m thick (Fig. 6A), representing the migration of the channel back and forth across the paleo-Hadar landscape. Conglomeratic components of DD-3 most likely represent the reworking of cobble material brought into the main axial fluvial system by braided streams from the western margin. The contribution of coarse material from braided streams at this time could also be related to the subsidence and change in topographic slope of the region following clay compaction. Exposures of the DD-3s at the eastern edge of Hadar contrast significantly compared to those in the west. In parts of the Unda Hadar Wadi, the DD-3s is still several meters thick but preserves tabular sets of cross-beds or bed sets of 10–30 cm that are sometimes separated by clay drapes. These structures are suggestive of a deltaic environment, but large carbonate nodules in the upper component may indicate that the DD-3 sand was at least partially subaerially exposed, such as on a delta plain, rather than a distributary channel or subaqueous delta lobe. This pattern mirrors the DD-1 and DD-2 west-east gradient from more terrestrial to more lacustrine conditions.

Earlier publications suggest that the DD-3 sand represents a climatically or tectonically induced deltaic or braided river system (Johanson et al., 1982; Tiercelin, 1986). However, the dramatic thickness and lateral extent of the DD-3s compared to other fluvial units of the Hadar Formation may be related to the underlying lacustrine and deltaic clays and silts. Compaction of these wet, fine-grained sediments could have resulted in significant nontectonic subsidence of the region. Such a drop in base level could have created a large accommodation space that the DD-3s deposits subsequently filled.

The DD-3 sand is consistently overlain by several meters of fining-upward floodplain deposits from fine sands to Vertisol clays with abundant carbonate nodules and rhizoliths (Fig. 6). A beige bentonite, the Kada Hadar Tuff (KHT), is usually preserved toward the top of this upward-fining sequence, and it marks the base of the Kada Hadar Member. Near the confluence of the Koborto Gaba and Unda Hadar Wadis, the 4-m-thick DD-3s is located just below the Confetti Clay and immediately above a bentonite, presumably the Kada Hadar Tuff (Fig. 7D). This stratigraphic position indicates the time-transgressive nature of the meandering DD-3 fluvial system and attests to the length of time over which this system impacted the Hadar landscape.

In addition to the fossil fauna preserved in the DD-3 submember, it also preserves a significant amount of fossilized wood in central and western Hadar, particularly in the upper components of the sand channels and near their interface with overbank silts. As observed in the modern Awash River system at Hadar, such vegetation is easily entrained and redeposited in the fluvial system as the river channel undercuts and erodes the heavily vegetated gallery forest located on the cutbank. In addition to the fossilized wood, the abundance of soil carbonate nodules and rhizoliths preserved in the overbank deposits indicates that the banks of this river system were well vegetated during DD-3 time.

Kada Hadar Member

The Kada Hadar Member is defined as those sediments between the base of the Kada Hadar Tuff (KHT) and the unconformity-related conglomerates that occur above BKT-2

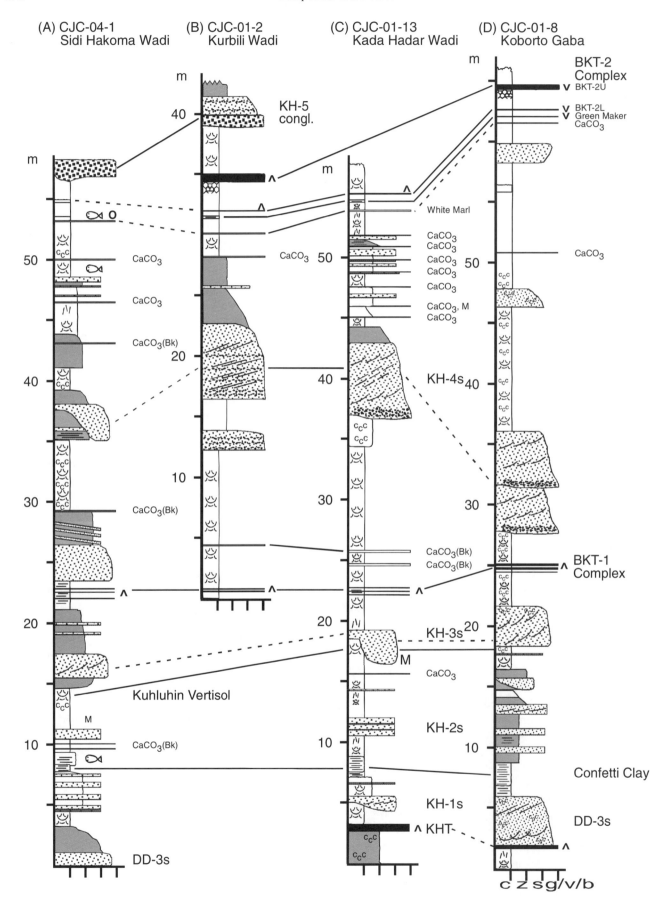

Figure 7. Stratigraphic sections of the Kada Hadar Member from Hadar. Sections progress left to right in a general eastward direction. Solid lines indicate known correlations; dashed lines indicate possible correlations. See Figure 2 for the geographic locations of sections. Approximate distances from base of sections: A–B = 2.8 km N; B–C = 0.9 km SE; C–D = 4.6 km E. KHT—Kada Hadar Tuff.

in the Hadar region. Because the top of this sequence is eroded by the unconformity-related conglomerates, the thickness of the Kada Hadar Member varies, but stratigraphic distances between the Kada Hadar Tuff and BKT-2 are typically between 50 and 60 m, with a slight thickening to the east (Fig. 7). The Kada Hadar Member has previously been informally divided into "lower" and "upper" components at the level of a major unconformity located just above BKT-2 (Vondra et al., 1996; Yemane, 1997). This unconformity, previously identified as a disconformity and referred to as the Main disconformity surface (MDS) (e.g., Hailemichael et al., 2002), is now recognized as an angular unconformity termed the Busidima unconformity surface (BUS) (Wynn et al., 2006, this volume). The Busidima unconformity surface marks a dramatic shift in facies character, stratigraphic completeness, and accumulation rates. Sediments previously referred to as the "upper" Kada Hadar Member are coeval with the lowermost strata of the newly designated Busidima Formation, which overlies the unconformity (Quade et al., 2004). The tephrostratigraphy and stratigraphic sequence of the Busidima Formation at Hadar are discussed in an accompanying paper in this volume (Campisano and Feibel, this volume, Chapter 6).

With the exception of the Gona archaeological sites and A.L. 288 (the "Lucy" site), the Kada Hadar Member received little attention prior to the 1990s because it was considered to be "essentially nonfossiliferous" (Gray, 1980). This lack of attention, combined with the lateral expanse of the Kada Hadar Member, left subdivisions of the member either poorly documented or undefined (e.g., Taieb et al., 1976; Johanson et al., 1978; Gray, 1980). Additionally, unlike the numbered sands of the Sidi Hakoma and Denen Dora Members, many of the sands of the Kada Hadar Member are discontinuous and/or difficult to correlate laterally. When work resumed in the 1990s, volcanic markers, as opposed to sand bodies, were used to subdivide the member into three submembers (Fig. 2).

Attaching formal names/numbers to the Kada Hadar sands is of limited utility, since a consensus in the nomenclature is found only in a few situations. Specifically, a clear agreement really only exists for three Kada Hadar sands/conglomerates: the KH-1 sand (between the Kada Hadar Tuff and the Confetti Clay), the KH-5 conglomerate (marking the Busidima unconformity surface), and the KH-7 conglomerate (Busidima Formation, beneath BKT-3 in the Maka'amitalu Basin). Comparisons of our stratigraphic sections with those of previous workers (e.g., Yemane, 1997; Hailemichael et al., 2002) demonstrate that the number of sands encountered in any particular Kada Hadar Member section is dependent upon where that section is measured. Although hesitantly, numbered sands/conglomerates are used in this study for the interval from the Kada Hadar Tuff to BKT-2. These assignments are not considered definitive but are used for reference purposes only. Assignments of the KH-1 sand to the KH-5 conglomerate are based upon sections along the Kada Hadar Wadi between the Farsita and Burukie Wadis (Fig. 7D).

KH-1 Submember

The KH-1 submember is defined as those sediments from the base of the Kada Hadar Tuff to the base of the BKT-1 Complex (Fig. 2), and it preserves both lacustrine and fluvial depositional environments.

The Kada Hadar Tuff is most commonly preserved as a beige, silty bentonite, 20–80 cm thick. In the Farsita Wadi, close to its confluence with the Kada Hadar, the Kada Hadar Tuff is preserved as a thick, vitric channel-fill deposit essentially lying immediately above the DD-3 sand. The channel fill is actually a series of four separate deposits that are preserved along four alternating cutbank surfaces over a distance of ~200 m (Fig. 4D). Using these Kada Hadar Tuff deposits, it is possible to reconstruct an ~26-m-wide, at least 5–6-m-deep channel flowing to the northeast (N72°E). A small channel-fill tuff preserved in the Middle Ledi region northeast of Hadar has been geochemically correlated to the Kada Hadar Tuff (DiMaggio et al., 2006; Dupont-Nivet et al., this volume), and Yemane (1997) also reported a 5–6-m-thick and 30–40-m-wide Kada Hadar Tuff channel-fill deposit at Merketa Negeria Koma (west of the Gona Wadi), originally discovered by J. Aronson. Sanidines collected from the Kada Hadar Tuff between the Gona and Sidiha Koma Wadis have been dated to ca. 3.20 Ma (Walter, 1994).

Based upon its size and degree of preservation, it is unlikely that this Kada Hadar Tuff channel represents the main axial fluvial system at the time. It is more likely that it represents a crevasse channel, probably a result of the influx of tephra into the axial system. Based on its observed relationship to the Kada Hadar Tuff channel, we interpret the bentonite floodplain deposits of the Kada Hadar Tuff as representing crevasse splays from the Kada Hadar Tuff channel, which may explain why it is not consistently recorded laterally across the Hadar region, as would be a typical sheet floodplain deposit.

Where the Kada Hadar Tuff is not preserved, a widespread green shale known as the Confetti Clay, ~4 m above the level of the Kada Hadar Tuff, provides an alternative marker bed near the base of the member (Fig. 7). Between the Kada Hadar Tuff and Confetti Clay, there are primarily pedogenically modified clays and silty clays, as well as a sand body typically up to 1 m thick, which is often, but not always preserved. This sand, known as the KH-1 sand, is noteworthy for the fact that the partial *A. afarensis* skeleton A.L. 288-1 ("Lucy") was discovered in this deposit (Johanson and Taieb, 1976). Yemane (1997) suggested that the ribbonlike sand bodies of KH-1s represent high-sinuosity meandering stream deposits, whereas the thin sheet-sand KH-1s deposits represent crevasse splays. However, the discontinuity of the KH-1s, as well as the multiple sand layers separated by clay drapes in some of the thicker KH-1s exposures, may be better explained as a network of crevasse channels and splays associated with the DD-3 fluvial system. Although the crevasse channels may still fit the description of high-sinuosity meanders, they probably do not represent the main axial fluvial system. The time-transgressive nature of the DD-3 sand is evident in its stratigraphic position just below the Confetti Clay in the Unda Hadar (and likely above the Kada Hadar Tuff), a position usually occupied by the KH-1s (Fig. 7D). Thus, the KH-1s would represent crevasse channels/splays from DD-3 meanders that, with the exception of the Unda Hadar, had already migrated out of the Hadar area.

The Confetti Clay is a laminated shale, pale olive to dark olive-gray in color, and it contains apatite nodules with a nucleus of fish debris. Its name derives from the fact that the weathering product of the shale is confetti-like in texture and appearance. With the exception of its lack of ostracods and a bentonite, the Confetti Clay is very similar in appearance to the TT-4 ostracod-bearing shale. In the Hurda, at the eastern edge of the Hadar project area, the Confetti Clay is lighter in color and diatomaceous (*Aulacoseira* sp. and ?*Stephanodiscus* sp.). The Confetti Clay continuously thickens from west to east, ranging from ~0.5 m thick in the Sidiha Koma Wadi to ~2.2 m in the Hurda (Fig. 7). The Confetti Clay is also present in the Gona and Dikika project areas (Tiercelin, 1986; Alemseged et al., 2005). Similar to the ostracod-bearing shale, the Confetti Clay represents the next major lacustrine transgression following the upper Sidi Hakoma and lower Denen Dora lacustrine interval. The eastward thickening of this lacustrine marker is also in line with previously mentioned information indicating a depocenter to the east or northeast.

The next major marker bed is the Bouroukie Tuff 1 (BKT-1), ~15 m above the Confetti Clay and the boundary between the KH-1 and KH-2 submembers (Fig. 2). Between the Confetti Clay and BKT-1, there are primarily brown clays and silty clays, usually laminated or with vertic structure, and two sand bodies (Fig. 7). The first sand body (here, KH-2s) is 1–1.5 m thick where present, composed of very fine to fine-grained sand, and in some areas, such as in the Kada Hadar Wadi, it is clearly seen to consist of three sheetlike sand layers separated by clay drapes (Fig. 7). The second sand body (here, KH-3s) is more distinct, as it is thicker and more laterally continuous. This upward-fining sand is as thin as 75 cm in some locations but displays channeling up to 3 m thick, such as in the Kada Hadar Wadi (Fig. 7). Typically preserved not far below the KH-3s, there lies a well-developed blue-gray to greenish gray Vertisol claystone containing abundant carbonate nodules, which we refer to as the Kuhluhin Vertisol (kuhlin = blue in Afar). This laterally extensive marker bed is occasionally channeled into or removed by KH-3s deposits and preserves a very localized gastropod coquina (~3 m diameter) in the Baruteita region.

Between the Confetti Clay and BKT-1 at Kuhluhin Koma (= Kahuli Koma of Yemane, 1997), there is a distinct conglomeratic delta facies that may be related to the KH-3s. Originally identified by C. Vondra and consistent with the description of Yemane (1997), the foreset beds of the conglomerate consist of ~4 m of steeply dipping planar cross-beds that are capped by ~2 m of nearly horizontal and laterally extensive conglomerate topset beds (Fig. 4E). No bottomset beds exist, but laminated brown clayey-silts and silts containing discontinuous horizons of ovate/discoid limestone nodules are preserved between the foreset and topset beds and lateral to the foresets. These laminated clayey-silts do not extend much farther laterally than the topset beds before they grade into a sequence of well-developed Vertisols. The topset beds at Kuhluhin Koma extend for more than 250 m and appear to be either just below or at the level of BKT-1 (BKT-1 is preserved lateral to the topset beds, but not above or below them). The delta conglomerate has also been reported at Merketa Negeria Koma (at Gona, southwest of Kuhluhin Koma) where it is several meters thicker (Yemane, 1997). In the Sidiha Koma region (northwest and north of Kuhluhin Koma), the sub–BKT-1 sequence preserves Vertisols and an extensive 2-m-thick sand with lateral accretion surfaces as opposed to a conglomerate (Fig. 7A).

The sediments between the Confetti Clay and BKT-1 are highly variable through time and across space and complicate the interpretation of depositional environments. In the western region of Hadar (Sidiha Koma Wadi and Kuhluhin Koma), lacustrine conditions continued after deposition of the Confetti Clay, as indicated by laminated sediments and sporadically preserved gastropod shells, but periods of exposure are noted in some horizons by varying degrees of Vertisol development and the presence of soil carbonates (Fig. 7A). The Kuhluhin Vertisol was originally interpreted as representing the basal component of the lacustrine interval that the conglomeratic delta prograded into as the bases of both the foreset beds and the Vertisol lie at the same level (Campisano, 2007). However, recent investigations have shown that although the clays of the Kuhluhin Vertisol may have been deposited under lacustrine conditions (based on its reduced color and locally preserved gastropods), it was already a developed soil when channeled into by the delta foresets. Instead, the laminated silty clays overlying the Kuhluhin Vertisol and lateral to the foresets represent the lacustrine interval during delta progradation.

The Confetti Clay to BKT-1 sequence in the central Kada Hadar region provides some evidence for continued lacustrine deposition, but very little that differentiates it from floodplain deposits. Most of the clays and clayey silts are pedogenically modified, sometimes with strong Vertisol development, including those olive-gray or olive-green in color, which are typically associated with lacustrine environments at Hadar. The KH-2 sand is represented by several thin sand sheets separated by clay drapes and could just as easily represent sheet sands on a fluvial floodplain as in a lacustrine setting. Possible lacustrine conditions may be indicated by two thin (2–3 cm) limestones preserved between the KH-2 and KH-3 sands in the Kada Hadar Wadi (Fig. 7C), a thin, localized ostracod silt in the Farsita Wadi, and a localized gastropod coquina preserved within the Kuhluhin

Vertisol in the Baruteita region. The KH-3s in the Kada Hadar Wadi fines upward and includes clear erosional channeling into the underlying Vertisol and bivalves concentrated at the bottom of the channel (Fig. 7C). However, it does not resemble the KH-3 conglomerate at Kuhluhin Koma or the context that surrounds it.

KH-1 submember deposits in the Koborto Gaba Wadi are more similar to those in the Kada Hadar Wadi than in the Sidiha Koma and Kuhluhin Koma region. Most fine-grained sediments show pedogenic modification, and there are three sands between the Confetti Clay and BKT-1, the lowest of which is very similar to the KH-2s from the Kada Hadar section. The upper sand, presumably KH-3s, is a thick, fining-upward sequence with lateral accretion surfaces that was clearly not deposited in a lacustrine setting (Fig. 7D).

Overall, the general pattern suggests a post–Confetti Clay continuation of fluctuating lacustrine conditions in at least some areas of Hadar, followed by a major fluvial episode associated with KH-3s. The distribution of more clearly defined lacustrine and deltaic deposits suggests that the post–Confetti Clay lake associated with the KH-3 delta at Kuhluhin Koma may have been centered more to the south than to the east, where previous lakes had been. Although lacustrine conditions existed at least long enough and to a depth sufficient to accommodate the Gilbert-type delta (~5 m), subaerial exposure and pedogenic modification of lacustrine sediments were frequent processes.

KH-2 Submember

The KH-2 submember is defined as those sediments from the base of the BKT-1 Complex (Fig. 2) to the top of the BKT-2 Complex, and it preserves both fluvial and lacustrine depositional environments. Excluding the tuff complexes themselves, the KH-2 submember consists of three major sedimentary packages.

Exposures of BKT-1, dated to ca. 3.12 Ma (Campisano, 2007), are consistently similar across Hadar. The BKT-1 "tuff" actually consists of several distinct units encased within a brown or olive-gray Vertisol claystone, which is referred to as the BKT-1 Complex in this study. The base of the BKT-1 Complex is a crystal-lithic volcaniclastic unit, ~3 cm thick, followed by 25–40 cm of massive or laminated claystone. This clay is followed by a double bentonite, where each bentonite is 5–10 cm thick, massive, white to yellow in color, and separated by 25–35 cm of laminated or massive brown claystone exhibiting conchoidal fracture. As the BKT-1 Complex is usually only 50–70 cm thick and does not display any distinctly lacustrine components beyond its lithologic structure, it is possible that this interval may represent a shallow, ephemeral floodplain lake. However, the lateral extent of the BKT-1 deposit across all of Hadar may more likely indicate a fully lacustrine transgression. It is worth noting that if it were not for the presence of the bentonites and the lithology of their immediate context, this thin lacustrine interval may have gone completely unnoticed.

In the Kada Hadar region, the first sedimentary package of the submember immediately overlies the BKT-1 Complex and consists of ~15 m of strongly developed Vertisol clays (Fig. 7C). The majority of the Vertisol thickness is composed of four brown Vertisol packages (1–4 m each) alternating with three, thinner (0.75–2 m each), grayish brown Vertisol packages that appear blue-gray in overall outcrop appearance. Whereas millimeter-sized carbonate nodules are intermittently dispersed throughout this Vertisol stack, there are two 20 cm Bk horizons in the first 3.5 m above BKT-1 that consist almost purely of unconsolidated 1–5 mm carbonate nodules. This impressive sequence of Vertisols and carbonate nodules is interpreted to represent a floodplain environment.

The Vertisol stack is followed by an extensive erosional sand body, up to 6 m thick, that forms a prominent sandstone bench in the Kada Hadar region (Fig. 7). The well-sorted sand (here, KH-4s) fines upward and in many areas has a pebble conglomerate base. It also has clearly visible cross-bedding and lateral accretion surfaces. The sand body is typically overlain by a fining-upward overbank sequence of fine sand to silt to clay with varying degrees of pedogenic modification. Although extensive, this fluvial sequence displays a high degree of lithologic variability. In the Sidiha Koma Wadi, two sand bodies are present, and the upper one is highly erosional with a pebble-conglomerate base in some areas (Fig. 7A). In the Kada Hadar region, the KH-4s is highly variable, ranging from conglomerate channel-lag deposits to thick sand bodies with erosional pebble-conglomerate bases and lateral accretion surfaces (Figs. 7B and 7C). Pebble-shaped carbonate nodules at the base of the KH-4s in the Kada Hadar Wadi near the Kurbili Wadi indicate that this meandering system was reworking soil carbonate nodules from the underlying paleosol as it aggraded across the landscape. Further up the Kada Hadar Wadi, at its confluence with the Burukie Wadi (near the A.L. 444 hominin site), a pebble-conglomerate channel facies of the KH-4s is clearly visible. Lateral to this channel (as at A.L. 444), there is a thick package of interbedded floodplain sands and silts that likely represent a levee deposit. Concentrations of fossil wood in the KH-4 fluvial system in the Kurbili Wadi may represent an undercut gallery-forest situation similar to that described for the DD-3 sand. Two ~4-m-thick fining-upward sand bodies with gravelly bases are also preserved in the Koborto Gaba Wadi (Fig. 7D). Lateral accretion surfaces in the lower sand indicate that this meandering fluvial system continued all the way to the eastern edge of Hadar.

Sediments between the KH-4 sand(s) and the BKT-2 Complex make up the final sedimentary package of the KH-2 submember. These deposits are several meters thick and are composed primarily of clays or sandy/silty clays, including several thin (10–30 cm), fine-grained sheet sands, laminated silts, and numerous (up to 7) thin (3–10 cm) carbonate horizons represented by either marls, limestone, or sparite, occasionally containing *Melanoides* shells or shell molds (Fig. 7). The uppermost carbonate in this series, defined here as the "White Marl," is a laterally extensive, 10-cm-thick marl that caps ~2 m of mudstone; it is a clearly visible marker bed in the region that aids in locating the overlying BKT-2 Complex.

The upper KH-2 submember sequence indicates a gradual transition from the KH-4 fluvial system to the BKT-2 lacustrine transgression. As with the Confetti Clay to BKT-1 sequence, the exact transition point from fluvial overbank to lacustrine deposits is obscured by pedogenic modification to both floodplain and lacustrine deposits. The marls and limestones indicate lacustrine or palustrine deposits. The fine-grained sediments that surround these limestones often exhibit degrees of pedogenic modification on top of relict horizontal lamination, which attest to periodic subaerial exposure of these sediments, probably a result of fluctuating lake-water levels. The thin sands represent an influx of terrigenous material from a fluvial system, possibly the KH-4 system, located outside Hadar. A localized, ~1.5 m sand body in the Koborto Gaba composed of poorly sorted, but coarsening-upward sand likely represents a subaqueous delta-channel fill (Fig. 7D).

Special attention has been paid to the Bouroukie Tuff 2 (BKT-2), the stratigraphically highest tuff of the Hadar Formation. It is a primary air-fall tuff complex with feldspar crystals suitable for dating, and it is laterally extensive throughout Hadar and adjacent regions, including the Ledi-Geraru, Dikika, and Gona study areas. Walter (1981) divided the BKT-2 into three different volcanic components, which are collectively referred to as the BKT-2 Complex in this study. Investigations by C. Campisano and colleagues in the Ledi-Geraru region, just east of Hadar, have identified as many as seven distinct volcanic units associated with the BKT-2 Complex encased in a laminated diatomite (DiMaggio et al., this volume).

The lowermost horizon of the BKT-2, BKT-2L_1, is also known as the Green Marker Tuff (GMT) or, perhaps more appropriately, the Green Marker Bed (GMB), as it does not always show evidence of being tuffaceous. In the Koborto Gaba region, the Green Marker Bed is a 5-cm-thick laminated dark green unit composed primarily of dark brown (basaltic) vitric tephra and augite crystals. It is encased in a brown, massive clay and underlies BKT-2L by ~50 cm. Although Walter (1981) described the unit as grading westward into a thin green tuffaceous sand as far as the Gona Wadi, it is more often preserved as a laminated olive-green sandy claystone grading into alternating laminated green and brown clays. The Green Marker Bed is commonly 10–20 cm thick, but it reaches a thickness of 85 cm in portions of the Burahin Dora and upper Kada Hadar Wadis.

The BKT-2L in the Koborto Gaba region is a 10-cm-thick, powdery, white to yellowish white, crystal-lithic rich unit with a small amount of rhyolitic vitric tephra. BKT-2L deposits in the central Hadar region (e.g., in the Farsita, Kurbili, Burukie, and Kada Hadar Wadis) are similar in appearance to deposits in the Koborto Gaba, but they are slightly thinner (5–7 cm). BKT-2L deposits in the central region also lack a vitric component, and feldspars are sometimes completely altered. Anorthoclase feldspars from BKT-2L from Gona, Hadar, and Ledi-Geraru have been dated to ca. 2.96 Ma (Semaw et al., 1997; Campisano, 2007; DiMaggio et al., this volume).

Depending on the location, BKT-2U (= BKT-2u) is separated from BKT-2L by 0.4–2.2 m of brown clays that typically preserve linked pentagonal and hexagonal desiccation cracks infilled with BKT-2U material (Figs. 4F and 7). BKT-2U in the Koborto Gaba is 45 cm thick and composed of three subunits, informally labeled here as BKT-2Ua, -b, and -c from bottom to top (Fig. 4F). The lowest unit is 15–20 cm thick and very similar in lithology and appearance to BKT-2L. The middle unit is also 15–20 cm thick and consists dominantly of crystal-lithic material and dark brown vesicular/scoriaceous basaltic glass, but it also includes rhyolitic glass and a minor quantity of small, light-colored pumices. The capping 2–4 cm is very similar in lithology to the basal BKT-2U unit, but it is exceptionally calcareous. Whereas the Green Marker Bed and BKT-2L appear to be continuously preserved across Hadar, BKT-2U is not consistently found in association with them (Fig. 7). Where preserved outside of the Koborto Gaba and Unda Hadar region, the thickness varies significantly, from 15 cm in the Farsita to 75 cm in the upper Kurbili. In these areas, BKT-2U is crystal-rich and yellowish white in appearance, like BKT-2L, but thicker, coarser grained, and contains reworked brown clay fragments, small tuff clasts, and occasional gastropod shells. As with BKT-2L, BKT-2U exposures in central and western Hadar preserve no vitric component. Anorthoclase feldspars from BKT-2U from Hadar and Ledi-Geraru have been dated to ca. 2.94 Ma (Campisano, 2007; DiMaggio et al., this volume; Fig. 1 in Walter, 1994).

As noted by others (e.g., Yemane, 1997; Hailemichael et al., 2002), the interval surrounding the BKT-2 Complex represents the last major lacustrine transgression recorded at Hadar. The laminated and/or massive clays in association with the White Marl and Green Marker represent the greatest extent of this southwestward transgression into Hadar during Kada Hadar times. It is uncertain whether this lacustrine transgression is climatic in nature or related to tectonic subsidence of the basin. Massive clays between the Green Marker Bed and BKT-2L were deposited under lacustrine conditions, but pedogenic overprinting evident in the Kada Hadar, but not Koborto Gaba, region suggest at least a brief period of lacustrine regression in western and central Hadar. Similarly, whereas BKT-2L and its overlying sediments were deposited under lacustrine conditions, Vertisol and/or desiccation crack development above BKT-2L at all locations indicates a lacustrine regression and subaerial exposure across all of Hadar. Localized pond basins may still have existed in some areas following the lacustrine regression, such as in the Burahin Dora Wadi, where a gastropod-packed tuffaceous sand is preserved at a level lateral to BKT-2. The sand is composed of crystal-lithic material including large feldspar crystals and scoriaceous basaltic glass, as found in BKT-2U. In association with the gastropods, this reworked deposit may represent a beach or nearshore environment.

The infilling of desiccation cracks by BKT-2U material may reflect deposition of the tephra on an already desiccated surface, such as a recently exposed delta plain or floodplain following a lacustrine regression. Alternatively, and perhaps more likely, the desiccation cracks may have formed in the clays while beneath the overlying tephra deposit, after which they were passively filled. A similar situation has been observed by one of us (C.S.F.)

on the modern landscape of the Turkana Basin, where sand has passively filled in an underlying, periodically desiccated soil. Compared to the eastern region, more terrestrial or fluctuating lacustrine conditions existed in the central and western parts of Hadar during BKT-2U times, as evident by reworked brown clay fragments, small tuff clasts, and occasional gastropod shells incorporated into the tuff. In contrast to the depositional environments at Hadar, BKT-2 exposures in the Ledi-Geraru region are fully lacustrine, encased in up to 8 m of diatomite or diatomaceous silts, and no pedogenic development or desiccation cracks underlie BKT-2L or -2U (DiMaggio et al., this volume).

KH-3 Submember

The KH-3 submember is defined as those sediments between the BKT-2 complex and the Busidima unconformity surface (Fig. 2). As the erosional extent of the unconformity varies considerably across the landscape, the thickness of this submember varies from 0 to 7.5 m at Hadar. Brown clays and silty clays represent the majority of the submember. These units are strongly vertic and show no clear indication of lacustrine conditions. In some areas, such as in the Burahin Dora Wadi, a thin sand is present that has produced a small quantity of fossils. In the upper Kada Hadar Wadi, an ~30-m-wide by 3-m-deep channel is lined with 10–20-cm-thick silty clay layers intercalated with 2–3-cm-thick carbonate evaporite horizons, indicating that the channels were subaerial.

PROVENANCE OF THE FOSSIL ASSEMBLAGES

The vast majority of the Hadar fossil specimens represent surface collections that, when possible, were assigned to a single stratigraphic submember (e.g., DD-3) or unit within a submember (e.g., DD-3s). These assignments were based on the stratigraphic unit(s) exposed at the locality, the adhering matrix, and the color and texture of the specimens when no matrix existed (Gray, 1980). The preponderance of vertebrate fossils in the collection is associated with the fluvial-deltaic sands and silts, particularly the channel and overbank deposits of the SH-2s, DD-2s, DD-3s, and KH-4s systems. Hominin remains have been collected from all of the Hadar Formation submembers except the stratigraphically thin and relatively unfossiliferous KH-3 submember. Although absent from the Basal Member at Hadar sensu stricto, hominin fossils have been recovered from Basal Member deposits across the Awash River in the Dikika study area (Alemseged et al., 2005). As with the other vertebrate remains, hominin fossils are also typically derived from the fluvial-deltaic sands and silts, particularly those within the SH-1 submember and from the KH-4s fluvial system of the KH-2 submember, where they reach their highest relative abundances (Campisano, 2007).

The pattern of fossil preservation at Hadar is likely a taphonomic artifact as opposed to representing the original distribution of faunal remains across the paleolandscape. The majority of the fine-grained floodplain deposits (clays and silty-clays) and many lacustrine deposits are pedogenically altered to Vertisols, which are typified by intense shrinking, swelling, and cracking (Soil Survey Staff, 1999). As evident in archaeological excavations at both Hadar and Gona, the pedogenic processes associated with Vertisol formation not only have the ability to reorient and distribute artifacts over tens of centimeters, but even abrade, shear/break, and polish artifacts (Hovers, 2003; Quade et al., 2004). These pedogenic processes clearly have implications for the preservation potential of bone on the paleolandscape, particularly if they were exposed on floodplain surfaces for an extended period of time. The vertical churning and leaching in Vertisols has also been proposed as an explanation for the differential preservation of bone at Gona archaeological sites (Quade et al., 2004). Mammalian remains from lacustrine-dominated submembers typically derive from delta or beach sand deposits as opposed to the laminated, massive, or pedogenically modified mudstone.

In addition to the temporally and spatially variable depositional environments that existed, a range of habitats was available to *A. afarensis*. Although Hadar habitats could broadly be defined as woodland mosaics, analyses of the Hadar faunal assemblages indicate the presence of open and closed woodlands, gallery forests, edaphic grasslands, and shrublands (Gray, 1980; Campisano, 2007; Campisano and Reed, 2007; Reed, 2008). Some of the variation observed in these faunal communities can be explained by the spatial distribution of fauna across the landscape, as well as by the depositional environments with which they were associated (Campisano, 2007; Campisano and Reed, 2007).

SUMMARY AND CONCLUSIONS

The strata of the Hadar Formation can be imagined as representing different components of a large-scale, but single fluvio-lacustrine system (Fig. 8). A related study has proposed that consistent oscillations observed between fluvial and lacustrine depositional environments at Hadar may be climatic in nature (Campisano and Feibel, 2007). As evident by some of the major sand bodies (e.g., DD-3s), some of the fluvial systems are clearly time-transgressive. As such, the extensive sand bodies preserved at different stratigraphic levels within the Hadar sequence may be interpreted as representing meanders of a single fluvial system entering the Hadar region at different times. Most fluvial deposits not represented directly by this meandering system are likely associated with it, either as crevasse splays or distributary channels. The most extensive and significant deposits of this fluvial system are represented by the SH-2, DD-3, and KH-4 sands (Fig. 8). Paleocurrent data from Hadar Formation sands show a wide dispersion of flow directions, consistent with a sinuous meandering system (Yemane, 1997). On average, paleocurrent directions trend to the NNE (Yemane, 1997), and a general eastward-fining trend is also observed in channel deposits. These patterns are consistent with more lacustrine-dominated sequences observed in exposures to the northeast and east (Dupont-Nivet et al., this volume). Hadar Formation conglomerate deposits are likely to have resulted from high-energy braided streams originating along

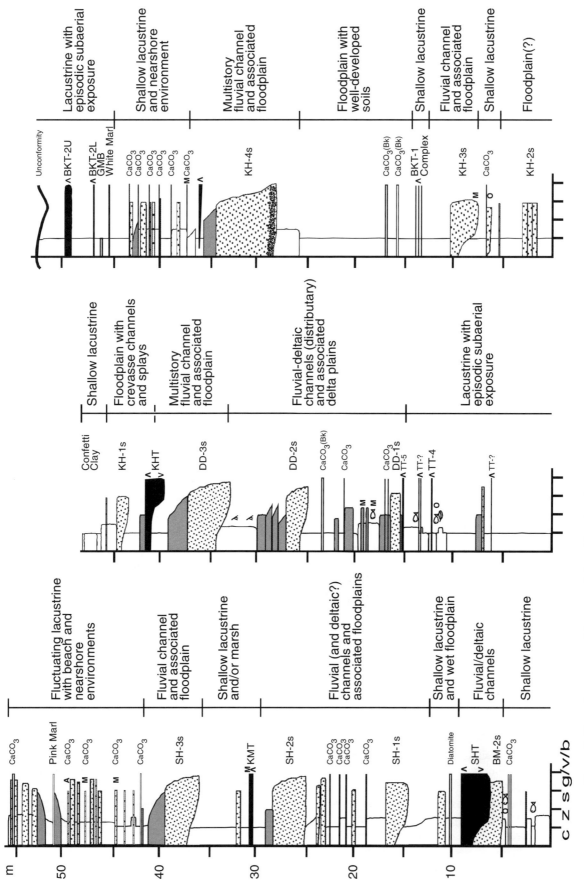

Figure 8. General summary of interpreted depositional environments for the Hadar Formation. Composite section was constructed primarily from sections CJC-02-1 (lower Sidi Hakoma Member), CJC-02-5 (upper Sidi Hakoma Member), CJC-02-7 (Denen Dora Member), and CJC-01-13 (Kada Hadar Member) located within the central region of the Hadar study area. GMB—Green Marker Bed; KHT—Kada Hadar Tuff; SHT—Sidi Hakoma Tuff.

the Ethiopian Escarpment, either entering directly into the Hadar region or intersecting with the axial fluvial system nearby.

As suggested by Leeder (1975), there is an inverse relationship between soil maturity and sediment accumulation rates on fluvial floodplains. Relatively immature paleosols develop adjacent to channel margins where sediment accumulates rapidly, whereas more mature paleosols develop under conditions of slow accumulation further from the channel. This relationship was expanded upon by Kraus and colleagues, who showed how soil maturity levels of overbank deposits can help predict avulsion patterns and the relative distance to fluvial channels (Bown and Kraus, 1987; Kraus, 1987, 2002; Kraus and Aslan, 1993; Kraus and Bown, 1988, 1993). The character of Hadar Formation paleosols suggests that fluvial avulsion/migration into and out of the Hadar region occurred relatively rapidly, as most paleosols are fine-grained and mature. However, the thickness of these paleosols also indicates a high rate of subsidence for the area. This avulsion pattern fits with other lines of evidence indicating that the paleolandscape at Hadar was topographically flat. Most of the Hadar paleosols are Vertisols, which are typically found in flat terrains, are characterized by strongly seasonal moisture regimes under semiarid climates (180–250 mm/yr of precipitation), and can form in only a few hundred years (Retallack, 2001). Additionally, most of the Hadar mudstones are dominated by smectite (Yemane, 1997), which also develops under a warm climate with alternating humid and arid seasons (Paquet, 1970).

Lacustrine transgressions into Hadar were a regular occurrence; the most significant and persistent episode spanned the upper Sidi Hakoma and lowermost Denen Dora Members (Fig. 8). At least seven transgressive episodes are clearly recorded in the Hadar record, and several more minor or ephemeral events are possible. Lacustrine deposits suggest a low-gradient shoreline composed mostly of mud with only localized evidence of true beach deposits. Similarly, the Hadar lakes were most likely relatively shallow (<10 m deep) and not very long-lasting, as most deposits show consistent evidence of subaerial exposure and pedogenic modification. The ubiquitous presence of *Cyprideis* from the lacustrine deposits (Peypouquet et al., 1983) also strongly suggests vegetated and/or marshy conditions (Balch et al., 2005). Lacustrine sediments in the Ledi-Geraru region, closer to the terminal depocenter, also show a high proportion of pedogenically modified to unmodified lacustrine sediments. Pedogenic modification has also made it difficult to identify what were once primary lacustrine deposits, particularly when other indicators of lacustrine conditions, such as gastropods, ostracods or fine laminations, are absent. In fact, it is likely that brief lacustrine transgressions or ephemeral floodplain lakes have gone unnoticed in this and other studies, instead being identified as pedogenically modified fluvial floodplain deposits. Similarly, it may not always be possible to distinguish between truly lacustrine conditions and ephemeral floodplain lakes in the lithostratigraphic record as the differences are often only an issue of spatial and temporal scale.

ACKNOWLEDGMENTS

This project was funded by grants from the National Science Foundation, the National Geographic Society, the Institute of Human Origins (ASU), and the Center for Human Evolutionary Studies (Rutgers University) to both the authors and directors of the Hadar Research Project. Our early fieldwork at Hadar benefited from an unpublished manuscript on sedimentary facies models of the Hadar Formation shared with us by James Aronson, Carl Vondra, and Tesfaye Yemane, as well as tours introducing us to the Hadar exposures by Carl Vondra and Gerry Eck in 2001. We would like to thank Kay Behrensmeyer and Jonathan Wynn for helpful discussions on depositional environments and Andy Cohen and Brian Currie for thoughtful and constructive reviews of this manuscript. For logistical and field support, we would like to thank Bill Kimbel, Don Johanson, Erella Hovers, Gerry Eck, the Institute of Human Origins, the National Museum of Ethiopia, the Ethiopian Authority for Research and Conservation of Cultural Heritage (A.R.C.C.H.), and especially our field crew and friends from the Eloaha region.

REFERENCES CITED

Alemseged, Z., Wynn, J.G., Kimbel, W.H., Reed, D.N., Geraads, D., and Bobe, R., 2005, A new hominin from the Basal Member of the Hadar Formation, Dikika, Ethiopia, and its geological context: Journal of Human Evolution, v. 49, p. 499–514, doi: 10.1016/j.jhevol.2005.06.001.

Aronson, J.L., and Taieb, M., 1981, Geology and paleogeography of the Hadar hominid site, Ethiopia, in Rapp, G., and Vondra, C.F., eds., Hominid Sites: Their Geologic Setting: Boulder, Colorado, Westview Press, p. 165–195.

Balch, D.P., Cohen, A.S., Schnurrenberger, D.W., Haskell, B.J., Valero Garces, B.L., Beck, J.W., Cheng, H., and Edwards, R.L., 2005, Ecosystem and paleohydrological response to Quaternary climate change in the Bonneville Basin, Utah: Palaeogeography, Palaeoclimatology, Palaeoecology, v. 221, p. 99–122, doi: 10.1016/j.palaeo.2005.01.013.

Behrensmeyer, A.K., 2008, this volume, Paleoenvironmental context of the Pliocene A.L. 333 "First Family" hominin locality, Hadar Formation, Ethiopia, in Quade, J., and Wynn, J.G., eds., The Geology of Early Humans in the Horn of Africa: Geological Society of America Special Paper 446, doi: 10.1130/2008.2446(09).

Behrensmeyer, A.K., Harmon, E.H., and Kimbel, W.H., 2003, Environmental context and taphonomy of the A.L. 333 locality, Hadar, Ethiopia: Tempe, Arizona, Paleoanthropology Society, Annual Meeting Abstracts, p. A2, www. paleoanthro.org/meeting.htm.

Beyene, A., and Abdelsalam, M.G., 2005, Tectonics of the Afar Depression: A review and synthesis: Journal of African Earth Sciences, v. 41, p. 41–59, doi: 10.1016/j.jafrearsci.2005.03.003.

Bown, T.M., and Kraus, M.J., 1987, Integration of channel and floodplain suites. I. Developmental sequence and lateral relations of alluvial paleosols: Journal of Sedimentary Petrology, v. 57, p. 587–601.

Campisano, C.J., 2007, Tephrostratigraphy and Hominin Paleoenvironments of the Hadar Formation, Afar Depression, Ethiopia [Ph.D. thesis]: New Brunswick, New Jersey, Rutgers University, 600 p.

Campisano, C.J., and Feibel, C.S., 2007, Connecting local environmental sequences to global climate patterns: Evidence from the hominin-bearing Hadar Formation, Ethiopia: Journal of Human Evolution, v. 53, p. 515–527, doi: 10.1016/j.jhevol.2007.05.015.

Campisano, C.J., and Feibel, C.S., 2008, this volume (Chapter 6), Tephrostratigraphy of the Hadar and Busidima Formations at Hadar, Afar Depression, Ethiopia, in Quade, J., and Wynn, J.G., eds., The Geology of Early Humans in the Horn of Africa: Geological Society of America Special Paper 446, doi: 10.1130/2008.2446(06).

Campisano, C.J., and Reed, K.E., 2007, Spatial and temporal patterns of *Australopithecus afarensis* habitats at Hadar, Ethiopia: Philadelphia,

Pennsylvania, Paleoanthropology Society, Annual Meeting Abstracts, p. A6, www.paleoanthro.org/meeting.htm.
de Heinzelin, J., and Haesaerts, P., 1983, The Shungura Formation, in de Heinzelin, J., ed., The Omo Group: Tervuren, Musée Royale de l'Afrique Central, p. 25–128.
DiMaggio, E.N., Campisano, C.J., Arrowsmith, J R., Reed, K.E., and Lockwood, C.A., 2006, Geochemistry and tephrochronology of the BKT-2 and Kada Hadar tephras in the Middle Ledi-Geraru region of Afar, Ethiopia: Eos (Transactions, American Geophysical Union), v. 87, p. 52.
DiMaggio, E.N., Campisano, C.J., Arrowsmith, J R., Reed, K.E., Swisher, C.C., III, and Lockwood, C.A., 2008, this volume, Correlation and stratigraphy of the BKT-2 volcanic complex in west-central Afar, Ethiopia, in Quade, J., and Wynn, J.G., eds., The Geology of Early Humans in the Horn of Africa: Geological Society of America Special Paper 446, doi: 10.1130/2008.2446(07).
Dupont-Nivet, G., Sier, M., Campisano, C.J., Arrowsmith, J R., DiMaggio, E., Reed, K., Lockwood, C., Franke, C., and Hüsing, S., 2008, this volume, Magnetostratigraphy of the eastern Hadar Basin (Ledi-Geraru research area, Ethiopia) and implications for hominin paleoenvironments, in Quade, J., and Wynn, J.G., eds., The Geology of Early Humans in the Horn of Africa: Geological Society of America Special Paper 446, doi: 10.1130/2008.2446(03).
Feibel, C.S., 1988, Paleoenvironments of the Koobi Fora Formation, Turkana Basin, Northern Kenya [Ph.D. thesis]: Salt Lake City, University of Utah, 330 p.
Gray, B.T., 1980, Environmental Reconstruction of the Hadar Formation (Afar, Ethiopia) [Ph.D. thesis]: Cleveland, Case Western Reserve University, 431 p.
Hailemichael, M., Aronson, J.L., Savin, S., Tevesz, M.J.S., and Carter, J.G., 2002, $\delta^{18}O$ in mollusk shells from Pliocene Lake Hadar and modern Ethiopian lakes: Implications for history of the Ethiopian monsoon: Palaeogeography, Palaeoclimatology, Palaeoecology, v. 186, p. 81–99, doi: 10.1016/S0031-0182(02)00445-5.
Harris, J.M., Brown, F.H., and Leakey, M.G., 1988, Stratigraphy and paleontology of Pliocene and Pleistocene localities west of Lake Turkana, Kenya: Contributions in Science, v. 399, p. 1–128.
Hecky, R.E., and Kilham, P., 1973, Diatoms in alkaline, saline lakes: Ecology and geochemical implications: Limnology and Oceanography, v. 18, p. 53–71.
Hounslow, M.W., 1997, Significance of localized pore pressures to the genesis of septarian concretions: Sedimentology, v. 44, p. 1133–1147, doi: 10.1046/j.1365-3091.1997.d01-64.x.
Hovers, E., 2003, Treading carefully; site formation processes and Pliocene lithic technology, in Martínez-Moreno, J.M., Mora, R., and de la Torre, I., eds., Oldowan: Rather More Than Smashing Stones: Barcelona, Universitat Autonoma de Barcelona, p. 145–164.
Johanson, D.C., and Taieb, M., 1976, Plio-Pleistocene hominid discoveries in Hadar, Ethiopia: Nature, v. 260, p. 293–297, doi: 10.1038/260293a0.
Johanson, D.C., Taieb, M., Gray, B.T., and Coppens, Y., 1978, Geological framework of the Pliocene Hadar Formation (Afar, Ethiopia) with notes on paleontology including hominids, in Bishop, W.W., ed., Geological Background to Fossil Man: Edinburgh, Scottish Academic Press, p. 549–564.
Johanson, D.C., Taieb, M., and Coppens, Y., 1982, Pliocene hominids from the Hadar Formation, Ethiopia (1973–1977): Stratigraphic, chronological, and paleoenvironmental contexts, with notes on hominid morphology and systematics: American Journal of Physical Anthropology, v. 57, p. 373–402, doi: 10.1002/ajpa.1330570402.
Kalb, J.E., Oswald, E.B., Tebedge, S., Mebrate, A., Tola, E., and Peak, D., 1982, Geology and stratigraphy of Neogene deposits, Middle Awash Valley, Ethiopia: Nature, v. 298, p. 17–25, doi: 10.1038/298017a0.
Kimbel, W.H., Johanson, D.C., and Rak, Y., 1994, The first skull and other new discoveries of Australopithecus afarensis at Hadar, Ethiopia: Nature, v. 368, p. 449–451, doi: 10.1038/368449a0.
Kimbel, W.H., Walter, R.C., Johanson, D.C., Reed, K.E., Aronson, J.L., Assefa, Z., Marean, C.W., Eck, G.G., Robe, R., Hovers, E., Rak, Y., Vondra, C., Yemane, T., York, D., Chen, Y., Evensen, N.M., and Smith, P.E., 1996, Late Pliocene Homo and Oldowan tools from the Hadar Formation (Kada Hadar Member), Ethiopia: Journal of Human Evolution, v. 31, p. 549–561, doi: 10.1006/jhev.1996.0079.
Kimbel, W.H., Johanson, D.C., and Rak, Y., 1997, Systematic assessment of a maxilla of Homo from Hadar, Ethiopia: American Journal of Physical Anthropology, v. 103, p. 235–262, doi: 10.1002/(SICI)1096-8644(199706)103:2<235::AID-AJPA8>3.0.CO;2-S.
Kimbel, W.H., Rak, Y., and Johanson, D.C., 2004, The Skull of Australopithecus afarensis: New York, Oxford University Press, 254 p.
Kraus, M.J., 1987, Integration of channel and floodplain suites: II. Lateral relations of alluvial paleosols: Journal of Sedimentary Petrology, v. 57, p. 602–612.
Kraus, M.J., 1999, Paleosols in clastic sedimentary rocks: Their geologic applications: Earth-Science Reviews, v. 47, p. 41–70, doi: 10.1016/S0012-8252(99)00026-4.
Kraus, M.J., 2002, Basin-scale changes in floodplain paleosols: Implications for interpreting alluvial architecture: Journal of Sedimentary Research, v. 72, p. 500–509, doi: 10.1306/121701720500.
Kraus, M.J., and Aslan, A., 1993, Eocene hydromorphic paleosols: Significance for interpreting ancient floodplain processes: Journal of Sedimentary Petrology, v. 63, p. 453–463.
Kraus, M.J., and Bown, T.M., 1988, Pedofacies analysis: A new approach to reconstructing ancient fluvial sequences, in Reinhardt, J., and Sigleo, W.R., eds., Paleosols and Weathering through Geologic Time: Principles and Applications: Geological Society of America Special Paper 216, p. 143–152.
Kraus, M.J., and Bown, T.M., 1993, Paleosols and sandbody prediction in alluvial sequences, in North, C.P., and Prosser, D.J., eds., Characterization of Fluvial and Aeolian Reservoirs: London, Geological Society of London, p. 23–32.
Leeder, M.R., 1975, Pedogenic carbonates and flood sediment accretion rates: A quantitative model for alluvial arid-zone lithofacies: Geological Magazine, v. 112, p. 257–270.
McKenzie, D.P., Davies, D., and Molnar, P., 1970, Plate tectonics of the Red Sea and East Africa: Nature, v. 226, p. 243–248, doi: 10.1038/226243a0.
Miall, A.D., 1977, A review of the braided river depositional environment: Earth-Science Reviews, v. 13, p. 1–62, doi: 10.1016/0012-8252(77)90055-1.
Miall, A.D., 1978, Lithofacies types and vertical profile models in braided river deposits: A summary, in Miall, A.D., ed., Fluvial Sedimentology: Calgary, Canadian Society of Petroleum Geologists, p. 597–604.
Mohr, P., 1975, Structural setting and evolution of Afar, in Pilger, A., and Rösler, A., eds., Afar Depression of Ethiopia: Inter-Union Commission on Geodynamics, Scientific Report No. 14: Stuttgart, Schweizerbart'sche Verlagsbuchhandlung, p. 27–37.
Paquet, H., 1970, Evolution géochimique des minéraux argileux dans les altérations et les sols des climats Méditérranéens et tropicaux à saisons contrastées: Memoires du Service de la Carte Geologique d'Alsace et de Lorraine, v. 30, p. 1–210.
Peypouquet, J.P., Carbonel, P., Taieb, M., Tiercelin, J.J., and Perinet, G., 1983, Ostracoda and evolution process of paleohydrologic environments in the Hadar Formation (the Afar Depression, Ethiopia), in Maddocks, R.F., ed., Applications of Ostracoda: Houston, Texas, University of Houston Geosciences, p. 277–285.
Pratt, B.R., 2001, Septarian concretions: Internal cracking caused by synsedimentary earthquakes: Sedimentology, v. 48, p. 189–213, doi: 10.1046/j.1365-3091.2001.00366.x.
Quade, J., Levin, N., Semaw, S., Stout, D., Renne, R., Rogers, M., and Simpson, S., 2004, Paleoenvironments of the earliest stone toolmakers, Gona, Ethiopia: Geological Society of America Bulletin, v. 116, p. 1529–1544, doi: 10.1130/B25358.1.
Quade, J., Levin, N.E., Simpson, S.W., Butler, R., McIntosh, W.C., Semaw, S., Kleinsasser, L., Dupont-Nivet, G., Renne, P., and Dunbar, N., 2008, this volume, The geology of Gona, Afar, Ethopia, in Quade, J., and Wynn, J.G., eds., The Geology of Early Humans in the Horn of Africa: Geological Society of America Special Paper 446, doi: 10.1130/2008.2446(01).
Radosevich, S.C., Retallack, G.J., and Taieb, M., 1992, Reassessment of the paleoenvironment and preservation of hominid fossils from Hadar, Ethiopia: American Journal of Physical Anthropology, v. 87, p. 15–27, doi: 10.1002/ajpa.1330870103.
Raiswell, R., and Fisher, Q.J., 2000, Mudrock-hosted carbonate concretions: A review of growth mechanisms and their influence on chemical and isotopic composition: Journal of the Geological Society of London, v. 157, p. 239–251.
Reed, K.E., 2008, Paleoecological patterns at the Hadar hominin site, Afar Regional State, Ethiopia: Journal of Human Evolution, v. 54, p. 743–768.
Renne, P., Walter, R., Verosub, K., Sweitzer, M., and Aronson, J., 1993, New data from Hadar (Ethiopia) support orbitally tuned time-scale to

3.3 Ma: Geophysical Research Letters, v. 20, p. 1067–1070, doi: 10.1029/93GL00733.

Renne, P.R., Swisher, C.C., Deino, A.L., Karner, D.B., Owens, T.L., and DePaolo, D.J., 1998, Intercalibration of standards, absolute ages and uncertainties in ^{40}Ar/^{39}Ar dating: Chemical Geology, v. 145, p. 117–152, doi: 10.1016/S0009-2541(97)00159-9.

Retallack, G.J., 2001, Soils of the Past: An Introduction to Paleopedology: Oxford, Blackwell Science, 404 p.

Schmitt, T.J., and Nairn, A.E.M., 1984, Interpretations of the magnetostratigraphy of the Hadar hominid site, Ethiopia: Nature, v. 309, p. 704–706, doi: 10.1038/309704a0.

Semaw, S., Renne, P., Harris, J.W.K., Feibel, C.S., Bernor, R.L., Fesseha, N., and Mowbray, K., 1997, 2.5-million-year-old stone tools from Gona, Ethiopia: Nature, v. 385, p. 333–336, doi: 10.1038/385333a0.

Soil Survey Staff, 1999, Soil Taxonomy: A Basic System of Soil Classification for Making and Interpreting Soil Surveys: U.S. Department of Agriculture–Natural Resources Conservation Service (NRCS) Agriculture Handbook Number 436, 871 p.

Taieb, M., and Tiercelin, J.J., 1979, Sédimentation Pliocène et paléoenvironnements de rift: Exemple de la formation à hominidés d'Hadar (Afar, Éthiopie): Bulletin de la Société Géologique de France, v. 21, p. 243–253.

Taieb, M., and Tiercelin, J.J., 1980, La stratigraphie et paléoenvironnements sédimentaires de la formation d'Hadar, depression de l'Afar, Éthiopie, in Leakey, R.E., and Ogot, B.A., eds., Proceedings of the 8th Panafrican Congress on Prehistory and Quaternary Studies, Nairobi, 1977: Nairobi, TILLMIAP, p. 109–114.

Taieb, M., Johanson, D.C., Coppens, Y., and Kalb, J.E., 1972, Dépôts sédimentaires et faunes du Plio-Pléistocène de la basse vallée de l'Awash (Afar central Ethiopia): Comptes Rendus de l'Académie des Sciences, série D, v. 275, p. 819–822.

Taieb, M., Johanson, D.C., Coppens, Y., and Aronson, J.L., 1976, Geological and palaeontological background of Hadar hominid site, Afar, Ethiopia: Nature, v. 260, p. 289–293, doi: 10.1038/260289a0.

Taieb, M., Johanson, D.C., Coppens, Y., and Tiercelin, J.J., 1978, Expédition internationale de l'Afar, Ethiopie (4e et 5e campagnes 1975–1977): Chronostratigraphie des gisements à hominidés Pliocènes d'Hadar et corrélations avec les sites préhistoriques du Kada Gona: Comptes Rendus de l'Académie des Sciences, série D, v. 287, p. 459–461.

Tesfaye, S., Harding, D.J., and Kusky, T.M., 2003, Early continental breakup boundary and migration of the Afar triple junction, Ethiopia: Geological Society of America Bulletin, v. 115, p. 1053–1067, doi: 10.1130/B25149.1.

Tiercelin, J.J., 1986, The Pliocene Hadar Formation, Afar Depression of Ethiopia, in Frostick, L.E., Renaut, R.W., Reid, I., and Tiercelin, J.J., eds., Sedimentation in the African Rifts: Geological Society of London Special Publication 25, p. 221–240.

Vondra, C.F., Yemane, T., Aronson, J.L., and Walter, R.C., 1996, A major disconformity within the Hadar Formation: Geological Society of America Abstracts with Programs, v. 28, no. 6, p. 69.

Walter, R.C., 1981, The Volcanic History of the Hadar Early-Man Site and the Surrounding Afar Region of Ethiopia [Ph.D. thesis]: Cleveland, Case Western Reserve University, 426 p.

Walter, R.C., 1994, Age of Lucy and the First Family: Single-crystal ^{40}Ar/^{39}Ar dating of the Denen Dora and lower Kada Hadar Members of the Hadar Formation, Ethiopia: Geology, v. 22, p. 6–10, doi: 10.1130/0091-7613(1994)022<0006:AOLATF>2.3.CO;2.

Walter, R.C., and Aronson, J.L., 1982, Revisions of K/Ar ages for the Hadar hominid site, Ethiopia: Nature, v. 296, p. 122–127, doi: 10.1038/296122a0.

Walter, R.C., and Aronson, J.L., 1993, Age and source of the Sidi Hakoma Tuff, Hadar Formation, Ethiopia: Journal of Human Evolution, v. 25, p. 229–240, doi: 10.1006/jhev.1993.1046.

Wolfenden, E., Ebinger, C., Yirgu, G., Deino, A., and Ayalew, D., 2004, Evolution of the northern Main Ethiopian rift: Birth of a triple junction: Earth and Planetary Science Letters, v. 224, p. 213–228, doi: 10.1016/j.epsl.2004.04.022.

Wynn, J.G., Alemseged, Z., Bobe, R., Geraads, D., Reed, D., and Roman, D.C., 2006, Geological and palaeontological context of a Pliocene juvenile hominin at Dikika, Ethiopia: Nature, v. 443, p. 332–336, doi: 10.1038/nature05048.

Wynn, J.G., Roman, D.C., Alemseged, Z., Reed, D., Geraads, D., and Munro, S., 2008, this volume, Stratigraphy, depositional environments, and basin structure of the Hadar and Busidima Formations at Dikika, Ethiopia, in Quade, J., and Wynn, J.G., eds., The Geology of Early Humans in the Horn of Africa: Geological Society of America Special Paper 446, doi: 10.1130/2008.2446(04).

Yemane, T., 1997, Stratigraphy and Sedimentology of the Hadar Formation [Ph.D. thesis]: Ames, Iowa State University, 182 p.

MANUSCRIPT ACCEPTED BY THE SOCIETY 17 JUNE 2008

Paleoenvironmental context of the Pliocene A.L. 333 "First Family" hominin locality, Hadar Formation, Ethiopia

Anna K. Behrensmeyer*
Department of Paleobiology, MRC 121, National Museum of Natural History,
Smithsonian Institution, P.O. Box 37012, Washington, DC 20013-7012, USA

ABSTRACT

Detailed lateral study of strata associated with the A.L. (Afar Locality) 333 hominin locality provides paleoenvironmental information at geographic scales of hundreds of meters to kilometers as well as insights regarding alluvial deposition and pedogenesis in the middle Denen Dora Member of the Hadar Formation. A.L. 333 is dated at ca. 3.2 Ma and has produced over 260 surface and excavated specimens of *Australopithecus afarensis*. It represents an unusual source of high-resolution information about the paleoenvironmental context of this hominin. The *in situ* hominin fossils are associated with the final stages of filling of a paleochannel and were buried prior to the formation of overlying paleosols. Preserved bedding structures in the fine-grained hominin-producing strata provide evidence that the abandoned channel continued to aggrade prior to the onset of sustained pedogenesis. Pedogenic carbonates associated with the hominin level thus postdate the death and burial of the hominins, possibly by centuries to millennia. The reconstructed paleodrainage of the DD-2 sandstone (DD-2s) is oriented south to north and consists of a trunk channel, ~40 m wide and 3–5 m deep, connecting a tributary system south of A.L. 333 to a distributary system to the north, which likely ended on the deltaic plain associated with the basin's depocenter. The hominin concentration occurs in the upper part of the fill of the trunk channel. The burial of the hominin remains involved fine-grained deposition indicating low-energy, seasonal flood events, and there is no sedimentological evidence for a high-energy, catastrophic flood that could have caused the demise of the hominins.

Keywords: Ethiopia, Pliocene, hominin, Hadar, paleoenvironment.

INTRODUCTION

This research focuses on the fluviolacustrine architecture and paleogeography of the Denen Dora Member of the Hadar Formation, Afar Depression, Ethiopia, in the vicinity of the unique fossil occurrence known as A.L. (Afar Locality) 333 or the "First Family" locality, which is preserved within this interval. Over 260 hominin specimens representing at least 17 individuals (Harmon et al., 2003; Behrensmeyer et al., 2003) occur at A.L. 333, making it arguably one of the most important and enigmatic concentrations of early hominin remains ever found. The Denen Dora Member represents a transition from lacustrine to fluvial environments between 3.26 and 3.20 Ma (Aronson and Taieb, 1981; Walter, 1994; Campisano and Feibel, this volume, Chapter 6; Campisano and Feibel, this volume, Chapter 8). The depositional history of A.L. 333

*behrensa@si.edu

has been debated since the locality was first discovered and specimens were collected in 1975–1976 (Johanson et al., 1978, 1982; White and Johanson, 1989).

Geological research targeting the sedimentary context of fossil concentrations such as A.L. 333 can address differing spatial and temporal scales, each of which provides information relating to depositional processes that contributed to the preservation of the fossil concentration. Studies of regional stratigraphy typically focus on placing fossil localities within an overall stratigraphic context over lateral scales of tens to hundreds of square kilometers, while the geologic setting of individual archaeological and paleontological sites may be documented at centimeter to meter scale. It is the purpose of this article to use intermediate scales of tens of meters to kilometers to reconstruct the fluvial architecture and paleoenvironments of the middle Denen Dora Member. This will provide a basis for interpreting the paleogeography of the physical landscape inhabited by *Australopithecus afarensis* and associated fauna at a well-constrained point of time. Research on the taphonomy and finer-scale microstratigraphy of this locality by the author and Elizabeth L. Harmon is continuing and will be published elsewhere.

BACKGROUND

The Hadar area of the northern Awash Basin in Ethiopia consists of an ~100 km² region centered on the Kada Hadar drainage at ~11°06′N, 40°35′E (Fig. 1); this region is well known for its fossil vertebrates, including many specimens of the hominin *Australopithecus afarensis* (Taieb et al., 1972; Johanson et al., 1978; Kimbel et al., 1994). The first reports of paleoenvironments of the Denen Dora (DD) Member were part of an overall assessment of the depositional context of the vertebrate faunal record throughout the Hadar Formation (Figs. 2 and 3). The Denen Dora Member was described as 30–40 m thick, with shallow-water lacustrine to deltaic plain deposits in the lower submember (DD-1) overlain by swamp and floodplain deposits (DD-2–3) (Gray, 1980; Aronson and Taieb, 1981; Johanson et al. 1982). The DD-2 submember represents a regressive phase of the paleolake and consists of the DD-2s (s = sandstone) and overlying fine-grained fluvial deposits with pedogenically modified units, $CaCO_3$ nodules, and root casts. The DD-3 submember is a major sand-dominated unit with abundant vertebrate fossils, interpreted originally as a network of distributary channels and floodplains (Aronson and Taieb, 1981). The paleogeographic reconstruction of the Denen Dora Member based on this earlier work posited high-energy streams from the highlands to the west that spread out into distributary networks upon entering the lower-gradient areas of the rift floor, forming marshes and small deltas (Johanson et al., 1982). Based on further field research, the lower part of the Denen Dora Member is now interpreted as regressive lake to lake-margin deposits, and DD-3s is interpreted as a single, large-scale meandering fluvial system that eroded into the upper part of the DD-2 submember and aggraded laterally over a wide area (Yemane, 1997; Campisano, 2007).

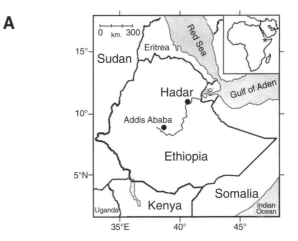

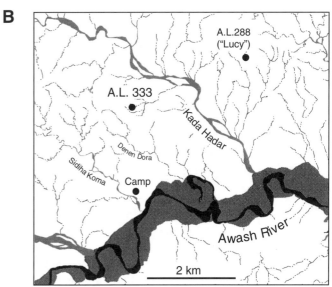

Figure 1. Maps showing the geographic location of (A) the Hadar area in Ethiopia, and (B) the A.L. 333 locality in relation to major drainages at Hadar, the Awash River, and A.L. 288, the fossil locality where the associated partial skeleton known as "Lucy" was found (map in B was modified from Eck, 2002). North is up on both maps; latitude and longitude reference points are shown on the perimeter of A.

In 1975–1976, large numbers of hominin specimens were discovered on the outcrop surfaces in a restricted area designated as A.L. 333, and 19 *in situ* hominin remains were subsequently excavated from carbonate-rich clayey silts near the base of a thick paleosol in the middle part of the Denen Dora Member (Johanson et al., 1982), below DD-3s (Figs. 2 and 3). The locality lies between the TT-4 (Triple Tuff 4), originally dated to 3.22 ± 0.01 Ma (Walter, 1994) and recently recalculated to ca. 3.24 ± 0.01 based on the revised age for the Fish Canyon sanidine standard (Table 5.19 *in* Campisano, 2007), and the overlying KHT (Kada Hadar Tuff) (3.18 ± 0.01 Ma; recalculated to 3.20 ± 0.01 Ma; Table 5.19 *in* Campisano, 2007). Recent analyses by Campisano (2007) indicate a slightly older date for TT-4 of 3.256 ± 0.016, but this will require further work to be confirmed.

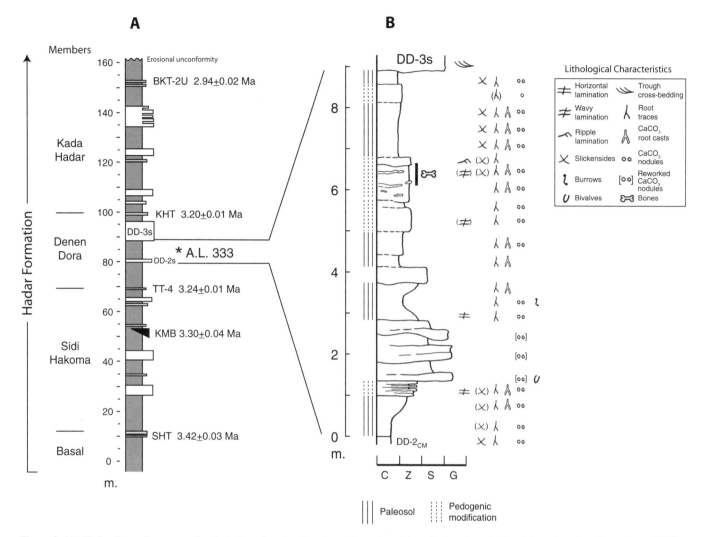

Figure 2. (A) Hadar Formation generalized stratigraphy, showing the major units with revised and recalculated dates based on Campisano (2007). Vertical scale in meters from base; BKT—Bouroukie Tuff, KHT—Kada Hadar Tuff, KMB—Kadada Moumou Basalt, SHT—Sidi Hakoma Tuff, TT-4—Triple Tuff 4. (B) Expansion of the portion of the Denen Dora Member that includes A.L. 333. Detailed section log (section 01/01) was measured at the main A.L. 333 excavation site and extends from the carbonate marker (DD-2_{CM}) into the base of the DD-3 sandstone (DD-3s). Lithologies and sedimentary structures show that the fossiliferous interval occurs in pedogenically altered clayey silts near the top of a fining-upward sequence associated with the DD-2 sandstone (DD-2s), a laterally discontinuous channel deposit. Solid vertical lines indicate well-developed paleosol features (slickensides, root traces, ped structure, calcium carbonate nodules, and root casts); dashed vertical lines indicate units with less-developed pedogenic features and relict bedding. Key to lithology scales below section logs: C—clay, Z—silt, S—sand, G—gravel.

The age of the A.L. 333 fossil locality thus is ca. 3.2 Ma. Most of the hominin fossils were collected along with other faunal remains from an area of ~40 m × 80 m (~3200 m²) on steep slopes and in small ravines up to the stratigraphic level of the excavated specimens. It has long been assumed that the surface hominin fossils were derived from the same sedimentary unit as the *in situ* remains (Aronson and Taieb, 1981).

The geological context of A.L. 333 has been described by a number of researchers, both at the scale of the locality itself (i.e., tens of meters) and within the overall lithostratigraphy of this area (Aronson and Taieb, 1981; Radosevich et al., 1992; Campisano, 2007). The Denen Dora Member is composed predominantly of clays and silty clays but includes three notable fluvial sandstone units (Campisano, 2007). The site occurs between the upper two of these, DD-2s and DD-3s. The DD-2s originally was characterized as a discontinuous but laterally persistent unit above a series of pedogenically modified mudstones (Fig. 2B) in the middle part of the Denen Dora Member (Aronson and Taieb, 1981). Where present, DD-2s is up to 5 m thick and internally complex, generally fining upward from basal, cross-stratified sands with intraclast conglomerates to well-sorted sandy and clayey silts. DD-3s is usually ~5.0 m in thickness but can reach 10 m locally and is a cross-stratified, multistoried sandstone with lateral accretion surfaces and an irregular, often deeply erosional base (Campisano, 2007). It is composed predominantly of coarse sand with lenses of allochthonous volcanic pebbles and cobbles, and it grades

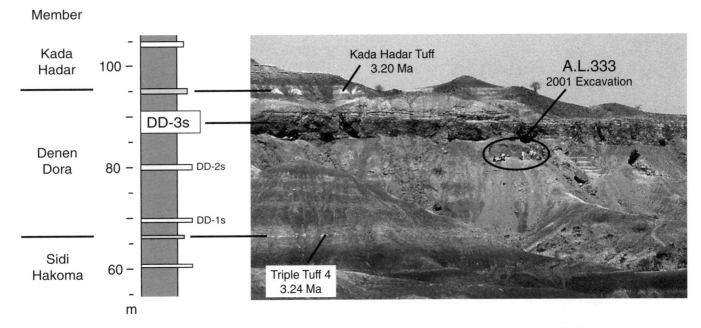

Figure 3. Photograph of outcrops associated with the A.L. 333 locality showing its position relative to the dated tuffs and DD-3s, a laterally extensive, 3–10-m-thick sheet sandstone. The *in situ* hominins were excavated from sediments 2.3–2.85 m below the local base of DD-3s, and surface fossils were collected on both sides of the low ridge and in the drainages below the excavation. Vertical scale shows meters above the base of the Hadar Formation (Campisano, 2007; Campisano and Feibel, 2008, this volume, Chapter 6); DD-1s and DD-2s are two lower, discontinuous sand bodies within the Denen Dora Member. Dates are based on Walter (1994) and Campisano (2007). For scale of photograph, note human figures at excavation (circled) and vertical stratigraphic scale to left.

upward to finer sands and silts below the KHT (see Aronson and Taieb [1981] and Campisano [2007] for further details). DD-3s forms a prominent, indurated sheet sand body throughout the areas of outcrop of the Denen Dora Member, and this helps to create steep slopes and good exposures of the underlying finer-grained DD-2 submember.

METHODS

The area targeted in this study is ~800 m wide (east-west) and ~2.5 km long (north-south), centered on A.L. 333 (Fig. 1). The locality occurs in an amphitheater of steeply sloping sediments extending from below TT-4 to DD-3s, which caps the local ridge top (over 30 m of relief); similar excellent exposures are present throughout the study area. Thirteen geological trenches were dug to document lithostratigraphy over an area of 300 × 500 m lateral to the main hominin-producing areas targeted by the 1976–1977 and 2000–2001 excavations. These, plus the walls of the two excavated areas, made it possible to describe vertical and lateral lithofacies patterns at decimeter scale vertically and 10^2–10^5 m scale laterally (Figs. 3 and 4; see also Supplementary Materials DR1[1]). Four additional trenches and two detailed stratigraphic logs provided documentation of the DD-2 channel and adjacent strata over a broader area (Figs. 5 and 6; Supplementary Materials DR1 [see footnote 1]); the farthest logged section was near A.L. 288—the site of the hominin fossil known as "Lucy"—which occurs at a higher stratigraphic level above DD-3s and just above the KHT.

The DD-2s unit was mapped throughout the study area, following the complex but roughly north-south orientation of the exposures (Fig. 7). Good exposures of the DD-2 submember occur under the continuous ridge-forming DD-3s, and it is straightforward to determine whether the DD-2s is present or not in these exposures. Where DD-2s is absent, there usually are slightly coarser, silty facies at the same level (e.g., sections 01/06 and 01/07 in Fig. 4), indicating overbank deposition from this channel. Abrupt lateral termination of DD-2s provides direct evidence of channel edges, which can be quite steep (Figs. 4–6). Mapping of these channel edges combined with the orientation of internal sedimentary structures indicate where a continuing channel should project onto adjacent outcrops; these predictions were tested by direct observation, trenching, and additional section logs (Figs. 4–7; section logs in Data Repository [see footnote 1]).

The panel diagrams (Figs. 4–6) provide cross-sectional views of the fluviolacustrine architecture at three different scales. These were constructed based on stratigraphic correlations between marker beds and unique sequences of beds. Correlation was straightforward over a distance of ~150 m in the A.L. 333

---

[1]GSA Data Repository item 2008196, consisting of scans of the author's original field logs documenting the stratigraphic context of the A.L. 333 locality, is available at www.geosociety.org/pubs/ft2008.htm, or on request from editing@geosociety.org, Documents Secretary, GSA, P.O. Box 9140, Boulder, CO 80301-9140, USA.

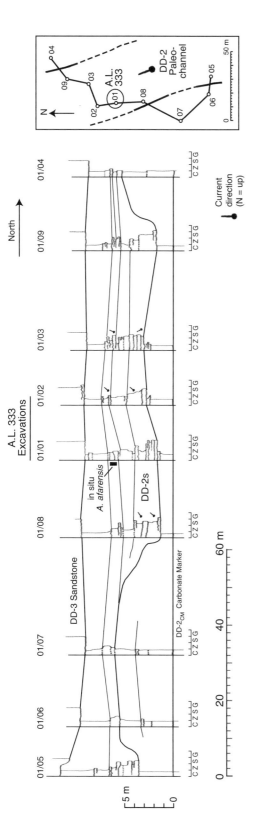

Figure 4. Two-dimensional panel showing the context of A.L. 333 within ~150 m of lateral exposures, based on geological step trenches and excavation walls. The top of a carbonate-rich unit in the Denen Dora Member, the carbonate marker (DD-2$_{CM}$), is the horizontal datum. Inset box shows the plan view of the section logs and the reconstructed boundaries of the DD-2 channel segment in the vicinity of A.L. 333; solid lines are documented channel margins, and dashed lines are reconstructed channel margins. Cross-stratification indicates a dominantly N to NE current direction. Key to lithology scales below section logs: C—clay, Z—silt, S—sand, G—gravel.

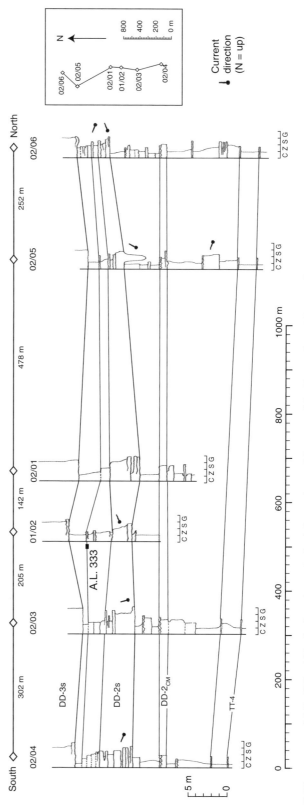

Figure 5. The context of A.L. 333 within the broader-scale fluviolacustrine architecture of the Denen Dora Member from TT-4 to the base of DD-3s. The panel represents a south-to-north cross section over ~1400 lateral meters (sections 02/04–02/06). Horizontal datum is the carbonate marker (CM). Inset box shows plan view of section positions. Key to lithology scales below section logs: C—clay, Z—silt, S—sand, G—gravel.

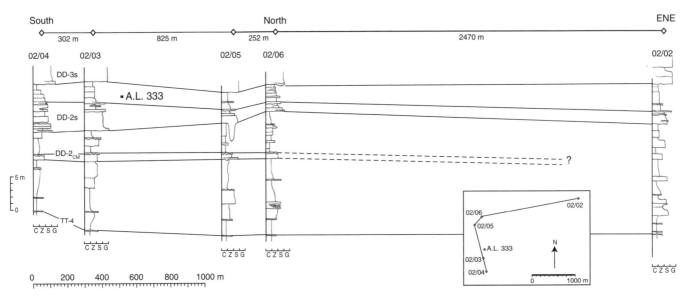

Figure 6. Panel diagram correlating Denen Dora Member strata from the area documented in Figure 5 to strata near the A.L. 288 "Lucy" site to the ENE. Total length of panel is ~4 km. Horizontal datum is the carbonate marker (CM), except between 02/06 and 02/02, where it is TT-4. Inset box shows plan view of section positions and A.L. 333 site. Key to lithology scales below section logs: C—clay, Z—silt, S—sand, G—gravel.

amphitheater, as individual beds could be traced along continuous outcrops (Figs. 3–4). The top of a prominent white-weathering, carbonate-rich claystone (see also Campisano, 2007), designated here as the DD-2 carbonate marker (DD-2_{CM}), was used as the horizontal datum in this area. The larger-scale panels also use this datum except between 02/06 and 02/02 (Fig. 6), where TT-4 is used because the correlation of DD-2_{CM} is no longer clear. Lateral correlations in Figure 5 are based on 1380 m of mostly continuous outcrops, and they show that DD-2s occurs throughout the study area. This scale oversimplifies the configuration of the channel system, however, which can be locally absent, as shown in Figure 4. There is marked thickening of the interval between TT-4 and DD-2s from south to north, indicating deepening of the basin in this direction. The nearly 4-km-long panel in Figure 6 extends the correlations to the vicinity of the A.L. 288 "Lucy" locality to the east, based on stratigraphic matching and marker units, including DD-3s and TT-4. This panel shows a consistent thickness of strata between TT-4 and DD-2s from A.L. 333 to A.L. 288, several kilometers to the ENE, suggesting that the basin was not deeper in that direction during the deposition of the clays and Vertisols of the DD-1 submember. This supports increased subsidence of the northern Ethiopian Rift basin toward the north between 3.8 and 2.9 Ma (Wynn et al., 2006) but is at odds with reconstructions of Hadar paleogeography during Denen Dora time (Yemane, 1997; Campisano, 2007), which suggest that the paleolake lay east of Hadar.

The Denen Dora Member is the most productive in terms of numbers of vertebrate fossils in the Hadar Formation (Taieb et al., 1976; Campisano, 2007), and most of these fossils occur in the DD-2 and DD-3 submembers. At A.L. 333, as well as elsewhere along the outcrops below DD-3s, many fossils found on the slopes of the DD-2 submember are clearly derived from overlying DD-3 strata based on adhering coarse sand matrix. Careful surface surveys of the DD-2 submember throughout the study area turned up few fossils that could be clearly associated with the A.L. 333 level (mainly fish, one crocodile tooth, a bovid horn core, an elephant mandible *in situ*). However, this may be in part because of previous intensive collecting in the Denen Dora Member near A.L. 333; farther north across the Kada Hadar drainage, the DD-2 channel sandstone becomes more fossiliferous where it broadens into a sheet sand body.

RESULTS

Strata between TT-4 and DD-3s support earlier interpretations of lake regression and the development of an emergent lake-margin plain (Taieb et al., 1976; Tiercelin, 1986; Aronson and Taieb, 1981; Campisano, 2007). Relatively pure, finely laminated clays at the level of TT-4 represent lacustrine sedimentation, and the upward increase in pedogenic features, including soil carbonate and well-developed slickensides, indicates wetlands that became progressively drier seasonally. These were incised by a paleochannel, the deposits of which formed the DD-2s submember. Continuing alluvial floodplain aggradation combined with a marked interval of paleosol formation followed the end of DD-2 channel deposition.

The DD-2 sand body is a ribbon sandstone (Friend et al., 1979), representing a paleochannel system that was oriented roughly south to north across the study area. It is well exposed in the A.L. 333 amphitheater (Figs. 2–4 and 7) and forms a prominent feature in the deposits of the emergent fluvial-deltaic plain. At A.L. 333, the channel deposit is ~40 m wide, 2–3 m deep, and extends north through the ridge that forms the amphitheater; its edges can be clearly seen on the ridge's northwest face (Figs. 2

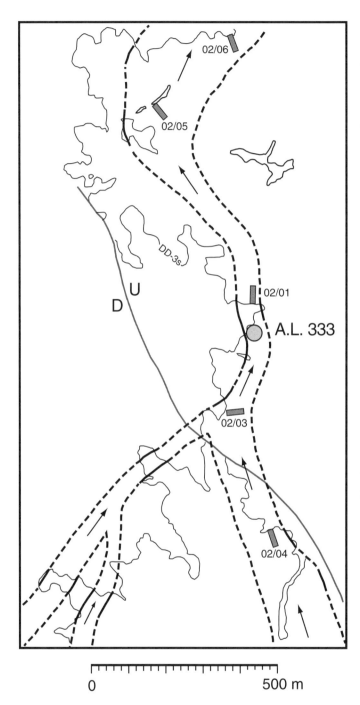

Figure 7. Reconstructed DD-2 channel system, based on documented channel facies and edges (solid line segments), current indicators within the channel, and intervening areas where channel deposits were absent. Thin irregular line marks the outcrop edge of DD-3s; all of the exposures under this sandstone were examined for the presence or absence of DD-2s. Gray rectangles show positions of geological trenches; north is up on the diagram. Normal fault (gray line) does not significantly affect exposures of the Denen Dora Member in this area.

and 4). The basal contact of the channel is irregular, with poorly sorted, locally derived mud and carbonate clasts generally less than 1 cm diameter and occasional articulated bivalves. The lower part of the channel has a sharp cutbank on its northwest edge just south of the excavation area (Fig. 4). Section 01/05 documents another, upstream portion of this same channel to the south (Fig. 4, inset plan view). Farther south (Figs. 5 and 7), exposures of channel segments at the same stratigraphic level are reconstructed as two branches feeding into the single channel at A.L. 333. Two parallel, shallower, and narrower U-shaped channel sand bodies toward the southwest (based on thickness and width of fill) are reconstructed as two smaller tributaries (Fig. 7). Toward the north, DD-2s broadens into a more sheet-like sand body, but internal features including current indicators are consistent with the continuation of the DD-2 channel system. The reconstructed channel in the vicinity of A.L. 333 is sinuous but not meandering, and it did not cut laterally to form a sheet sand. This is evidence that the channel was incised into the alluvial plain and then filled and abandoned in a relatively short period of time without cutting laterally into preexisting floodplain sediments (in contrast to DD-3s).

Aronson and Taieb (1981) described some channels in the Hadar Formation as U-shaped and noted that this implied pauses in sediment accumulation with intervals of channelized erosion into preexisting alluvium. The DD-2 channel provides evidence of such localized incision. In addition to the channel cut-and-fill documented at A.L. 333 (Fig. 4), three steep-sided, 2.5-m-deep "runnel" features at the base of DD-2s north of A.L. 333 (two at the 02/05 trench and one documented across the Kada Hadar drainage to the north) indicate marked erosional incision followed by rapid channel filling (Fig. 5). This suggests either temporary regression of the paleolake and headward erosion from the north, or tectonically controlled variability in local subsidence rates, e.g., local syndepositional flexure or faulting that caused some areas of the depositional basin to be slightly uplifted relative to others and subject to greater fluvial downcutting (Aronson and Taieb, 1981). Short-term climate cycling also could have affected the erosive power of flow into the lake, e.g., if the alluvial substrate was stabilized by vegetation, thereby reducing sediment load in the A.L. 333 channel and increasing its tendency to erode downward rather than expand laterally.

The DD-2 channel fill at A.L. 333 is sand and fine gravel at the base but includes interbedded sand and pedogenically modified sandy silts; these were regarded as separate small channels by Aronson and Taieb (1981), but they actually form parts of the fill of the larger channel feature shown in Figure 4. Deposits of silty clay with $CaCO_3$ root casts and pedogenic nodules cap the lower part of the channel fill (Fig. 2B, meters 2.8–3.7, Fig. 4). This represents an intermediate phase of fine-grained channel fill when flow energy had temporarily decreased. Above this paleosol, 40 cm silty sand indicate reactivated channel flow, followed by ~2.5 m of silts with variable degrees of pedogenic modification. The top of this sequence is formed by ~1.0 m of well-sorted gray silt, which extends throughout the exposures of the channel in the

A.L. 333 amphitheater at this level but is thickest in the vicinity of the main hominin excavation. Based on original maps, stratigraphic logs, and photographs from the 1976–1977 excavations, reexcavation of the locality in 2000–2001 by W.H. Kimbel (2001, personal commun.), and the author's personal experience at the site, the upper part of this gray clayey silt is the stratum in which the *in situ* hominin remains were originally found. Relative to the horizontal datum provided by DD-2_{CM} (Figs. 4 and 5), the channel underlying these silts forms a slight depression in the area where the hominins were concentrated, and the gray silt unit also is thickest there. This implies that the hominin remains were buried during the final infilling of a shallow ($\leq$~0.5 m), abandoned channel swale, with deposition overlapping onto the adjacent floodplain.

Radosevich et al. (1992) noted that carbonate precipitation at the A.L. 333 site may have occurred in loose substrates shortly after burial as well as during subsequent soil formation. Some of the hominin remains found on the surface up to the level of the excavation had thick $CaCO_3$ encrustations and also matrix consisting of medium-grained sand and reworked $CaCO_3$ nodules (Aronson and Taieb, 1981). The latter suggests that some of the fossil remains were buried by moderately high-energy flow and that these coarser facies of the original hominin-bearing deposit have been lost to erosion. The channel swale could have been temporarily reactivated during a wet-season flooding event, depositing coarser sediment along with the hominin remains as well as finer silts and clays prior to its final abandonment. Alternatively, the surface-collected hominins at A.L. 333 may be derived from successive strata representing different depositional events that combined to form the upper portions of the DD-2 sand body.

The gray silts at the excavated hominin level are cross-stratified at section 01/02, and current directions are oriented 50° east, similar to the current indicators in the lower, sandier DD-2 channel deposits. This suggests that the flow that buried the hominin remains was partly constrained and directed by the linear swale provided by the preexisting channel. The fact that cross-bedding and horizontal stratification are preserved in this unit implies that the silt deposit was a one time event, or part of a succession of events that were thick enough to protect some of the stratified sediments from later bioturbation and other homogenizing processes of pedogenesis associated with the overlying, well-developed clayey Vertisol. This 1.3 m Vertisol has abundant $CaCO_3$ nodules and root casts (Fig. 2). The soil's parent material accumulated during overbank sedimentation that preceded or was partially contemporaneous with pedogenesis (i.e., an accretionary soil). All of the sediments above the gray silts postdate the burial of the hominin remains, and it is possible that some (though probably not all) of the carbonate precipitation and nodule formation associated with the hominin bones occurred in the subsurface as the Vertisol was forming above the clayey silts. Regardless of the precise timing of the carbonate precipitation, the Vertisol near the top of the DD-2 submember postdates the hominin burial event.

The geological history of the interval between TT-4 and DD-3s, summarized next, is based on documentation of strata in the vicinity of A.L. 333 and is in general agreement with that of Aronson and Taieb (1981) as well as Campisano (2007); it provides additional details regarding the DD-2 channel and its relationship to the hominin remains at A.L. 333. This history can be separated into 14 stages:

1. lacustrine sedimentation (laminated beds) in a relatively deep lake with clay-grade clastic input plus occasional airfall tuffs (Triple Tuff sequence);
2. shallowing of the lake, with localized influxes of silts and sands, likely from distributary lobes of a nearby delta;
3. development of lake-margin wetlands with continuing lake regression, seasonal emergence, and pedogenesis;
4. increased development of pedogenic carbonate, indicating more pronounced periods of emergence; continuing lake regression;
5. continuing fluctuation of fine-grained clastic deposition and paleosol formation; more pronounced carbonates (Soil Carbonate Stage 2 of Gile et al., 1966);
6. formation of drainage channel (DD-2) in the study area and variable but locally pronounced downcutting resulting in narrow basal erosion gullies incised into earlier fine-grained deposits;
7. filling of first channel phase, up to approximately the level of the initial erosional incision;
8. temporary cessation of sand-grade clastic deposition and evidence of inactive channel and paleosol formation on the initial channel fill;
9. channel reactivation, with sand followed by silt deposition gradually filling in the channel and aggrading laterally beyond the margins of the earlier incised channel;
10. deposition of hominin remains in a shallow swale near the top of DD-2 channel fill;
11. final phase of deposition in the channel, burying the A.L. 333 hominin assemblage;
12. continued fine-grained deposition on an alluvial plain, where the DD-2 channel system had died or avulsed away from the study area; calcium carbonate precipitation in and around some of the hominin remains and other buried organic materials;
13. formation of a clay-rich accretionary Vertisol, with occasional influxes of fine silt and sand-grade sediment; pronounced development of pedogenic carbonate and prismatic ped structure indicating marked seasonal water-table fluctuations; and
14. erosion of the DD-3 channel system into the upper part of DD-2 submember, including the capping Vertisol, and deposition of a multistoried sheet sand over the area.

DISCUSSION

The DD-2s, although mostly below the level of the excavated hominins, provides the paleoenvironmental context for strata at the excavation site and a rationale for the depositional events that buried the hominin remains. All of the DD-2 channel fill deposits are similar in lithology, bedding features, and current direction, and this evidence plus the consistent stratigraphic position above a widespread green, $CaCO_3$-rich clay ($DD-2_{CM}$) and below DD-3s support the interpretation that this was single-channel system that flowed across a flat, emergent plain toward the Hadar paleolake to the north. This reconstructed channel system, which can be traced as a single entity over ~1.5 km along its axis (Figs. 5 and 7), is north-directed and has a variable preserved width (40–~80 m) with smaller and shallower tributaries feeding in from the south. Near the modern Kada Hadar drainage and across it to the north, the channel is well defined but wider and appears to diverge into multiple channels to form a braided distributary system.

The excavated hominin fossils occurred in clayey silt at the top of the channel fill and below a well-developed Vertisol. Preservation of vertebrate remains in the upper parts of abandoned channel fills is common in the terrestrial fossil record (Behrensmeyer, 1988). Previous work by Radosevich et al. (1992) has provided a thorough characterization of paleosols associated with the original A.L. 333 excavation area, based on samples collected by M. Taieb and analyzed by S. Radosevich and G. Retallack. The designation on their Figure 2 (Radosevich et al., 1992, p. 22) indicating "Hominids in Place" at 1.1–1.2 m below DD-3s, however, is in error, based on 1970s diagrams and photographs (W. Kimbel, 2004, personal commun.), relocation of the hominin-bearing level in 2000, and Figure 6 in Aronson and Taieb (1981, p. 188). The actual level is 2.3–2.85 m below DD-3s. Diagrams and measurements of Radosevich et al. (1992) show that all paleosol samples analyzed were above the hominin level except the lowest sample, F-2, at 2.45 m below the base of DD-3s. Thus, their inferences regarding the paleoenvironmental context of the hominin remains, based on the analysis of the paleosols at A.L. 333, relate to environmental conditions that postdate the time of hominin death and burial. The upper part of the hominin-bearing level is in the base of the "Type Fo" paleosol, which Radosevich et al. (1992) characterized as very weakly developed. The F-2 sample is dominantly clay, with ~16% silt and sand, and it differs from most other samples in having relatively more organic carbon and strontium, suggesting unique features of this layer.

As noted by Radosevich et al. (1992), the A.L. 333 hominin remains could have been buried with the parent material or incorporated into the sediment during a subsequent period of pedogenesis. They suggested a catastrophic flood event as a cause of the death and burial of the hominins, supporting the first alternative and implying that paleosol development was subsequent to the burial event. Aronson and Taieb (1981) also attributed the hominin concentration to mass death during a flood event. The results of the 2002–2003 research provide support for burial of the remains during a fluvial depositional event and later pedogenesis superimposed on the original parent sediment. Given the context within the top of a dying channel, however, in a gently sloping swale that was ≤0.5 m deep, as well as the fragmentary nature of the buried hominin remains, it is unlikely that this depositional event—i.e., a flood—also caused their death. Hominin body parts or individual bones could have been affected by this flow, but the fine grain size of the enclosing sediment and lack of basal erosion features indicate that the depositional event was primarily aggradational, gently covering rather than scouring preexisting sediment and organic remains, at least at the site of the excavated hominins. Possible transport of the hominin remains from upstream in the channel swale or from a wider scatter on the adjacent floodplain can be tested using the hominin fossils themselves; taphonomic study bearing on such hypotheses will be presented in subsequent publications.

Although there is no direct record of vegetation at the A.L. 333 site, other than root casts associated with pedogenesis, palynological research in the lower Denen Dora Member (DD-1 submember) indicates that the regional habitat was predominantly a dry grassland (Bonnefille et al., 2004). Gray (1980) noted that, locally, fossils of the genus *Kobus* (waterbuck) and other reduncines are common in the Denen Dora Member, which indicates moist substrates with "fresh grass" forage (Reed, 1998). Recent detailed analysis of the depositional environments and mammalian fauna of the Hadar Formation by Campisano (2007) shows paleogeographic differences in the DD-2 submember, with edaphic grasslands and marshy conditions to the north and east of A.L. 333 and more closed bush or woodland habitats in the vicinity of and west of this locality. Stable isotope analyses of pedogenic carbonates at A.L. 333 are in general agreement with Campisano's faunal evidence, indicating 30%–34% C_4 grassland (Hailemichael, 2000), which is a relatively low proportion of grass compared with other samples from the Denen Dora Member. However, it should be noted that since pedogenic carbonate at the site formed after the burial of the hominin fossils, this habitat signature may not relate to conditions when living hominins were actually present.

The combined evidence indicates that both closed and open habitats were present in the DD-2 submember; the gradient went from more closed to the west to more open, edaphic grasslands to the east (Campisano, 2007). The north-directed DD-2 paleochannel in the immediate area of A.L. 333 likely was one of many drainages that carried water and sediment from the areas to the south and west across the deltaic plain to the north and northeast. The overall landscape that the A.L. 333 hominins inhabited, or perhaps occasionally traversed, would have been a relatively featureless, seasonally dry, grass-dominated plain (on a scale of kilometers) where the slight depression created by the dying channel may have hosted a different type of vegetation, including bushes or trees that were able to grow under conditions of more stable soil moisture (available to plant roots in the silty and sandy channel deposits underlying the channel swale). It is possible that hominins and other animals moved along such linear depressions left by old channels when they ventured into the more open grassland environments or used such areas as sheltering places. The scale of root casts in the brown paleosols lateral to the DD-2 channel indicates grass or small shrubs, and the root casts in the silts and clays in the upper channel fill indicate grass, shrubs, and bushes, but probably not large trees. The immediate vicinity of the A.L. 333 excavation site has some of the densest concentrations of $CaCO_3$ root casts and nodules in the upper DD-2 channel deposit, suggesting seasonal fluctuations in moisture (i.e., due to seasonally elevated periods of soil evaporation and plant evapotranspiration) relating to more intense biological activity in this particular spot. The narrow channel between the tributary and distributary systems (Fig. 7) could have provided increased groundwater availability in the underlying channel sands, thus helping to focus more plant growth and deeper root systems in this linear paleogeographic feature.

Although estimates of the time represented by deposition versus pedogenesis versus erosional hiatuses are problematic at the scale of individual geological strata, the DD-2 submember provides an unusual opportunity to examine depositional history between two well-dated tuffs that are close in absolute age. The total amount of elapsed time, based on revised and recalculated dates for the TT-4 and KHT (Walter, 1994; Campisano, 2007) indicates that the ~30 m of strata represent ~40 k.y. (Fig. 2A) for a sediment accumulation rate of 75 cm/k.y., which corresponds in time to a period of increased basin subsidence between 3.4 and 3.2 Ma (Walter, 1994; Dupont-Nivet et al., this volume). The deposits of this interval include DD-3s, which is up to 10 m thick with a deeply erosional base. It seems unlikely that this sheet sand body could have been emplaced in less than 10,000 yr, leaving ~30 k.y. for the ~20 m below DD-3. There are at least eight paleosols in the DD-2 submember, for an average maximum of ~4000 yr per paleosol. The Vertisol above the DD-2 channel is thicker, more internally complex, and more extensively developed than earlier clay-rich soils, and it likely represents a relatively longer period of time, perhaps on the order of 5000 yr. The period of DD-2s channel incision and filling itself should represent no less than several thousand years and likely is closer to 5000 yr, particularly since it includes at least one period of inactive flow and soil formation. This leaves an estimated remaining time span of 20 k.y. below DD-2 for deposition of seven distinct intervals of fine-grained lacustrine to wetland sediments and their subsequent modification by pedogenesis, or ~3000 yr per interval. If Campisano's date of 3.256 ± 0.016 Ma for TT-4 (Campisano, 2007) is used, the total interval would increase from 40 k.y. to 56 k.y., increasing these estimates by ~30%. In spite of the uncertainty of the time estimates, it is clear that the DD-2 submember preserves a high-resolution history of short-term environmental change on a scale of millennia, superimposed on the longer period of lake regression that spanned tens of thousands of years. Against this backdrop of change in the physical environment, the final burial of the concentration of hominin remains at A.L. 333 occurred over much shorter period of time, perhaps in minutes or hours but likely no more than a few years, based on the geological evidence.

Among the East African localities where *A. afarensis* has been documented, only two provide high-resolution temporal and spatial information about the geological and paleoenvironmental context of this species, Hadar and Laetoli. The hominin trackway at Laetoli in Tanzania and the numerous autochthonous fossil bones and teeth from hominins and associated fauna preserved in volcaniclastic silts and paleosols at 3.5–3.8 Ma indicate bush, woodland, and open habitats (Leakey and Harris, 1987; Agnew et al., 1996; Kappelman et al., 1997; Su and Harrison, 2007). The *in situ* hominin remains at A.L. 333 can be related to a death—and likely life—association of multiple hominin individuals with an abandoned channel swale that crossed a flat, extensive alluvial plain several kilometers from a paleolake to the north. The "Lucy" skeleton and other *A. afarensis* specimens from throughout the Hadar Formation are derived from fluvial and lake-margin deposits but lack more detailed evidence concerning sedimentary context, mainly because most are surface finds. Other East African localities with *A. afarensis* provide only general spatial and temporal information regarding its association with particular habitats. The Dikika *A. afarensis* child from deposits east of Hadar was found in a sandstone block associated with channel sandstones (Alemseged et al., 2006), and the remains may have been transported by fluvial processes prior to final burial in this sand. Teeth and other surface remains from Turkana Basin localities in Kenya are generally associated with fluvial or fluvial-deltaic deposits or are not complete enough to be certainly identified as *A. afarensis* (Feibel et al., 1989; Leakey and Harris, 2003; Campisano et al., 2004). Thus, the A.L. 333 *in situ* hominins and the Laetoli footprints provide the highest resolution contextual evidence for *A. afarensis* ecology currently available and indicate that this hominin occupied a mix of open and closed habitats in aggrading rift basin fluvial plains (Hadar) as well as more "upland" volcaniclastic terrain (Laetoli).

CONCLUSIONS

The study of the paleoenvironmental context of the A.L. 333 locality presented in this paper involves spatial and temporal scales ranging from those of the excavated site itself to decimeter to kilometer cross sections of the aggrading fluvial-deltaic system of the Denen Dora Member of the Hadar Formation. Each scale of inquiry provides information bearing on the depositional history of the locality itself as well as the adjacent strata.

The *in situ* hominin concentration at A.L. 333 is associated with the final stages of filling of the DD-2 paleochannel, and remains were buried prior to the formation of the overlying Vertisol. The burial event(s) that interred the hominin remains at the top of the channel likely occurred during seasonal flooding, and the resulting sediment was subsequently modified by pedogenic processes over centuries to millennia. Preserved bedding structures in the hominin-producing strata suggest that the immediate area of the excavated site continued to aggrade for some time prior to the development of the overlying paleosol(s). Thus, although pedogenic carbonates occur in the hominin level and the fossils have been described as occurring in a paleosol (Aronson and Taieb, 1981; Radosevich et al., 1992), the evidence presented here shows that they were buried by alluvial parent sediment that was later modified by pedogenic processes, not incorporated into the paleosol after it had begun to form.

The reconstructed DD-2 paleodrainage in the study area consists of a trunk channel connecting a tributary system within ~0.5 km to the south of A.L. 333 with a distributary system ~0.5 km to the north; the latter continued northeast for at least several kilometers and likely fed the deltaic plain formed by the retreating paleolake to the north. The hominin concentration is associated with the relatively narrow single channel between the branching distributary and tributary portions of the drainage system, which raises the possibility of a behavioral as well as a taphonomic cause for the paleogeographic position of A.L. 333, i.e., hominins and other animals frequenting a narrow strip of habitat formed by the abandoned channel fill. This channel may have been one of many of similar scale that combined to form the discontinuous sandstone outcrops of DD-2s across the broader extent of the Denen Dora Member. The burial of the excavated hominin remains involved fine-grained deposition, probably due to a shallow, seasonal flood event, and there is no sedimentological evidence for a high-energy, catastrophic flood that could have caused the demise of the hominins.

Based on revised age dates for the TT-4 and KHT, the DD-2 channel system is estimated to have formed and filled in ~5000 yr, and the overlying fine-grained floodplain sediments and Vertisol development may represent ~10,000 yr. The preservation of the hominins occurred during an interval of overall rapid aggradation of the Hadar Formation (5.5 m in 5 k.y. or ~110 cm/1000 yr for the DD-2 channel fill based on section 01/01). The link between well-preserved vertebrate fossils and high rates of sediment accumulation has been previously noted (Campisano et al., 2004) and underscores the paleontological importance of tectonic subsidence in the Afar Depression rift system during the mid-Pliocene. Correlations of sections in the area surrounding A.L. 333 indicate that the depocenter of the Hadar Formation during Denen Dora Member time was toward the north, but faunal gradients (Campisano, 2007) appear to have been transverse rather than parallel to this direction. This suggests that rift-margin climatic gradients could have interacted with axial river and lake systems in complex ways to create habitat variability over scales of tens of kilometers in the Pliocene Ethiopian Rift. Channel systems such as that represented by the DD-2s could have been important as corridors for animals moving among the different habitats and utilizing diverse resources on the Pliocene paleolandscape, as well as providing taphonomically favorable sites for fossil preservation.

ACKNOWLEDGMENTS

The author offers sincere thanks to Abebaw Ejigu and Mohammed Ahmadin, who helped with supervision and translation in the field, Mesfin Mekonen and the Hadar camp staff for superb field support, and all the Afar people who guarded us and helped with the geological trenching. I am grateful to Don Johanson and Bill Kimbel for welcoming me as a member of the Hadar team and for their support and generous sharing of unpublished information about A.L. 333, Gerry Eck for assistance with the Hadar maps, and Erella Hovers for advice and encouragement. I thank Elizabeth Harmon for her expertise in documenting the excavations and geological trenches in the A.L. 333 amphitheater and for her ongoing contributions as a collaborator in the overall A.L. 333 project. Chris Campisano provided essential help and advice during the course of this study, including constructive manuscript reviews, and Jim Aronson also provided advice and encouragement. I thank Mamitu Yilma, Director of the National Museum of Ethiopia, for her gracious assistance with the research. Permission for field research was granted by the Authority for Research and Conservation of Cultural Heritage (ARCCH) of the Ministry of Youth, Sport, and Culture and the National Museum of Ethiopia, and funding support was provided by the National Geographic Society and the National Science Foundation (BCS-0080378 to Bill Kimbel). Finally, I thank Jay Quade and Jonathan Wynn for their patience and encouragement with the production of this manuscript.

REFERENCES CITED

Agnew, N., Demas, M., and Leakey, M.D., 1996, The Laetoli footprints: Science, v. 271, p. 1651–1652, doi: 10.1126/science.271.5256.1651b.

Alemseged, Z., Spoor, F., Kimbel, W.H., Geraads, D., Reed, D., and Wynn, J.G., 2006, A juvenile early hominin skeleton from Dikika, Ethiopia: Nature, v. 443, p. 296–301, doi: 10.1038/nature05047.

Aronson, J.L., and Taieb, M., 1981, Geology and paleogeography of the Hadar hominid site, Ethiopia, in Rapp, G., Jr., and Vondra, C.F., eds., Hominid Sites: Their Geologic Settings, American Association for the Advancement of Science Selected Symposium 63: Boulder, Colorado, Westview Press, p. 165–195.

Behrensmeyer, A.K., 1988, Vertebrate preservation in fluvial channels, in Behrensmeyer, A.K., and Kidwell, S.M., eds., Ecological and Evolutionary Implications of Taphonomic Processes: Palaeogeography, Palaeoclimatology, Palaeoecology, v. 63, no. 1–3, p. 183–199.

Behrensmeyer, A.K., Harmon, E.H., and Kimbel, W.H., 2003, Environmental context and taphonomy of the First Family hominid locality, Hadar, Ethiopia: Journal of Vertebrate Paleontology, v. 23, supplement to no. 3, p. 33A.

Bonnefille, R., Potts, R., Chalie, F., Jolly, D., and Peyron, O., 2004, High-resolution vegetation and climate change associated with Pliocene *Australopithecus afarensis*: Proceedings of the National Academy of Sciences of the United States of America, v. 101, no. 33, p. 12,125–12,129, doi: 10.1073/pnas.0401709101.

Campisano, C.J., 2007, Tephrostratigraphy and Hominin Paleoenvironments of the Hadar Formation, Afar Depression, Ethiopia [Ph.D. thesis]: New Brunswick, New Jersey, Rutgers University, 620 p.

Campisano, C.J., and Feibel, C.S., 2008, this volume (Chapter 8), Depositional environments and stratigraphic summary of the Pliocene Hadar Formation at Hadar, Afar Depression, Ethiopia, in Quade, J., and Wynn, J.G., eds., The Geology of Early Humans in the Horn of Africa: Geological Society of America Special Paper 446, doi: 10.1130/2008.2446(08).

Campisano, C.J., and Feibel, C.S., 2008, this volume (Chapter 6), Tephrostratigraphy of the Hadar and Busidima Formations at Hadar, Afar Depression, Ethiopia, in Quade, J., and Wynn, J.G., eds., The Geology of Early Humans in the Horn of Africa: Geological Society of America Special Paper 446, doi: 10.1130/2008.2446(06).

Campisano, C.J., Behrensmeyer, A.K., Bobe, R., and Levin, N., 2004, High-resolution paleoenvironmental comparisons between Hadar and Koobi Fora: Preliminary results of a combined geological and paleontological approach: Paleoanthropology Society Abstracts, p. A34; http://www.paleoanthro.org/abst2004.htm.

Dupont-Nivet, G., Sier, M., Campisano, C.J., Arrowsmith, J.R., DiMaggio, E., Reed, K., Lockwood, C., Franke, C., and Hüsing, S., 2008, this volume (Chapter 3), Magnetostratigraphy of the eastern Hadar Basin (Ledi-Geraru research area, Ethiopia) and implications for hominin paleoenvironments, in Quade, J., and Wynn, J.G., eds., The Geology of Early Humans in the Horn of Africa: Geological Society of America Special Paper 446, doi: 10.1130/2008.2446(03).

Eck, G.G., 2002, An Atlas of Hadar: Tempe, Arizona, Institute of Human Origins at Arizona State University.

Feibel, C.S., Brown, F.H., and McDougall, I., 1989, Stratigraphic context of fossil hominids from the Omo Group deposits, northern Turkana Basin, Kenya and Ethiopia: American Journal of Physical Anthropology, v. 78, p. 595–622, doi: 10.1002/ajpa.1330780412.

Friend, P.F., Slater, M.J., and Williams, R.C., 1979, Vertical and lateral building of river sandstone bodies, Ebro Basin, Spain: Journal of the Geological Society of London, v. 136, p. 39–46, doi: 10.1144/gsjgs.136.1.0039.

Gile, L.H., Peterson, F.F., and Grossman, R.B., 1966, Morphological and genetic sequences of carbonate accumulation in desert soils: Soil Science, v. 101, p. 347–360, doi: 10.1097/00010694-196605000-00001.

Gray, B.T., 1980, Environmental Reconstruction of the Hadar Formation (Afar, Ethiopia) [Ph.D. thesis]: Cleveland, Ohio, Case Western Reserve University, 431 p.

Hailemichael, M., 2000, The Pliocene Environment of Hadar, Ethiopia: A Comparative Isotopic Study of Paleosol Carbonates and Lacustrine Mollusk Shells of the Hadar Formation [Ph.D. thesis]: Cleveland, Ohio, Case Western Reserve University, 239 p.

Harmon, E.H., Behrensmeyer, A.K., Kimbel, W.H., and Johanson, D.C., 2003, Preliminary taphonomic analysis of hominin remains from A.L. 333, Hadar Formation, Ethiopia: Paleoanthropology Society Abstracts; http://www.paleoanthro.org/abst2003.htm.

Johanson, D.C., Taieb, M., Gray, B.T., and Coppens, Y., 1978, Geological framework of the Pliocene Hadar Formation (Afar, Ethiopia), in Bishop, W.W., ed., Geological Background to Fossil Man: Edinburgh, Scottish Academic Press, p. 549–564.

Johanson, D.C., Taieb, M., and Coppens, Y., 1982, Pliocene hominids from the Hadar Formation, Ethiopia (1973–1977): Stratigraphic, chronologic, and paleoenvironmental contexts, with notes on hominid morphology and systematics: American Journal of Physical Anthropology, v. 57, p. 373–402, doi: 10.1002/ajpa.1330570402.

Kappelman, J., Plummer, T., Bishop, L., Duncan, A., and Appleton, S., 1997, Bovids as indicators of Plio-Pleistocene paleoenvironments in East Africa: Journal of Human Evolution, v. 32, p. 229–256, doi: 10.1006/jhev.1996.0105.

Kimbel, W.H., Johanson, D.C., and Rak, Y., 1994, The first skull and other new discoveries of *Australopithecus afarensis* at Hadar, Ethiopia: Nature, v. 368, p. 449–451, doi: 10.1038/368449a0.

Leakey, M.D., and Harris, J.M., eds., 1987, Laetoli: A Pliocene Site in Northern Tanzania: Oxford, Oxford University Press, 584 p.

Leakey, M.G., and Harris, J.M., eds., 2003, Lothagam: The Dawn of Humanity in Eastern Africa: New York, Columbia University Press, 678 p.

Radosevich, S.C., Retallack, G.J., and Taieb, M., 1992, Reassessment of the paleoenvironment and preservation of hominid fossils from Hadar, Ethiopia: American Journal of Physical Anthropology, v. 87, p. 15–27, doi: 10.1002/ajpa.1330870103.

Reed, K.E., 1998, Using large mammal communities to examine ecological and taxonomic structure and predict vegetation in extant and extinct assemblages: Paleobiology, v. 24, p. 384–408.

Su, D., and Harrison, T., 2007, The paleoecology of the Upper Laetoli Beds at Laetoli: A reconsideration of the large mammal evidence, in Bobe, R., Alemseged, Z., and Behrensmeyer, A.K., eds., Hominin Environments in the East African Pliocene: An Assessment of the Faunal Evidence: New York, Springer, Vertebrate Paleobiology and Paleoanthropology Series, v. 1, p. 279–314.

Taieb, M., Coppens, Y., Johanson, D.C., and Kalb, J., 1972, Dépôts sédimentaires et faunes du Plio-Pléistocène de la basse vallée de l'Awash (Afar central Ethiopia): Comptes Rendus de l'Académie des Sciences (Paris), Série D, v. 275, p. 819–822.

Taieb, M., Johanson, D.C., Coppens, Y., and Aronson, J.L., 1976, Geological and paleontological background of Hadar hominid site, Afar, Ethiopia: Nature, v. 260, p. 289–293, doi: 10.1038/260289a0.

Tiercelin, J.J., 1986, The Pliocene Hadar Formation, Afar Depression of Ethiopia, in Frostick, L.E., Renaut, R.W., Reid, I., and Tiercelin, J.J., eds., Sedimentation in the African Rifts: Oxford, Blackwell Scientific, p. 221–240.

Walter, R.C., 1994, Age of Lucy and the First Family: Single-crystal $^{40}Ar/^{39}Ar$ dating of the Denen Dora and lower Kada Hadar Members of the Hadar Formation, Ethiopia: Geology, v. 22, no. 1, p. 6–10, doi: 10.1130/0091-7613(1994)022<0006:AOLATF>2.3.CO;2.

White, T.D., and Johanson, D., 1989, The hominid composition of Afar Locality 333: Some preliminary observations, in Hominidae: Proceedings of the 2nd International Congress of Human Paleontology: Milan, Editoriale Jaca Book, p. 97–101.

Wynn, J.G., Alemseged, Z., Bobe, R., Reed, D., and Roman, D.C., 2006, Geological and palaeontological context of a Pliocene juvenile hominin at Dikika, Ethiopia: Nature, v. 443, no. 7109, p. 332–336, doi: 10.1038/nature05048.

Yemane, T., 1997, Stratigraphy and Sedimentology of the Hadar Formation [Ph.D. thesis]: Ames, Iowa, Iowa State University, 182 p.

MANUSCRIPT ACCEPTED BY THE SOCIETY 17 JUNE 2008

Herbivore enamel carbon isotopic composition and the environmental context of Ardipithecus at Gona, Ethiopia

Naomi E. Levin*†
Department of Geology and Geophysics, University of Utah, Salt Lake City, Utah 84112, USA

Scott W. Simpson*
Department of Anatomy, Case Western Reserve University School of Medicine, Cleveland, Ohio 44106, USA, and
Laboratory of Physical Anthropology, Cleveland Museum of Natural History, Cleveland, Ohio 44106, USA

Jay Quade*
Department of Geosciences, University of Arizona, Tucson, Arizona 85721, USA

Thure E. Cerling*
Department of Geology and Geophysics, University of Utah, Salt Lake City, Utah 84112, USA

Stephen R. Frost*
Department of Anthropology, University of Oregon, Eugene, Oregon 97403, USA

ABSTRACT

Ardipithecus fossils found in late Miocene and early Pliocene deposits in the Afar region of Ethiopia, along with *Sahelanthropus tchadensis* from Chad and *Orrorin tugenensis* from Kenya, are among the earliest known human ancestors and are considered to be the predecessors to the subsequent australopithecines (*Australopithecus anamensis* and *Australopithecus afarensis*). Current paleoenvironmental reconstructions suggest a wooded habitat for both *Ardipithecus kadabba* and *Ardipithecus ramidus* but more open and varied environments for other hominids living in Africa during the late Miocene and early Pliocene. To further evaluate the environmental context of *Ardipithecus*, we present stable carbon isotope data of 182 fossil herbivore teeth from *Ardipithecus*-bearing fossil deposits in the Gona Paleoanthropological Research Project area, in the Afar region of Ethiopia. The sampled teeth include representatives of all major fossil herbivore taxa and the majority of the mammalian biomass that lived in the same time and place as the hominids. When compared to extant herbivores from East Africa, the spectra of isotopic results from herbivores found in late Miocene *Ar. kadabba* and early Pliocene *Ar. ramidus* sites at Gona are most similar to isotopic values from extant herbivores living in bushland and grassland regions and

---

*E-mails: Levin: naomi@gps.caltech.edu; Simpson: sws3@cwru.edu; Quade: jquade@geo.arizona.edu; Cerling: thure.cerling@utah.edu; Frost: sfrost@uoregon.edu.
†Current address: Division of Geological and Planetary Sciences, California Institute of Technology, MC100-23, 1200 E. California Blvd., Pasadena, California 91125, USA.

Levin, N.E., Simpson, S.W., Quade, J., Cerling, T.E., and Frost, S.R., 2008, Herbivore enamel carbon isotopic composition and the environmental context of *Ardipithecus* at Gona, Ethiopia, *in* Quade, J., and Wynn, J.G., eds., The Geology of Early Humans in the Horn of Africa: Geological Society of America Special Paper 446, p. 215–234, doi: 10.1130/2008.2446(10). For permission to copy, contact editing@geosociety.org. ©2008 The Geological Society of America. All rights reserved.

dissimilar to those from herbivores living in closed-canopy forests, montane forests, and high-elevation grasslands. The tooth enamel isotopic data from fossil herbivores make it clear that *Ardipithecus* at Gona lived among a guild of animals whose diet was dominated by C_4 grass, and where there is no record of closed-canopy vegetation.

Keywords: Gona, Ethiopia, carbon isotopes, tooth enamel, *Ardipithecus*.

INTRODUCTION

Ecological change is considered to be a key component to understanding the tempo and mode of human evolution (Laporte and Zihlman, 1983; Potts, 1998; White et al., 2006). The few late Miocene and early Pliocene deposits that do contain hominid fossils (Toros-Menalla in Chad, the Middle Awash, Gona, and Galili in Ethiopia, and the Tugen Hills, Kanapoi, and Lothagam in Kenya) suggest that these early hominids were associated with a range of environments, including shrublands, heterogeneous woodlands, closed woodland/forests, and rain forests (Wolde-Gabriel et al., 1994, 2001; Wynn, 2000; Haile-Selassie, 2001; Vignaud et al., 2002; Cerling et al., 2003a; Haile-Selassie et al., 2004; Pickford et al., 2004; Semaw et al., 2005; Urbanek et al., 2005; White et al., 2006). Late Miocene and early Pliocene hominid fossils, *Ardipithecus kadabba* and *Ardipithecus ramidus*, occur in the Middle Awash and Gona study areas, in the Afar region of Ethiopia (White et al., 1994; Haile-Selassie, 2001; Semaw et al., 2005; Simpson et al., 2007) (Fig. 1). The presence of *Ardipithecus* fossils from coeval deposits less than 100 km apart provides an opportunity to evaluate the environmental context of *Ardipithecus* beyond the scale of a single paleoanthropological research project area. In the Middle Awash study area, *Ar. kadabba* and *Ar. ramidus* specimens are associated with other fossils that suggest that they were restricted to wooded environments, avoided open environments, and likely lived in highland settings (WoldeGabriel et al., 1994, 2001). There is considerable overlap in the faunal composition of fossils found associated with late Miocene *Ar. kadabba* and early Pliocene *Ar. ramidus* in the Middle Awash and Gona study areas. However, fossil assemblages from Gona display distinct differences in faunal proportions, including a lower relative percentage of tragelaphine bovids and colobine primates (Semaw et al., 2005), suggesting that *Ar. ramidus* may have accessed a broader range of environments than documented in the Middle Awash study area.

Stable isotopes provide an additional perspective on the environmental context of *Ardipithecus* (WoldeGabriel et al., 2001; Semaw et al., 2005; White et al., 2006). The carbon isotopic composition of herbivore tooth enamel is a reliable dietary indicator that can be used to identify the relative contribution of grasses, trees and shrubs, and closed-canopy resources in an animal's diet. The carbon isotopic composition of teeth can be ecosystem specific (Cerling et al., 2003b, 2004) and, within a fossil context, provide information on the dietary preferences of the animals among which hominids lived. The carbon isotopic composition of tooth enamel, which reflects diet in an animal's early adult years, strongly complements morphological data, which reflect long-term evolutionary trends.

In this paper, we compare fossil tooth enamel isotopic results from *Ardipithecus*-bearing deposits at Gona to similar data from extant mammals in Ethiopia and across eastern Africa. This comparison makes it clear that early hominids at Gona, in both the late Miocene and early Pliocene, lived in an ecologically diverse environment that included abundant grasses.

BACKGROUND

Geology and Fossil Fauna

Late Miocene and early Pliocene fluvial and lacustrine deposits from the Gona Paleoanthropological Research Project (GPRP) study area, in the Afar region of Ethiopia, contain abundant vertebrate fossil remains, including hominids (Semaw et al., 2005; Simpson et al., 2007; Kleinsasser et al., this volume; Quade et al., this volume). The deposits containing *Ar. kadabba*, assigned to the 6.5–5.2 Ma Adu-Asa Formation, occur in small packages of sediment between stacked basalt flows (Simpson et al., 2007; Kleinsasser et al., this volume). The *Ar. ramidus*–bearing deposits in the Gona study area come from the Sagantole Formation (5.2–3.9 Ma), which is composed of basalts and fluviolacustrine sediments that rest conformably on the Adu-Asa Formation and are bound on the east by the As Duma fault. East of the As Duma fault, Pliocene and Pleistocene sediments of the Hadar and Busidima Formations have filled the basin, which formed as the result of movement along the As Duma fault (Quade et al., this volume) (Fig. 2).

In the Adu-Asa Formation, *Ar. kadabba* and other mammalian fauna are found in three fossiliferous horizons, above the Bodele Tuff (6.48 ± 0.22 Ma), above the Sifi Tuff, and at the level of the Kobo'o Tuff (5.44 ± 0.06 Ma) (Kleinsasser et al., this volume). Herbivore fossil fauna associated with *Ar. kadabba* include specimens of Colobinae and Cercopithecinae, *Nyanzachoerus syrticus*, Hippopotamidae, multiple bovid taxa, Giraffidae, *Eurygnathohippus* cf. *feibeli*, Rhinocerotidae, and *Anancus*. This assemblage bears similarities to that found in coeval deposits of the Adu-Asa Formation in the Middle Awash area (WoldeGabriel et al., 2001; Haile-Selassie et al., 2004). The fossil-bearing sediments in the Adu-Asa Formation at Gona include laminated siltstones, sands, and conglomerates indicative of both lacustrine and fluvial environments (Kleinsasser et al., this volume; Quade et al., this volume). High sedimentation rates, the lack of well-developed paleosols, and extensive basalts in the

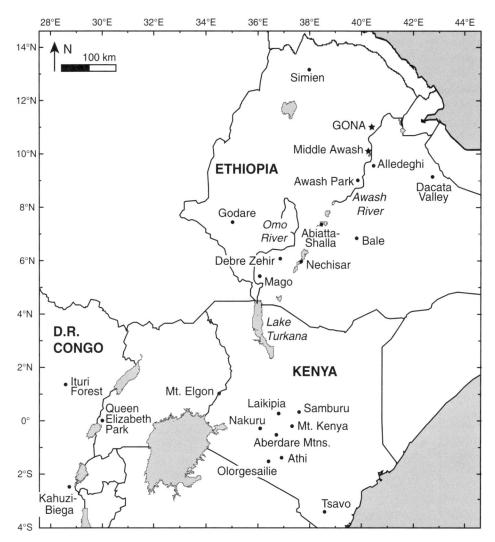

Figure 1. Map of sampling localities for modern teeth. Fossil localities discussed in text at length are marked with stars.

Adu-Asa Formation suggest active deposition in narrow floodplains confined by large expanses of basalt flows.

Ar. ramidus and other mammalian fossils are found predominantly in marginal lacustrine or marshy settings within the Sagantole Formation (Semaw et al., 2005; Quade et al., this volume). The herbivore fossil fauna associated with *Ar. ramidus* include Colobinae and Cercopithecinae, three suid species, Hippopotamidae, multiple bovid taxa, Giraffidae, Equidae, at least three proboscidean taxa, and Rhinocerotidae. To date, more than 130 fossil localities have been identified in the Sagantole Formation at Gona. Most of these localities can be confidently assigned to the Segala Noumou (4.6–4.2 Ma) or As Duma (<5.2–4.6 Ma) Members. The Segala Noumou and As Duma Members have been intensively studied at their type locations between the Sifi and Busidima Rivers (Quade et al., this volume). The fossil sites in the As Duma Member all occur at the same stratigraphic level and paleoenvironmental setting, the shallow margins of a lake where it lapped onto an active basaltic cinder cone. Mammalian fossils from the Segala Noumou Member are found in gastropod-rich tufas formed in shallow lakes or marshes and in bedded siltstones that represent the overbank deposits of small rivers. The faunal compositions from sites in the As Duma and Segala Noumou Members are indistinguishable biostratigraphically. Sediments containing the Gona Western Margin South (GWMS) sites are exposed south of the Sifi River (Fig. 2). Unlike areas north of the Sifi, the geology of this area is not known in detail. Sediments in this area strongly resemble those of the Segala Noumou Member, with which we provisionally correlate them. The stratigraphic position of other sites, including GWM10 and 11 (part of the same sedimentary package), GWM45, and GWM18, is less clear, except that they lie stratigraphically above the top of the Adu-Asa Formation (<5.2 Ma) and below the top of the Sagantole Formation (>3.9 Ma).

Carbon Isotopes in Vegetation

Carbon isotope ratios ($^{13}C/^{12}C$) of C_3 and C_4 plants are distinct because they use different photosynthetic pathways to metabolize CO_2 (Farquhar et al., 1989). In East Africa, the majority of shrubs and trees uses the C_3 pathway, whereas most grasses use the C_4 pathway, except for those growing at high elevations (above ~3000 m) (Tieszen et al., 1979; Young and Young, 1983). Stable carbon isotope ratios are commonly reported relative to the isotopic standard Vienna Peedee belemnite (VPDB) and presented as δ values in per mil (‰) units, wherein $\delta^{13}C = ([^{13}C/^{12}C]_{sample}/[^{13}C/^{12}C]_{standard} - 1) \times 1000‰$. The $\delta^{13}C$ values of C_3 plants range from –36.5‰ to –22‰ and between –15‰ and –11‰ for C_4 vegetation in equatorial Africa (Cerling and Harris, 1999; Cerling et al., 2004). Plants that use the Crassulacean Acid Metabolism (CAM) photosynthetic pathway have $\delta^{13}C$ values intermediate to C_3 and C_4 plants, but because CAM plants (e.g., cacti and succulents) do not make a large contribution to the diets of most extant mammals, they are not usually considered in the interpretation of $\delta^{13}C$ values of herbivore tooth enamel. Cerling et al. (2003c) reported that $\delta^{13}C$ values of C_3 plants from savannas and bushlands in Kenya average –27.0 ± 0.2‰, whereas plants from the subcanopy in the Ituri Forest average –34.0 ± 1.5‰. The $\delta^{13}C$ values of C_3 plants growing in an open-canopy forest average –27.8 ± 0.3‰, and they more closely resemble isotopic values of C_3 plants in savannas and bushlands than those in closed-canopy ecosystems. The ^{13}C-depleted values of C_3 plants in the Ituri Forest are due to the "canopy effect" (van der Merwe, 1991), wherein

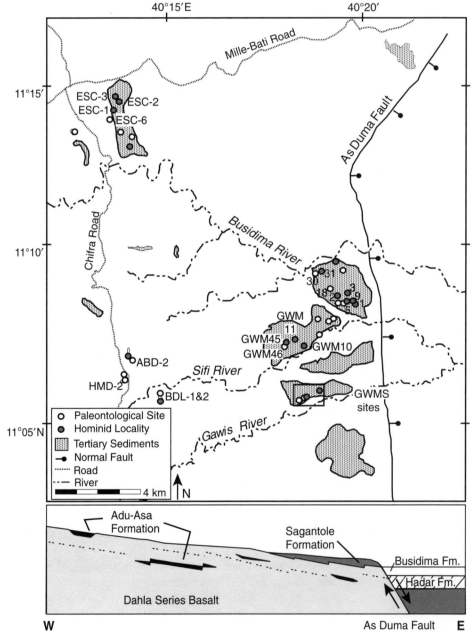

Figure 2. Map and schematic cross section of fossil localities in the western portion of the Gona Paleoanthropological Research Project study area where fossil teeth were sampled. Sites from the As Duma and Segala Noumou Members in the Sagantole Formation are labeled with site number. All other sites are labeled with site prefix and number. The prefixes for site names Gona Western Margin (GWM) and Gona Western Margin South (GWMS) are abbreviated accordingly. For detailed geological maps and cross sections of the Adu-Asa and Sagantole Formations at Gona, see Kleinsasser et al. (this volume) and Quade et al. (this volume).

$\delta^{13}C$ values of plants are reduced by recycling of CO_2 within the forest canopy. This distinction within C_3 plants is important because the dietary composition of herbivores living in a closed-canopy ecosystem can be detected in the fossil record (MacFadden and Higgins, 2004). The $\delta^{13}C$ values of C_3 plants can increase by several per mil when water-stressed (Farquhar et al., 1989). The $\delta^{13}C$ values of C_4 grasses in East Africa also vary with water stress. The three main subpathways of C_4 photosynthesis include nicotinamide adenine dinucleotide phosphate (NADP), nicotinamide adenine dinucleotide (NAD), and phosphoenolpyruvate carboxykinase (PCK). The $\delta^{13}C$ values of mesic (NADP subpathway) grasses average -11.8 ± 0.2‰, whereas xeric (NAD, PCK subpathways) grasses have a lower average $\delta^{13}C$ value, -13.1 ± 0.3‰ (Cerling et al., 2003c).

Interpretation of Fossil $\delta^{13}C$ Values

The reconstruction of herbivore paleodiet and paleovegetation from $\delta^{13}C$ values of fossil teeth and paleosol carbonate involves several steps. One must estimate three values: $\delta^{13}C$ of atmospheric CO_2, the carbon isotope enrichment factor between atmospheric CO_2 and plants, and the carbon isotope enrichment factor between diet and tooth enamel or between vegetation and paleosol carbonate. The enrichment factor is the isotopic difference between two phases, and it is represented by ε^*_{A-B}, where $\varepsilon^*_{A-B} = ([\delta^{13}C_{PhaseB} + 1000]/[\delta^{13}C_{PhaseA} + 1000] - 1) \times 1000$. ε^* is used instead of the fractionation factor ($\alpha_{A-B} = [\delta^{13}C_{PhaseB} + 1000]/[\delta^{13}C_{PhaseA} + 1000]$), in order to present isotopic differences in per mil (‰). ε^* refers to isotopic differences between two phases that may not be in isotopic equilibrium (Cerling and Harris, 1999).

The $\delta^{13}C$ values of vegetation, teeth, and soil carbonates must be considered with respect to the $\delta^{13}C$ value of the atmosphere in which they formed. The $\delta^{13}C$ values of Atlantic Ocean benthic foraminifera can be used as recorders of $\delta^{13}C$ values of atmospheric CO_2 ($\delta^{13}C_{atmCO_2}$), assuming that the $\delta^{13}C$ values of foraminifera reflect surface dissolved inorganic carbon, which is in equilibrium with atmospheric CO_2. The ε^* value between atmospheric CO_2 and benthic foraminifera ($\varepsilon^*_{CO_2\text{-foram}}$) can be estimated for the past 30 k.y. from benthic foraminifera $\delta^{13}C$ values (Zachos et al., 2001) and the $\delta^{13}C_{atmCO_2}$ value of air trapped in ice cores (Indermühle et al., 1999; Smith et al., 1999). With the Cenozoic record of benthic foraminifera $\delta^{13}C$ values from Zachos et al. (2001), the calculated $\varepsilon^*_{CO_2\text{-foram}}$ can be used to determine $\delta^{13}C_{atmCO_2}$ values for the late Miocene and early Pliocene fossils deposits at Gona (Table 1). These estimated values are more ^{13}C-enriched than today's $\delta^{13}C_{atmCO_2}$ value (-8‰), which has decreased over the past 150 yr due to the combustion of ^{13}C-depleted fossil fuels (Keeling et al., 1979, 1995; Francey et al., 1999). Assuming that modern plants fix carbon from the present atmosphere with consistent $\varepsilon^*_{atmosphere-plant}$ values as in the past, $\varepsilon^*_{atmosphere-plant}$ can be estimated and used to estimate plant $\delta^{13}C$ values for time periods when $\delta^{13}C_{atmCO_2}$ values were different from today's values (Table 1).

The carbon isotope enrichment factor between herbivore diet and animal tissue ($\varepsilon^*_{diet-tissue}$) has been determined experimentally for domestic and laboratory animals and estimated for wild animals (Cerling and Harris, 1999; Passey et al., 2005). For

TABLE 1. ESTIMATED $\delta^{13}C$ VALUES FOR ATMOSPHERIC CO_2, VEGETATION, TOOTH ENAMEL, AND SOIL CARBONATE

	Present	0–30 ka	Sagantole Formation	Adu-Asa Formation
Time Interval (Ma)	0.0	0–0.03	5.2–3.9	6.5–5.2
$\delta^{13}C$ benthic foraminifera* (‰, VPDB)	—	0.6	0.4	1.0
$\delta^{13}C$ atmospheric CO_2† (‰, VPDB)	–8.0	–6.5	–6.7	–6.1
$\delta^{13}C$ vegetation§ (‰, VPDB)				
Closed-canopy C_3	–31.4	–30.0	–30.1	–29.6
C_3	–27.0	–25.5	–25.7	–25.1
C_4	–12.0	–10.5	–10.6	–10.0
$\delta^{13}C$ tooth enamel# (‰, VPDB)				
Closed-canopy C_3	–17.6	–16.1	–16.3	–15.7
C_3	–13.1	–11.6	–11.7	–11.1
C_4	2.2	3.7	3.5	4.1
$\delta^{13}C$ soil carbonate** (‰, VPDB)				
Closed-canopy C_3	–17.7	–16.2	–16.4	–15.8
C_3	–13.2	–11.7	–11.8	–11.2
C_4	2.1	3.6	3.4	4.0

Note: VPDB is an abbreviation for the isotopic standard Vienna Peedee belemnite.
*Average $\delta^{13}C$ values of Atlantic benthic foraminifera were calculated from Zachos et al. (2001).
†Present $\delta^{13}C$ value of atmospheric CO_2 is from Francey et al. (1999), and the 0–30 ka value is from air trapped in ice cores from Smith et al. (1999) and Indermuhle et al. (1999). $\delta^{13}C$ values of atmospheric CO_2 for the fossil time intervals at Gona were calculated from benthic foraminifera $\delta^{13}C$ values, assuming that the enrichment factor between atmospheric CO_2 and benthic foraminifera was the same for these time intervals as it was for the 0–30 ka interval.
§Average values for modern plants from Cerling et al. (2003c). Values for plants from other time intervals were calculated using the observed enrichment factor between $\delta^{13}C$ values of atmospheric CO_2 and of each vegetation type for the present time interval.
#Expected enamel $\delta^{13}C$ values for animals consuming only the respective vegetation type, assuming a diet-enamel enrichment of +14.1‰, as discussed in the text.
**Expected $\delta^{13}C$ values of soil carbonates formed under different vegetation types, assuming plant to soil carbonate enrichment factor of +14‰.

large ruminant mammals, Cerling and Harris (1999) estimated a $\varepsilon^*_{diet-tissue}$ of 14.1 ± 0.5‰ for tooth enamel. This enrichment factor is consistent with the experimental results from Passey et al. (2005) for domestic cows, which show a $\varepsilon^*_{diet-enamel}$ value of 14.6 ± 0.3‰, but it is distinct from the $\varepsilon^*_{diet-enamel}$ value of 13.3 ± 0.3‰ found for the modern domestic pig, *Sus scrofa*. In this paper, we use the $\varepsilon^*_{diet-enamel}$ value of 14.1 ± 0.5‰ (Cerling and Harris, 1999) to interpret tooth enamel $\delta^{13}C$ values for all herbivore taxa because we do not know enough about the digestive physiologies of the fossil species to adjust the enrichment factors accordingly.

Carbonate-rich soils form in seasonally dry environments. The $\delta^{13}C$ values of soil carbonates that form at depths greater than 30 cm record the carbon isotope composition of vegetation growing in the soil during carbonate formation (Cerling, 1999). The carbon isotope enrichment factor between vegetation and soil carbonate depends on a combination of equilibrium and kinetic fractionation processes and can range between 13‰ and 16‰, depending on plant productivity, atmospheric CO_2 concentration, and soil properties (Cerling, 1999).

A summary of our estimates for $\delta^{13}C$ values of atmospheric CO_2, plants in different ecosystems, herbivore enamel, and soil carbonate is listed in Table 1. This approach to adjusting expected $\delta^{13}C$ values of enamel and paleosol carbonates according to changing $\delta^{13}C_{atmCO_2}$ values through time is similar to the approach used by Passey et al. (2002).

METHODS

Sample Collection

Tooth enamel was sampled from existing collections of herbivore skulls and isolated mandibles in several national parks and reserves in Ethiopia (Abiatta-Shalla, Alledeghi, Awash, Mago, Nechisar, Bale, and Simien) (Fig. 1), the Natural History Museum at the University of Addis Ababa, and the headquarters of the Ethiopian Wildlife Conservation Organization in Addis Ababa. These collections were supplemented by cranio-dental remains found on survey in the national parks and the Gona Paleoanthropological Research Project study area, which increased the sample size to 142 teeth from 122 individuals. The $\delta^{13}C$ values of tooth enamel from herbivores found in modern Ethiopian environments were compared to carbon isotopic data of teeth from similar taxa living in the mountain regions and rift valleys in Kenya and Uganda, and forests in the Democratic Republic of the Congo (D.R.C.) (Fig. 1). Modern collection localities are characterized using the ecoregion classification system defined by Olson et al. (2001) (Table 2). Ecoregions are large areas of land (~150,000 km²) that contain geographically distinct assemblages of natural communities that share a large majority of species, ecosystem dynamics, and environmental conditions (Olson et al., 2001).

For isotopic analysis, 182 fossil teeth from paleontological localities in the Adu-Asa and Sagantole Formations at Gona were sampled for isotopic analysis. This sample includes all mammalian taxa available except for hominids, micromammals, and carnivores (e.g., cercopithecids, suids, hippopotamids, bovids, giraffids, equids, rhinocerotids, and proboscideans). Bulk samples (GONBULK) of fragmented large herbivore teeth (e.g., suid, bovid, proboscidean, etc.) and complete teeth accessioned by the National Museums of Ethiopia were sampled. When possible, we sampled third molars, teeth that have formed after the adult diet has been adopted. Teeth formed prior to weaning may have ^{13}C-depleted $\delta^{13}C_{enamel}$ values compared to teeth formed with only an adult diet because milk contains a high proportion of lipids, which are depleted in ^{13}C relative to carbohydrates and proteins (DeNiro and Epstein, 1978).

Enamel from fragmentary GONBULK samples was separated from dentine and ground using a mortal and pestle, yielding up to 20 mg of powder. Accessioned specimens were sampled in the National Museum of Ethiopia in Addis Ababa, using a diamond bit and a hand-held drill. These teeth were sampled in a shallow vertical groove or pit removing 1–3 mg of enamel powder. All teeth were cleaned or wiped with ethanol before sampling, and great care was taken to avoid dentine, cementum, and matrix in the sample powder. Only one sample was taken from each fossil tooth. Modern teeth were sampled using a similar approach, but, in some cases, several teeth were sampled from a single individual, and single teeth were sampled in multiple places.

When present in stratigraphic section with fossil-bearing sediments at Gona, paleosol carbonates were collected from carbonate (Bk) horizons in paleo-Vertisols at least 30 cm below the upper boundary of the paleosol. Paleosol carbonates were collected at these depths to avoid the influence of ^{13}C-enriched atmospheric CO_2 on carbonate $\delta^{13}C$ values. Paleosol carbonates sampled for isotopic analysis in this study represent different substrates than the fossil teeth that were sampled, which come from nonpedogenic horizons.

Sample Treatment

All GONBULK sample powders were treated with 3% H_2O_2 for 30 min and 1 M acetic acid for 15–30 min, rinsed after each treatment with deionized water, and then dried at 60 °C before analysis. For the majority of modern teeth and accessioned fossil teeth, untreated powders were analyzed first. If enough powder remained after the initial isotopic analysis of the untreated portion, the sample was treated in the same manner as the GONBULK samples, except they were treated with 0.1 M acetic acid instead of 1 M acetic acid.

Isotopic Analysis and Notation

All GONBULK sample powders were reacted in 100% H_3PO_4 in a constant temperature bath at 50 °C. The resultant CO_2 was extracted offline and then analyzed on a Finnigan Delta S mass spectrometer at the University of Arizona.

TABLE 2. LOCATION AND DESCRIPTION OF MODERN ENVIRONMENTS SAMPLED

Location	Ecoregion	Latitude (°N)	Longitude (°E)	Elevation (m)	Country	Locations included
Abiatta-Shalla	Somali acacia-commiphora bushlands and thickets	7.5	38.6	1580–1700	Ethiopia	Abiatta-Shalla NP* and Bulbula town
Alledeghi	Somali acacia-commiphora bushlands and thickets	9.2	40.4	840	Ethiopia	
Athi Plains	Northern acacia-commiphora bushlands and thickets	−1.4	36.9	1640	Kenya	Athi Game Ranch, Nairobi NP*, Kitengela, Ngong, Lukenya Hills, Athi River, Sukari Ranch, Kikuyu, Embu Kjabe, Limuru
Dacata Valley	Ethiopian montane forests	9.1	42.4	1350	Ethiopia	Near Babille town
Bale	Ethiopian montane moorland & montane grasslands/woodlands	6.8	39.8	3000–4000	Ethiopia	
Debre Zehir	Somali acacia-commiphora bushlands and thickets	6.1	36.9	1550	Ethiopia	
Godare	Ethiopian montane forests	7.4	35.0	900	Ethiopia	
Gona	Somali acacia-commiphora bushlands and thickets	11.1	40.3	500–900	Ethiopia	
Ituri Forest	Northeastern Congolian lowland forests	1.4	28.6	770	D.R.C.	
Kahuzi-Biega	Albertine Rift montane forests	−2.5	28.8	1800–2500	D.R.C.	Mountain and lowland sectors
Kenya Mountains	East African montane forests	−0.4	36.6	2000–4000	Kenya	Mt. Kenya NP, Mt. Elgon NP, and Aberdare NP*
Laikipia	Northern acacia-commiphora bushlands and thickets	0.3	36.8	1700	Kenya	
Mago	Somali acacia-commiphora bushlands and thickets	5.5	36.0	400	Ethiopia	
Nakuru	Northern acacia-commiphora bushlands and thickets	−0.3	36.1	1870	Kenya	
Nechisar	Somali acacia-commiphora bushlands and thickets	6.0	37.6	1250	Ethiopia	
Olorgesailie	Northern acacia-commiphora bushlands and thickets	−1.5	36.4	619	Kenya	
Queen Elizabeth Park	Victoria Basin forest-savanna mosaic	0.0	30.0	950	Uganda	
Samburu	Northern acacia-commiphora bushlands and thickets	0.4	37.6	1100	Kenya	
Simien	Ethiopian montane moorland & montane grasslands/woodlands	13.2	38.0	3100–3800	Ethiopia	
Tsavo	Northern acacia-commiphora bushlands and thickets	−3.4	38.6	560	Kenya	
Turkana	Masai xeric grasslands and shrublands	4.0	36.0	360–450	Kenya	

Note: Ecoregion descriptions are from Olson et al. (2001). D.R.C.—Democratic Republic of Congo.
*NP—National Park.

Corrections were based on internal Carrara Marble standards calibrated to the NBS-19 calcite standard. Some of the GONBULK samples and all other tooth enamel and carbonates were analyzed using an online carbonate device, the Finnigan Carboflo, reacted in 100% H_3PO_4 with silver capsules at 90 °C and analyzed on a Finnigan MAT 252 at the University of Utah. Internal Carrara Marble and tooth enamel standards (MRS and MHS), calibrated to NBS-19, were used for the University of Utah analyses.

Statistics were performed using SYSTAT10. Significance (alpha) for statistics was set at $p < 0.05$. The post-hoc Scheffe analysis of variance (ANOVA) was used for all of the multipair comparisons. When comparisons were made between modern and fossil $\delta^{13}C_{enamel}$ values and between fossil time periods, the fossil data were adjusted to be compatible with modern $\delta^{13}C_{atm-CO_2}$ values using enrichment factors of −1.4‰ and −1.9‰ for the Sagantole and Adu-Asa data respectively (Table 1).

RESULTS

Analytical Precision

Repeat analyses of tooth enamel yielded consistent $\delta^{13}C$ values within and between the laboratories at which they were analyzed. Among fossil samples that were analyzed in different runs at the Utah laboratory, absolute differences between repeat analyses averaged 0.2 ± 0.2‰ for $\delta^{13}C$ ($n = 33$). The $\delta^{13}C$ values of enamel standards had a standard deviation of 0.15‰ over the course of the analyses performed for this study. For the Arizona laboratory, $\delta^{13}C$ values of the Carrara marble analyzed with the enamel samples had a standard deviation of 0.07‰, and their average was not distinct from the long-term averages of the analyses of these standards in that laboratory.

Some of the GONBULK samples originally analyzed at Arizona and published in Semaw et al. (2005) were reanalyzed at the Utah laboratory to test interlaboratory precision and data compatibility. The average absolute difference in δ¹³C values for samples analyzed in both laboratories was 0.2 ± 0.2‰ (n = 40).

Effects of Treatment on Isotopic Results

In general, there is a strong 1:1 relationship between δ¹³C values of treated and untreated powders in both fossil and modern tooth enamel (Fig. 3). The average residual between the treated and untreated δ¹³C values is 0.1 ± 0.6‰ for modern teeth and 0.4 ± 0.7‰ for fossil teeth. Among different fossil assemblages sampled (Adu-Asa Formation, As Duma Member, Segala Noumou Member, and the GWMS sites), there are no differences in treated versus untreated regressions for $\delta^{13}C_{enamel}$ values.

There is a substantial body of literature on the effects of treating tooth enamel powder prior to isotopic analysis (e.g., Lee-Thorp and van der Merwe, 1991; Koch et al., 1997; Passey et al., 2002). Although treatment of tooth enamel is intended to eliminate organic material and nonstructural carbonate, it can also produce new compounds or cause isotopic fractionation in existing ones (Lee-Thorp, 2000). From the treatment data presented in this study, it is clear that there are systematic differences in the ways that the fossil and modern enamel isotopic values respond to the same treatment (Fig. 3); however, it is unclear which value, treated or untreated, best represents animal diet.

A substantial amount of sample powder can be lost during sample treatment, leaving insufficient amounts of powder for analysis when the initial sample powder is small (≤1mg). Sample loss during treatment makes some of the smaller specimens inaccessible for isotopic analysis. Given the uncertainty in the value of treatment, we decided to collect small enamel powder samples from some specimens and leave them untreated for analysis.

Out of 182 fossil teeth sampled for this study, 32 remained untreated. All δ¹³C values from the untreated samples fall within the range of the treated values for each taxon at each site, and they do not affect the interpretation of any of the values. None of the primate tooth enamel powders was treated. The untreated primate values (in addition to the other untreated values) could be "converted" to be equivalent to treated values using the regression equations established from the Gona fossil samples (Fig. 3). However, for this study, none of the untreated values has been adjusted to be compatible with treated values, or vice versa.

Diagenesis in Tooth Enamel

In general, tooth enamel δ¹³C ($\delta^{13}C_{enamel}$) values have been shown to be resistant to diagenetic overprinting (Wang and Cerling, 1994). Among fossil teeth from the Adu-Asa and Sagantole Formations at Gona, $\delta^{13}C_{enamel}$ values span and are restricted to the spectrum of expected $\delta^{13}C_{enamel}$ values. Analyses of teeth from fossil animals known to be obligate browsers and grazers, like the deinothere and gomphotheres (Cerling et al., 1999; Zazzo et al., 2000; Cerling et al., 2003a; Kingston and Harrison, 2007), yield $\delta^{13}C_{enamel}$ values indicative of strict

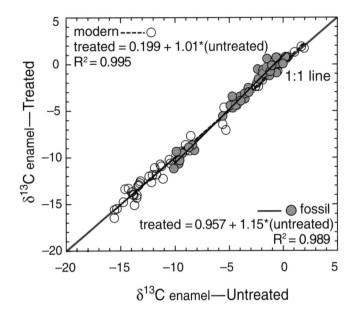

Figure 3. Scatter plot of δ¹³C values (VPBD) of untreated and treated enamel for the same sample, in ‰ units, for modern and fossil tooth enamel. Linear regression equations for modern and fossil enamel are reported on plot.

C_3 and C_4 diets, as shown in a later section of this paper. These data provide confidence that the dietary signature from $\delta^{13}C_{enamel}$ values is preserved in the fossil record.

General Patterns: Modern Teeth

The $\delta^{13}C_{enamel}$ values from modern East African environments (Fig. 4) provide a framework for interpreting $\delta^{13}C_{enamel}$ values of fossil taxa at Gona. We first present $\delta^{13}C_{enamel}$ values from modern and fossil assemblages, grouped by location, to demonstrate how $\delta^{13}C_{enamel}$ values vary with environment, and, in later sections, we present the isotopic data grouped by taxon. The $\delta^{13}C_{enamel}$ values of tooth enamel (modern and fossil) sampled from Ethiopia are available in the GSA Data Repository.[1]

The $\delta^{13}C_{enamel}$ values of modern teeth were obtained from locations that encompass the spectrum of environments found in East Africa today, including bushlands, grasslands, lowland forests, montane forests, and montane moorlands (Table 2; Fig. 4). The $\delta^{13}C_{enamel}$ values from animals sampled from the Ituri Forest, in the northeastern Congolian lowland forest ecoregion, range from −26.0‰ to −14.1‰ (n = 35) and have a median value of −16.2‰, which is close to the expected average $\delta^{13}C_{enamel}$ value for animals consuming vegetation in a C_3 closed-canopy forest (Fig. 4A; Table 1). The $\delta^{13}C_{enamel}$ values from animals living in the Albertine Rift montane forests at Kahuzi-Biega have a median value of −14.6‰ and range from −17.8‰ to

---
[1]GSA Data Repository item 2008197, tables listing isotopic data from modern and fossil tooth enamel and pedogenic carbonates, is available online at www.geosociety.org/pubs/ft2008.htm, or on request from editing@geosociety.org or Documents Secretary, GSA, P.O. Box 9140, Boulder, CO 80301, USA.

−11.3‰ ($n = 73$) (Fig. 4B). The $\delta^{13}C_{enamel}$ values from mammals in the Ethiopian highlands (Bale and Simien National Parks) range between −15.5‰ and −6.5‰ ($n = 21$), and, although they are higher than $\delta^{13}C_{enamel}$ values from the lowland and montane forests at Ituri and Kahuzi-Biega, they still are centered near the expected value for animals on a strict C_3 diet (median $\delta^{13}C_{enamel}$ = −12.0‰) (Fig. 4C). Herbivores sampled from all other bushlands, shrublands, grasslands, and the Kenyan mountains (Aberdare Mountains, Mount Kenya, and Mount Elgon) have more positive $\delta^{13}C_{enamel}$ values that span the full spectrum of values expected for animals consuming C_3 and C_4 plants, with median values ranging from −12.0‰ in the Kenyan mountains ($n = 37$) to +0.2‰ in the grasslands of the Athi Plains ($n = 65$) (Figs. 4D–4H).

Carbon isotope data from extant herbivore teeth show distinct distributions of $\delta^{13}C_{enamel}$ values among East African ecoregions. Mean $\delta^{13}C_{enamel}$ values from the Ituri Forest are consistent with a C_3 diet (−16‰) (Fig. 4A), but a tail of ^{13}C-depeleted values indicates the reliance by some animals on subcanopy vegetation, as documented in Cerling et al. (2004). There is no carbon isotope evidence for C_4 grasses in the diets of the animals living in the closed-canopy Ituri Forest. Mean $\delta^{13}C_{enamel}$ values of herbivores from Kahuzi-Biega indicate a C_3 diet, with some of the ^{13}C-depleted values nearing the expected value for closed-canopy diets (Fig. 4B). The shape and range of the Kahuzi-Biega histogram appear to be similar to the histogram from animals that lived in the montane moorlands, woodlands, and grasslands within Bale and Simien National Parks. Grazing animals at Bale and Simien are expected to yield $\delta^{13}C_{enamel}$ values indicative of C_3 diets because high-elevation grasses use the C_3 pathway. However, some $\delta^{13}C_{enamel}$ values from the Ethiopian highlands reach up to −6.5‰ and indicate that these animals may migrate to lower elevations where they have access to C_4 grass. No animals sampled from Ituri, Kahuzi-Biega, or the Ethiopian highlands rely solely on C_4 vegetation. Bovids are the only herbivores from the Kenyan mountains with $\delta^{13}C_{enamel}$ values that indicate a significant dietary intake of C_4 grass.

The $\delta^{13}C_{enamel}$ values from Nechisar Park, the Athi Plains, the Awash region, and Turkana indicate significant consumption of C_4 plants, consistent with the ecoregion classification of these areas as grasslands, shrublands, and bushlands (Figs. 4E–4H). A near bimodal distribution of $\delta^{13}C_{enamel}$ values of herbivores at Turkana indicates that most of the herbivores are either obligate browsers (C_3 consumers) or grazers (C_4 consumers), with the exception of the hippopotamids and some bovids, which yield intermediate $\delta^{13}C_{enamel}$ values. The distribution of $\delta^{13}C_{enamel}$ values from herbivores living in grassland, shrubland, and bushland ecoregions indicates the predominant consumption of C_4 plants,

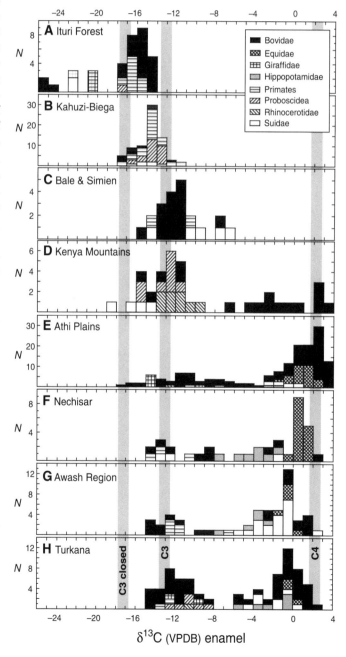

Figure 4. Histograms of $\delta^{13}C$ values in ‰ of tooth enamel from extant herbivores in East Africa plotted by location: (A) Ituri Forest, Democratic Republic of Congo (D.R.C.) (Cerling et al., 2004); (B) Kahuzi-Biega Forest, D.R.C.; (C) Bale and Simien National Parks in the Ethiopian highlands; (D) Kenya Mountains, including data from the Aberdare Mountains, Mt. Kenya, and Mt. Elgon (Cerling et al., 1999; Harris and Cerling, 2002; Cerling et al., 2003b, 2003c); (E) Athi Plains, which includes samples from the grasslands near the Athi River and in the Nairobi vicinity, Kenya (Cerling et al., 1999, 2003b, 2003c; Harris and Cerling, 2002); (F) Nechisar Park, Ethiopia; (G) Awash region of Ethiopia, including samples from Awash National Park, the Alledeghi Wildlife Reserve, and the Gona Paleoanthropological Research Project study area; and (H) Turkana, including samples collected from Mago National Park in Ethiopia (this study) and from regions around Lake Turkana in Kenya (Cerling et al., 2003b, 2003c). Gray vertical bars represent averages of expected $\delta^{13}C_{enamel}$ values for herbivores eating strict diets of C_3 plants from a closed-canopy forest, C_3 plants from a forest or bushland, and C_4 grasses, based on calculations described in Table 1.

and it is distinct from the distribution of $\delta^{13}C_{enamel}$ values of herbivores living in high-elevation or forested ecoregions, where diets are dominated by C_3 plants.

General Patterns: Fossil Teeth

Fossil teeth were sampled from four stratigraphic groupings at Gona, the late Miocene Adu-Asa Formation, the early Pliocene Segala Noumou and As Duma Members of the Sagantole Formation, and the early Pliocene Gona Western Margin South (GWMS) sites in the Sagantole Formation (Fig. 2). Eleven teeth were analyzed from other sites in the Sagantole Formation (GWM10, 11, 45, and 18), but we cannot place them confidently in any of these groupings. The data from these teeth will only be presented in the section that reviews $\delta^{13}C_{enamel}$ values by taxon. Fossils sampled from the Adu-Asa Formation are from sites in three stratigraphic horizons: one in lacustrine sediments above the Bodele Tuff (6.48 ± 0.22 Ma), another in fluvial sediments above the Sifi Tuff, and the highest level is associated with the Kobo'o Tuff (5.44 ± 0.06 Ma) (Kleinsasser et al., this volume). Most fossils sampled for isotopic analysis are associated with the Kobo'o Tuff and are considered together with the fossils sampled from the lower stratigraphic horizons.

The $\delta^{13}C_{enamel}$ values from the Adu-Asa fossils range between −15.7‰ to +0.8‰ and have a median value of −5.1‰ ($n = 35$) (Fig. 5D). Median $\delta^{13}C_{enamel}$ values for fossil teeth from the As Duma ($n = 76$), Segala Noumou ($n = 32$), and GWMS ($n = 28$) assemblages are −2.4‰, −2.8‰, and −1.6‰, respectively (Figs. 5A–5C). Fossil teeth from the As Duma Member include the most teeth sampled from one fossil assemblage at Gona, and they have the largest range in $\delta^{13}C_{enamel}$ values (−11.9‰ to +1.5‰) among fossil assemblages from the Sagantole Formation.

The $\delta^{13}C_{enamel}$ values of teeth from the Adu-Asa Formation suggest that the diets of some herbivores (suids, bovids, and giraffids) were dominated by C_3 plants, whereas other herbivores (hippopotamids, equids, and proboscidea) relied heavily on C_4 grasses. The presence of herbivores with C_4-dominated diets distinguishes fossil taxa of the Adu-Asa Formation from herbivores living in modern forests, where they do not eat C_4 grass (Figs. 4A–4B).

The distributions of $\delta^{13}C_{enamel}$ values from the Sagantole Formation fossil assemblages most resemble $\delta^{13}C_{enamel}$ values of extant herbivores living in bushlands, thickets, and grassland ecoregions, where there is a mix of C_3 and C_4 vegetation (Figs. 4E–4H). The fossil taxa from the Sagantole Formation with ^{13}C-depleted $\delta^{13}C_{enamel}$ values indicative of C_3 diets (colobines, bovids, giraffids, and rhinocerotids) are the same taxa that display ^{13}C-depleted $\delta^{13}C_{enamel}$ values in bushland and grassland ecoregions today. The suids are an exception because *Kolpochoerus deheinzelini* from the Sagantole Formation yields $\delta^{13}C_{enamel}$ values as low as −10.5‰, indicative of a C_3-dominated diet, whereas most extant suids living in bushlands and thickets today have diets dominated by C_4 grass. The details of $\delta^{13}C_{enamel}$ values from the fossil suids at Gona are discussed in a later section of this paper.

Carbon Isotope Results by Taxa

Cercopithecidae

Modern primate teeth sampled from Ethiopia include the papionins, *Papio hamadryas hamadryas* (sacred baboon), *P. h. anubis* (olive baboon), and *Theropithecus gelada* (gelada baboon), and the cercopithecins, *Chlorocebus aethiops* (vervet) and *Cercopithecus neglectus* (de Brazza's monkey). We compared the $\delta^{13}C_{enamel}$ values of these extant Ethiopian cercopithecids to those of similar taxa from Kenya and the D.R.C. They were also compared to the papionin *Cercocebus agilis* (agile mangabey), the cercopithecin *Cercopithecus mitis* (the diadem or blue monkey), the colobines *Procolobus badius* (red colobus) and *Colobus angolensis* (Angolan black-and-white colobus), as well as hominoids *Gorilla gorilla* (gorilla) and *Pan troglodytes* (chimpanzee). Teeth tentatively allocated to the fossil papionin *Pliopapio alemui* and the colobine monkey *Kuseracolobus aramisi* were sampled from the As Duma and Segala Noumou Members of the Sagantole Formation. Fossils of *P. alemui* are

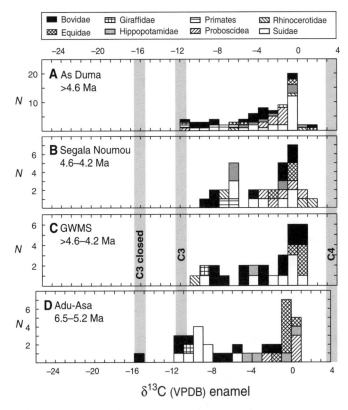

Figure 5. Histograms of $\delta^{13}C$ values in ‰ of tooth enamel for fossil herbivores from (A) the As Duma Member, (B) the Segala Noumou Member, and (C) Gona Western Margin South (GWMS) sites in the Sagantole Formation, and (D) the Adu-Asa Formation. The axis for the Adu-Asa plot (D) is shifted −0.6‰ to account for changes in $\delta^{13}C$ values of atmospheric CO_2 (see Table 1). Gray vertical bars represent averages of expected $\delta^{13}C_{enamel}$ values for herbivores eating strict diets of C_3 plants from a closed-canopy forest, C_3 plants from a forest or bushland, and C_4 grasses.

twice as common as *K. aramisi* fossils in the Sagantole Formation at Gona (Semaw et al., 2005).

The $\delta^{13}C_{enamel}$ values of modern primate teeth span a large range (>13‰) of values expected for herbivore tooth enamel (Fig. 6). The $\delta^{13}C_{enamel}$ values of primates from the Ituri Forest are less than −15‰ and are lower than $\delta^{13}C_{enamel}$ values from the majority of extant primates living elsewhere in eastern Africa. Baboons (*Papio anubis*, *P. hamadryas*, and *T. gelada*), which live in many different environments today, exhibit a large range in $\delta^{13}C_{enamel}$ values, between −16.7 ± 0.5‰ ($n = 2$) in the Ituri Forest and −5.2 ± 1.4‰ ($n = 3$) at Laikipia. A single tooth from *T. gelada*, a grass-eating baboon, sampled from Simien yields a $\delta^{13}C_{enamel}$ value of −10.2‰. The $\delta^{13}C_{enamel}$ values of extant colobine monkeys sampled from the forests at Ituri and Kahuzi-Biega in the D.R.C. average −15.5 ± 1.3‰ ($n = 5$) (Fig. 6).

The $\delta^{13}C_{enamel}$ values of fossil primates *P. alemui* and *K. aramisi* are enriched in ^{13}C relative to $\delta^{13}C_{enamel}$ values of extant East African primates. The $\delta^{13}C_{enamel}$ values for *P. alemui* average −7.3 ± 1.2‰ ($n = 7$), and those from *K. aramisi* average −9.4 ± 1.5‰ ($n = 5$). There are no significant distinctions in $\delta^{13}C_{enamel}$ values between specimens found in the As Duma or Segala Noumou Members for either *P. alemui* or *K. aramisi*. When compared to each other, $\delta^{13}C_{enamel}$ values of *P. alemui* are significantly higher than those of *K. aramisi*. The $\delta^{13}C_{enamel}$ values of *P. alemui* are most similar to $\delta^{13}C_{enamel}$ values of extant papionins than to any of the other primates, but they are significantly higher than $\delta^{13}C_{enamel}$ values of the baboons from Ituri and all non-papionin primates in the comparison. The $\delta^{13}C_{enamel}$ values of *K. aramisi* do not vary from $\delta^{13}C_{enamel}$ values of any of the extant primates, except they have higher $\delta^{13}C_{enamel}$ values than those measured for the cercopithecines from Ituri and Kahuzi-Biega, gorillas at Kahuzi-Biega, and the colobine *P. badius* at Ituri.

There are two useful observations from the extant primate $\delta^{13}C_{enamel}$ data that help us interpret $\delta^{13}C_{enamel}$ values from fossil primates at Gona: (1) primates living in closed-canopy systems are isotopically distinctive, and (2) papionin diet can be variable. The ^{13}C-depleted $\delta^{13}C_{enamel}$ values (<−15‰) of primates from the Ituri Forest are indicative of animals living in a closed-canopy forest and are consistent with other primate isotopic data from closed-canopy ecosystems (Schoeninger et al., 1998; Carter, 2001). All of the $\delta^{13}C_{enamel}$ values from the fossil primates at Gona are more enriched in ^{13}C than those from primates living in closed-canopy forests, precluding the possibility that the fossil primates lived in a closed-canopy environment. The large range in $\delta^{13}C_{enamel}$ values from extant papionins indicates that their diet is variable and can include some C_4 vegetation, as also observed by Codron et al. (2006). The overlap in $\delta^{13}C_{enamel}$ values between the extant and fossil papionins suggests that *P. alemui*, the more prevalent non-hominoid primate at Gona, may have had dietary behaviors similar to extant baboons and consumed varying amounts of C_4 plants, including grass leaves, seeds, or rhizomes. Lower $\delta^{13}C_{enamel}$ values for *K. aramisi* show that these fossil colobines ate proportionately more C_3 vegetation than *P. alemui*.

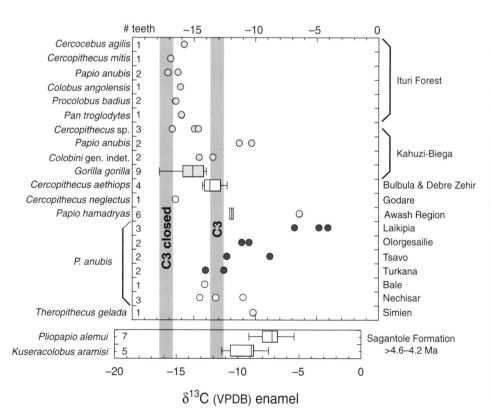

Figure 6. Box plot of $\delta^{13}C$ values in ‰ of extant and fossil primate teeth. In the box plot, median values are marked by a vertical line within the box, the edges of the boxes represent quartile values, the horizontal lines indicate the range, and outliers are plotted as circles. The $\delta^{13}C_{enamel}$ values for taxa with three or fewer samples are plotted as circles. Axes for the plots of fossil $\delta^{13}C_{enamel}$ values are offset from modern $\delta^{13}C_{enamel}$ values based on calculations for changes in $\delta^{13}C_{atm-CO_2}$ values described in Table 1. The $\delta^{13}C_{enamel}$ values from the Democratic Republic of Congo (D.R.C.) (Cerling et al., 2004), Kenya (T.E. Cerling et al., 2007, personal commun.), and Ethiopia are plotted in light gray, dark gray, and white, respectively. Gray vertical bars represent averages of expected $\delta^{13}C_{enamel}$ values for herbivores eating strict diets of C_3 plants from a closed-canopy forest and C_3 plants from a forest or bushland. The estimate for $\delta^{13}C_{enamel}$ values formed from a diet of C_4 plants is excluded because it lies off the plot.

Suidae

Extant suid teeth sampled include *Phacochoerus* sp. (warthog) and *Potomachoerus porcus* (bush pig) from Ethiopia and *Phacochoerus* sp., *Pot. porcus*, and *Hylochoerus meinertzhageni* (forest hog) from Kenya, Uganda, and the D.R.C. (Harris and Cerling, 2002). Warthogs are grouped on the genus level because morphologic distinctions between the common warthog (*Phacochoerus africanus*) and the desert warthog (*Phacochoerus aethiopicus*) (d'Huart and Grubb, 2001; Randi et al., 2002) were not made for this study. Fossil suidae sampled from Gona include *Nyanzachoerus syrticus* from the late Miocene and three sympatric taxa from the early Pliocene deposits, *Nyanzachoerus jaegeri*, *Nyanzachoerus kanamensis*, and *Kolpochoerus deheinzelini*. Of the early Pliocene suids found in the Sagantole Formation at Gona, *Ny. jaegeri* is the most common, followed by *K. deheinzelini* and then by *Ny. kanamensis*. Subtle morphological distinctions between *Ny. kanamensis* and *Ny. jaegeri* limit some of the identifications of isolated teeth to the genus level.

Extant suids from East Africa display the full spectrum of $\delta^{13}C_{enamel}$ values expected for herbivores (Fig. 7). Average $\delta^{13}C_{enamel}$ values for warthogs range from $-8.4 \pm 2.0‰$ ($n = 4$) at Bale to $-0.1 \pm 0.6‰$ ($n = 12$) at Nakuru. Within the Awash region alone, $\delta^{13}C_{enamel}$ values of warthogs range between $-5.6‰$ and $+2.2‰$ (average $-1.7 \pm 1.8‰$, $n = 18$). The $\delta^{13}C_{enamel}$ values of extant bush pigs range between $-16.1‰$ ($n = 1$) in the Ituri Forest and $-3.0 \pm 3.7‰$ ($n = 6$) in the Athi Plains. Forest hogs have the lowest $\delta^{13}C_{enamel}$ values among extant suids, averaging $-22.6 \pm 0.4‰$ ($n = 2$) at Ituri, $-16.3 \pm 1.8‰$ ($n = 4$) in the Kenyan mountains, and $-14.5‰$ ($n = 1$) in Queen Elizabeth Park.

The $\delta^{13}C_{enamel}$ values of fossil suids point to different dietary behaviors between the late Miocene and early Pliocene taxa (Fig. 7). The $\delta^{13}C_{enamel}$ values of *Ny. syrticus*, the late Miocene suid at Gona, average $-9.2 \pm 1.5‰$ ($n = 9$) and are responsible for a large portion of the C_3 signal in the histogram of $\delta^{13}C_{enamel}$ values of teeth from the Adu-Asa Formation (Fig. 5D). The $\delta^{13}C_{enamel}$ values of the small suid from the Sagantole Formation, *K. deheinzelini*, average $-6.2 \pm 2.3‰$ ($n = 7$), whereas $\delta^{13}C_{enamel}$ values of the nyanzachoeres from the Sagantole Formation average $-1.5 \pm 1.5‰$ ($n = 22$), $-1.4 \pm 1.6‰$ ($n = 8$), and $-2.1 \pm 1.3‰$ ($n = 5$) for *Ny. jaegeri*, *Ny. kanamensis*, and *Ny.* sp. indet., respectively. The $\delta^{13}C_{enamel}$ values of *Ny. syrticus* from the Adu-Asa Formation are significantly lower than $\delta^{13}C_{enamel}$ values of all suid species from the Sagantole Formation. Among suids from the Sagantole Formation, $\delta^{13}C_{enamel}$ values of *K. deheinzelini* are lower than $\delta^{13}C_{enamel}$ values of the nyanzachoeres.

The $\delta^{13}C_{enamel}$ values of the extant suids show that warthogs eat primarily C_4 plants, that bush pigs can have variable diets and will opportunistically feed on C_3 or C_4 vegetation, and that forest hogs are hyperbrowsers (Harris and Cerling, 2002). The $\delta^{13}C_{enamel}$ values of warthogs from Ethiopia extend the known isotopic range for extant warthogs. Low $\delta^{13}C_{enamel}$ values ($-10.8‰$ to $-6.5‰$) of warthogs at Bale National Park indicate diets with a significant C_3 plant component, which may be explained by consumption of C_3 grasses prevalent at high elevations in the Bale Mountains (>3000 m). Low $\delta^{13}C_{enamel}$ values ($-13‰$) of two adult molar teeth from a single warthog at Nechisar (Fig. 7) are unexpected, given the 1250 m elevation grassland setting at Nechisar. The large range of $\delta^{13}C_{enamel}$ values among warthogs from the Awash region suggests that

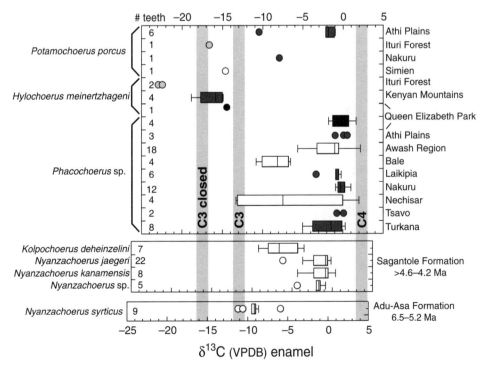

Figure 7. Box plot of $\delta^{13}C$ values in ‰ of extant and fossil suid teeth, plotted in same manner as Figure 6. Axes for the plots of fossil $\delta^{13}C_{enamel}$ values are offset from each other and from modern $\delta^{13}C_{enamel}$ values based on calculations for changes in $\delta^{13}C_{atm-CO_2}$ values described in Table 1. The $\delta^{13}C_{enamel}$ values from Uganda, the Democratic Republic of Congo (D.R.C.), Kenya (Harris and Cerling, 2002), and Ethiopia are plotted in black, light gray, dark gray, and white, respectively. Gray vertical bars represent averages of expected $\delta^{13}C_{enamel}$ values for herbivores eating strict diets of C_3 plants from a closed-canopy forest, C_3 plants from a forest or bushland, and C_4 grasses.

warthogs can have variable diets, even in environments where C_4 plants are the only available grass resources.

The $\delta^{13}C_{enamel}$ values of fossil suids at Gona indicate a dietary shift between the late Miocene, when suid diet was dominated by C_3 plants, and the early Pliocene, when suids had mixed C_3-C_4 and C_4-dominated diets. The isotope data suggest that *Ny. syrticus* was a browser living in a forest, woodland, or bushland, but not in a closed-canopy ecosystem. In contrast, the early Pliocene nyanzachoeres were grazers with $\delta^{13}C_{enamel}$ values indicative of diets similar to those of extant East African warthogs living at lower elevations (<2000 m). These results are consistent with isotopic data from *Ny. syrticus* and *Ny. jaegeri* from late Miocene and early Pliocene deposits at Kenya and Chad (Zazzo et al., 2000; Harris and Cerling, 2002).

While there is no significant difference in $\delta^{13}C_{enamel}$ values between *Ny. jaegeri* and *Ny. kanamensis*, there is a distinction between the nyanzachoeres and *K. deheinzelini* in the Sagantole Formation. The $\delta^{13}C_{enamel}$ values for *K. deheinzelini* are most similar to those of extant bush pigs, and the large range in values suggests a cosmopolitan diet. There are no published isotopic data from *K. deheinzelini* for comparison, but $\delta^{13}C_{enamel}$ values from younger taxa in the kolpochoere lineage (Zazzo et al., 2000; Harris and Cerling, 2002; Kingston and Harrison, 2007) indicate a shift toward an increased dependence on C_4 vegetation through the Pliocene.

Hippopotamidae

Hippopotamid teeth sampled from Ethiopia include the extant hippopotamus, *Hippopotamus amphibius*, from Nechisar and Awash Parks and fossil Hippopotamidae from the Adu-Asa and Sagantole Formations (Fig. 8A). The $\delta^{13}C_{enamel}$ values of *H. amphibius* average $-4.2 \pm 1.7‰$ ($n = 5$) at Awash Park and $-4.1 \pm 2.0‰$ ($n = 7$) at Nechisar Park. Among the fossils, $\delta^{13}C_{enamel}$ values average $-2.0 \pm 2.5‰$ ($n = 4$) and $-2.8 \pm 2.5‰$ ($n = 11$) for the late Miocene and early Pliocene hippopotamidae, respectively. The $\delta^{13}C_{enamel}$ values from fossil hippopotamids are indistinguishable from values of modern Ethiopian *H. amphibius*.

The carbon isotope data suggest that the diets of extant and fossil hippopotamids are dominated by C_4 plants but include a C_3 component. Mixed C_3-C_4 diets in extant and fossil hippopotamidae have also been observed elsewhere in Africa (Boisserie et al., 2005; Cerling et al., 2008; Harris et al., 2008).

Bovidae

Extant bovids sampled from Ethiopia include Alcelaphini, Bovini, Caprini, Cephalophini, Hippotragini, Neotragini, Reduncini, and Tragelaphini. The $\delta^{13}C_{enamel}$ values of extant bovids from Ethiopia were compared to $\delta^{13}C_{enamel}$ values of bovids from Kenya, Uganda, and the D.R.C (Cerling et al., 2003c, 2004) and used as references for interpreting $\delta^{13}C_{enamel}$ values of fossil bovids from Gona, which include Aepycerotini, Bovini, Reduncini, and Tragelaphini (Fig. 9). In the fossil deposits of the Sagantole Formation, Tragelaphini and Aepycerotini are equally common and together comprise ~80% of the bovid sample.

The $\delta^{13}C_{enamel}$ values of extant Ethiopian bovids plot within the ranges of corresponding tribes from elsewhere in East Africa (Fig. 9). An outlying $\delta^{13}C_{enamel}$ value among the Reduncini serves as one exception. Although $\delta^{13}C_{enamel}$ values of most Reduncini are indicative of diets of C_4 graze, the one *Redunca redunca* (Bohor reedbuck) sampled from the Bale Mountains yielded a $\delta^{13}C_{enamel}$ value of $-7.1‰$, suggesting the presence of C_3 plants in its diet. Like the other grazing animals sampled from Bale and Simien, in the Ethiopian highlands, the intermediate $\delta^{13}C_{enamel}$ value from this reedbuck indicates that it ate both C_3 and C_4 grass. The ^{13}C-depleted $\delta^{13}C_{enamel}$ values among Neotragini are from the Ituri Forest (Fig. 9) and indicate that these Neotragini (*Neotragus batesi*) eat ^{13}C-depleted plants from a closed-canopy setting. Cerling et al. (2004) classified these animals as subcanopy folivores.

In general, the $\delta^{13}C_{enamel}$ values of the late Miocene bovids from the Adu-Asa Formation lie outside the ranges of their modern analogs. The $\delta^{13}C_{enamel}$ value from cf. Hippotragini ($-10.2‰$) is considerably lower than values of modern grazing Hippotragini, and the $\delta^{13}C_{enamel}$ value of the cf. Reduncini specimen ($-5.4‰$) is similar only to the low outlier value from the extant Reduncini. The $\delta^{13}C_{enamel}$ value of the Antilopini ($-6.6‰$) from the Adu-Asa Formation does fall within the range of extant Antilopini. The $\delta^{13}C_{enamel}$ values of cf. *Tragelaphus* average $-13.6 \pm 2.9‰$ ($n = 2$) and are similar to those of extant Tragelaphini, whereas $\delta^{13}C_{enamel}$ values of *Tragelaphus* sp. are significantly more positive (average $-2.2 \pm 0.9‰$, $n = 2$) (Fig. 9).

The $\delta^{13}C_{enamel}$ values of bovids from the Sagantole Formation generally correspond to those of their extant counterparts. The $\delta^{13}C_{enamel}$ values of Aepycerotini range from $-11.4‰$ to $-0.3‰$ (average $-4.8 \pm 4.1‰$, $n = 6$) and are indistinct from $\delta^{13}C_{enamel}$ values of extant Aepycerotini. The $\delta^{13}C_{enamel}$ values of Bovini average $-3.9 \pm 1.3‰$ ($n = 4$) and lie within the extant range, except for those from the Ituri Forest and the Kenyan mountain region, which include data from the forest buffalo. Reduncini $\delta^{13}C_{enamel}$ values average $-0.7 \pm 1.1‰$ ($n = 2$) and are not distinct from those of extant Reduncini. Among $\delta^{13}C_{enamel}$ values of *Tragelaphus* sp., there are no distinctions between the different fossil assemblages within the Sagantole Formation, and when pooled together, they average $-6.8 \pm 3.4‰$ ($n = 20$) and range from $-11.0‰$ to $-0.1‰$. Although some $\delta^{13}C_{enamel}$ values of Tragelaphini from the Sagantole Formation overlap with the range of $\delta^{13}C_{enamel}$ values of extant Tragelaphini, they extend to values much higher ($-0.1‰$) than the high values measured in extant Tragelaphini ($-7‰$). The Tragelaphini are the only group of fossil bovids from the Sagantole Formation that do not have a range similar to their modern analogs.

We can only make general observations from the fossil bovid isotope data, given (1) the difficulty of making taxonomic identifications from isolated teeth, (2) the still-developing taxonomy of bovids from the late Miocene and early Pliocene in Ethiopia, and (3) the small number of isotopic analyses for some of the taxa from each fossil assemblage. All of the isotopic data from the fossil bovid teeth

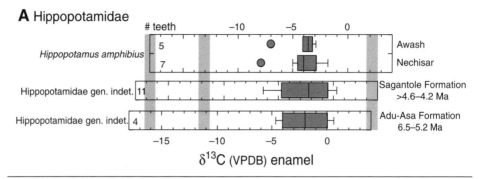

Figure 8. Box plot of $\delta^{13}C_{enamel}$ values in ‰ of large extant herbivores and their fossil forms from the Adu-Asa and Sagantole Formations, plotted in a similar manner as Figure 6. Axes for the plots of fossil $\delta^{13}C_{enamel}$ values are offset from each other and from modern $\delta^{13}C_{enamel}$ values based on calculations for changes in $\delta^{13}C_{atm-CO_2}$ values described in Table 1. Gray vertical bars represent averages of expected $\delta^{13}C_{enamel}$ values for herbivores eating strict diets of C_3 plants from a closed-canopy forest, C_3 plants from a forest or bushland, and C_4 grasses.

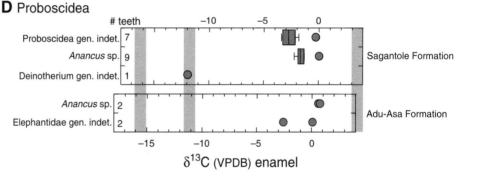

at Gona plot within the expected range of $\delta^{13}C_{enamel}$ values from animals eating C_3 and C_4 plants (Fig. 9), but most of the $\delta^{13}C$ values are intermediate and do not represent diets composed solely of C_3 or C_4 vegetation. The $\delta^{13}C_{enamel}$ values of bovids from the Adu-Asa Formation indicate that these animals relied more heavily on C_3 vegetation, whereas $\delta^{13}C_{enamel}$ values of bovids from the Sagantole Formation suggest an increased reliance on C_4 vegetation in the early Pliocene. The $\delta^{13}C_{enamel}$ values from the Sagantole bovids generally match those of their modern counterparts. However, $\delta^{13}C_{enamel}$ values of *Tragelaphus* sp. indicate a broad range of C_3 and C_4 diets, unlike extant tragelaphines, which exclusively eat C_3 plants. If these teeth represent one bovid species, then it had a very diverse diet, like extant Antilopini and Aepycerotini. Alternatively, the large range in $\delta^{13}C_{enamel}$ values among Tragelaphini from the Sagantole Formation may indicate that more than one taxon is represented by *Tragelaphus* sp.

Giraffidae

The sample of giraffid teeth from the Adu-Asa and Sagantole Formations includes those identified to family. The $\delta^{13}C_{enamel}$ values of giraffids from the Adu-Asa Formation average -10.9 ± 1.0‰ ($n = 2$), and those from the Sagantole Formation average -9.8 ± 1.7‰ ($n = 3$), with one outlier value of -1.4‰ (Fig. 8B). The isotopic data from all giraffids, except for the outlier value, indicate diets dominated by C_3 plants, consis-

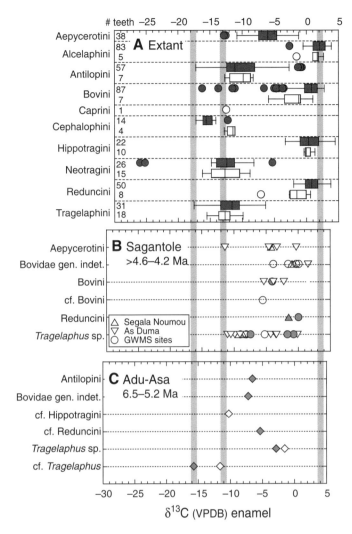

Figure 9. Bovid $\delta^{13}C_{enamel}$ values in ‰. (A) Box plot of $\delta^{13}C_{enamel}$ values from extant East African bovids grouped by tribe. The $\delta^{13}C_{enamel}$ values from bovids in Kenya, the Democratic Republic of Congo (D.R.C.), and Uganda (Cerling et al., 2003c, 2004) are plotted in dark gray, whereas white boxes represent the data from Ethiopia. (B–C) Point plots of $\delta^{13}C_{enamel}$ values from the Sagantole and Adu-Asa fossil assemblages. The $\delta^{13}C_{enamel}$ values from untreated fossil samples are plotted as filled symbols. Axes for the plots of fossil $\delta^{13}C_{enamel}$ values are offset from each other and from modern $\delta^{13}C_{enamel}$ values based on calculations for changes in $\delta^{13}C_{atm-CO_2}$ values described in Table 1. Gray vertical bars represent averages of expected $\delta^{13}C_{enamel}$ values for herbivores eating strict diets of C_3 plants from a closed-canopy forest, C_3 plants from a forest or bushland, and C_4 grasses.

tent with isotopic data from late Miocene giraffids from Chad (Zazzo et al., 2000) and early Pliocene giraffids from Tanzania (Kingston and Harrison, 2007).

Equidae

Extant equid teeth sampled from Ethiopia include *Equus grevyi* (Grevy's zebra), *E. burchelli* (Burchell's zebra), and *E. africanus somalicus* (Somali wild ass). All fossil equid teeth sampled from the Adu-Asa Formation were assigned to *Eurygnathohippus* cf. *feibeli*, and those from the Sagantole Formation are classified as *Eurygnathohippus* sp.

The $\delta^{13}C_{enamel}$ values of most East African equids, ancient and modern, overlap and fall within a narrow range of –2‰ to +2‰ (Fig. 10). More subtle, but statistically significant, distinctions include *E. burchelli* from the Athi Plains and Nechisar, which have higher $\delta^{13}C_{enamel}$ values compared to those at Samburu and Nakuru. Among the fossil material from Gona, there are no significant differences among equid $\delta^{13}C_{enamel}$ values from the different fossil assemblages. The $\delta^{13}C_{enamel}$ values of *E.* cf. *feibeli* teeth from the late Miocene average –0.6 ± 0.5‰ ($n = 8$), and the $\delta^{13}C_{enamel}$ values of early Pliocene *Eurygnathohippus* sp. average –1.2 ± 2.4‰ ($n = 13$). The $\delta^{13}C_{enamel}$ values of two specimens from the As Duma Member (*Eurygnathohippus* sp.) average –6.1 ± 0.5‰ and are significantly lower than the $\delta^{13}C_{enamel}$ values of other equids from the As Duma Member and from other sites in the deposits of the Sagantole Formation (Fig. 10).

These isotopic results demonstrate that most fossil equidae from Gona, in both the late Miocene and early Pliocene deposits, were grazers and fed primarily on C_4 grass, like extant *Equus* in eastern Africa today. The low (–6‰) $\delta^{13}C_{enamel}$ values from two *Eurygnathohippus* sp. teeth in the As Duma Member suggest a diet with a large input of C_3 vegetation. However, given the fragmentary nature of these two teeth (both are unaccessioned GONBULK specimens), we must consider the possibility that these teeth might have formed early in life and represent a pre-adult diet. In bovids, the influence of mother's milk depletes $\delta^{13}C_{enamel}$ values by less than 1‰ (Balasse, 2002; Zazzo et al., 2002). The magnitude of this effect is likely similar for equids, which, like bovids, have milk with a low fat content (Oftedal, 1984). Consequently, the pre-weaning effect cannot account for the entire offset (~4‰) in $\delta^{13}C_{enamel}$ values of these equid teeth; a diet consisting of some C_3 plants likely explains the relatively ^{13}C-depleted $\delta^{13}C_{enamel}$ values.

Rhinocerotidae

The rhinoceros teeth sampled from the Sagantole Formation are fragmentary enamel pieces and are identified to family, except for one specimen identified as cf. *Diceros praecox*. No rhinoceros teeth sampled from the Adu-Asa Formation. The $\delta^{13}C_{enamel}$ values of the rhinocerotids from the Sagantole Formation average –2.0 ± 3.7‰ ($n = 6$), but within this sample, two teeth have $\delta^{13}C_{enamel}$ values less than –6‰ (Fig. 8C). The specimen identified as cf. *Diceros praecox* has a $\delta^{13}C_{enamel}$ value of –10.3‰.

The isotopic data suggest that the diets of rhinocerotids contained varying amounts of C_3 and C_4 plants. The diets of the fossil rhinocerotids from Gona may have had similar diets to extant rhinocerotids, which include both a browsing and a grazing species, but it is unclear how the Pliocene fossil rhinocerotids are related to the extant forms (Geraads, 2005). Other isotopic studies have found a similar range in diets of rhinocerotids from the early Pliocene in Africa (Zazzo et al., 2000; Cerling et al., 2003a; Kingston and Harrison, 2007).

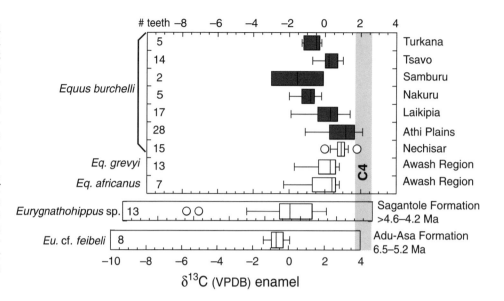

Figure 10. Box plot of $\delta^{13}C$ values in ‰ of extant and fossil equids. Plotted in the same manner as Figure 6, except note the different scale. Axes for the plots of fossil $\delta^{13}C_{enamel}$ values are offset from each other and from modern $\delta^{13}C_{enamel}$ values based on calculations for changes in $\delta^{13}C_{atm-CO_2}$ values described in Table 1. The $\delta^{13}C_{enamel}$ values from Kenya (T.E. Cerling, 2007, personal commun.) and Ethiopia are plotted in dark gray and white, respectively. Gray vertical bars represent averages of expected $\delta^{13}C_{enamel}$ values for herbivores eating strict diets of C_4 plants. Estimates for $\delta^{13}C_{enamel}$ values formed from a strict diet of C_3 plants from a closed-canopy forest and C_3 plants from a forest or bushland are excluded because they lie off the plot.

Proboscidea

The fossil proboscidea from Gona include gomphotheres (*Anancus* sp.) and elephantids from the Adu-Asa Formation, and gomphotheres (*Anancus* sp.), deinotheres (*Deinotherium* sp.), and unclassified proboscidea (Proboscidea gen. indet.) from the Sagantole Formation. Most of the proboscidean teeth sampled from the Sagantole Formation are enamel fragments and are difficult to identify in more detail. The $\delta^{13}C_{enamel}$ values of *Anancus* sp. and elephantids from the Adu-Asa Formation average +0.8 ± 0.1‰ ($n = 2$) and –1.2 ± 1.9‰ ($n = 2$), respectively (Fig. 8D). The $\delta^{13}C_{enamel}$ values of *Anancus* sp. and Proboscidea gen. indet. from the Sagantole Formation average –1.5 ± 0.7‰ ($n = 9$) and –2.4 ± 1.1‰ ($n = 7$). The $\delta^{13}C_{enamel}$ value of the sole deinothere analyzed from the Sagantole Formation is –11.9‰.

The $\delta^{13}C_{enamel}$ values of proboscidea from the Adu-Asa and Sagantole Formations at Gona indicate that the diets of these large-bodied herbivores were dominated by C_4 grass in both the late Miocene and the early Pliocene, except for the dentally specialized deinothere, which had a strict diet of C_3 plants. These results are consistent with isotopic data of proboscidea from fossil sites in Kenya, Tanzania, and Chad (Cerling et al., 1999, 2003a; Zazzo et al., 2000; Kingston and Harrison, 2007).

Summary

Examination of the herbivore tooth enamel isotopic data by each taxon confirms the perspective on late Miocene and early Pliocene environments at Gona provided by the $\delta^{13}C_{enamel}$ histograms in Figure 5. There is no isotopic evidence for a closed-canopy forest at Gona in the late Miocene or early Pliocene. The presence of suids, hippopotamids, bovids, equids, rhinocerotids, and proboscideans with C_4-dominated diets makes the fossil sites at Gona very different from the modern closed-canopy Ituri Forest, the forests at Kahuzi-Biega, and the Ethiopian highlands, where there is little to no C_4 component in herbivore diets.

Although some fossil taxa from the Adu-Asa and Sagantole Formations at Gona (e.g., primates, suids, and bovids) have $\delta^{13}C_{enamel}$ values that are similar to their extant forms in forest and highland environments, they are also similar to $\delta^{13}C_{enamel}$ values of analog taxa living in woodland and bushland ecoregions. The $\delta^{13}C_{enamel}$ values of fossil herbivores from the Adu-Asa and Sagantole Formations at Gona suggest that the vegetation at Gona during the late Miocene and early Pliocene was most similar to mixed C_3-C_4 woodlands or bushlands today.

Paleosol Carbonates

Paleosols in the Sagantole Formation at Gona are dark brown and clay rich with occasional manganese and iron-oxide staining, and they have either blocky ped structures or are massive in outcrop. Carbonate nodules collected from paleosols are 0.5–2 cm in diameter and are from distinct carbonate (Bk) horizons. In general, the early Pliocene paleosols at Gona are not as mature as the paleosols found in the younger Busidima Formation (Levin et al., 2004; Quade et al., 2004, this volume). There are no well-developed soils in the sedimentary sequences that contain fossil sites in the As Duma Member of the Sagantole Formation or in the Adu-Asa Formation at Gona.

Pedogenic carbonates were sampled from paleosols in the Sagantole Formation within <2 km radii of paleontological localities in the Segala Noumou Member, the GWMS fossil sites, and the fossil locality GWM45. Pedogenic carbonate nodules sampled from 10 paleosols in the Segala Noumou Member within a 2 km radius of GWM3, an *Ar. ramidus* locality, yield $\delta^{13}C$ values that average –7.5 ± 1.6‰ and range from –10.4‰ to –3.9‰ (Fig. 11). Pedogenic carbonate nodules from four paleosols associated with GWMS fossil sites (<1 m directly above or a below a fossil horizon) yield $\delta^{13}C$ values that average –5.5 ± 1.1‰. Three nodules from a paleosol directly underlying a fossiliferous gravel

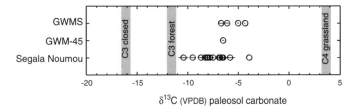

Figure 11. The δ¹³C values in ‰ of paleosol carbonates associated with paleontological localities in the Sagantole Formation from the Gona Western Margin South (GWMS) sites, the GWM-45 site, and in the Segala Noumou Member. Plotted values are averages of multiple nodules from a single paleosol outcrop, and horizontal lines represent standard deviations. Gray vertical bars indicate approximate values that would be expected for soil carbonates forming beneath a canopy forest, C_3 plants in a forest or bushland, and a C_4 grassland during deposition of the Sagantole Formation when $δ^{13}C_{atm-CO_2}$ was more enriched in ¹³C.

and hominid locality, GWM45, have δ¹³C values that average −6.5 ± 0.1‰. A summary of the δ¹³C values of soil carbonate from paleosols in the Sagantole Formation is presented in GSA Data Repository Table DR3 (see footnote 1). The δ¹³C values of pedogenic nodules from the Sagantole Formation indicate the presence of both C_3 and C_4 vegetation but, generally, a greater proportion of C_3 plants (Fig. 11).

The C_3-dominated landscapes in the Sagantole Formation, as inferred from the soil carbonate record, contrast with the tooth enamel isotopic record, which suggests an abundance of C_4 vegetation. The δ¹³C record from soil carbonates may represent just one end of the spectrum of environments in which animals lived. Differences between the records can be reconciled if C_3 plants grew in the well-watered areas where there was active deposition and soil formation, and if C_4 plants grew on slopes or topographic highs where depositional rates were lower. Active volcanic cones and basalt flows adjacent to the depositional environments in both the Adu-Asa and Sagantole Formations would have generated topographically and lithologically diverse landscapes that could support a variety of soil types and vegetation communities, including C_4 grasses.

Paleoelevation

In addition to identifying the degree of forestation and the presence of grasslands, δ¹³C values from soil carbonates and teeth can also be indicators of paleoelevation. The altitudinal effect on $δ^{13}C_{enamel}$ values is clear in modern data from the Ethiopian highlands, where only a few taxa have C_4 grass in their diet, despite the abundance of grasses (Fig. 4C). The presence of C_4 plants growing in soils and in the diets of most herbivores in the late Miocene and early Pliocene fossil assemblages at Gona precludes the possibility that these environments were above the altitudinal limit for growth of C_4 plants, which is between 2800 and 3200 m in East Africa today (Young and Young, 1983). However, this limit depends on climatic conditions (e.g., effective moisture and growing-season temperatures), and there is evidence that it has moved upward in the past, during the Last Glacial Maximum, on Mt. Kenya, when C_4 grasses were abundant at 3080 m (Wooller et al., 2003). We do not know the location of this limit in East Africa in the late Miocene and early Pliocene, before the onset of Northern Hemisphere glaciation, but if C_4 grasses did exist at higher elevations during these periods, then they were accompanied by the environmental conditions, like warm growing-season temperatures, in which C_4 grasses thrive at lower elevations today.

The presence of fossil *Tachyoryctes*, a genus of mole rat, has been used as an indicator of high elevation for *Ardipithecus*-bearing deposits in the Middle Awash and Gona study areas (WoldeGabriel et al., 1994, 2001; Haile-Selassie et al., 2004; Semaw et al., 2005). *T. macrocephalus*, the giant mole rat, is currently found at Bale and Simien in the Ethiopian highlands and is restricted to 3200–4150 m (Yalden, 1985), above the elevation limit for C_4 grasses. However, another species, *T. splendens*, is found at lower elevations (1200–3200 m; Yalden et al., 1996), where C_4 vegetation thrives. Assuming that *Tachyoryctes* in the fossil record was restricted to similar elevations as the extant forms, and that C_4 plants in the late Miocene and early Pliocene were limited to environments below 3200 m, then *Ardipithecus* must have lived at elevations between 1200 and 3200 m. If the high-elevation limit of C_4 plants was above 3200 m in the late Miocene or early Pliocene, then the environments would have to be similar to those below that limit today, given the clear isotopic indicators of C_4 vegetation in the fossil record from the Adu-Asa and Sagantole Formations at Gona. Either of these interpretations would place late Miocene and early Pliocene *Ardipithecus* environments at elevations higher than the present elevation of *Ardipithecus*-bearing deposits in the Gona Paleoanthropological Research Project study area, which ranges between 800 and 950 m. Alternatively, the fossil *Tachyoryctes* may have had a lower elevation limit than its present forms, in which case there might have been little elevation change since the late Miocene or early Pliocene. This latter hypothesis is consistent with a kinematic model proposed by Redfield et al. (2003), which suggests little change in Afar topography since the late Miocene.

DISCUSSION

Regional Environments of *Ardipithecus*

The isotopic data from large herbivores and paleosol carbonates associated with *Ardipithecus* fossils at Gona indicate a different landscape from that described for either the *Ardipithecus*-bearing deposits in the Middle Awash, Ethiopia, or at Tabarin, Kenya. WoldeGabriel et al. (1994, 2001) and White et al. (2006) have suggested that *Ardipithecus* lived in wooded biomes and avoided open environments. Isotopic data from soil carbonates in these deposits suggest the presence of both C_3 and C_4 vegetation on the landscape, but the paleoenvironmental reconstructions are based primarily on the presence or absence of fossils with extant forms usually found in wooded or open environments, respectively. For example, the characterization of *Ar. ramidus* habitat at Aramis as "varying from closed to grassy woodlands" is based on the presence of colobine monkeys and tragelaphine bovids (White et al., 2006; WoldeGabriel et al., 1994). Pickford and colleagues (2004) reconstructed the habitat at the Tabarin *Ar. ramidus* site as a rain forest from the presence of peafowl and tragulids. However, assignments of the habitat and diet of extant taxa to their fossil forms can be tenuous.

The carbon isotope data from Gona show that there were abundant C_4 grasses associated with paleohabitats where *Ardipithecus* lived and that the majority of the large mammalian herbivores with which *Ardipithecus* fossils are found sought C_4 plants for food. Considering the isotopic data from Gona, the assertion that there is a "scarcity of basal Pliocene hominid remains in non-woodland settings in the Middle Awash and beyond" (WoldeGabriel et al., 1994, p. 333) must be revised. However, this does not preclude the possibility that less than 100 km to the south, in the Middle Awash study area, *Ardipithecus* lived in closed, forested habitats. *Ardipithecus* fossils found in the Middle Awash study area may have been deposited in a more axial part of the basin than the Gona sediments, which were closer to the basin margin (Quade et al., this volume). The different paleoenvironmental reconstructions indicate that *Ardipithecus* may have inhabited a variety of landscapes and was not as ecologically restricted as previous studies have suggested (WoldeGabriel et al., 1994, 2001). Active tectonism, frequent volcanism, and the segmented nature of basalt-bounded grabens in this part of the Afar Rift during the late Miocene and early Pliocene could explain variation in *Ardipithecus* habitat within the 100 km that separates the Middle Awash and Gona project areas. Direct comparisons of the paleontological records and geological setting in the future will further clarify the paleobiology of *Ardipithecus*.

CONCLUSIONS

Carbon isotopic data show that extant East African herbivores living in closed-canopy forests, forests, and montane regions have $\delta^{13}C_{enamel}$ values that are distinct from those of herbivores living in bushlands and grasslands. When compared to data from modern environments, isotopic data of fossil herbivores from the *Ardipithecus*-bearing deposits at Gona most closely resemble the isotopic data from herbivores living in bushlands today. The $\delta^{13}C_{enamel}$ values of the fauna from the Miocene-Pliocene fossil deposits at Gona do not match $\delta^{13}C_{enamel}$ values of herbivores living in closed-canopy forests like the Ituri Forest, the forests in Kahuzi-Biega, or the mountainous regions in Kenya and Ethiopia. Examination of individual taxa confirms this and shows that the diets of the majority of taxa in the late Miocene at Gona were composed of a mixture of C_3 and C_4 plants, and, in the early Pliocene, they were dominated by C_4 grass. The carbon isotopic record from fossil teeth and soil carbonates shows that *Ardipithecus* at Gona was part of ecosystems in which C_4 grasses were a major component of the floral biomass.

ACKNOWLEDGMENTS

The help of many people has made the different stages of this study possible. We foremost thank S. Semaw for his ongoing support of isotopic studies of Gona material, and we are grateful to K. Schick and N. Toth for their overall support of research at Gona. At the Authority for Research and Conservation of Cultural Heritage (ARCCH) of the Ministry of Culture and Tourism, we thank J.H. Mariam, M. Yilma, Y. Beyene, A. Admasu, B. Tadesse, S. Kebede, H. Habtemichael, and M. Bitew for their assistance with research permits and general support of this work. We appreciate the hospitality of the Bureau of Tourism and Culture of the Afar Regional State at Semera and the Mille and Eloha Kebeles. The Ministry of Geology and Mines in Ethiopia was critical for exporting geologic samples. L. Kleinsasser, M. Everett, M. Rogers, D. Stout, T. Kidane, L. Harlacker, W. McIntosh, and N. Dunbar are all warmly acknowledged for their help and discussions in the field. Our special thanks go to A. Humet and the many other Afar colleagues who facilitate fieldwork at Gona. For the sampling of modern teeth from Ethiopia, we thank the Ethiopian Wildlife Conservation Department, especially T. Hailu, A. Kebede, and F. Kebede. We also thank S. Yirga for making it possible to sample the collections at the University of Addis Ababa Natural History Museum. A.K. Behrensmeyer, F.H. Brown, J.M. Harris, P. Kaleme, S. Kebede, Z. Kubsa, M.G. Leakey, L.N. Leakey, L. Swedell, the Kenya Wildlife Service, and the National Museums of Kenya have all been critical for ongoing access to modern samples. We thank C. Cook, D. Dettman, M. Lott, B. Passey, J. Pigati, and L. Roe for their analytical help and advice. The thoughtful comments of A.K. Behrensmeyer, N. Tabor, J.G. Wynn, and an anonymous reviewer greatly improved this manuscript. The Leakey Foundation, National Geographic Society, Packard Foundation, Wenner-Gren Foundation, Geological Society of America, Sigma Xi, and the National Science Foundation (EAR-0617010, SBR-9910974) funded this research, as did Tim White and the late Clark Howell through the National Science Foundation's Revealing Hominid Origins Initiative (RHOI) (BCS-0321893) program.

REFERENCES CITED

Balasse, M., 2002, Reconstructing dietary and environmental history from enamel isotopic analysis: Time resolution of intra-tooth sequential sampling: International Journal of Osteoarchaeology, v. 12, p. 155–165, doi: 10.1002/oa.601.

Boisserie, J.-R., Zazzo, A., Merceron, G., Blondel, C., Vignaud, P., Likius, A., Mackaye, H.T., and Brunet, M., 2005, Diets of modern and late Miocene hippopotamids: Evidence from carbon isotope composition and microwear of tooth enamel: Palaeogeography, Palaeoclimatology, Palaeoecology, v. 221, p. 153–174, doi: 10.1016/j.palaeo.2005.02.010.

Carter, M.L., 2001, Sensitivity of Stable Isotopes (^{13}C, ^{15}N and ^{18}O) in Bone to Dietary Specialization and Niche Separation among Sympatric Primates in Kibale National Park, Uganda [Ph.D. thesis]: Chicago, University of Chicago, 288 p.

Cerling, T.E., 1999, Stable carbon isotopes in palaeosol carbonates, *in* Thiry, M., and Simon-Coincon, R., eds., Palaeoweathering, Palaeosurfaces and Related Continental Deposits: International Association of Sedimentology Special Publication 27, p. 43–60.

Cerling, T.E., and Harris, J.M., 1999, Carbon isotope fractionation between diet and bioapatite in ungulate mammals and implications for ecological and paleoecological studies: Oecologia, v. 120, p. 347–363, doi: 10.1007/s004420050868.

Cerling, T.E., Harris, J.M., and Leakey, M.G., 1999, Browsing and grazing in elephants: The isotope record of modern and fossil proboscideans: Oecologia, v. 120, p. 364–374, doi: 10.1007/s004420050869.

Cerling, T.E., Harris, J.M., and Leakey, M.G., 2003a, Isotope paleoecology of the Nawata and Nachukui Formations at Lothagam, Turkana Basin, Kenya, in Leakey, M.G., and Harris, J.M., eds., Lothagam: The Dawn of Humanity: New York, Columbia University Press, p. 605–614.

Cerling, T.E., Harris, J.M., Leakey, M.G., and Mudida, N., 2003b, Stable isotope ecology of northern Kenya with emphasis on the Turkana Basin, in Leakey, M.G., and Harris, J.M., eds., Lothagam: The Dawn of Humanity: New York, Columbia University Press, p. 583–594.

Cerling, T.E., Harris, J.M., and Passey, B.H., 2003c, Diets of East African Bovidae based on stable isotope analysis: Journal of Mammalogy, v. 84, no. 2, p. 456–470, doi: 10.1644/1545-1542(2003)084<0456:DOEABB>2.0.CO;2.

Cerling, T.E., Hart, J.A., and Hart, T.B., 2004, Stable isotope ecology in the Ituri Forest: Oecologia, v. 138, p. 5–12, doi: 10.1007/s00442-003-1375-4.

Cerling, T.E., Harris, J.M., Hart, J.A., Kaleme, P., Klingel, H., Leakey, M.G., Levin, N.E., Lewison, R.L., and Passey, B.H., 2008, Stable isotope ecology of the common hippopotamus: Journal of Zoology, doi: 10.1111/j.1469-7998.2008.00450.x.

Codron, D., Lee-Thorp, J.A., Sponheimer, M., de Ruiter, D., and Codron, J., 2006, Inter- and intrahabitat dietary variability of Chacma baboons (*Papio ursinus*) in South African savannas based on fecal $\delta^{13}C$, $\delta^{15}N$, and %N: American Journal of Physical Anthropology, v. 129, p. 204–214, doi: 10.1002/ajpa.20253.

DeNiro, M.J., and Epstein, S., 1978, Influence of diet on the distribution of carbon isotopes in animals: Geochimica et Cosmochimica Acta, v. 42, p. 495–506, doi: 10.1016/0016-7037(78)90199-0.

d'Huart, J.-P., and Grubb, P., 2001, Distribution of the common warthog (*Phacochoerus africanus*) and the desert warthog (*Phacochoerus aethiopicus*) in the Horn of Africa: African Journal of Ecology, v. 39, p. 156–169, doi: 10.1046/j.0141-6707.2000.00298.x.

Farquhar, G.D., Ehleringer, J.R., and Hubick, K.T., 1989, Carbon isotope discrimination and photosynthesis: Annual Review of Plant Physiology and Plant Molecular Biology, v. 40, p. 503–537, doi: 10.1146/annurev.pp.40.060189.002443.

Francey, R.J., Allison, C.E., Etheridge, D.M., Trudinger, C.M., Enting, I.G., Leuenberger, M., Langenfelds, R.L., Michel, E., and Steele, L.P., 1999, A 1000-year high precision record of $\delta^{13}C$ in atmospheric CO_2: Tellus, v. 51B, p. 170–193.

Geraads, D., 2005, Pliocene Rhinocerotidae (Mammalia) from Hadar and Dikika (Lower Awash, Ethiopia), and a revision of the origin of modern African rhinos: Journal of Vertebrate Paleontology, v. 25, no. 2, p. 451–461, doi: 10.1671/0272-4634(2005)025[0451:PRMFHA]2.0.CO;2.

Haile-Selassie, Y., 2001, Late Miocene hominids from the Middle Awash, Ethiopia: Nature, v. 412, p. 178–181, doi: 10.1038/35084063.

Haile-Selassie, Y., WoldeGabriel, G., White, T.D., Bernor, R.L., Degusta, D., Renne, P.R., Hart, W.K., Ambrose, S., and Howell, F.C., 2004, Mio-Pliocene mammals from the Middle Awash, Ethiopia: Geobios, v. 37, p. 536–552, doi: 10.1016/j.geobios.2003.03.012.

Harris, J.M., and Cerling, T.E., 2002, Dietary adaptations of extant and Neogene African suids: Journal of Zoology, v. 256, p. 45–54.

Harris, J.M., Cerling, T.E., Leakey, M.G., and Passey, B.H., 2008, Stable isotope ecology of fossil hippopotamids from the Lake Turkana Basin of East Africa: Journal of Zoology, v. 275, p. 323–331, doi: 10.1111/j.1469-7998.2008.00444.x.

Indermühle, A., Stocker, T.F., Joos, F., Fischer, H., Smith, H.J., Wahlen, M., Deck, B., Mastroianni, D., Tschumi, J., Blunier, T., Meyer, R., and Stauffer, B., 1999, Holocene carbon-cycle dynamics based on CO_2 trapped in ice at Taylor Dome, Antarctica: Nature, v. 398, p. 121–126, doi: 10.1038/18158.

Keeling, C.D., Mook, W.G., and Tans, P.P., 1979, Recent trends in the $^{13}C/^{12}C$ ratio of atmospheric carbon dioxide: Nature, v. 277, p. 121–123, doi: 10.1038/277121a0.

Keeling, C.D., Whorf, T.P., Wahlen, M., and van der Plicht, J., 1995, Interannual extremes in the rate of rise of atmospheric carbon dioxide since 1980: Nature, v. 375, p. 666–670, doi: 10.1038/375666a0.

Kingston, J.D., and Harrison, T., 2007, Isotopic dietary reconstructions of Pliocene herbivores at Laetoli: Implications for early hominin palaeoecology: Palaeogeography, Palaeoclimatology, Palaeoecology, v. 243, p. 272–306, doi: 10.1016/j.palaeo.2006.08.002.

Kleinsasser, L.L., Quade, J., McIntosh, W.C., Levin, N.E., Simpson, S.W., and Semaw, S., 2008, this volume, Stratigraphy and geochronology of the late Miocene Adu-Asa Formation at Gona, Ethiopia, in Quade, J., and Wynn, J.G., eds., The Geology of Early Humans in the Horn of Africa: Geological Society of America Special Paper 446, doi: 10.1130/2008.2446(02).

Koch, P.L., Tuross, N., and Fogel, M.L., 1997, The effects of sample treatment and diagenesis on the isotopic integrity of carbonate in biogenic hydroxylapatite: Journal of Archaeological Science, v. 24, p. 417–429, doi: 10.1006/jasc.1996.0126.

Laporte, L.F., and Zihlman, A.L., 1983, Plates, climate and hominoid evolution: South African Journal of Science, v. 79, p. 96–110.

Lee-Thorp, J.A., 2000, Preservation of biogenic carbon isotope signals in Plio-Pleistocene bone and tooth mineral, in Ambrose, S.H., and Katzenberg, M.A., eds., Biogeochemical Approaches to Paleodietary Analysis: New York, Kluwer Academic/Plenum Press, p. 89–115.

Lee-Thorp, J.A., and van der Merwe, N.J., 1991, Aspects of the chemistry of modern and fossil biological apatites: Journal of Archaeological Science, v. 18, p. 343–354, doi: 10.1016/0305-4403(91)90070-6.

Levin, N.E., Quade, J., Simpson, S.W., Semaw, S., and Rogers, M.J., 2004, Isotopic evidence for Plio-Pleistocene environmental change at Gona, Ethiopia: Earth and Planetary Science Letters, v. 219, p. 93–110, doi: 10.1016/S0012-821X(03)00707-6.

MacFadden, B.J., and Higgins, P., 2004, Ancient ecology of 15-million-year-old browsing mammals within C_3 plant communities from Panama: Oecologia, v. 140, p. 169–182, doi: 10.1007/s00442-004-1571-x.

Oftedal, O.T., 1984, Milk composition, milk yield and energy output at peak lactation: A comparative review: Symposia of the Zoological Society of London, v. 51, p. 33–85.

Olson, D.M., Dinerstein, E., Wikramanayake, E.D., Burgess, N.D., Powell, G.V.N., Underwood, E.C., D'Amico, J.A., Itoua, I., Strand, H.E., Morrison, J.C., Loucks, C.J., Allnutt, T.F., Ricketts, T.H., Kura, Y., Lamoureux, J.F., Wettengel, W.W., Hedao, P., and Kassem, K.R., 2001, Terrestrial ecoregions of the world: A new map of life on Earth: BioScience, v. 51, no. 11, p. 933–938.

Passey, B.H., Cerling, T.E., Perkins, M.E., Voorhies, M.R., Harris, J.M., and Tucker, S.T., 2002, Environmental change in the Great Plains: An isotopic record from fossil horses: The Journal of Geology, v. 110, p. 123–140, doi: 10.1086/338280.

Passey, B.H., Robinson, T.F., Ayliffe, L.K., Cerling, T.E., Sponheimer, M., Dearing, M.D., Roeder, B.L., and Ehleringer, J.R., 2005, Carbon isotope fractionation between diet, breath CO_2, and bioapatite in different mammals: Journal of Archaeological Science, v. 32, p. 1459–1470, doi: 10.1016/j.jas.2005.03.015.

Pickford, M., Senut, B., and Mourer-Chauviré, C., 2004, Early Pliocene Tragulidae and peafowls in the Rift Valley, Kenya: Evidence for rainforest in East Africa: Comptes Rendus Palevol, v. 3, p. 179–189, doi: 10.1016/j.crpv.2004.01.004.

Potts, R., 1998, Environmental hypotheses of hominin evolution: Yearbook of Physical Anthropology, v. 41, p. 93–136, doi: 10.1002/(SICI)1096-8644(1998)107:27+<93::AID-AJPA5>3.0.CO;2-X.

Quade, J., Levin, N., Semaw, S., Stout, D., Renne, P.R., Rogers, M., and Simpson, S., 2004, Paleoenvironments of the earliest stone toolmakers, Gona, Ethiopia: Geological Society of America Bulletin, v. 116, no. 11–12, p. 1529–1544, doi: 10.1130/B25358.1.

Quade, J., Levin, N.E., Simpson, S.W., Butler, R., McIntosh, W.C., Semaw, S., Kleinsasser, L., Dupont-Nivet, G., Renne, P., and Dunbar, N., 2008, this volume, The geology of Gona, Afar, Ethiopia, in Quade, J., and Wynn, J.G., eds., The Geology of Early Humans in the Horn of Africa: Geological Society of America Special Paper 446, doi: 10.1130/2008.2446(01).

Randi, E., d'Huart, J.-P., Lucchini, V., and Aman, R., 2002, Evidence of two genetically deeply divergent species of warthog, *Phacochoerus africanus* and *P. aethiopicus* (Artiodactyla: Suiformes) in East Africa: Mammalian Biology, v. 67, p. 91–96, doi: 10.1078/1616-5047-00013.

Redfield, T.F., Wheeler, W.H., and Often, M., 2003, A kinematic model for the development of the Afar Depression and its paleogeographic implications: Earth and Planetary Science Letters, v. 216, p. 383–398, doi: 10.1016/S0012-821X(03)00488-6.

Schoeninger, M.J., Iwaniec, U.T., and Nash, L.T., 1998, Ecological attributes recorded in stable isotope ratios of arboreal prosimian hair: Oecologia, v. 113, p. 222–230, doi: 10.1007/s004420050372.

Semaw, S., Simpson, S.W., Quade, J., Renne, P.R., Butler, R.F., McIntosh, W.C., Levin, N., Dominguez-Rodrigo, M., and Rogers, M.J., 2005, Early Pliocene hominids from Gona, Ethiopia: Nature, v. 433, p. 301–305, doi: 10.1038/nature03177.

Simpson, S.W., Quade, J., Kleinsasser, L., Levin, N., McIntosh, W., Dunbar, N., and Semaw, S., 2007, Late Miocene hominid teeth from Gona project area, Ethiopia: American Journal of Physical Anthropology, v. 132, no. S44, p. 219.

Smith, H.J., Fischer, H., Wahlen, M., Mastroianni, D., and Deck, B., 1999, Dual modes of the carbon cycle since the Last Glacial Maximum: Nature, v. 400, p. 248–250, doi: 10.1038/22291.

Tieszen, L.L., Senyimba, M.M., Imbamba, S.K., and Troughton, J.H., 1979, The distribution of C_3 and C_4 grasses and carbon isotope discrimination along an altitudinal and moisture gradient in Kenya: Oecologia, v. 37, p. 337–350.

Urbanek, C., Faupl, P., Hujer, W., Ntaflos, T., Richter, W., Weber, G., Schaefer, K., Viola, B., Gunz, P., Neubauer, S., Stadlmayr, A., Kullmer, O., Sandrock, O., Nagel, D., Conroy, G., Falk, D., Woldearegay, K., Said, H., Assefa, G., and Seidler, H., 2005, Geology, paleontology and paleoanthropology of the Mount Galili Formation in the southern Afar Depression, Ethiopia—Preliminary results: Joannea Geologie und Paläontologie, v. 6, p. 29–43.

van der Merwe, N.J., 1991, The canopy effect: Carbon isotope ratios and food-webs in Amazonia: Journal of Archaeological Science, v. 18, p. 249–259, doi: 10.1016/0305-4403(91)90064-V.

Vignaud, P., Duringer, P., Mackaye, H.T., Likius, A., Blondel, C., Boisserie, J.-R., de Bonis, L., Eisenmann, V., Etienne, M.-E., Geraads, D., Guy, F., Lehmann, T., Lihoreau, F., Lopez-Martinez, N., Mourer-Chauviré, C., Otero, O., Rage, J.-C., Schuster, M., Viriot, L., Zazzo, A., and Brunet, M., 2002, Geology and palaeontology of the Upper Miocene Toros-Menalla hominid locality, Chad: Nature, v. 418, p. 152–155, doi: 10.1038/nature00880.

Wang, Y., and Cerling, T.E., 1994, A model of fossil tooth and bone diagenesis: Implications for paleodiet reconstruction from stable isotopes: Palaeogeography, Palaeoclimatology, Palaeoecology, v. 107, p. 281–289, doi: 10.1016/0031-0182(94)90100-7.

White, T., Suwa, G., and Asfaw, B., 1994, *Australopithecus ramidus*, a new species of early hominid from Aramis, Ethiopia: Nature, v. 371, p. 306–312, doi: 10.1038/371306a0.

White, T.D., WoldeGabriel, G., Asfaw, B., Ambrose, S., Beyene, Y., Bernor, R.L., Boisserie, J.-R., Currie, B., Gilbert, H., Haile-Selassie, Y., Hart, W.K., Hlusko, L.J., Howell, F.C., Kono, R.T., Lehmann, T., Louchart, A., Lovejoy, C.O., Renne, P.R., Saegusa, H., Vrba, E.S., Wesselman, H., and Suwa, G., 2006, Asa Issie, Aramis, and the origin of *Australopithecus*: Nature, v. 440, p. 883–889, doi: 10.1038/nature04629.

WoldeGabriel, G., White, T.D., Suwa, G., Renne, P., de Heinzelin, J., Hart, W.K., and Heiken, G., 1994, Ecological and temporal placement of early Pliocene hominids at Aramis, Ethiopia: Nature, v. 371, p. 330–333, doi: 10.1038/371330a0.

WoldeGabriel, G., Haile-Selassie, Y., Renne, P.R., Hart, W.K., Ambrose, S.H., Asfaw, B., Heiken, G., and White, T., 2001, Geology and palaeontology of the late Miocene Middle Awash Valley, Afar Rift, Ethiopia: Nature, v. 412, p. 175–178, doi: 10.1038/35084058.

Wooller, M.J., Swain, D.L., Ficken, K.J., Agnew, A.D.Q., Street-Perrott, F.A., and Eglinton, G., 2003, Late Quaternary vegetation changes around Lake Rutundu, Mount Kenya, East Africa: Evidence from grass cuticles, pollen and stable carbon isotopes: Journal of Quaternary Science, v. 18, no. 1, p. 3–15, doi: 10.1002/jqs.725.

Wynn, J.G., 2000, Paleosols, stable carbon isotopes, and paleoenvironmental interpretation of Kanapoi, northern Kenya: Journal of Human Evolution, v. 39, p. 411–432, doi: 10.1006/jhev.2000.0431.

Yalden, D.W., 1985, *Tachyoryctes macrocephalus*: Mammalian Species, v. 237, p. 1–3, doi: 10.2307/3503827.

Yalden, D.W., Largen, M.J., Kock, D., and Hillman, J.C., 1996, Catalogue of the mammals of Ethiopia and Eritrea. 7. Revised checklist, zoogeography and conservation: Tropical Zoology, v. 9, p. 73–164.

Young, H.J., and Young, T.P., 1983, Local distribution of C_3 and C_4 grasses in sites of overlap on Mount Kenya: Oecologia, v. 58, p. 373–377, doi: 10.1007/BF00385238.

Zachos, J., Pagani, M., Sloan, L., Thomas, E., and Billups, K., 2001, Trends, rhythms, and aberrations in global climate 65 Ma to present: Science, v. 292, p. 686–693, doi: 10.1126/science.1059412.

Zazzo, A., Bocherens, H., Brunet, M., Beauvilain, A., Billiou, D., Mackaye, H.T., Vignaud, P., and Mariotti, A., 2000, Herbivore paleodiet and paleoenvironmental changes in Chad during the Pliocene using stable isotope ratios of tooth enamel carbonate: Paleobiology, v. 26, no. 2, p. 294–309, doi: 10.1666/0094-8373(2000)026<0294:HPAPCI>2.0.CO;2.

Zazzo, A., Mariotti, A., Lécuyer, C., and Heintz, E., 2002, Intra-tooth isotope variations in late Miocene bovid enamel from Afghanistan: Paleobiological, taphonomic, and climatic implications: Palaeogeography, Palaeoclimatology, Palaeoecology, v. 186, p. 145–161, doi: 10.1016/S0031-0182(02)00449-2.

MANUSCRIPT ACCEPTED BY THE SOCIETY 17 JUNE 2008